# Infrared and Raman Spectra of Inorganic and Coordination Compounds

# Infrared and Raman Spectra of Inorganic and Coordination Compounds

THIRD EDITION

*KAZUO NAKAMOTO*

Wehr Professor of Chemistry
Marquette University

A Wiley-Interscience Publication
JOHN WILEY & SONS, *New York* • *Chichester* • *Brisbane* • *Toronto*

The first and second editions were entitled
*Infrared Spectra of Inorganic and Coordination Compounds.*



***Library of Congress Cataloging in Publication Data:***

Nakamoto, Kazuo, 1922–

Infrared and Raman spectra of inorganic and coordination compounds.

Second ed. published in 1970 under title: Infrared spectra of inorganic and coordination compounds.

"A Wiley-Interscience publication."

Includes bibliographical references and index.

1. Infra-red spectrometry. 2. Raman spectroscopy.
I. Title.

QD96.15N33 1977 544′.63 77-15107

ISBN 0-471-62979-0

Printed in the United States of America

10 9 8 7 6 5 4 3 2 1

# Preface to the First Edition

Since 1945, the volume of literature on the infrared spectra of inorganic and coordination compounds has grown with ever-increasing rapidity. As a result, it is becoming more and more difficult for any one individual to read all pertinent articles. Excellent books on the theory of vibrational spectra and on the infrared spectra of organic compounds have been published, but relatively few comprehensive reviews of the vibrational spectra of inorganic or coordination compounds are available. This situation has prompted me to write this book.

In reviewing the literature, I have attempted to interpret the experimental results in terms of normal vibrations, as far as it is possible. Consequently, I have been most interested in normal coordinate analysis, in fundamental frequencies, in band assignments and in structural considerations, while omitting any consideration of band intensities or rotational fine structure. The experimental aspects of vibrational spectra have also been omitted.

Part I is devoted to the minimum amount of theory necessary for an understanding of the concept of the normal vibration and of the method of normal coordinate analysis. For a more detailed discussion of the theory, the reader may consult the references cited in Part I. An application of the method of normal coordinate analysis to a complex system is given in detail in Appendix III. In Parts II and III, the observed fundamental vibrational frequencies of molecules are discussed in terms of their relation to molecular structure. Although most of the data are from infrared spectra, Raman spectral data have been quoted wherever pertinent or necessary. This volume is intended as a critical review of the infrared spectra of inorganic and coordination compounds; but it was clearly impossible to refer to all work which has been reported in these fields. I have, however, tried to give a broad and reasonably complete coverage of the fields. But in the discussion I have had to select examples which I considered more illustrative, more important or more interesting. Although this book is intended to cover all inorganic and coordination compounds, the border line between these and organic compounds is a difficult one to draw. Therefore I have chosen to omit most of the metallo-organic compounds.

I wish to express my sincere thanks to Professor Arthur E. Martell whose discussions inspired me to begin this book. A special word of gratitude is due to Professors S. Mizushima, T. Shimanouchi and I. Nakagawa whose publications and personal communications have greatly helped me in writing on normal coordinate analysis. I am also deeply indebted to Dr. J. L. Bethune, Rev. Paul J. McCarthy, S. J., and Sister M. Paulita, C.S.J., who read the whole manuscript and gave many valuable comments. Thanks are also due to Dr. Junnosuke Fujita and Mr. Yukiyoshi Morimoto, who assisted in the preparation of this book, and to Mrs. Helen Kwan and Mrs. Jeannette Lynch, who typed the manuscript.

KAZUO NAKAMOTO

*Chicago, Illinois*
*December* 1962

# Preface to the Second Edition

A reference book is doomed to become outdated rapidly because of the ever-increasing volume of new literature. I began to feel a need for a new edition of this book several years ago, but it was not until the spring of 1968 that I finished my survey of the literature and started to rewrite the chapters. This task was finally completed in the summer of 1969.

In the second edition, I have expanded Part I by adding several new sections concerning the theories of Raman spectra, crystal spectra, and infrared intensities. In Parts II and III, I have attempted to include all important and significant results published before the end of 1967, while omitting those proved to be erroneous by later studies. Some results obtained after 1967 were included rather arbitrarily simply because they were easily accessible to me. It was clearly not possible to cover all the work on infrared and Raman spectra of inorganic and coordination compounds. I have tried, however, to present topics and results that I consider important or interesting in a proper and reasonable balance. Finally, I have replaced several appendices by new ones that I consider to be more useful and convenient to the reader.

I express my sincere thanks to all persons who gave valuable comments on the first edition through private communications or book reviews. All these comments were taken into consideration in preparing the second edition. I am particularly indebted to Dr. Marcia Cordes who read the whole manuscript of the second edition. Thanks are also due to Professor I. Nakagawa (Part I), Dr. A. Fadini and Professor A. Müller (Part II), and Professor D. F. Shriver (Part III) who helped me in writing some sections. Most of the literature search and writing for the second edition were done at the Department of Chemistry, Illinois Institute of Technology, whose library facilities greatly helped me in the preparation of this volume. The second edition includes some recent results we obtained during its preparation. Most of these projects were supported by the ACS-PRF unrestricted research grant, and I would like to thank the American Chemical Society for giving me opportunities to pursue these, rather uncoordinated, research projects.

KAZUO NAKAMOTO

*Milwaukee, Wisconsin*
*October* 1969

# Preface to the Third Edition

Since I wrote the second edition of my book in 1969, the field of inorganic vibrational spectroscopy has been expanded by a variety of research techniques, leading to the production of many new and interesting results. The most notable developments during this period have been the wide applications of laser–Raman spectroscopy and matrix isolation spectroscopy to inorganic and coordination compounds. This situation necessitated the addition of several new sections on the theories of Raman spectroscopy (Part I) and a large volume of new data obtained by these and other new techniques (Parts II–IV). The title of the book has also been changed to emphasize the importance of Raman spectroscopy. Part IV has been added to give a brief review of vibrational studies in the ever-growing field of organometallic chemistry.

As I emphasized in the Prefaces of the preceding editions, this book is intended to describe fundamental theories of vibrational spectroscopy in condensed form and to illustrate their applications to inorganic, coordination, and organometallic compounds, using typical examples. Although I have tried to give broad and reasonably complete coverage of the field, it is clearly impossible to include all published work in the limited space available. To compensate for this deficiency, I have cited reference books and review articles whenever possible.

I would like to express my sincere thanks to all those who helped me in preparing this volume. I am particularly indebted to professors D. P. Strommen (Carthage College), D. F. Shriver (Northwestern University), A. Müller (University of Dortmund), R. J. H. Clark (University College, London) and E. Maslowsky, Jr. (Loras College), who read my manuscript and gave me many valuable comments. This book could not have been written without the encouragement of these and many other colleagues. Special thanks are due also to the Alexander von Humboldt Stiftung, which gave me an opportunity to visit German colleagues in 1974, and to the American Chemical Society–Petroleum Research Fund, which provided continuing support for my own research as described in this book.

KAZUO NAKAMOTO

*Milwaukee, Wisconsin*
*June* 1977

# Abbreviations

IR, infrared; R, Raman; *p*, polarized; *dp*, depolarized; *ip*, inverse polarization.

$\nu$, stretching; $\delta$, in-plane bending or defomation; $\rho_w$, wagging; $\rho_r$, rocking; $\rho_t$, twisting; $\pi$, out-of-plane bending. Subscripts *a*, *s*, and *d* denote antisymmetric, symmetric, and degenerate modes, respectively. Approximate normal modes of vibration corresponding to these vibrations are given in Figs. III-2 and III-13.

GVF, generalized valence force field; UBF, Urey–Bradley force field.

M, metal; L, ligand; X, halogen; R, alkyl group or cyclopentadienyl or other ring compound.

g, gas; l, liquid; s, solid; m or mat, matrix; sol'n or sl, solution.

Me, methyl; Et, ethyl; Bu, butyl; OAc, acetate ion. Abbreviations of the ligands are given when they appear in the text.

In the tables of observed frequencies given in Parts II–IV, values in parentheses are calculated or estimated unless otherwise stated.

# Contents

**Part I. Theory of Normal Vibration**

## Part II. Inorganic Compounds



## Part III. Coordination Compounds

**Part IV. Organometallic Compounds**



**Appendices**

# Theory of Normal Vibration

## *Part I*

## I-1. ORIGIN OF MOLECULAR SPECTRA

As a first approximation, it is possible to separate the energy of a molecule into three additive components associated with (1) the rotation of the molecule as a whole, (2) the vibrations of the constituent atoms, and (3) the motion of the electrons in the molecule.* The translational energy of the molecule may be ignored in this discussion. The basis for this separation lies in the fact that the velocity of electrons is much greater than the vibrational velocity of nuclei, which is again much greater than the velocity of molecular rotation. If a molecule is placed in an electromagnetic field (e.g., light), a transfer of energy from the field to the molecule will occur only when Bohr's frequency condition is satisfied:

$$\Delta E = h\nu \tag{1.1}$$

where $\Delta E$ is the difference in energy between two quantized states, $h$ is Planck's constant, and $\nu$ is the frequency of the light.† If

$$\Delta E = E'' - E' \tag{1.2}$$

where $E''$ is a quantized state of higher energy than $E'$, the molecule *absorbs* radiation when it is excited from $E'$ to $E''$ and *emits* radiation of the same frequency as given by Eq. 1.1 when it reverts from $E''$ to $E'$.

Because rotational levels are relatively close to each other, transitions between these levels occur at low frequencies (long wavelengths). In fact, pure rotational spectra appear in the range between 1 $cm^{-1}$ ($10^4$ $\mu$) and $10^2$ $cm^{-1}$ ($10^2$ $\mu$). The separation of vibrational energy levels is greater,

---

* Hereafter the word *molecule* may also represent an *ion*.

† The frequency, $\nu$, is converted to the wave number, $\tilde{\nu}$, or the wavelength, $\lambda_\omega$, through the relation

$$\nu = c\tilde{\nu} = \frac{c}{\lambda_\omega}$$

where $c$ is the velocity of light. For theoretical discussion, $\nu$ and $\tilde{\nu}$ are more convenient than $\lambda_\omega$, since they are proportional to the energy of radiation. More explicit relations between these three units are given in the following table for the region in which vibrational spectra occur.

| Frequency ($sec^{-1}$) | Wave Number ($cm^{-1}$) | Wavelength ($\mu$) |
|---|---|---|
| $3 \cdot 10^{14}$ | $10^4$ | 1 |
| $3 \cdot 10^{13}$ | $10^3$ | 10 |
| $3 \cdot 10^{12}$ | $10^2$ | $10^2$ |

Although the dimensions of $\nu$ and $\tilde{\nu}$ differ from one another, it is conventional to use them interchangeably. For example, a phrase such as "a frequency shift of 25 $cm^{-1}$" is often employed. All the spectral data in this book are given in terms of $\tilde{\nu}$ ($cm^{-1}$).

and the transitions occur at higher frequencies (shorter wavelengths) than do the rotational transitions. As a result, pure vibrational spectra are observed in the range between $10^2$ cm$^{-1}$ ($10^2$ $\mu$) and $10^4$ cm$^{-1}$ (1 $\mu$). Finally, electronic energy levels are usually far apart, and electronic spectra are observed in the range between $10^4$ cm$^{-1}$ (1 $\mu$) and $10^5$ cm$^{-1}$ ($10^{-1}$ $\mu$). Thus pure rotational, vibrational, and electronic spectra are usually observed in the microwave and far-infrared, the infrared, and the visible and ultraviolet regions, respectively. This division into three regions, however, is to some extent arbitrary, for pure rotational spectra may appear in the near-infrared region ($1.5 \sim 0.5 \times 10^4$ cm$^{-1}$) if transitions to higher excited states are involved, and pure electronic transitions may appear in the near-infrared region if the levels are closely spaced.

Figure I-1 illustrates transitions of the three types mentioned for a diatomic molecule. As the figure shows, rotational intervals tend to increase as the rotational quantum number $J$ increases, whereas vibrational intervals tend to decrease as the vibrational quantum number $v$ increases. The dotted line below each electronic level indicates the zero point energy that exists even at a temperature of absolute zero as a result of nuclear vibration. It should be emphasized that not all transitions between these levels are possible. To see whether the transition is *allowed* or *forbidden*, the relevant selection rule must be examined. This, in turn, is determined by the symmetry of the molecule. As will be seen later, vibrational problems like those mentioned above can be solved for polyatomic molecules in an elegant manner by the use of group theory.

Since this book is concerned only with vibrational spectra, no description of electronic and rotational spectra is given. Although vibrational spectra are observed experimentally as infrared or Raman spectra, the physical origins of these two types of spectra are different. Infrared spectra originate in transitions between two vibrational levels of the molecule in the electronic ground state and are usually observed as *absorption spectra* in the infrared region. On the other hand, Raman spectra originate in the electronic polarization caused by ultraviolet or visible light. If a molecule is irradiated by monochromatic light of frequency $\nu$, then, because of electronic polarization induced in the molecule by this incident light, light of frequency $\nu$ (*Rayleigh scattering*) as well as of $\nu \pm \nu_i$ (*Raman scattering*) is emitted ($\nu_i$ represents a vibrational frequency). Thus the vibrational frequencies are observed as Raman shifts from the incident frequency $\nu$ in the ultraviolet or visible region.

Although Raman scattering is much weaker than Rayleigh scattering (by a factor of $10^{-3}$ to $10^{-4}$), it is possible to observe the former by using a strong exciting source. In the past, the mercury lines at 435.8 nm

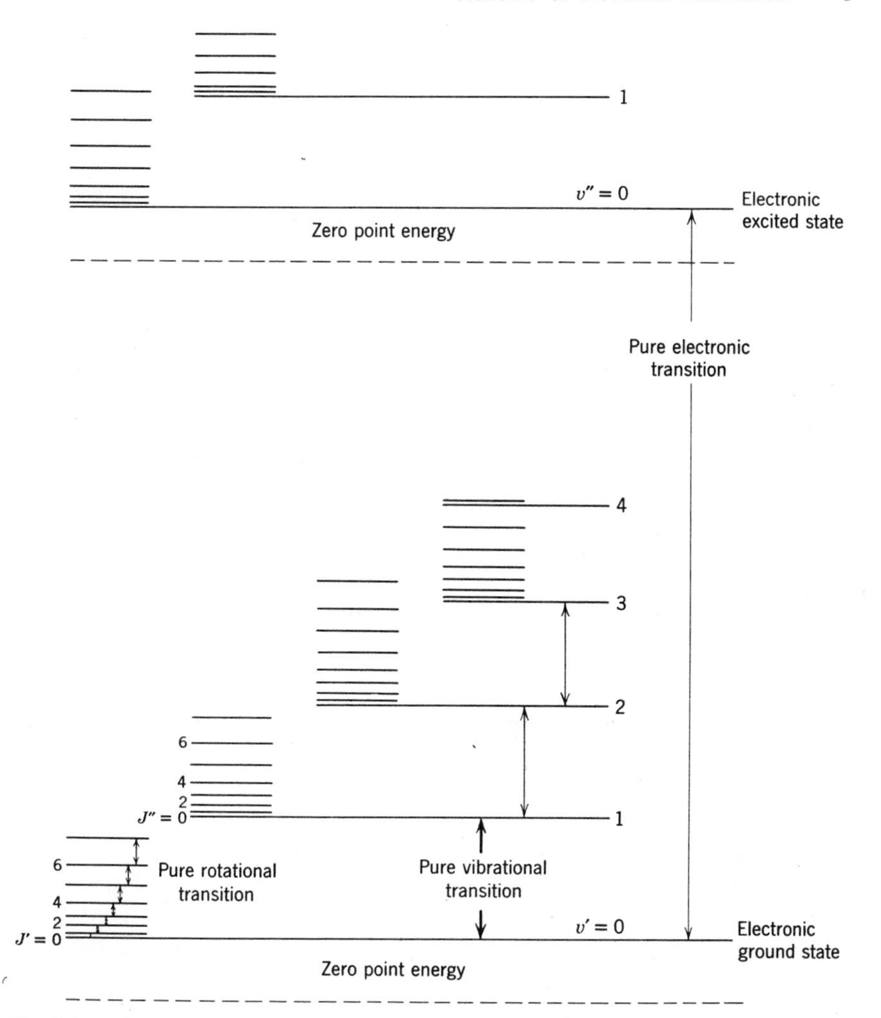


**Fig. I-1.** Energy levels of a diatomic molecule (the actual spacings of electronic levels are much larger, and those of rotational levels much smaller, than those shown in the figure).

(22,938 cm$^{-1}$) and 404.7 nm (24,705 cm$^{-1}$) from a low pressure mercury arc were used to observe Raman scattering. These lines, however, may be absorbed by a number of compounds which have an electronic absorption band in this region. Recently, several lasers which provide strong monochromatic beams at other frequencies have been developed. Typical laser lines are as follows: $Kr^+$ (647.1 nm, 15,450 cm$^{-1}$, red), He–Ne (632.8 nm, 15,798 cm$^{-1}$, red), $Ar^+$ (514.5 nm, 19,430 cm$^{-1}$, green), and $Ar^+$ (488.0 nm, 20,487 cm$^{-1}$, blue). With these and other lines, it is possible to measure Raman spectra outside the electronic absorption band. In the case of resonance Raman spectroscopy (Sec. I-20), the exciting frequency is chosen so as to fall inside the electronic band. The degree of resonance enhancement changes as a function of the exciting frequency and reaches a maximum when the frequency approximately coincides with that of the electronic absorption maximum. Recently developed tunable dye-lasers provide a valuable means of studying resonance Raman spectra.

The origin of Raman spectra can be explained by an elementary classical theory. Consider a light wave of frequency $\nu$ with an electric field strength $E$. Since $E$ fluctuates at frequency $\nu$, we can write

$$E = E_0 \cos 2\pi\nu t \tag{1.3}$$

where $E_0$ is the amplitude and $t$ the time. If a diatomic molecule is irradiated by this light, the dipole moment $P$ given by

$$P = \alpha E = \alpha E_0 \cos 2\pi\nu t \tag{1.4}$$

is induced. Here $\alpha$ is a proportionality constant and is called the *polarizability*. If the molecule is vibrating with frequency $\nu_1$, the nuclear displacement $q$ is written as

$$q = q_0 \cos 2\pi\nu_1 t \tag{1.5}$$

where $q_0$ is the vibrational amplitude. For small amplitudes of vibration, $\alpha$ is a linear function of $q$. Thus we can write

$$\alpha = \alpha_0 + \left(\frac{\partial\alpha}{\partial q}\right)_0 q \tag{1.6}$$

Here $\alpha_0$ is the polarizability at the equilibrium position, and $(\partial\alpha/\partial q)_0$ is the rate of change of $\alpha$ with respect to the change in $q$, evaluated at the equilibrium position. If we combine Eqs. 1.4, 1.5, and 1.6, we have

$$\begin{aligned} P &= \alpha E_0 \cos 2\pi\nu t \\ &= \alpha_0 E_0 \cos 2\pi\nu t + \left(\frac{\partial\alpha}{\partial q}\right)_0 q_0 E_0 \cos 2\pi\nu t \cdot \cos 2\pi\nu_1 t \\ &= \alpha_0 E_0 \cos 2\pi\nu t \\ &\quad + \frac{1}{2}\left(\frac{\partial\alpha}{\partial q}\right)_0 q_0 E_0 [\cos\{2\pi(\nu+\nu_1)t\} + \cos\{2\pi(\nu-\nu_1)t\}] \end{aligned} \tag{1.7}$$

According to classical theory, the first term describes an oscillating dipole which radiates light of frequency $\nu$ (Rayleigh scattering). The second term gives the Raman scattering of frequencies $\nu+\nu_1$ (*anti-Stokes*) and $\nu-\nu_1$ (*Stokes*). If $(\partial\alpha/\partial q)_0$ is zero, the second term vanishes. Thus the vibration

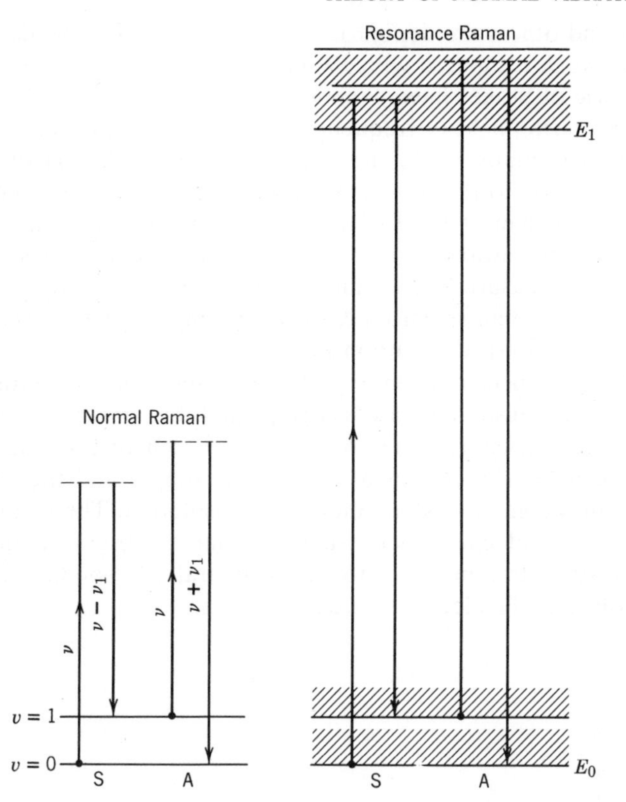


**Fig. I-2.** Mechanisms of normal and resonance Raman scattering. S, Stokes; A, anti-Stokes. The dotted lines represent the virtual state. The shaded areas indicate the broadening of rotational–vibrational levels.

is not Raman active unless the polarizability changes during the vibration.

Figure I-2 illustrates the mechanisms of *normal* and *resonance Raman* (*RR*) *scattering.* In the former, the energy of the exciting line falls far below that required to excite the first electronic transition. In the latter, the energy of the exciting line coincides with that of an electronic transition.* If the photon is absorbed and then emitted during the process, it is called *resonance fluorescence* (*RF*). Although the conceptual difference between resonance Raman scattering and resonance fluroescence is subtle, there are several experimental differences which can be used to distinguish between these two phenomena. For example, in RF spectra all lines are depolarized, whereas in RR spectra some are

* If the exciting line is close to but not inside an electronic absorption band, the process is called "preresonance Raman scattering."

polarized and others are depolarized. Additionally, RR bands tend to be broad and weak compared with RF bands.

In the case of Stokes lines, the molecule at $\upsilon = 0$ is excited to the $\upsilon = 1$ state by scattering light of frequency $\nu - \nu_1$. Anti-Stokes lines arise when the molecule initially in the $\upsilon = 1$ state scatters radiation of frequency $\nu + \nu_1$ and reverts to the $\upsilon = 0$ state. Since the population of molecules is larger at $\upsilon = 0$ than at $\upsilon = 1$ (*Maxwell–Boltzmann distribution law*), the Stokes lines are always stronger than the anti-Stokes lines. Thus it is customary to measure Stokes lines in Raman spectroscopy. Figure I-3 illustrates the Raman spectrum (below 500 $cm^{-1}$) of $CCl_4$ excited by the blue line (488.0 nm) of an argon-ion laser.

It is to be expected from Fig. I-1 that electronic spectra are very complicated because they are accompanied by vibrational as well as rotational fine structure. The rotational fine structure in the electronic spectrum can be observed if a molecule is simple and the spectrum is measured in the gaseous state under high resolution. The vibrational fine structure of the electronic spectrum is easier to observe than the rotational fine structure, and can provide structural and bonding information about molecules in electronic excited states.

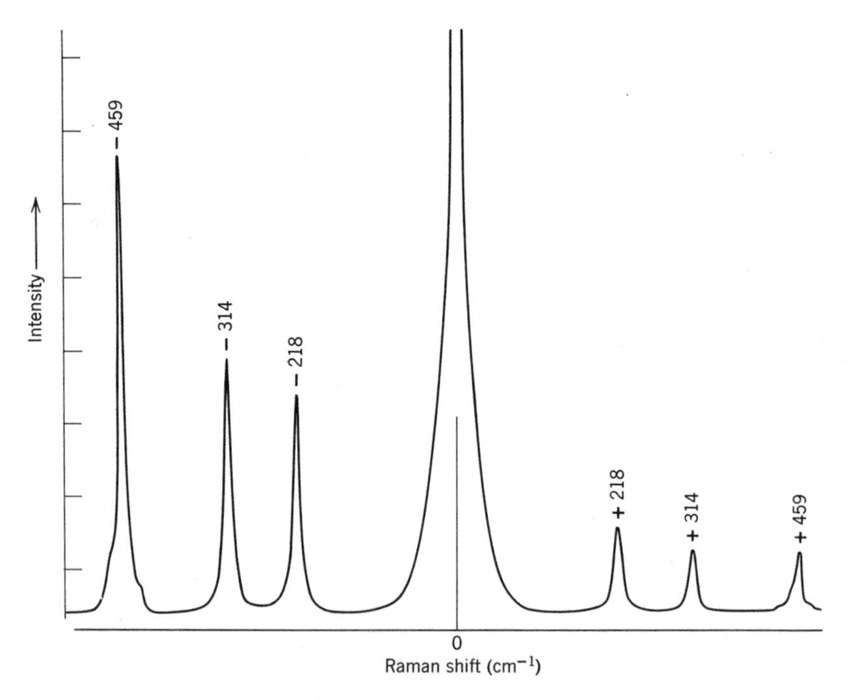


**Fig. I-3.** Raman spectrum of $CCl_4$ (488.0 nm excitation).

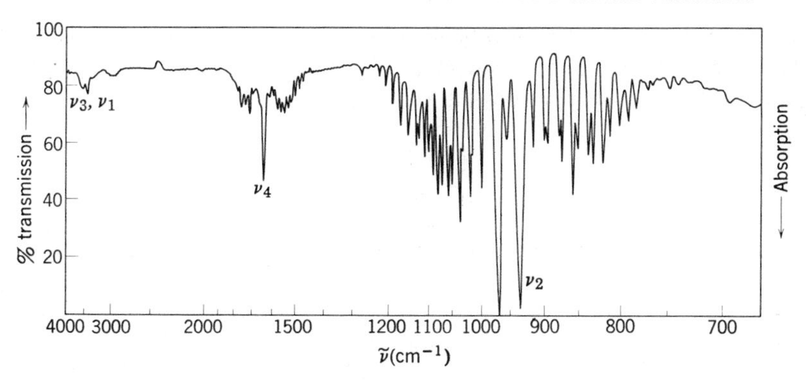


**Fig. I-4.** Rotational fine structure of gaseous $NH_3$.

Vibrational spectra are accompanied by rotational transitions. Figure I-4 shows the rotational fine structure observed for the gaseous ammonia molecule. In most polyatomic molecules, however, such a rotational fine structure is not observed because the rotational levels are closely spaced as a result of relatively large moments of inertia. Vibrational spectra obtained in solution do not exhibit rotational fine structure, since molecular collisions occur before a rotation is completed and the levels of the individual molecules are perturbed differently. Since Raman spectra are often obtained in liquid state, they do not exhibit rotational fine structure.

According to the selection rule for the harmonic oscillator, any transitions corresponding to $\Delta \upsilon = \pm 1$ are allowed (Sec. I-2). Under ordinary conditions, however, only the *fundamentals* that originate in the transition from $\upsilon = 0$ to $\upsilon = 1$ in the electronic ground state can be observed because of the Maxwell–Boltzmann distribution law. In addition to the selection rule for the harmonic oscillator, another restriction results from the symmetry of the molecule (Sec. I-9). Thus the number of allowed transitions in polyatomic molecules is greatly reduced. The *overtones and combination bands** of these fundamentals are forbidden by the selection rule of the harmonic oscillator. However, they are weakly observed in the spectrum because of the anharmonicity of the vibration (Sec. I-2). Since they are less important than the fundamentals, they will be discussed only when necessary.

## I-2. VIBRATION OF A DIATOMIC MOLECULE

Through quantum mechanical considerations,[4,5] the vibration of two nuclei in a diatomic molecule can be reduced to the motion of a single

* Overtones represent multiples of some fundamental, whereas combination bands arise from the sum or difference of two or more fundamentals.

particle of mass $\mu$, whose displacement $q$ from its equilibrium position is equal to the change of the internuclear distance. The mass $\mu$ is called the *reduced mass* and is represented by

$$\frac{1}{\mu}=\frac{1}{m_1}+\frac{1}{m_2} \tag{2.1}$$

where $m_1$ and $m_2$ are the masses of the two nuclei. The kinetic energy is then

$$T=\tfrac{1}{2}\mu\dot{q}^2=\frac{1}{2\mu}p^2 \tag{2.2}$$

where $p$ is the conjugate momentum, $\mu\dot{q}$. If a simple parabolic potential function such as that shown in Fig. I-5 is assumed, the system represents a *harmonic oscillator*, and the potential energy is simply given by

$$V=\tfrac{1}{2}Kq^2 \tag{2.3}$$

Here $K$ is the force constant for the vibration. Then the Schrödinger wave equation becomes

$$\frac{d^2\psi}{dq^2}+\frac{8\pi^2\mu}{h^2}(E-\tfrac{1}{2}Kq^2)\psi=0 \tag{2.4}$$

If this equation is solved with the condition that $\psi$ must be single valued,

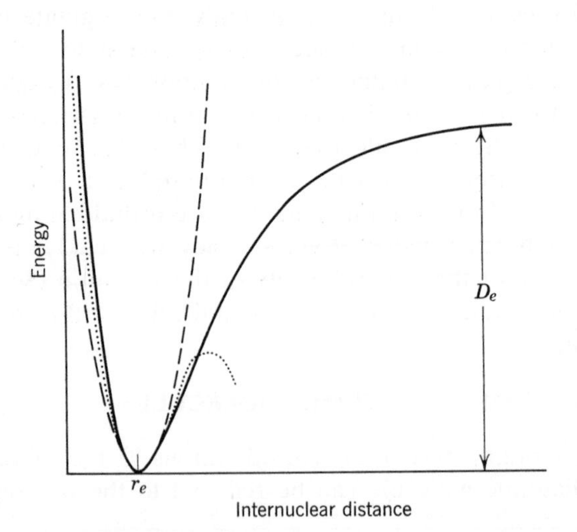


**Fig. I-5.** Potential curve for a diatomic molecule. Actual potential, solid line; parabola, broken line; cubic parabola, dotted line.

finite, and continuous, the eigenvalues are

$$E_{\upsilon} = h\nu(\upsilon + \tfrac{1}{2}) = hc\tilde{\nu}(\upsilon + \tfrac{1}{2}) \tag{2.5}$$

with the frequency of vibration

$$\nu = \frac{1}{2\pi}\sqrt{\frac{K}{\mu}} \quad \text{or} \quad \tilde{\nu} = \frac{1}{2\pi c}\sqrt{\frac{K}{\mu}} \tag{2.6}$$

Here $\upsilon$ is the vibrational quantum number, and it can have the values 0, 1, 2, 3, . . . .

The corresponding eigenfunctions are

$$\psi_{\upsilon} = \frac{(\alpha/\pi)^{1/4}}{\sqrt{2^{\upsilon}\upsilon!}} e^{-\alpha q^2/2} H_{\upsilon}(\sqrt{\alpha}q) \tag{2.7}$$

where $\alpha = 2\pi\sqrt{\mu K}/h = 4\pi^2\mu\nu/h$, and $H_{\upsilon}(\sqrt{\alpha}q)$ is a Hermite polynomial of the $\upsilon$th degree. Thus the eigenvalues and the corresponding eigenfunctions are

$$\begin{array}{ll} E_0 = \tfrac{1}{2}h\nu & \psi_0 = (\alpha/\pi)^{1/4} e^{-\alpha q^2/2} \\ E_1 = \tfrac{3}{2}h\nu & \psi_1 = (\alpha/\pi)^{1/4} 2^{1/2} q e^{-\alpha q^2/2} \\ \vdots & \vdots \end{array} \tag{2.8}$$

As Fig. I-5 shows, actual potential curves can be approximated more exactly by adding a cubic term:[2]

$$V = \tfrac{1}{2}Kq^2 - Gq^3 \qquad (K \gg G) \tag{2.9}$$

Then the eigenvalues are

$$E_{\upsilon} = hc\omega_e(\upsilon + \tfrac{1}{2}) - hcx_e\omega_e(\upsilon + \tfrac{1}{2})^2 + \cdots \tag{2.10}$$

where $\omega_e$ is the wave number corrected for *anharmonicity*, and $x_e\omega_e$ indicates the magnitude of anharmonicity. Table II-1*a* of Part II lists $\omega_e$ and $x_e\omega_e$ for a number of diatomic molecules. Equation 2.10 shows that the energy levels of the anharmonic oscillator are not equidistant, and the separation decreases slowly as $\upsilon$ increases. This anharmonicity is responsible for the appearance of overtones and combination vibrations, which are forbidden in the harmonic oscillator. Since the anharmonicity correction has not been made for most polyatomic molecules, in large part because of the complexity of the calculation, the frequencies given in Parts II to IV are not corrected for anharmonicity (except those given in Table II-1*a*).

According to Eq. 2.6, the frequency of the vibration in a diatomic

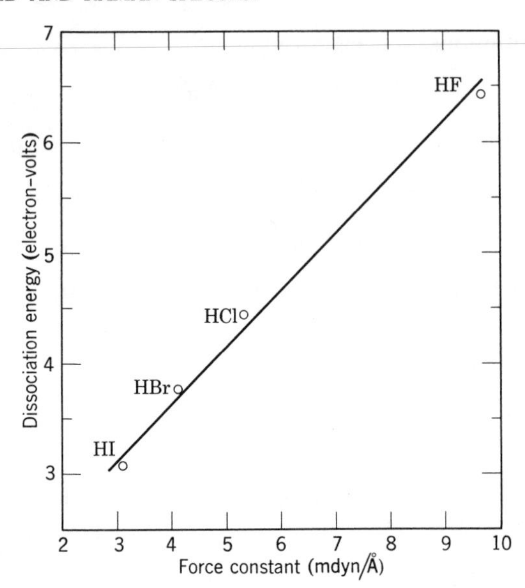


**Fig. I-6.** Relation between force constant and dissociation energy in hydrogen halides.

molecule is proportional to the square root of $K/\mu$. If $K$ is approximately the same for a series of diatomic molecules, the frequency is inversely proportional to the square root of $\mu$. This point is illustrated by the series $H_2$, HD, and $D_2$ shown in Table II-1*a*. If $\mu$ is approximately the same for a series of diatomic molecules, the frequency is proportional to the square root of $K$. This point is illustrated by the series HF, HCl, HBr, and HI. These simple rules, obtained for a diatomic molecule, are helpful in understanding the vibrational spectra of polyatomic molecules.

Figure I-6 indicates the relation between the force constant $K$, calculated from Eq. 2.6, and the dissociation energy in a series of hydrogen halides. Evidently, the bond becomes stronger as the force constant becomes larger. It should be noted, however, that a general theoretical relation between these two quantities is difficult to derive even for a diatomic molecule. The force constant is a measure of the curvature of the potential well near the equilibrium position:

$$K = \left(\frac{d^2V}{dq^2}\right)_{q\to 0} \tag{2.11}$$

whereas the dissociation energy $D_e$ is given by the depth of the potential energy curve (Fig. I-5). Thus a large force constant means sharp curvature of the potential well near the bottom but does not necessarily indicate a deep potential well. Usually, however, a larger force constant is an indication of a stronger bond in a series of molecules belonging to the same type (Fig. I-6).

In the case of small molecules, attempts have been made to calculate the force constants by quantum mechanics. The principle of the method is to express the total electronic energy of a molecule as a function of nuclear displacements near the equilibrium position and to calculate its second derivatives, $\partial^2 V/\partial q_i^2$, and so on for each displacement coordinate $q_i$. Thus far, *ab initio* calculations of force constants have been made for molecules such as HF, $H_2O$, and $NH_3$.[37] The force constants thus obtained are in good agreement with those calculated from the analysis of vibrational spectra.

## I-3. NORMAL COORDINATES AND NORMAL VIBRATIONS

In diatomic molecules, the vibration of the nuclei occurs only along the line connecting two nuclei. In polyatomic molecules, however, the situation is much more complicated because all the nuclei perform their own harmonic oscillations. It can be shown, however, that any of these extremely complicated vibrations of the molecule may be represented as a superposition of a number of *normal vibrations.*

Let the displacement of each nucleus be expressed in terms of rectangular coordinate systems with the origin of each system at the equilibrium position of each nucleus. Then the kinetic energy of an $N$-atom molecule would be expressed as

$$T=\frac{1}{2}\sum_N m_N\left[\left(\frac{d\,\Delta x_N}{dt}\right)^2+\left(\frac{d\,\Delta y_N}{dt}\right)^2+\left(\frac{d\,\Delta z_N}{dt}\right)^2\right] \tag{3.1}$$

If generalized coordinates such as

$$q_1=\sqrt{m_1}\,\Delta x_1, \qquad q_2=\sqrt{m_1}\,\Delta y_1, \qquad q_3=\sqrt{m_1}\,\Delta z_1, \qquad q_4=\sqrt{m_2}\,\Delta x_2, \cdots \tag{3.2}$$

are used, the kinetic energy is simply written as

$$T=\frac{1}{2}\sum_i^{3N}\dot{q}_i^2 \tag{3.3}$$

The potential energy of the system is a complex function of all the coordinates involved. For small values of the displacements, it may be expanded in a Taylor's series as

$$V(q_1, q_2, \ldots, q_{3N})=V_0+\sum_i^{3N}\left(\frac{\partial V}{\partial q_i}\right)_0 q_i+\frac{1}{2}\sum_{i,j}^{3N}\left(\frac{\partial^2 V}{\partial q_i\,\partial q_j}\right)_0 q_iq_j+\cdots \tag{3.4}$$

where the derivatives are evaluated at $q_i = 0$, the equilibrium position. The constant term $V_0$ can be taken as zero if the potential energy at $q_i = 0$ is taken as a standard. The $(\partial V/\partial q_i)_0$ terms also become zero, since $V$ must be a minimum at $q_i = 0$. Thus $V$ may be represented by

$$V = \frac{1}{2}\sum_{i,j}^{3N}\left(\frac{\partial^2 V}{\partial q_i\,\partial q_j}\right)_0 q_i q_j = \frac{1}{2}\sum_{i,j}^{3N} b_{ij} q_i q_j \tag{3.5}$$

neglecting higher order terms.

If the potential energy given by Eq. 3.5 did not include any cross products such as $q_i q_j$, the problem could be solved directly by using Newton's equation:

$$\frac{d}{dt}\left(\frac{\partial T}{\partial \dot{q}_i}\right) + \frac{\partial V}{\partial q_i} = 0 \qquad i = 1, 2, \ldots, 3N \tag{3.6}$$

From Eqs. 3.3 and 3.5, Eq. 3.6 is written as

$$\ddot{q}_i + \sum_j b_{ij} q_j = 0 \qquad j = 1, 2, \ldots, 3N \tag{3.7}$$

If $b_{ij} = 0$ for $i \neq j$, Eq. 3.7 becomes

$$\ddot{q}_i + b_{ii} q_i = 0 \tag{3.8}$$

and the solution is given by

$$q_i = q_i^0 \sin(\sqrt{b_{ii}}\,t + \delta_i) \tag{3.9}$$

where $q_i^0$ and $\delta_i$ are the amplitude and the phase constant, respectively.

Since, in general, this simplification is not applicable, the coordinates $q_i$ must be transformed into a set of new coordinates $Q_i$ through the relations

$$\begin{aligned} q_1 &= \sum_i B_{1i} Q_i \\ q_2 &= \sum_i B_{2i} Q_i \\ &\;\;\vdots \\ q_k &= \sum_i B_{ki} Q_i \end{aligned} \tag{3.10}$$

The $Q_i$ are called *normal coordinates* for the system. By appropriate choice of the coefficients $B_{ki}$, both the potential and the kinetic energies can be written as

$$T = \frac{1}{2} \sum_i \dot{Q}_i^2 \tag{3.11}$$

$$V = \frac{1}{2} \sum_i \lambda_i Q_i^2 \tag{3.12}$$

without any cross products.

If Eqs. 3.11 and 3.12 are combined with Newton's equation (3.6), there results

$$\ddot{Q}_i + \lambda_i Q_i = 0 \tag{3.13}$$

The solution of this equation is given by

$$Q_i = Q_i^0 \sin(\sqrt{\lambda_i} t + \delta_i) \tag{3.14}$$

and the frequency is

$$\nu_i = \frac{1}{2\pi} \sqrt{\lambda_i} \tag{3.15}$$

Such a vibration is called a *normal vibration.*

For the general $N$-atom molecule, it is obvious that the number of the normal vibrations is only $3N-6$, since six coordinates are required to describe the translational and rotational motion of the molecule as a whole. Linear molecules have $3N-5$ normal vibrations, as no rotational freedom exists around the molecular axis. Thus the general form of the molecular vibration is a superposition of the $3N-6$ (or $3N-5$) normal vibrations given by Eq. 3.14.

The physical meaning of the normal vibration may be demonstrated in the following way. As shown in Eq. 3.10, the original displacement coordinate is related to the normal coordinate by

$$q_k = \sum_i B_{ki} Q_i \tag{3.10}$$

Since all the normal vibrations are independent of each other, consideration may be limited to a special case in which only one normal vibration, subscripted by 1, is excited (i.e., $Q_1^0 \neq 0$, $Q_2^0 = Q_3^0 = \cdots = 0$). Then it follows from Eqs. 3.10 and 3.14 that

$$\begin{aligned} q_k = B_{k1} Q_1 &= B_{k1} Q_1^0 \sin(\sqrt{\lambda_1} t + \delta_1) \\ &= A_{k1} \sin(\sqrt{\lambda_1} t + \delta_1) \end{aligned} \tag{3.16}$$

This relation holds for all $k$. Thus it is seen that the excitation of one normal vibration of the system causes vibrations, given by Eq. 3.16, of all the nuclei in the system. In other words, in the normal vibration, all the nuclei move with the same frequency and in phase.

This is true for any other normal vibration. Thus Eq. 3.16 may be written in the more general form

$$q_k = A_k \sin(\sqrt{\lambda}t + \delta) \tag{3.17}$$

If Eq. 3.17 is combined with Eq. 3.7, there results

$$-\lambda A_k + \sum_j b_{kj} A_j = 0 \tag{3.18}$$

This is a system of first-order simultaneous equations with respect to $A$. In order for all the $A$'s to be nonzero,

$$\begin{vmatrix} b_{11}-\lambda & b_{12} & b_{13} & \cdots \\ b_{21} & b_{22}-\lambda & b_{23} & \cdots \\ b_{31} & b_{32} & b_{33}-\lambda & \cdots \\ \cdot & \cdot & \cdot & \\ \cdot & \cdot & \cdot & \\ \cdot & \cdot & \cdot & \end{vmatrix} = 0 \tag{3.19}$$

The order of this secular equation is equal to $3N$. Suppose that one root, $\lambda_1$, is found for Eq. 3.19. If it is inserted in Eq. 3.18, $A_{k1}$, $A_{k2}, \ldots$ are obtained for all the nuclei. The same is true for the other roots of Eq. 3.19. Thus the most general solution may be written as a superposition of all the normal vibrations:

$$q_k = \sum_l B_{kl} Q_l^0 \sin(\sqrt{\lambda_l}t + \delta_l) \tag{3.20}$$

The general discussion developed above may be understood more easily if we apply it to a simple molecule such as $CO_2$, which is constrained to move in only one direction. If the mass and the displacement of each atom are defined as follows:

$m_1$ $\Delta x_1$ $m_2$ $\Delta x_2$ $m_3$ $\Delta x_3$ $x$

the potential energy is given by

$$V = \tfrac{1}{2}k[(\Delta x_1 - \Delta x_2)^2 + (\Delta x_2 - \Delta x_3)^2] \tag{3.21}$$

Considering that $m_1 = m_3$, we find that the kinetic energy is written as

$$T = \tfrac{1}{2}m_1(\Delta\dot{x}_1^2 + \Delta\dot{x}_3^2) + \tfrac{1}{2}m_2\,\Delta\dot{x}_2^2 \tag{3.22}$$

Using the generalized coordinates defined by Eq. 3.2, we may rewrite these energies as

$$2V = k\left[\left(\frac{q_1}{\sqrt{m_1}} - \frac{q_2}{\sqrt{m_2}}\right)^2 + \left(\frac{q_2}{\sqrt{m_2}} - \frac{q_3}{\sqrt{m_1}}\right)^2\right] \tag{3.23}$$

$$2T = \sum \dot{q}_i^2 \tag{3.24}$$

From comparison of Eq. 3.23 with Eq. 3.5, we obtain

$$b_{11} = \frac{k}{m_1} \qquad b_{22} = \frac{2k}{m_2}$$

$$b_{12} = b_{21} = -\frac{k}{\sqrt{m_1 m_2}} \qquad b_{23} = b_{32} = -\frac{k}{\sqrt{m_1 m_2}}$$

$$b_{13} = b_{31} = 0 \qquad b_{33} = \frac{k}{m_1}$$

If these terms are inserted in Eq. 3.19, we obtain the following result:

$$\begin{vmatrix} \frac{k}{m_1} - \lambda & -\frac{k}{\sqrt{m_1 m_2}} & 0 \\ -\frac{k}{\sqrt{m_1 m_2}} & \frac{2k}{m_2} - \lambda & -\frac{k}{\sqrt{m_1 m_2}} \\ 0 & -\frac{k}{\sqrt{m_1 m_2}} & \frac{k}{m_1} - \lambda \end{vmatrix} = 0 \tag{3.25}$$

By solving this secular equation, we obtain three roots:

$$\lambda_1 = \frac{k}{m_1}, \qquad \lambda_2 = k\mu, \qquad \lambda_3 = 0$$

where

$$\mu = \frac{2m_1 + m_2}{m_1 m_2}$$

Equation 3.18 gives the following three equations:

$$-\lambda A_1 + b_{11}A_1 + b_{12}A_2 + b_{13}A_3 = 0$$
$$-\lambda A_2 + b_{21}A_1 + b_{22}A_2 + b_{23}A_3 = 0$$
$$-\lambda A_3 + b_{31}A + b_{32}A_2 + b_{33}A_3 = 0$$

Using Eq. 3.17, we rewrite these as

$$(b_{11}-\lambda)q_1+b_{12}q_2+b_{13}q_3=0$$
$$b_{21}q_1+(b_{22}-\lambda)q_2+b_{23}q_3=0$$
$$b_{31}q_1+b_{32}q_2+(b_{33}-\lambda)q_3=0$$

If $\lambda_1=k/m_1$ is inserted in the above simultaneous equations, we obtain

$$q_1=-q_3, \qquad q_2=0$$

Similar calculations give

$$q_1=q_3, \qquad q_2=-2\sqrt{\frac{m_1}{m_2}}\,q_1 \qquad \text{for } \lambda_2=k\mu$$

$$q_1=q_3, \qquad q_2=\sqrt{\frac{m_2}{m_1}}\,q_1 \qquad \text{for } \lambda_3=0$$

The relative displacements are depicted in the following figure:

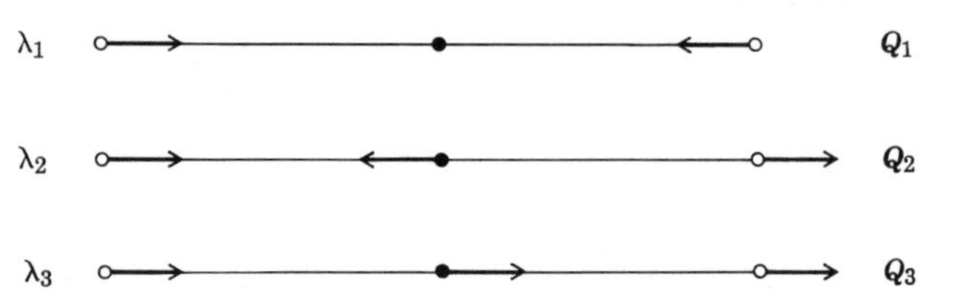


It is easy to see that $\lambda_3$ corresponds to the translational mode ($\Delta x_1=\Delta x_2=\Delta x_3$). The inclusion of $\lambda_3$ could be avoided if we consider the restriction that the center of gravity does not move ; $m_1(\Delta x_1+\Delta x_3)+m_2\,\Delta x_2=0$.

The relationships between the generalized coordinates and the normal coordinates are given by Eq. 3.10. In the present case, we have

$$q_1=B_{11}Q_1+B_{12}Q_2+B_{13}Q_3$$
$$q_2=B_{21}Q_1+B_{22}Q_2+B_{23}Q_3$$
$$q_3=B_{31}Q_1+B_{32}Q_2+B_{33}Q_3$$

In the normal vibration whose normal coordinate is $Q_1$, $B_{11}:B_{21}:B_{31}$ gives the ratio of the displacements. From the previous calculation, it is obvious that $B_{11}:B_{21}:B_{31}=1:0:-1$. Similarly, $B_{12}:B_{22}:B_{32}=1:-2\sqrt{m_1/m_2}:1$ gives the ratio of the displacements in the normal vibration whose normal coordinate is $Q_2$. Thus the mode of a normal vibration can be drawn if the normal coordinate is translated into a set of rectangular coordinates, as is shown above.

So far, we have discussed only the vibrations whose displacements occur along the molecular axis. There are, however, two other normal vibrations in which the displacements occur in the direction perpendicular to the molecular axis. They are not treated here, since the calculation is not simple. It is clear that the method described above will become more complicated as a molecule becomes larger. In this respect, the **GF** matrix method described in Sec. I-11 is important in the vibrational analysis of complex molecules.

By using the normal coordinates, the Schrödinger wave equation for the system can be written as

$$\sum_i \frac{\partial^2 \psi_n}{\partial Q_i^2} + \frac{8\pi^2}{h^2}\left(E - \tfrac{1}{2}\sum_i \lambda_i Q_i^2\right)\psi_n = 0 \tag{3.26}$$

Since the normal coordinates are independent of each other, it is possible to write

$$\psi_n = \psi_1(Q_1)\cdot\psi_2(Q_2)\cdot\cdots \tag{3.27}$$

and solve the simpler one-dimensional problem.

If Eq. 3.27 is substituted in Eq. 3.26, there results

$$\frac{d^2\psi_i}{dQ_i^2} + \frac{8\pi^2}{h^2}(E_i - \tfrac{1}{2}\lambda_i Q_i^2)\psi_i = 0 \tag{3.28}$$

where

$$E = E_1 + E_2 + \cdots$$

with

$$E_i = h\nu_i(v_i + \tfrac{1}{2})$$

$$\nu_i = \frac{1}{2\pi}\sqrt{\lambda_i} \tag{3.29}$$

## I-4. SYMMETRY ELEMENTS AND POINT GROUPS[11–15]

As noted before, polyatomic molecules have $3N-6$ or, if linear, $3N-5$ normal vibrations. For any given molecule, however, only vibrations that are permitted by the selection rule for that molecule appear in the infrared and Raman spectra. Since the selection rule is determined by the symmetry of the molecule, this must first be studied.

The spatial geometrical arrangement of the nuclei constituting the molecule determines its symmetry. If a coordinate transformation (a reflection or a rotation or a combination of both) produces a configuration of the nuclei indistinguishable from the original one, this transformation is called a *symmetry operation*, and the molecule is said to have a corresponding *symmetry element*. Molecules may have the following symmetry elements.

**(1) Identity, *I***

This symmetry element is possessed by every molecule no matter how unsymmetrical it is, the corresponding operation being to leave the molecule unchanged. The inclusion of this element is necessitated by mathematical reasons which will be discussed in Sec. I-6.

**(2) A Plane of Symmetry, $\sigma$**

If reflection of a molecule with respect to some plane produces a configuration indistinguishable from the original one, the plane is called a plane of symmetry.

**(3) A Center of Symmetry, *i***

If reflection at the center, that is, inversion, produces a configuration indistinguishable from the original one, the center is called a center of symmetry. This operation changes the signs of all the coordinates involved, $x_i \rightarrow -x_i$, $y_i \rightarrow -y_i$, $z_i \rightarrow -z_i$.

**(4) A *p*-fold Axis of Symmetry, $C_p$**

If rotation through an angle $360°/p$ about an axis produces a configuration indistinguishable from the original one, the axis is called a $p$-fold axis of symmetry, $C_p$. For example, a twofold axis, $C_2$, implies that a rotation of 180° about the axis reproduces the original configuration. A molecule may have a two-, three-, four-, five-, or sixfold, or higher axis. A linear molecule has an infinite-fold (denoted by $\infty$-fold) axis of symmetry, $C_\infty$, since a rotation of $360°/\infty$, that is an infinitely small angle, transforms the molecule into one indistinguishable from the original.

**(5) A *p*-fold Rotation–Reflection Axis, $S_p$**

If rotation by $360°/p$ about the axis, followed by reflection at a plane perpendicular to the axis, produces a configuration indistinguishable from the original one, the axis is called a $p$-fold rotation–reflection axis. A molecule may have a two-, three-, four-, five, or sixfold, or higher, rotation–reflection axis. A symmetrical linear molecule has an $S_\infty$ axis. It is easily seen that the presence of $S_p$ always means the presence of $C_p$ as well as $\sigma$ when $p$ is odd.

A molecule may have more than one of these symmetry elements. Combination of more and more of these elements produces systems of higher and higher symmetry. Not all combinations of symmetry elements, however, are possible. For example, it is highly improbable that a molecule will have a $C_3$ and $C_4$ axis in the same direction because this requires the existence of a twelvefold axis in the molecule. It should also be noted that the presence of some symmetry elements often implies the presence of other elements. For example, if a molecule has two $\sigma$-planes at right angles to each other, the line of intersection of these two planes must be a $C_2$ axis. A possible combination of symmetry operations whose axes intersect at a point is called a *point group.**

Theoretically, an infinite number of point groups exist, since there is no restriction on the order ($p$) of rotation axes which may exist in an isolated molecule. Practically, however, there are few molecules and ions that possess a rotation axis higher than $C_6$. Thus most of the compounds discussed in this book belong to the following point groups.

1. $\mathbf{C}_p$. Molecules having only a $C_p$ and no other elements of symmetry: $\mathbf{C}_1$, $\mathbf{C}_2$, $\mathbf{C}_3$, etc.
2. $\mathbf{C}_{ph}$. Molecules having a $C_p$ and a $\sigma_h$ perpendicular to it: $\mathbf{C}_{1h} \equiv \mathbf{C}_s$, $\mathbf{C}_{2h}$, $\mathbf{C}_{3h}$, etc.
3. $\mathbf{C}_{pv}$. Molecules having a $C_p$ and $p\,\sigma_v$ through it: $\mathbf{C}_{1v} \equiv \mathbf{C}_s$, $\mathbf{C}_{2v}$, $\mathbf{C}_{3v}$, $\mathbf{C}_{4v}, \ldots, \mathbf{C}_{\infty v}$.
4. $\mathbf{D}_p$. Molecules having a $C_p$ and $p\,C_2$ perpendicular to the $C_p$ and at equal angles to one another: $\mathbf{D}_1 \equiv \mathbf{C}_2$, $\mathbf{D}_2 \equiv \mathbf{V}$, $\mathbf{D}_3$, $\mathbf{D}_4$, etc.
5. $\mathbf{D}_{ph}$. Molecules having a $C_p$, $p\sigma_v$ through it at angles of 360°/2$p$ to one another, and a $\sigma_h$ perpendicular to the $C_p$: $\mathbf{D}_{1h} \equiv \mathbf{C}_{2v}$, $\mathbf{D}_{2h} \equiv \mathbf{V}_h$, $\mathbf{D}_{3h}$, $\mathbf{D}_{4h}$, $\mathbf{D}_{5h}$, $\mathbf{D}_{6h}, \ldots, \mathbf{D}_{\infty h}$.
6. $\mathbf{D}_{pd}$. Molecules having a $C_p$, $p\,C_2$ perpendicular to it, and $p\,\sigma_d$ which go through the $C_p$ and bisect the angles between two successive $C_2$ axes: $\mathbf{D}_{2d} \equiv \mathbf{V}_d$, $\mathbf{D}_{3d}$, $\mathbf{D}_{4d}$, $\mathbf{D}_{5d}$, etc.
7. $\mathbf{S}_p$. Molecules having only a $S_p$ ($p$ even). For $p$ odd, $S_p$ is equivalent to $C_p \times \sigma_h$, for which other notations such as $\mathbf{C}_{3h}$ are used: $\mathbf{S}_2 \equiv \mathbf{C}_i$, $\mathbf{S}_4$, $\mathbf{S}_6$, etc.
8. $\mathbf{T}_d$. Molecules having three mutually perpendicular $C_2$ axes, four $C_3$ axes, and a $\sigma_d$ through each pair of $C_3$ axes: regular tetrahedral molecules.
9. $\mathbf{O}_h$. Molecules having three mutually perpendicular $C_4$ axes, four $C_3$ axes, and a center of symmetry, $i$: regular octahedral and cubic molecules.

Complete listings of the symmetry elements present in each point group are given in the character tables in Appendix I. From the symmetry point of view, molecules belonging to the $\mathbf{C}_1$, $\mathbf{C}_2$, $\mathbf{C}_3$, $\mathbf{D}_2 \equiv \mathbf{V}$, and $\mathbf{D}_3$ groups are optically active since they lack an $S_p$ axis.

---

* In this respect, point groups differ from space groups, which involve translations and rotations about nonintersecting axes (see Sec. I-22).

## I-5. SYMMETRY OF NORMAL VIBRATIONS AND SELECTION RULES

Figure I-7 indicates the normal modes of vibration in $CO_2$ and $H_2O$ molecules. In each normal vibration, the individual nuclei carry out a simple harmonic motion in the direction indicated by the arrow, and all the nuclei have the same frequency of oscillation (i.e., the frequency of the normal vibration) and are moving in the same phase. Furthermore, the relative lengths of the arrows indicate the relative velocities and the amplitudes for each nucleus.* The $\nu_2$ vibrations in $CO_2$ are worth comment, since they differ from the others in that two vibrations ($\nu_{2a}$ and $\nu_{2b}$) have exactly the same frequency. Apparently, there are an infinite number of normal vibrations of this type, which differ only in their directions perpendicular to the molecular axis. Any of them, however, can be resolved into two vibrations such as $\nu_{2a}$ and $\nu_{2b}$, which are perpendicular to each other. In this respect, the $\nu_2$ vibrations in $CO_2$ are called *doubly degenerate vibrations.* Doubly degenerate vibrations occur only when a molecule has an axis higher than twofold. *Triply degenerate vibrations* also occur in molecules having more than one $C_3$ axis.

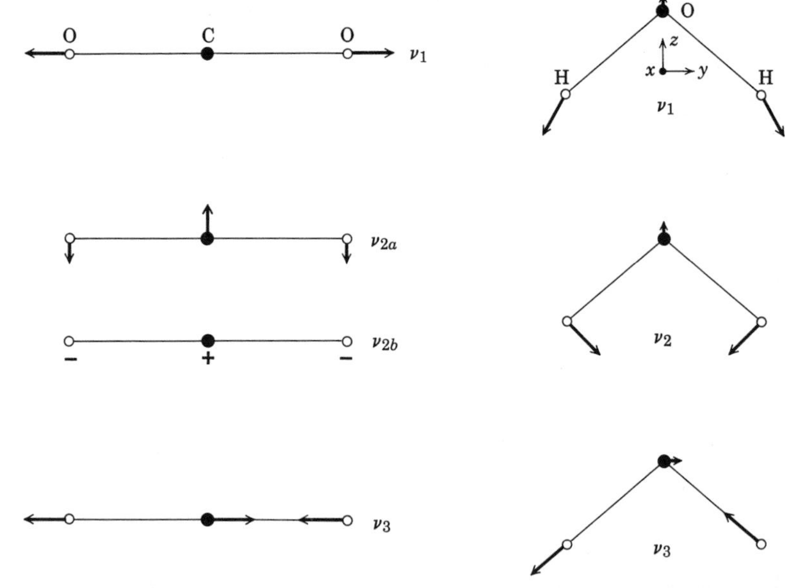


**Fig. I-7.** Normal modes of vibration in $CO_2$ and $H_2O$ molecules (+ and − denote vibrations going upward and downward, respectively, in the direction perpendicular to the paper plane).

---

* In this respect, all the normal modes of vibration shown in this book are only approximate.

To determine the symmetry of a normal vibration, it is necessary to begin by considering the kinetic and potential energies of the system. These were discussed in Sec. I-3.

$$T = \frac{1}{2} \sum_i \dot{Q}_i^2 \tag{5.1}$$

$$V = \frac{1}{2} \sum_i \lambda_i Q_i^2 \tag{5.2}$$

Consider a case in which a molecule performs only one normal vibration, $Q_i$. Then $T = \frac{1}{2}\dot{Q}_i^2$ and $V = \frac{1}{2}\lambda_i Q_i^2$. These energies must be invariant when a symmetry operation, $R$, changes $Q_i$ to $RQ_i$. Thus

$$T = \tfrac{1}{2}\dot{Q}_i^2 = \tfrac{1}{2}(R\dot{Q}_i)^2$$
$$V = \tfrac{1}{2}\lambda_i Q_i^2 = \tfrac{1}{2}\lambda_i (RQ_i)^2$$

For these relations to hold, it is necessary that

$$(RQ_i)^2 = Q_i^2 \qquad \text{or} \qquad RQ_i = \pm Q_i \tag{5.3}$$

Thus the normal coordinate must change either into itself or into its negative. If $Q_i = RQ_i$, the vibration is said to be *symmetric*. If $Q_i = -RQ_i$, it is said to be *antisymmetric*.

If the vibration is doubly degenerate, we have

$$T = \tfrac{1}{2}\dot{Q}_{ia}^2 + \tfrac{1}{2}\dot{Q}_{ib}^2$$
$$V = \tfrac{1}{2}\lambda_i (Q_{ia})^2 + \tfrac{1}{2}\lambda_i (Q_{ib})^2$$

In this case, a relation such as

$$(RQ_{ia})^2 + (RQ_{ib})^2 = Q_{ia}^2 + Q_{ib}^2 \tag{5.4}$$

must hold. As will be shown later, such a relationship is expressed more conveniently by using a matrix form:

$$R\begin{bmatrix} Q_{ia} \\ Q_{ib} \end{bmatrix} = \begin{bmatrix} A & B \\ C & D \end{bmatrix}\begin{bmatrix} Q_{ia} \\ Q_{ib} \end{bmatrix}$$

where the values of $A$, $B$, $C$, and $D$ depend on the symmetry operation, $R$. In any case, the normal vibration must be either symmetric or antisymmetric or degenerate for each symmetry operation.

The symmetry properties of the normal vibrations of the $H_2O$ molecule shown in Fig. I-7 are classified as indicated in Table I-1. Here, $+1$ and $-1$ denote symmetric and antisymmetric, respectively. In the $\nu_1$ and $\nu_2$ vibrations, all the symmetry properties are preserved during the vibration. Therefore they are *symmetric vibrations* and are called, in particular, *totally symmetric vibrations*. In the $\nu_3$ vibration, however, symmetry

TABLE I-1

| $\mathbf{C}_{2v}$ | $I$ | $C_2(z)$ | $\sigma_v(xz)$[a] | $\sigma_v(yz)$[a] |
|---|---|---|---|---|
| $Q_1, Q_2$ | +1 | +1 | +1 | +1 |
| $Q_3$ | +1 | −1 | −1 | +1 |

[a] $\sigma_v$ : vertical plane of symmetry.

elements such as $C_2$ and $\sigma_v(xz)$ are lost. Thus it is called a *nonsymmetric vibration.* If a molecule has a number of symmetry elements the normal vibrations are classified as various species according to the number and the kind of symmetry elements preserved during the vibration.

To determine the activity of the vibrations in the infrared and Raman spectra, the selection rule must be applied to each normal vibration. From a quantum mechanical point of view, *a vibration is active in the infrared spectrum if the dipole moment of the molecule is changed during the vibration, and is active in the Raman spectrum if the polarizability of the molecule is changed during the vibration.* As stated in Sec. I-1, the induced dipole moment $P$ is related to the strength of the electric field $E$ by the relation

$$P = \alpha E$$

where $\alpha$ is called the *polarizability*. If we resolve $P$, $\alpha$, and $E$ in the $x$, $y$, and $z$ directions, simple relationships such as

$$P_x = \alpha_x E_x, \qquad P_y = \alpha_y E_y, \qquad \text{and} \qquad P_z = \alpha_z E_z \tag{5.6}$$

do not hold, since the direction of polarization does not coincide with the direction of the applied field. This is so because the direction of chemical bonds in the molecule also affects the direction of polarization. Thus, instead of Eq. 5.6, we have the relationships:

$$\begin{aligned} P_x &= \alpha_{xx}E_x + \alpha_{xy}E_y + \alpha_{xz}E_z \\ P_y &= \alpha_{yx}E_x + \alpha_{yy}E_y + \alpha_{yz}E_z \\ P_z &= \alpha_{zx}E_x + \alpha_{zy}E_y + \alpha_{zz}E_z \end{aligned} \tag{5.7}$$

In matrix form, Eq. 5.7 is written as

$$\begin{bmatrix} P_x \\ P_y \\ P_z \end{bmatrix} = \begin{bmatrix} \alpha_{xx} & \alpha_{xy} & \alpha_{xz} \\ \alpha_{yx} & \alpha_{yy} & \alpha_{yz} \\ \alpha_{zx} & \alpha_{zy} & \alpha_{zz} \end{bmatrix} \begin{bmatrix} E_x \\ E_y \\ E_z \end{bmatrix} \tag{5.8}$$

and the first matrix on the right-hand side is called the *polarizability tensor.* In normal Raman scattering, the tensor is symmetric; $\alpha_{xy} = \alpha_{yx}$,

$\alpha_{yz} = \alpha_{zy}$, and $\alpha_{xz} = \alpha_{zx}$. This is not so, however, in the case of resonance Raman scattering (Sec. I-20).

According to quantum mechanics, the vibration is Raman active if one of these six components of the polarizability changes during the vibration. Similarly, it is infrared active if one of the three components of the dipole moment ($\mu_x$, $\mu_y$, and $\mu_z$) changes during the vibration. Changes in dipole moment or polarizability are not obvious from inspection of the normal modes of vibration in most polyatomic molecules. As will be shown later, application of group theory gives a clear-cut solution to this problem.

In simple molecules, however, the activity of a vibration may be determined by inspection of the normal mode. For example, it is obvious that the vibration in a homopolar diatomic molecule is not infrared active but is Raman active, whereas the vibration in a heteropolar diatomic molecule is both infrared and Raman active. It is also obvious that all three vibrations of $H_2O$ and $\nu_2$ and $\nu_3$ of $CO_2$ are infrared active. Except for $\nu_1$ of $CO_2$, the Raman activity is not easy to predict even for such simple molecules.

The polarizability tensor can be visualized easily if we draw a *polarizability ellipsoid* by plotting $1/\sqrt{\alpha}$ in any direction from the origin. This gives a three-dimensional surface such as is shown in Fig. I-8. If we orient this ellipsoid with its principal axes along the $X$, $Y$, $Z$ axes of the coordinate system, Eq. 5.8 is simplified to

$$\begin{bmatrix} P_X \\ P_Y \\ P_Z \end{bmatrix} = \begin{bmatrix} \alpha_{XX} & 0 & 0 \\ 0 & \alpha_{YY} & 0 \\ 0 & 0 & \alpha_{ZZ} \end{bmatrix} \begin{bmatrix} E_X \\ E_Y \\ E_Z \end{bmatrix} \tag{5.9}$$

These three axes are called the *principal axes of polarizability*. In terms of the polarizability ellipsoid, the Raman selection rule can be stated as follows: *The vibration is Raman active if the ellipsoid changes in size, shape, or orientation during the vibration.* Consider the $\nu_1$ vibration of $CO_2$. As shown in Fig. I-8, the ellipsoid changes its size during this vibration (or $\alpha_{xx}$, $\alpha_{yy}$, and $\alpha_{zz}$ change during the vibration). Thus it is Raman active. Although the size of the ellipsoid changes during the $\nu_3$ vibration, they are identical in two extreme positions, as shown in Fig. 1–8. If we consider a limiting case where the nuclei undergo very small displacements, there is effectively no change in the polarizability; hence the $\nu_3$ vibration is not Raman active. The same is true for the $\nu_2$ vibration.

In the case of $H_2O$, both the $\nu_1$ and $\nu_2$ vibrations are Raman active because the size and the shape of the ellipsoid change during these vibrations. The $\nu_3$ vibration of $H_2O$ is different from the other vibrations

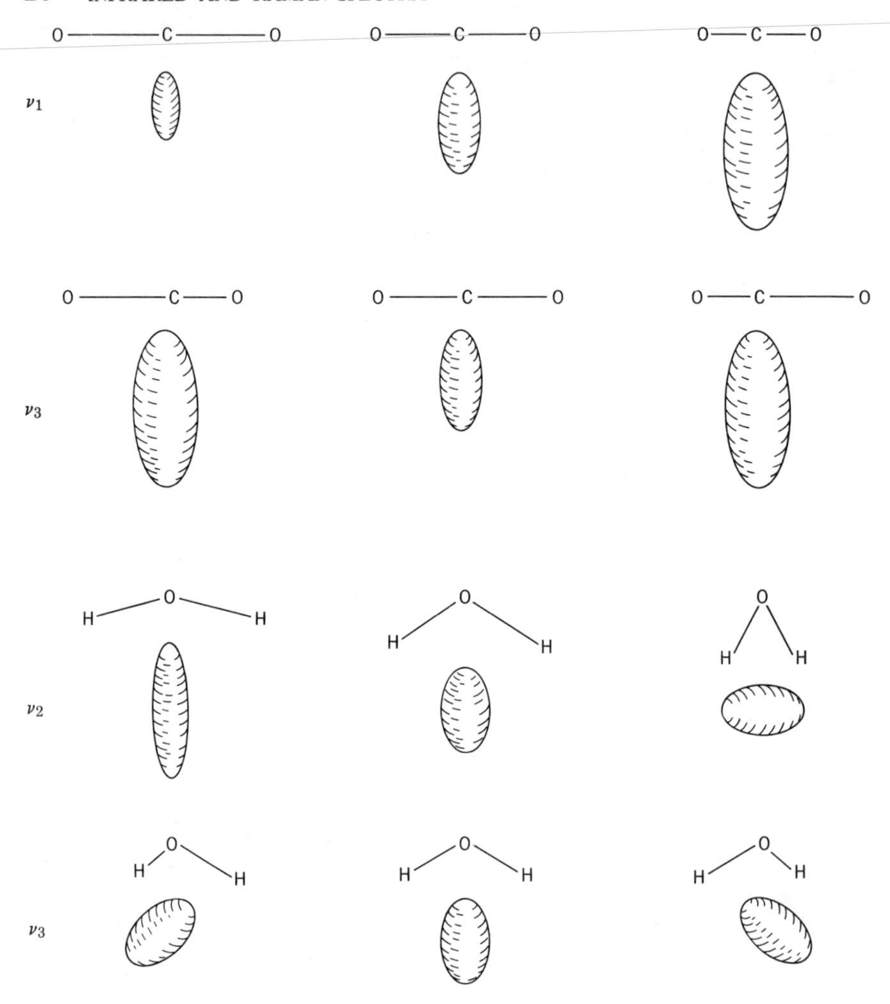


**Fig. I-8.** Changes of polarizability ellipsoids during the normal vibrations of $CO_2$ and $H_2O$.

in that the orientation of the ellipsoid changes ($\alpha_{yz}$ changes) during the vibration. Thus all three normal vibrations of $H_2O$ are Raman active.

It should be noted that in $CO_2$ the vibration symmetric with respect to the center of symmetry ($\nu_1$) is Raman active and not infrared active, whereas the vibrations antisymmetric with respect to the center of symmetry ($\nu_2$ and $\nu_3$) are infrared active but not Raman active. In a polyatomic molecule having a center of symmetry, the vibrations symmetric with respect to the center of symmetry ($g$ vibrations*) are Raman active and not infrared active, but the vibrations antisymmetric with respect to the center of symmetry ($u$ vibrations*) are infrared active and not Raman active. This rule is called the *mutual exclusion rule.* It should be noted, however, that in polyatomic molecules having several symmetry elements in addition to the center of symmetry, the vibrations that should be active according to this rule may not necessarily be active, because of the presence of other symmetry elements. An example is seen in a square-planar $XY_4$ type molecule of $\mathbf{D}_{4h}$ symmetry, where the $A_{2g}$ vibrations are not Raman active and the $A_{1u}$, $B_{1u}$, and $B_{2u}$ vibrations are not infrared active (see Sec. II-6 and Appendix I).

## I-6. INTRODUCTION TO GROUP THEORY

In Sec. I-4, the symmetry and the point group allocation of a given molecule were discussed. To understand the symmetry and selection rules of normal vibrations in polyatomic molecules, however, a knowledge of group theory is required. The minimum amount of group theory needed for this purpose is given here.†

Consider a pyramidal $XY_3$ molecule (Fig. I-9) for which the symmetry operations $I$, $C_3^+$, $C_3^-$, $\sigma_1$, $\sigma_2$, and $\sigma_3$ are applicable. Here $C_3^+$ and $C_3^-$ denote rotation through 120° in the clockwise and counterclockwise directions, respectively, and $\sigma_1$, $\sigma_2$, and $\sigma_3$ indicate the symmetry planes that pass through X and $Y_1$, X and $Y_2$, and X and $Y_3$, respectively. For simplicity, let these symmetry operations be denoted by $I$, $A$, $B$, $C$, $D$, and $E$, respectively. Other symmetry operations are possible, but each is equivalent to some one of the operations mentioned. For instance, a

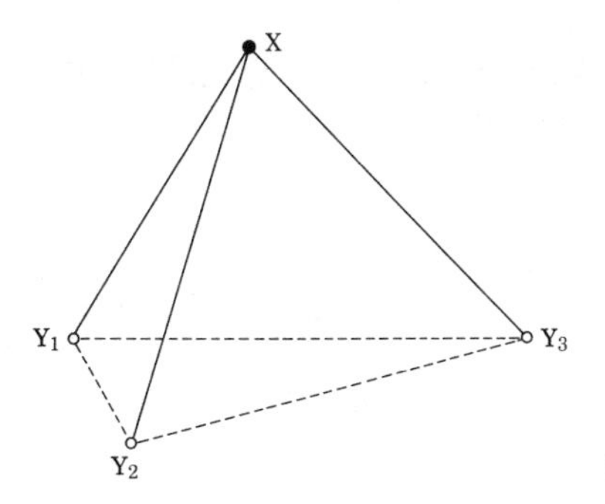


**Fig. I-9.** Pyramidal $XY_3$ molecule.

---

* The symbols $g$ and $u$ stand for *gerade* and *ungerade* (German), respectively.
† For details on group theory and matrix theory, see Refs. 11–15.

TABLE I-2

| | I | A | B | C | D | E |
|---|---|---|---|---|---|---|
| I | I | A | B | C | D | E |
| A | A | B | I | D | E | C |
| B | B | I | A | E | C | D |
| C | C | E | D | I | B | A |
| D | D | C | E | A | I | B |
| E | E | D | C | B | A | I |

clockwise rotation through 240° is identical with operation *B*. It may also be shown that two successive applications of any one of these operations is equivalent to some single operation of the group mentioned. Let operation *C* be applied to the original figure. This interchanges $Y_2$ and $Y_3$. If operation *A* is applied to the resulting figure, the net result is the same as application of the single operation *D* to the original figure. This is written as $CA = D$. If all the possible multiplicative combinations are made, Table I-2, in which the operation applied first is written across the top, is obtained. This is called the *multiplication table* of the group.

It is seen that a group consisting of the mathematical elements (symmetry operations) *I*, *A*, *B*, *C*, *D*, and *E* satisfies the following conditions:

1. The product of any two elements in the set is another element in the set.
2. The set contains the identity operation that satisfies the relation $IP = PI = P$, where *P* is any element in the set.
3. The associative law holds for all the elements in the set, that is, $(CB)A = C(BA)$, for example.
4. Every element in the set has its reciprocal, *X*, which satisfies the relation $XP = PX = I$, where *P* is any element in the set. This reciprocal is usually denoted by $P^{-1}$.

These are necessary and sufficient conditions for a set of elements to form a *group*. It is evident that operations *I*, *A*, *B*, *C*, *D*, and *E* form a group in this sense. It should be noted that the commutative law of multiplication does not necessarily hold. For example, Table I-2 shows that $CD \neq DC$.

The six elements can be classified into three types of operations: the identity operation, *I*; the rotations, $C_3^+$ and $C_3^-$; and the reflections, $\sigma_1$, $\sigma_2$, and $\sigma_3$. Each of these sets of operations is said to form a *class*. More precisely, two operations, *P* and *Q*, which satisfy the relation $X^{-1}PX = P$

or $Q$, where $X$ is any operation of the group and $X^{-1}$ is its reciprocal, are said to belong to the same class. It can easily be shown that $C_3^+$ and $C_3^-$, for example, satisfy the relation. Thus the six elements of the point group $\mathbf{C}_{3v}$ are usually abbreviated as $I$, $2C_3$, and $3\sigma_v$.

The relations between the elements of the group are shown in the multiplication table, Table I-2. Such a tabulation of a group is, however, awkward to handle. The essential features of the table may be abstracted by replacing the elements by some analytical function that reproduces the multiplication table. Such an analytical expression may be composed of a simple integer, an exponential function, or a matrix. Any set of such expressions that satisfies the relations given by the multiplication table is called a *representation* of the group and is designated by $\Gamma$. The representations of the point group $\mathbf{C}_{3v}$ discussed above are indicated in Table I-3. It is easily proved that each representation in the table satisfies the multiplication table.

In addition to the three representations in Table I-3, it is possible to write an infinite number of other representations of the group. If a set of six matrices of the type $S^{-1}R(K)S$ is chosen, where $R(K)$ is a representation of the element $K$ given in Table I-3, $S(|S| \neq 0)$ is any matrix of the same order as $R$, and $S^{-1}$ is the reciprocal of $S$, this set also satisfies the relations given by the multiplication table. The reason is obvious from the relation

$$S^{-1}R(K)SS^{-1}R(L)S = S^{-1}R(K)R(L)S = S^{-1}R(KL)S$$

Such a transformation is called a *similarity transformation.* Thus it is possible to make an infinite number of representations by means of similarity transformations.

On the other hand, this statement suggests that a given representation may be broken into simpler ones. If each representation of the symmetry

TABLE I-3

| $\mathbf{C}_{3v}$ | $I$ | $A$ | $B$ | $C$ | $D$ | $E$ |
|---|---|---|---|---|---|---|
| $A_1(\Gamma_1)$ | 1 | 1 | 1 | 1 | 1 | 1 |
| $A_2(\Gamma_2)$ | 1 | 1 | 1 | −1 | −1 | −1 |
| $E(\Gamma_3)$ | $\begin{pmatrix} 1 & 0 \\ 0 & 1 \end{pmatrix}$ | $\begin{pmatrix} -\frac{1}{2} & \frac{\sqrt{3}}{2} \\ -\frac{\sqrt{3}}{2} & -\frac{1}{2} \end{pmatrix}$ | $\begin{pmatrix} -\frac{1}{2} & -\frac{\sqrt{3}}{2} \\ \frac{\sqrt{3}}{2} & -\frac{1}{2} \end{pmatrix}$ | $\begin{pmatrix} -1 & 0 \\ 0 & 1 \end{pmatrix}$ | $\begin{pmatrix} \frac{1}{2} & -\frac{\sqrt{3}}{2} \\ -\frac{\sqrt{3}}{2} & -\frac{1}{2} \end{pmatrix}$ | $\begin{pmatrix} \frac{1}{2} & \frac{\sqrt{3}}{2} \\ \frac{\sqrt{3}}{2} & -\frac{1}{2} \end{pmatrix}$ |

element $K$ is transformed into the form

$$R(K) = \begin{vmatrix} Q_1(K) & 0 & 0 & 0 \\ 0 & Q_2(K) & 0 & 0 \\ 0 & 0 & Q_3(K) & 0 \\ 0 & 0 & 0 & Q_3(K) \end{vmatrix} \tag{6.1}$$

by a similarity transformation, $Q_1(K)$, $Q_2(K)$, ... are simpler representations. In such a case, $R(K)$ is called *reducible.* If a representation cannot be simplified any further, it is said to be *irreducible.* The representations $\Gamma_1$, $\Gamma_2$, and $\Gamma_3$ in Table I-3 are all irreducible representations. It can be shown generally that the number of irreducible representations is equal to the number of classes. Thus only three irreducible representations exist for the point group $\mathbf{C}_{3v}$. These representations are entirely independent of each other. Furthermore, the sum of the squares of the dimensions ($l$) of the irreducible representations of a group is always equal to the total number of the symmetry elements, namely, the *order of the group* ($h$). Thus

$$\sum l_i^2 = l_1^2 + l_2^2 + \cdots = h \tag{6.2}$$

In the point group $\mathbf{C}_{3v}$, it is seen that

$$1^2 + 1^2 + 2^2 = 6$$

A point group is classified into *species* according to its irreducible representations. In the point group $\mathbf{C}_{3v}$, the species having the irreducible representations $\Gamma_1$, $\Gamma_2$, and $\Gamma_3$ are called the $A_1$, $A_2$, and $E$ species, respectively.*

The sum of the diagonal elements of a matrix is called the *character* of the matrix and is denoted by $\chi$. It is to be noted in Table I-3 that the character of each of the elements belonging to the same class is the same. Thus, using the character, Table I-3 can be simplified to Table I-4. Such a table is called the *character table* of the point group $\mathbf{C}_{3v}$.

TABLE I-4. THE CHARACTER TABLE OF THE POINT GROUP $\mathbf{C}_{3v}$

| $\mathbf{C}_{3v}$ | $I$ | $2C_3(z)$ | $3\sigma_v$ |
|---|---|---|---|
| $A_1(\chi_1)$ | 1 | 1 | 1 |
| $A_2(\chi_2)$ | 1 | 1 | −1 |
| $E(\chi_3)$ | 2 | −1 | 0 |

* For the labeling of the irreducible representations (species), see Appendix I.

That the *character* of a matrix is not changed by a similarity transformation can be proved as follows. If a similarity transformation is expressed by $T = S^{-1}RS$, then

$$\chi_T = \sum_i (S^{-1}RS)_{ii} = \sum_{i,j,k} (S^{-1})_{ij}R_{jk}S_{ki} = \sum_{j,k,i} S_{ki}(S^{-1})_{ij}R_{jk}$$

$$= \sum_{j,k} \delta_{kj}R_{jk} = \sum_k R_{kk} = \chi_R$$

where $\delta_{kj}$ is Kronecker's delta (0 for $k \neq j$ and 1 for $k = j$). Thus any reducible representation can be reduced to its irreducible representations by a similarity transformation that leaves the character unchanged. Therefore the character of the reducible representation, $\chi(K)$, is written as

$$\chi(K) = \sum_m a_m \chi_m(K) \tag{6.3}$$

where $\chi_m(K)$ is the character of $Q_m(K)$, and $a_m$ is a positive integer that indicates the number of times $Q_m(K)$ appears in the matrix of Eq. 6.1. Hereafter the character will be used rather than the corresponding representation because a 1:1 correspondence exists between these two, and the former is sufficient for vibrational problems.

It is important to note that the following relation holds in Table I-4:

$$\sum_K \chi_i(K)\chi_j(K) = h\delta_{ij} \tag{6.4}$$

If Eq. 6.3 is multiplied by $\chi_i(K)$ on both sides, and the summation is taken over all the symmetry operations, then

$$\sum_K \chi(K)\chi_i(K) = \sum_K \sum_m a_m \chi_m(K)\chi_i(K)$$

$$= \sum_m \sum_K a_m \chi_m(K)\chi_i(K)$$

For a fixed $m$, we have

$$\sum_K a_m \chi_m(K)\chi_i(K) = a_m \sum_K \chi_m(K)\chi_i(K) = a_m h\delta_{im}$$

If we consider the sum of such a term over $m$, only the sum in which $m = i$ remains. Thus

$$\sum_K \chi(K)\chi_m(K) = ha_m$$

or

$$a_m = \frac{1}{h}\sum_K \chi(K)\chi_m(K) \tag{6.5}$$

This formula is written more conveniently as

$$a_m = \frac{1}{h} \sum n\chi(K)\chi_m(K) \tag{6.6}$$

where $n$ is the number of symmetry elements in any one class, and the summation is made over the different classes. As Sec. I-7 will show, this formula is very useful in determining the number of normal vibrations belonging to each species.

## I-7. THE NUMBER OF NORMAL VIBRATIONS FOR EACH SPECIES

As shown in Sec. I-5, the $3N-6$ (or $3N-5$) normal vibrations of an $N$-atom molecule can be classified into various species according to their symmetry properties. The number of normal vibrations in each species can be calculated by using the general equations given in Appendix II. These equations were derived from consideration of the vibrational degrees of freedom contributed by each set of identical nuclei for each symmetry species.[1] As an example, let us consider the $NH_3$ molecule belonging to the $\mathbf{C}_{3v}$ point group. The general equations are as follows:

$A_1$ species: $3m + 2m_v + m_0 - 1$

$A_2$ species: $3m + m_v - 1$

$E$ species: $6m + 3m_v + m_0 - 2$

$N$ (total number of atoms) $= 6m + 3m_v + m_0$

From the definitions given in the footnotes of Appendix II, it is obvious that $m = 0$, $m_0 = 1$, and $m_v = 1$ in this case. To check these numbers, we calculate the total number of atoms from the equation for $N$ given above. Since the result is 4, these assigned numbers are correct. Then the number of normal vibrations in each species can be calculated by inserting these numbers in the general equations given above: 2, 0, and 2, respectively, for the $A_1$, $A_2$, and $E$ species. Since the $E$ species is doubly degenerate, the total number of vibrations is counted as 6, which is expected from the $3N-6$ rule.

A more general method of finding the number of normal vibrations in each species can be developed by using group theory. The principle of the method is that all the representations are irreducible if normal coordinates are used as the basis for the representations. For example, the representations for the symmetry operations based on three normal coordinates, $Q_1$, $Q_2$, and $Q_3$, which correspond to the $\nu_1$, $\nu_2$, and $\nu_3$ vibrations in the $H_2O$ molecule of Fig. I-5, are as follows:

$$I\begin{bmatrix}Q_1\\Q_2\\Q_3\end{bmatrix}=\begin{bmatrix}1&0&0\\0&1&0\\0&0&1\end{bmatrix}\begin{bmatrix}Q_1\\Q_2\\Q_3\end{bmatrix}\qquad C_2(z)\begin{bmatrix}Q_1\\Q_2\\Q_3\end{bmatrix}=\begin{bmatrix}1&0&0\\0&1&0\\0&0&-1\end{bmatrix}\begin{bmatrix}Q_1\\Q_2\\Q_3\end{bmatrix}$$

$$\sigma_v(xz)\begin{bmatrix}Q_1\\Q_2\\Q_3\end{bmatrix}=\begin{bmatrix}1&0&0\\0&1&0\\0&0&-1\end{bmatrix}\begin{bmatrix}Q_1\\Q_2\\Q_3\end{bmatrix}\qquad \sigma_v(yz)\begin{bmatrix}Q_1\\Q_2\\Q_3\end{bmatrix}=\begin{bmatrix}1&0&0\\0&1&0\\0&0&1\end{bmatrix}\begin{bmatrix}Q_1\\Q_2\\Q_3\end{bmatrix}$$

Let a representation be written with the $3N$ rectangular coordinates of an $N$-atom molecule as its basis. If it is decomposed into its irreducible components, the basis for these irreducible representations must be the normal coordinates, and the number of appearances of the same irreducible representation must be equal to the number of normal vibrations belonging to the species represented by this irreducible representation. As stated previously, however, the $3N$ rectangular coordinates involve six (or five) coordinates, which correspond to the translational and rotational motions of the molecule as a whole. Therefore the representations that have such coordinates as their basis must be subtracted from the result obtained above. Use of the character of the representation, rather than the representation itself, yields the same result.

For example, consider a pyramidal $XY_3$ molecule that has six normal vibrations. At first, the representations for the various symmetry operations must be written with the 12 rectangular coordinates in Fig. I-10 as their basis. Consider pure rotation, $C_p^+$. If the clockwise rotation of the point $(x, y, z)$ around the $z$ axis by the angle $\theta$ brings it to the point denoted by the coordinates $(x', y', z')$, the relations between these two sets of coordinates are given by

$$\begin{aligned}x'&=x\cos\theta+y\sin\theta\\y'&=-x\sin\theta+y\cos\theta\\z'&=z\end{aligned}\tag{7.1}$$

By using matrix notation, this can be written as

$$\begin{bmatrix}x'\\y'\\z'\end{bmatrix}=C_\theta^+\begin{bmatrix}x\\y\\z\end{bmatrix}=\begin{bmatrix}\cos\theta&\sin\theta&0\\-\sin\theta&\cos\theta&0\\0&0&1\end{bmatrix}\begin{bmatrix}x\\y\\z\end{bmatrix}\tag{7.2}$$

Then the character of the matrix is given by

$$\chi(C_\theta^+)=1+2\cos\theta\tag{7.3}$$

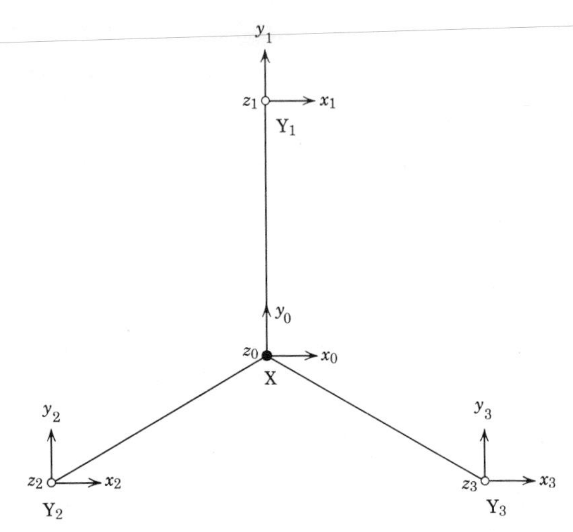


**Fig. I-10.** Rectangular coordinates in a pyramidal $XY_3$ molecule ($z$ axis is perpendicular to the paper plane).

The same result is obtained for $\chi(C_\theta^-)$. If this symmetry operation is applied to all the coordinates of the $XY_3$ molecule, the result is

$$C_\theta \begin{bmatrix} x_0 \\ y_0 \\ z_0 \\ x_1 \\ y_1 \\ z_1 \\ x_2 \\ y_2 \\ z_2 \\ x_3 \\ y_3 \\ z_3 \end{bmatrix} = \left[\begin{array}{c:c:c:c} \mathbf{A} & 0 & 0 & 0 \\ \hdashline 0 & 0 & 0 & \mathbf{A} \\ \hdashline 0 & \mathbf{A} & 0 & 0 \\ \hdashline 0 & 0 & \mathbf{A} & 0 \end{array}\right] \begin{bmatrix} x_0 \\ y_0 \\ z_0 \\ x_1 \\ y_1 \\ z_1 \\ x_2 \\ y_2 \\ z_2 \\ x_3 \\ y_3 \\ z_3 \end{bmatrix} \tag{7.4}$$

where **A** denotes the small square matrix given by Eq. 7.2. Thus the character of this representation is simply given by Eq. 7.3. It should be noted in Eq. 7.4 that only the small matrix **A**, related to the nuclei unchanged by the symmetry operation, appears as a diagonal element. Thus a more general form of the character of the representation for rotation around the axis by $\theta$ is

$$\boxed{\chi(R) = N_R(1 + 2\cos\theta)} \tag{7.5}$$

where $N_R$ is the number of nuclei unchanged by the rotation. In the present case, $N_R = 1$ and $\theta = 120°$. Therefore

$$\chi(C_3) = 0 \tag{7.6}$$

Identity ($I$) can be regarded as a special case of Eq. 7.5 in which $N_R = 4$ and $\theta = 0°$. The character of the representation is

$$\chi(I) = 12 \tag{7.7}$$

Pure rotation and identity are called *proper rotation.*

It is evident from Fig. I-10 that a symmetry plane such as $\sigma_1$ changes the coordinates from $(x_i, y_i, z_i)$ to $(-x_i, y_i, z_i)$. The corresponding representation is therefore written as

$$\sigma_1 \begin{bmatrix} x \\ y \\ z \end{bmatrix} = \begin{bmatrix} -1 & 0 & 0 \\ 0 & 1 & 0 \\ 0 & 0 & 1 \end{bmatrix} \begin{bmatrix} x \\ y \\ z \end{bmatrix} \tag{7.8}$$

The result of such an operation on all the coordinates is

$$\sigma_1 \begin{bmatrix} x_0 \\ y_0 \\ z_0 \\ x_1 \\ y_1 \\ z_1 \\ x_2 \\ y_2 \\ z_2 \\ x_3 \\ y_3 \\ z_3 \end{bmatrix} = \left[\begin{array}{c:c:c:c} \mathbf{B} & 0 & 0 & 0 \\ \hdashline 0 & \mathbf{B} & 0 & 0 \\ \hdashline 0 & 0 & 0 & \mathbf{B} \\ \hdashline 0 & 0 & \mathbf{B} & 0 \end{array}\right] \begin{bmatrix} x_0 \\ y_0 \\ z_0 \\ x_1 \\ y_1 \\ z_1 \\ x_2 \\ y_2 \\ z_2 \\ x_3 \\ y_3 \\ z_3 \end{bmatrix} \tag{7.9}$$

where $\mathbf{B}$ denotes the small square matrix of Eq. 7.8. Thus the character of this representation is calculated as $2 \times 1 = 2$. It is noted again that the matrix on the diagonal is nonzero only for the nuclei unchanged by the operation.

More generally, a reflection at a plane ($\sigma$) is regarded as $\sigma = i \times C_2$. Thus the general form of Eq. 7.8 may be written as

$$\begin{bmatrix} -1 & 0 & 0 \\ 0 & -1 & 0 \\ 0 & 0 & -1 \end{bmatrix} \begin{bmatrix} \cos\theta & \sin\theta & 0 \\ -\sin\theta & \cos\theta & 0 \\ 0 & 0 & 1 \end{bmatrix} = \begin{bmatrix} -\cos\theta & -\sin\theta & 0 \\ \sin\theta & -\cos\theta & 0 \\ 0 & 0 & -1 \end{bmatrix}$$

Then

$$\chi(\sigma) = -(1 + 2\cos\theta)$$

As a result, the character of the large matrix shown in Eq. 7.9 is given by

$$\boxed{\chi(R) = -N_R(1 + 2\cos\theta)} \quad (7.10)$$

In the present case, $N_R = 2$ and $\theta = 180°$. This gives

$$\chi(\sigma_v) = 2 \quad (7.11)$$

Symmetry operations such as $i$ and $S_p$ are regarded as

$$\begin{aligned} i &= i \times I & \theta &= 0° \\ S_3 &= i \times C_6 & \theta &= 60° \\ S_4 &= i \times C_4 & \theta &= 90° \\ S_6 &= i \times C_3 & \theta &= 120° \end{aligned}$$

Therefore the characters of these symmetry operations can be calculated by Eq. 7.10 with the values of $\theta$ defined above. Operations such as $\sigma$, $i$, and $S_p$ are called *improper rotations.* Thus the character of the representation based on 12 rectangular coordinates is as follows:

| $I$ | $2C_3$ | $3\sigma_v$ |
|---|---|---|
| 12 | 0 | 2 |

(7.12)

To determine the number of normal vibrations belonging to each species, the $\chi(R)$ thus obtained must be resolved into the $\chi_i(R)$ of the irreducible representations of each species in Table I-4. First, however, the characters corresponding to the translational and rotational motions of the molecule must be subtracted from the result shown in Eq. 7.12.

The characters for the translational motion of the molecule in the $x$, $y$, and $z$ directions (denoted by $T_x$, $T_y$, and $T_z$) are the same as those obtained in Eqs. 7.5 and 7.10. They are as follows:

$$\boxed{\chi_t(R) = \pm(1 + 2\cos\theta)} \quad (7.13)$$

where the + and − signs are for proper and improper rotations, respectively. The characters for the rotations around the $x$, $y$, and $z$ axes (denoted by $R_x$, $R_y$, and $R_z$) are given by

$$\boxed{\chi_r(R) = +(1 + 2\cos\theta)} \qquad (7.14)$$

for both proper and improper rotations. This is due to the fact that a rotation of the vectors in the plane perpendicular to the $x$, $y$, and $z$ axes can be regarded as a rotation of the components of angular momentum, $M_x$, $M_y$, and $M_z$, about the given axes. If $p_x$, $p_y$, and $p_z$ are the components of linear momentum in the $x$, $y$, and $z$ directions, the following relations hold:

$$M_x = yp_z - zp_y$$
$$M_y = zp_x - xp_z$$
$$M_z = xp_y - yp_x$$

Since $(x, y, z)$ and $(p_x, p_y, p_z)$ transform as shown in Eq. 7.2, it follows that

$$C_\theta \begin{bmatrix} M_x \\ M_y \\ M_z \end{bmatrix} = \begin{bmatrix} \cos\theta & \sin\theta & 0 \\ -\sin\theta & \cos\theta & 0 \\ 0 & 0 & 1 \end{bmatrix} \begin{bmatrix} M_x \\ M_y \\ M_z \end{bmatrix}$$

Then a similar relation holds for $R_x$, $R_y$, and $R_z$:

$$C_\theta \begin{bmatrix} R_x \\ R_y \\ R_z \end{bmatrix} = \begin{bmatrix} \cos\theta & \sin\theta & 0 \\ -\sin\theta & \cos\theta & 0 \\ 0 & 0 & 1 \end{bmatrix} \begin{bmatrix} R_x \\ R_y \\ R_z \end{bmatrix}$$

Thus the characters for the proper rotations are given by Eq. 7.14. The same result is obtained for the improper rotation if the latter is regarded as $i \times$(proper rotation). Therefore the character for the vibration is obtained from

$$\boxed{\chi_v(R) = \chi(R) - \chi_t(R) - \chi_r(R)} \qquad (7.15)$$

It is convenient to tabulate the foregoing calculations as in Table I-5. By using the formula in Eq. 6.6 and the character of the irreducible representations in Table I-4, $a_m$ can be calculated as follows:

$$a_m(A_1) = \tfrac{1}{6}[(1)(6)(1) + (2)(0)(1) + (3)(2)(1)] = 2$$
$$a_m(A_2) = \tfrac{1}{6}[(1)(6)(1) + (2)(0)(1) + (3)(2)(-1)] = 0$$
$$a_m(E) = \tfrac{1}{6}[(1)(6)(2) + (2)(0)(-1) + (3)(2)(0)] = 2$$

TABLE I-5

| Symmetry operation | $I$ | $2C_3$ | $3\sigma_v$ |
|---|---|---|---|
| Kind of rotation | proper | | improper |
| $\theta$ | 0° | 120° | 180° |
| $\cos\theta$ | 1 | $-\frac{1}{2}$ | $-1$ |
| $1+2\cos\theta$ | 3 | 0 | $-1$ |
| $N_R$ | 4 | 1 | 2 |
| $\chi$, $\pm N_R(1+2\cos\theta)$ | 12 | 0 | 2 |
| $\chi_t$, $\pm(1+2\cos\theta)$ | 3 | 0 | 1 |
| $\chi_r$, $+(1+2\cos\theta)$ | 3 | 0 | $-1$ |
| $\chi_v$, $\chi-\chi_t-\chi_r$ | 6 | 0 | 2 |

and

$$\chi_v = 2\chi_{A_1} + 2\chi_E \tag{7.16}$$

In other words, the six normal vibrations of a pyramidal $XY_3$ molecule are classified into two $A_1$ and two $E$ species.

This procedure is applicable to any molecule. As another example, a similar calculation is shown in Table I-6 for an octahedral $XY_6$ molecule. By use of Eq. 6.6 and the character table in Appendix I, the $a_m$ are obtained as

$$\begin{aligned} a_m(A_{1g}) &= \tfrac{1}{48}[(1)(15)(1)+(8)(0)(1)+(6)(1)(1)+(6)(1)(1) \\ &\quad +(3)(-1)(1)+(1)(-3)(1)+(6)(-1)(1)+(8)(0)(1) \\ &\quad +(3)(5)(1)+(6)(3)(1)] \\ &= 1 \end{aligned}$$

$$\begin{aligned} a_m(A_{1u}) &= \tfrac{1}{48}[(1)(15)(1)+(8)(0)(1)+(6)(1)(1)+(6)(1)(1) \\ &\quad +(3)(-1)(1)+(1)(-3)(-1)+(6)(-1)(-1)+(8)(0)(-1) \\ &\quad +(3)(5)(-1)+(6)(3)(-1)] \\ &= 0 \end{aligned}$$

..................................................

and therefore

$$\chi_v = \chi_{A_{1g}} + \chi_{E_g} + 2\chi_{F_{1u}} + \chi_{F_{2g}} + \chi_{F_{2u}}$$

TABLE I-6

| Symmetry operation | $I$ | $8C_3$ | $6C_2$ | $6C_4$ | $3C_4^2 \equiv C_2''$ | $S_2 \equiv i$ | $6S_4$ | $8S_6$ | $3\sigma_h$[a] | $6\sigma_d$[a] |
|---|---|---|---|---|---|---|---|---|---|---|
| Kind of rotation | proper | | | | | improper | | | | |
| $\theta$ | 0° | 120° | 180° | 90° | 180° | 0° | 90° | 120° | 180° | 180° |
| $\cos\theta$ | 1 | $-\frac{1}{2}$ | $-1$ | 0 | $-1$ | 1 | 0 | $-\frac{1}{2}$ | $-1$ | $-1$ |
| $1+2\cos\theta$ | 3 | 0 | $-1$ | 1 | $-1$ | 3 | 1 | 0 | $-1$ | $-1$ |
| $N_R$ | 7 | 1 | 1 | 3 | 3 | 1 | 1 | 1 | 5 | 3 |
| $\chi, \pm N_R(1+2\cos\theta)$ | 21 | 0 | $-1$ | 3 | $-3$ | $-3$ | $-1$ | 0 | 5 | 3 |
| $\chi_t, \pm(1+2\cos\theta)$ | 3 | 0 | $-1$ | 1 | $-1$ | $-3$ | $-1$ | 0 | 1 | 1 |
| $\chi_r, +(1+2\cos\theta)$ | 3 | 0 | $-1$ | 1 | $-1$ | 3 | 1 | 0 | $-1$ | $-1$ |
| $\chi_v, \chi-\chi_t-\chi_r$ | 15 | 0 | 1 | 1 | $-1$ | $-3$ | $-1$ | 0 | 5 | 3 |

[a] $\sigma_h$ = horizontal plane of symmetry; $\sigma_d$ = diagonal plane of symmetry

## I-8. INTERNAL COORDINATES

In Sec. I-3, the potential and the kinetic energies were expressed in terms of rectangular coordinates. If, instead, these energies are expressed in terms of *internal coordinates* such as increments of the bond length and bond angle, the corresponding force constants have clearer physical meanings than those expressed in terms of rectangular coordinates, since these force constants are characteristic of the bond stretching and the angle deformation involved. The number of internal coordinates must be equal to, or greater than, $3N-6$ (or $3N-5$), the degrees of vibrational freedom of an $N$-atom molecule. If more than $3N-6$ (or $3N-5$) coordinates are selected as the internal coordinates, this means that these coordinates are not independent of each other. Figure I-11 illustrates the internal coordinates for various types of molecules.

In linear XYZ (*a*), bent $XY_2$ (*b*), and pyramidal $XY_3$ (*c*) molecules, the number of internal coordinates is the same as the number of normal vibrations. In a nonplanar $X_2Y_2$ molecule (*d*) such as $H_2O_2$, the number of internal coordinates is the same as the number of vibrations if the twisting angle around the central bond ($\Delta\tau$) is considered. In a tetrahedral $XY_4$ molecule (*e*), however, the number of internal coordinates exceeds the number of normal vibrations by one. This is due to the fact that the

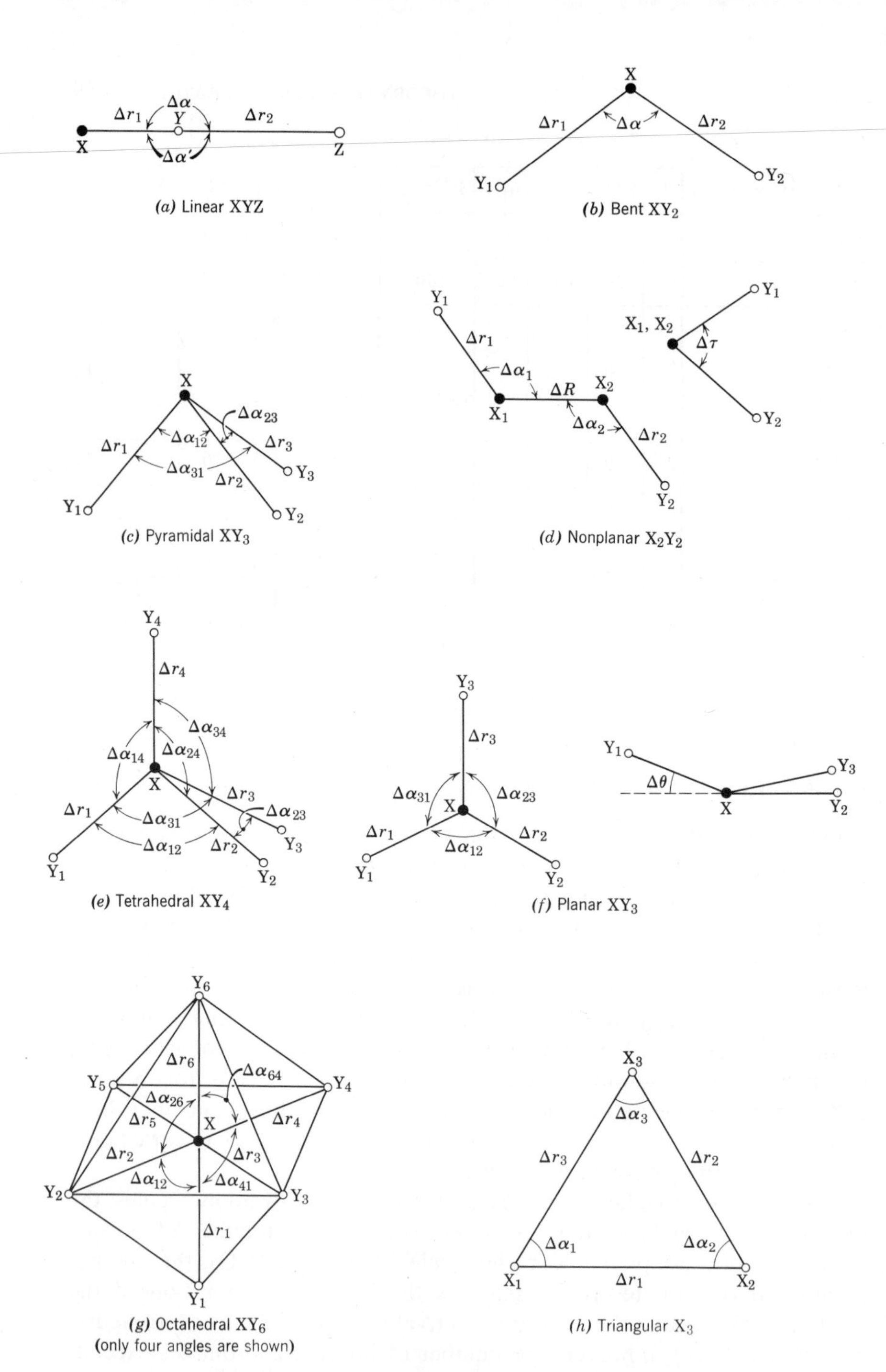


**Fig. I-11.** Internal coordinates for various molecules.

six angle coordinates around the central atom are not independent of each other, that is, they must satisfy the relation

$$\Delta\alpha_{12}+\Delta\alpha_{23}+\Delta\alpha_{31}+\Delta\alpha_{41}+\Delta\alpha_{42}+\Delta\alpha_{43}=0 \tag{8.1}$$

This is called a *redundant condition.* In planar $XY_3$ molecule (*f*), the number of internal coordinates is seven when the coordinate $\Delta\theta$, which represents the deviation from planarity, is considered. Since the number of vibrations is six, one redundant condition such as

$$\Delta\alpha_{12}+\Delta\alpha_{23}+\Delta\alpha_{31}=0 \tag{8.2}$$

must be involved. Such redundant conditions always exist for the angle coordinates around the central atom. In an octahedral $XY_6$ molecule (*g*), the number of internal coordinates exceeds the number of normal vibrations by three. This means that, of the 12 angle coordinates around the central atom, three redundant conditions are involved:

$$\begin{aligned}\Delta\alpha_{12}+\Delta\alpha_{26}+\Delta\alpha_{64}+\Delta\alpha_{41}&=0\\ \Delta\alpha_{15}+\Delta\alpha_{56}+\Delta\alpha_{63}+\Delta\alpha_{31}&=0\\ \Delta\alpha_{23}+\Delta\alpha_{34}+\Delta\alpha_{45}+\Delta\alpha_{52}&=0\end{aligned} \tag{8.3}$$

The redundant conditions are more complex in ring compounds. For example, the number of internal coordinates in a triangular $X_3$ molecule (*h*) exceeds the number of vibrations by three. One of these redundant conditions ($A_1'$ species) is

$$\Delta\alpha_1+\Delta\alpha_2+\Delta\alpha_3=0 \tag{8.4}$$

The other two redundant conditions ($E'$ species) involve bond stretching and angle deformation coordinates such as

$$\begin{aligned}(2\Delta r_1-\Delta r_2-\Delta r_3)+\frac{r}{\sqrt{3}}(\Delta\alpha_1+\Delta\alpha_2-2\Delta\alpha_3)&=0\\ (\Delta r_2-\Delta r_3)-\frac{r}{\sqrt{3}}(\Delta\alpha_1-\Delta\alpha_2)&=0\end{aligned} \tag{8.5}$$

where $r$ is the equilibrium length of the X–X bond. The redundant conditions mentioned above can be derived by using the method described in Sec. I-11.

The procedure for finding the number of normal vibrations in each species was described in Sec. I-7. This procedure is, however, considerably simplified if internal coordinates are used. Again consider a pyramidal $XY_3$ molecule. Using the internal coordinates shown in Fig. I-11$c$, we can write the representation for the $C_3^+$ operation as

$$C_3^+ \begin{bmatrix} \Delta r_1 \\ \Delta r_2 \\ \Delta r_3 \\ \Delta\alpha_{12} \\ \Delta\alpha_{23} \\ \Delta\alpha_{31} \end{bmatrix} = \begin{bmatrix} 0 & 0 & 1 & 0 & 0 & 0 \\ 1 & 0 & 0 & 0 & 0 & 0 \\ 0 & 1 & 0 & 0 & 0 & 0 \\ 0 & 0 & 0 & 0 & 0 & 1 \\ 0 & 0 & 0 & 1 & 0 & 0 \\ 0 & 0 & 0 & 0 & 1 & 0 \end{bmatrix} \begin{bmatrix} \Delta r_1 \\ \Delta r_2 \\ \Delta r_3 \\ \Delta\alpha_{12} \\ \Delta\alpha_{23} \\ \Delta\alpha_{31} \end{bmatrix} \tag{8.6}$$

Thus $\chi(C_3^+)=0$, as does $\chi(C_3^-)$. Similarly, $\chi(I)=6$ and $\chi(\sigma_v)=2$. This result is exactly the same as that obtained in Table I-5 using rectangular coordinates. *When using internal coordinates, however, the character of the representation is simply given by the number of internal coordinates unchanged by each symmetry operation.*

If this procedure is made separately for stretching ($\Delta r$) and bending ($\Delta\alpha$) coordinates, it is readily seen that

$$\begin{aligned} \chi^r(R) &= \chi_{A_1} + \chi_E \\ \chi^\alpha(R) &= \chi_{A_1} + \chi_E \end{aligned} \tag{8.7}$$

Thus it is found that both $A_1$ and $E$ species have one stretching and one bending vibration, respectively. No consideration of the translational and rotational motions is necessary if the internal coordinates are taken as the basis for the representation.

Another example, for an octahedral $XY_6$ molecule, is given in Table I-7. Using Eq. 6.6 and the character table in Appendix I, we find that these characters are resolved into

$$\chi^r(R) = \chi_{A_{1g}} + \chi_{E_g} + \chi_{F_{1u}} \tag{8.8}$$

$$\chi^\alpha(R) = \chi_{A_{1g}} + \chi_{E_g} + \chi_{F_{1u}} + \chi_{F_{2g}} + \chi_{F_{2u}} \tag{8.9}$$

Comparison of this result with that obtained in Sec. I-7 immediately suggests that three redundant conditions are included in these bending vibrations (one in $A_{1g}$ and one in $E_g$). Therefore $\chi^\alpha(R)$ for genuine vibrations becomes

$$\chi^\alpha(R) = \chi_{F_{1u}} + \chi_{F_{2g}} + \chi_{F_{2u}} \tag{8.10}$$

TABLE I-7

| | $I$ | $8C_3$ | $6C_2$ | $6C_4$ | $3C_4^2 \equiv C_2''$ | $S_2 \equiv i$ | $6S_4$ | $8S_6$ | $3\sigma_h$ | $6\sigma_d$ |
|---|---|---|---|---|---|---|---|---|---|---|
| $\chi^r(R)$ | 6 | 0 | 0 | 2 | 2 | 0 | 0 | 0 | 4 | 2 |
| $\chi^\alpha(R)$ | 12 | 0 | 2 | 0 | 0 | 0 | 0 | 0 | 4 | 2 |

Thus it is concluded that six stretching and nine bending vibrations are distributed as indicated in Eqs. 8.8 and 8.10, respectively. Although the method given above is simpler than that of Sec. I-7, caution must be exercised with respect to the bending vibrations whenever redundancy is involved. In such a case, comparison of the results obtained from both methods is useful in finding the species of redundancy.

## I-9. SELECTION RULES FOR INFRARED AND RAMAN SPECTRA

According to quantum mechanics,[4,5] the selection rule for the infrared spectrum is determined by the integral:

$$[\mu]_{v'v''} = \int \psi_{v'}(Q_a)\mu\psi_{v''}(Q_a)\, dQ_a \tag{9.1}$$

Here $\mu$ is the dipole moment in the electronic ground state, $\psi$ is the vibrational eigenfunction given by Eq. 2.7, and $v'$ and $v''$ are the vibrational quantum numbers before and after the transition, respectively. The activity of the normal vibration whose normal coordinate is $Q_a$ is being determined. By resolving the dipole moment into the three components in the $x$, $y$, and $z$ directions, we obtain the result

$$\begin{aligned} [\mu_x]_{v'v''} &= \int \psi_{v'}(Q_a)\mu_x\psi_{v''}(Q_a)\, dQ_a \\ [\mu_y]_{v'v''} &= \int \psi_{v'}(Q_a)\mu_y\psi_{v''}(Q_a)\, dQ_a \\ [\mu_z]_{v'v''} &= \int \psi_{v'}(Q_a)\mu_z\psi_{v''}(Q_a)\, dQ_a \end{aligned} \tag{9.2}$$

If one of these integrals is not zero, the normal vibration associated with $Q_a$ is infrared active. If all the integrals are zero, the vibration is infrared inactive.

Similarly, the selection rule for the Raman spectrum is determined by the integral:

$$[\alpha]_{v'v''} = \int \psi_{v'}(Q_a)\alpha\psi_{v''}(Q_n)\, dQ_a \tag{9.3}$$

As shown in Sec. I-5, $\alpha$ consists of six components, $\alpha_{xx}$, $\alpha_{yy}$, $\alpha_{zz}$, $\alpha_{xy}$, $\alpha_{yz}$, and $\alpha_{xz}$. Thus Eq. 9.3 may be resolved into six components:

$$
\begin{aligned}
[\alpha_{xx}]_{v'v''} &= \int \psi_{v'}(Q_a)\alpha_{xx}\psi_{v''}(Q_a)\, dQ_a \\
[\alpha_{yy}]_{v'v''} &= \int \psi_{v'}(Q_a)\alpha_{yy}\psi_{v''}(Q_a)\, dQ_a \\
&\cdots\cdots\cdots\cdots\cdots \\
&\cdots\cdots\cdots\cdots\cdots
\end{aligned}
\tag{9.4}
$$

If one of these integrals is not zero, the normal vibration associated with $Q_a$ is Raman active. If all the integrals are zero, the vibration is Raman inactive.

It is possible to decide whether the integrals of Eqs. 9.2 and 9.4 are zero or nonzero from a consideration of symmetry. As stated in Sec. I-1, the vibrations of interest are the fundamentals in which transitions occur from $v'=0$ to $v''=1$. It is evident from the form of the vibrational eigenfunction (Eq. 2.8) that $\psi_0(Q_a)$ is invariant under any symmetry operation, whereas the symmetry of $\psi_1(Q_a)$ is the same as that of $Q_a$. Thus the integral does not vanish when the symmetry of $\mu_x$, for example, is the same as that of $Q_a$. If the symmetry properties of $\mu_x$ and $Q_a$ differ in even one symmetry element of the group, the integral becomes zero. In other words, for the integral to be nonzero, $Q_a$ must belong to the same species as $\mu_x$. More generally, the normal vibration associated with $Q_a$ becomes infrared active when at least one of the components of the dipole moment belongs to the same species as $Q_a$. Similar conclusions are obtained for the Raman spectrum.

Since the species of the normal vibration can be determined by the methods described in Secs. I-7 and I-8, it is necessary only to determine the species of the components of the dipole moment and polarizability of the molecule. This can be done as follows. The components of the dipole moment, $\mu_x$, $\mu_y$, and $\mu_z$, transform as do those of translational motion, $T_x$, $T_y$, and $T_z$, respectively. These were discussed in Sec. I-7. Thus the character of the dipole moment is given by Eq. 7.13, which is

$$
\boxed{\chi_\mu(R) = \pm(1+2\cos\theta)} \tag{9.5}
$$

where + and − have the same meaning as before. In a pyramidal $XY_3$ molecule, Eq. 9.5 gives

| | $I$ | $2C_3$ | $3\sigma_v$ |
|---|---|---|---|
| $\chi_\mu(R)$ | 3 | 0 | 1 |

Using Eq. 6.6, we resolve this into $A_1+E$. It is obvious that $\mu_z$ belongs to $A_1$. Then $\mu_x$ and $\mu_y$ must belong to $E$. In fact, the pair, $\mu_x$ and $\mu_y$, transforms as follows.

$$I\begin{bmatrix}\mu_x\\ \mu_y\end{bmatrix}=\begin{bmatrix}1 & 0\\ 0 & 1\end{bmatrix}\begin{bmatrix}\mu_x\\ \mu_y\end{bmatrix} \qquad C_3^+\begin{bmatrix}\mu_x\\ \mu_y\end{bmatrix}=\begin{bmatrix}-\frac{1}{2} & \frac{\sqrt{3}}{2}\\ -\frac{\sqrt{3}}{2} & -\frac{1}{2}\end{bmatrix}\begin{bmatrix}\mu_x\\ \mu_y\end{bmatrix}$$

$$\chi(I)=2 \qquad \chi(C_3^+)=-1$$

$$\sigma_1\begin{bmatrix}\mu_x\\ \mu_y\end{bmatrix}=\begin{bmatrix}-1 & 0\\ 0 & 1\end{bmatrix}\begin{bmatrix}\mu_x\\ \mu_y\end{bmatrix}$$

$$\chi(\sigma_1)=0$$

Thus it is found that $\mu_z$ belongs to $A_1$ and $(\mu_x, \mu_y)$ belong to $E$.

The character of the representation of the polarizability is given by

$$\boxed{\chi_\alpha(R)=2\cos\theta(1+2\cos\theta)} \tag{9.6}$$

for both proper and improper rotations. This can be derived as follows. The polarizability in the $x$, $y$, and $z$ directions is related to that in $X$, $Y$, and $Z$ coordinates by

$$\begin{bmatrix}\alpha_{XX} & \alpha_{XY} & \alpha_{XZ}\\ \alpha_{YX} & \alpha_{YY} & \alpha_{YZ}\\ \alpha_{ZX} & \alpha_{ZY} & \alpha_{ZZ}\end{bmatrix}$$

$$=\begin{bmatrix}C_{Xx} & C_{Xy} & C_{Xz}\\ C_{Yx} & C_{Yy} & C_{Yz}\\ C_{Zx} & C_{Zy} & C_{Zz}\end{bmatrix}\begin{bmatrix}\alpha_{xx} & \alpha_{xy} & \alpha_{xz}\\ \alpha_{yx} & \alpha_{yy} & \alpha_{yz}\\ \alpha_{zx} & \alpha_{zy} & \alpha_{zz}\end{bmatrix}\begin{bmatrix}C_{Xx} & C_{Yx} & C_{Zx}\\ C_{Xy} & C_{Yy} & C_{Zy}\\ C_{Xz} & C_{Yz} & C_{Zz}\end{bmatrix}$$

where $C_{Xx}$, $C_{Xy}$, and so forth, denote the direction cosines between the two axes subscripted. If a rotation through $\theta$ around the $Z$ axis superimposes the $X$, $Y$, and $Z$ axes on the $x$, $y$, and $z$ axes, the preceding relation becomes

$$C_\theta\begin{bmatrix}\alpha_{xx} & \alpha_{xy} & \alpha_{xz}\\ \alpha_{yx} & \alpha_{yy} & \alpha_{yz}\\ \alpha_{zx} & \alpha_{zy} & \alpha_{zz}\end{bmatrix}$$

$$=\begin{bmatrix}\cos\theta & \sin\theta & 0\\ -\sin\theta & \cos\theta & 0\\ 0 & 0 & 1\end{bmatrix}\begin{bmatrix}\alpha_{xx} & \alpha_{xy} & \alpha_{xz}\\ \alpha_{yx} & \alpha_{yy} & \alpha_{yz}\\ \alpha_{zx} & \alpha_{zy} & \alpha_{zz}\end{bmatrix}\begin{bmatrix}\cos\theta & -\sin\theta & 0\\ \sin\theta & \cos\theta & 0\\ 0 & 0 & 1\end{bmatrix}$$

This can be written as

$$C_\theta \begin{bmatrix} \alpha_{xx} \\ \alpha_{yy} \\ \alpha_{zz} \\ \alpha_{xy} \\ \alpha_{xz} \\ \alpha_{yz} \end{bmatrix} = \begin{bmatrix} \cos^2\theta & \sin^2\theta & 0 & 2\sin\theta\cos\theta & 0 & 0 \\ \sin^2\theta & \cos^2\theta & 0 & -2\sin\theta\cos\theta & 0 & 0 \\ 0 & 0 & 1 & 0 & 0 & 0 \\ -\sin\theta\cos\theta & \sin\theta\cos\theta & 0 & 2\cos^2\theta-1 & 0 & 0 \\ 0 & 0 & 0 & 0 & \cos\theta & \sin\theta \\ 0 & 0 & 0 & 0 & -\sin\theta & \cos\theta \end{bmatrix} \begin{bmatrix} \alpha_{xx} \\ \alpha_{yy} \\ \alpha_{zz} \\ \alpha_{xy} \\ \alpha_{xz} \\ \alpha_{yz} \end{bmatrix}$$

Thus the character of this representation is given by Eq. 9.6. The same results are obtained for improper rotations if they are regarded as the product $i \times$(proper rotation). For a pyramidal $XY_3$ molecule, Eq. 9.6 gives

| | $I$ | $2C_3$ | $3\sigma_v$ |
|---|---|---|---|
| $\chi_\alpha(R)$ | 6 | 0 | 2 |

Using Eq. 6.6, this is resolved into $2A_1+2E$. Again, it is immediately seen that the component $\alpha_{zz}$ belongs to $A_1$, and the pair $\alpha_{zx}$ and $\alpha_{zy}$ belongs to $E$ since

$$\begin{bmatrix} zx \\ zy \end{bmatrix} = z\begin{bmatrix} x \\ y \end{bmatrix} \approx A_1 \times E = E$$

It is more convenient to consider the components $\alpha_{xx}+\alpha_{yy}$ and $\alpha_{xx}-\alpha_{yy}$ than $\alpha_{xx}$ and $\alpha_{yy}$. If a vector of unit length is considered, the relation

$$x^2+y^2+z^2=1$$

holds. Since $\alpha_{zz}$ belongs to $A_1$, $\alpha_{xx}+\alpha_{yy}$ must belong to $A_1$. Then the pair $\alpha_{xx}-\alpha_{yy}$ and $\alpha_{xy}$ must belong to $E$. As a result, the character table of the point group $\mathbf{C}_{3v}$ is completed as in Table I-8. Thus it is concluded that, in the point group $\mathbf{C}_{3v}$, both the $A_1$ and the $E$ vibrations are infrared as well as Raman active, while the $A_2$ vibrations are inactive.

Complete character tables like Table I-8 have already been worked out for all the point groups. Therefore no elaborate treatment such as that described in this section is necessary in practice. Appendix I gives

TABLE I-8. CHARACTER TABLE OF THE POINT GROUP $\mathbf{C}_{3v}$

| $\mathbf{C}_{3v}$ | $I$ | $2C_3$ | $3\sigma_v$ | | |
|---|---|---|---|---|---|
| $A_1$ | +1 | +1 | +1 | $\mu_z$ | $\alpha_{xx} + \alpha_{yy}, \alpha_{zz}$ |
| $A_2$ | +1 | +1 | −1 | | |
| $E$ | +2 | −1 | 0 | $(\mu_x, \mu_y)^a$ | $(\alpha_{xz}, \alpha_{yz}),^a (\alpha_{xx} - \alpha_{yy}, \alpha_{xy})^a$ |

[a] A doubly degenerate pair is represented by two terms in parentheses.

complete character tables for the point groups that appear frequently in this book. From these tables, the selection rules for the infrared and Raman spectra are obtained immediately: *The vibration is infrared or Raman active if it belongs to the same species as one of the components of the dipole moment or polarizability, respectively*. For example, the character table of the point group $\mathbf{O}_h$ signifies immediately that only the $F_{1u}$ vibrations are infrared active and only the $A_{1g}$, $E_g$, and $F_{2g}$ vibrations are Raman active, for the components of the dipole moment or the polarizability belong to these species in this point group. It is to be noted in these character tables that (1) a totally symmetric vibration is Raman active in any point group, and (2) the infrared and Raman active vibrations always belong to $u$ and $g$ types, respectively, in point groups having a center of symmetry.

As stated in Sec. I-2, some overtones and combination bands are observed weakly because the actual vibrations are not harmonic and some of them are allowed by symmetry selection rules. For the symmetry selection rules of these nonfundamental vibrations, see Refs. 3, 7, and 8.

## I-10. STRUCTURE DETERMINATION

Suppose that a molecule has several probable structures each of which belongs to a different point group. Then the number of infrared and Raman active fundamentals should be different for each structure. Therefore the most probable model can be selected by comparing the observed number of infrared and Raman active fundamentals with that predicted theoretically for each model.

Consider the $XeF_4$ molecule as an example. It may be tetrahedral or square-planar. By use of the methods described in the preceding sections, the number of infrared or Raman active fundamentals can be found easily for each structure. Tables I-9*a* and I-9*b* summarize the results. It is seen that the tetrahedral structure predicts two infrared active fundamentals (one stretching and one bending), whereas the square-planar structure predicts three infrared active fundamentals (one stretching and two bendings). The infrared spectrum of $XeF_4$ in the vapor phase exhibits one XeF stretching at 586 $cm^{-1}$ and two FXeF bendings at 291 and 161 $cm^{-1}$ (Ref. II-558). Thus the square-planar structure of $\mathbf{D}_{4h}$ symmetry is preferable to the tetrahedral structure of $\mathbf{T}_d$ symmetry*. As is seen in Tables I-9*a* and I-9*b*, group theory predicts two Raman active XeF stretchings for both structures, but two Raman active bendings for the tetrahedral and one Raman active bending for the square-planar structure. The observed Raman spectrum (554, 524, and 218 $cm^{-1}$) again confirms the square-planar structure.

TABLE I-9*a*. NUMBER OF FUNDAMENTALS FOR TETRAHEDRAL $XeF_4$

| $\mathbf{T}_d$ | Activity | Number of Fundamentals | XeF Stretching | FXeF Bending |
|---|---|---|---|---|
| $A_1$ | R | 1 | 1 | 0 |
| $A_2$ | ia | 0 | 0 | 0 |
| $E$ | R | 1 | 0 | 1 |
| $F_1$ | ia | 0 | 0 | 0 |
| $F_2$ | IR, R | 2 | 1 | 1 |
| Total | IR | 2 | 1 | 1 |
| | R | 4 | 2 | 2 |

TABLE I-9*b*. NUMBER OF FUNDAMENTALS FOR SQUARE-PLANAR $XeF_4$

| $\mathbf{D}_{4h}$ | Activity | Number of Fundamentals | XeF Stretching | FXeF Bending |
|---|---|---|---|---|
| $A_{1g}$ | R | 1 | 1 | 0 |
| $A_{1u}$ | ia | 0 | 0 | 0 |
| $A_{2g}$ | ia | 0 | 0 | 0 |
| $A_{2u}$ | IR | 1 | 0 | 1 |
| $B_{1g}$ | R | 1 | 1 | 0 |
| $B_{1u}$ | ia | 0 | 0 | 0 |
| $B_{2g}$ | R | 1 | 0 | 1 |
| $B_{2u}$ | ia | 1 | 0 | 1 |
| $E_g$ | R | 0 | 0 | 0 |
| $E_u$ | IR | 2 | 1 | 1 |
| Total | IR | 3 | 1 | 2 |
| | R | 3 | 2 | 1 |

* This conclusion may be drawn directly from observation of the mutual exclusion rule, which holds for $\mathbf{D}_{4h}$ but not for $\mathbf{T}_d$.

This method is widely used for elucidation of the molecular structure of inorganic, organic, and coordination compounds. In Part II and Appendix III, the number of infrared and Raman active fundamentals is compared for $XY_3$ (planar, $\mathbf{D}_{3h}$, and pyramidal, $\mathbf{C}_{3v}$), $XY_4$ (square-planar, $\mathbf{D}_{4h}$, and tetrahedral, $\mathbf{T}_d$), $XY_5$ (trigonal-bipyramidal, $\mathbf{D}_{3h}$, and tetragonal-pyramidal, $\mathbf{C}_{4v}$) and other molecules. Recently, the structures of various metal carbonyl compounds (Sec. III-12) have been determined by this simple technique.

It should be noted, however, that this method does not give a clear-cut answer if the predicted numbers of infrared and Raman active fundamentals are similar for various probable structures. Furthermore, a practical difficulty arises in determining the number of fundamentals from the observed spectrum, since the intensities of overtone and combination bands are sometimes comparable to those of fundamentals when they appear as satellite bands of the fundamental. This is particularly true when overtone and combination bands are enhanced anomalously by *Fermi resonance* (accidental degeneracy). For example, the frequency of the first overtone of the $\nu_2$ vibration of $CO_2$ (667 $cm^{-1}$) is very close to that of the $\nu_1$ vibration (1337 $cm^{-1}$). Since these two vibrations belong to the same symmetry species ($\Sigma_g^+$), they interact with each other and give rise to two strong Raman lines at 1388 and 1286 $cm^{-1}$. Fermi resonances similar to the resonance observed for $CO_2$ may occur for a number of other molecules. It is to be noted also that the number of observed bands depends on the resolving power of the instrument used. Finally, the molecular symmetry in the isolated state is not necessarily the same as that in the crystalline state (Sec. I-22). Therefore this method must be applied with caution to spectra obtained for compounds in the crystalline state.

## I-11. PRINCIPLE OF THE GF MATRIX METHOD*

As described in Sec. I-3, the frequency of the normal vibration is determined by the kinetic and potential energies of the system. The kinetic energy is determined by the masses of the individual atoms and their geometrical arrangement in the molecule. On the other hand, the potential energy arises from interaction between the individual atoms and is described in terms of the force constants. Since the potential energy provides valuable information about the nature of interatomic forces, it is

---

* For details, see Refs. 3–8. The term "normal coordinate analysis" is almost synonymous with the **GF** matrix method, since most of the normal coordinate calculations are carried out by using this method.

highly desirable to obtain the force constants from the observed frequencies. This is usually done by calculating the frequencies, assuming a suitable set of force constants. If the agreement between the calculated and observed frequencies is satisfactory, this particular set of the force constants is adopted as a representation of the potential energy of the system.

To calculate the vibrational frequencies, it is necessary first to express both the potential and the kinetic energies in terms of some common coordinates (Sec. I-3). Internal coordinates (Sec. I-8) are more suitable for this purpose than rectangular coordinates, since (1) force constants expressed in terms of internal coordinates have clearer physical meanings than those expressed in terms of rectangular coordinates, and (2) a set of internal coordinates does not involve translational and rotational motion of the molecule as a whole.

Using the internal coordinates, $R_i$, we write the potential energy as

$$2V = \tilde{\mathbf{R}}\mathbf{F}\mathbf{R} \tag{11.1}$$

For a bent $Y_1XY_2$ molecule such as that in Fig. I-11*b*, **R** is a column matrix of the form

$$\mathbf{R} = \begin{bmatrix} \Delta r_1 \\ \Delta r_2 \\ \Delta\alpha \end{bmatrix}$$

$\tilde{\mathbf{R}}$ is its transpose:

$$\tilde{\mathbf{R}} = [\Delta r_1 \quad \Delta r_2 \quad \Delta\alpha]$$

and **F** is a matrix whose components are the force constants:

$$\mathbf{F} = \begin{bmatrix} f_{11} & f_{12} & r_1 f_{13} \\ f_{21} & f_{22} & r_2 f_{23} \\ r_1 f_{31} & r_2 f_{32} & r_1 r_2 f_{33} \end{bmatrix} \equiv \begin{bmatrix} F_{11} & F_{12} & F_{13} \\ F_{21} & F_{22} & F_{23} \\ F_{31} & F_{32} & F_{33} \end{bmatrix} \tag{11.2}^*$$

Here $r_1$ and $r_2$ are the equilibrium lengths of the $X{-}Y_1$ and $X{-}Y_2$ bonds, respectively.

The kinetic energy is not easily expressed in terms of the same internal coordinates. Wilson[38] has shown, however, that the kinetic energy can be written as

$$2T = \tilde{\dot{\mathbf{R}}}\mathbf{G}^{-1}\dot{\mathbf{R}} \tag{11.3}^\dagger$$

---

* Here $f_{11}$ and $f_{22}$ are the stretching force constants of the $X{-}Y_1$ and $X{-}Y_2$ bonds, respectively, and $f_{33}$ is the bending force constant of the $Y_1XY_2$ angle. The other symbols represent interaction force constants between stretching and stretching or between stretching and bending vibrations. To make the dimensions of all the force constants the same, $f_{13}$ (or $f_{31}$), $f_{23}$ (or $f_{32}$), and $f_{33}$ are multiplied by $r_1$, $r_2$, and $r_1r_2$, respectively.

† Appendix IV gives the derivation of Eq. 11.3.

where $\mathbf{G}^{-1}$ is the reciprocal of the $\mathbf{G}$ matrix, which will be defined later.
If Eqs. 11.1 and 11.3 are combined with Newton's equation,

$$\frac{d}{dt}\left(\frac{\partial T}{\partial \dot{R}_k}\right)+\frac{\partial V}{\partial R_k}=0 \tag{3.6}$$

the following secular equation, which is similar to Eq. 3.19, is obtained:[3,7]

$$\begin{vmatrix} F_{11}-(G^{-1})_{11}\lambda & F_{12}-(G^{-1})_{12}\lambda & \cdots \\ F_{21}-(G^{-1})_{21}\lambda & F_{22}-(G^{-1})_{22}\lambda & \cdots \\ \cdot & \cdot & \\ \cdot & \cdot & \\ \cdot & \cdot & \end{vmatrix} \equiv |\mathbf{F}-\mathbf{G}^{-1}\lambda|=0 \tag{11.4}$$

By multiplying by the determinant of $\mathbf{G}$

$$\begin{vmatrix} G_{11} & G_{12} & \cdots \\ G_{21} & G_{22} & \cdots \\ \cdot & \cdot & \\ \cdot & \cdot & \\ \cdot & \cdot & \end{vmatrix} \equiv |\mathbf{G}| \tag{11.5}$$

from the left of Eq. 11.4, the following equation is obtained:

$$\begin{vmatrix} \sum G_{1t}F_{t1}-\lambda & \sum G_{1t}F_{t2} & \cdots \\ \sum G_{2t}F_{t1} & \sum G_{2t}F_{t2}-\lambda & \cdots \\ \cdot & \cdot & \\ \cdot & \cdot & \\ \cdot & \cdot & \end{vmatrix} \equiv |\mathbf{GF}-\mathbf{E}\lambda|=0 \tag{11.6}$$

Here $\mathbf{E}$ is the unit matrix, and $\lambda$ is related to the wave number, $\tilde{\nu}$, by the relation $\lambda=4\pi^2c^2\tilde{\nu}^2$.* The order of the equation is equal to the number of internal coordinates used.

The $\mathbf{F}$ matrix can be written by assuming a suitable set of force constants. If the $\mathbf{G}$ matrix is constructed by the following method, the vibrational frequencies are obtained by solving Eq. 11.6. The $\mathbf{G}$ matrix is defined as

$$\mathbf{G}=\mathbf{BM}^{-1}\tilde{\mathbf{B}} \tag{11.7}$$

* Here $\lambda$ should not be confused with $\lambda_w$ (wavelength).

Here $\mathbf{M}^{-1}$ is a diagonal matrix whose components are $\mu_i$, where $\mu_i$ is the reciprocal of the mass of the $i$th atom. For a bent $XY_2$ molecule,

$$\mathbf{M}^{-1} = \begin{bmatrix} \mu_1 & & & & & & 0 \\ & \mu_1 & & & & & \\ & & \mu_1 & & & & \\ & & & \cdot & & & \\ & & & & \cdot & & \\ & & & & & \cdot & \\ 0 & & & & & & \mu_3 \end{bmatrix}$$

where $\mu_3$ and $\mu_1$ are the reciprocals of the masses of the X and Y atoms, respectively. The **B** matrix is defined as

$$\mathbf{R} = \mathbf{B}\mathbf{X} \tag{11.8}$$

where **R** and **X** are column matrices whose components are the internal and rectangular coordinates, respectively. For a bent $XY_2$ molecule, Eq. 11.8 is written as

$$\begin{bmatrix} \Delta r_1 \\ \Delta r_2 \\ \Delta\alpha \end{bmatrix} = \left[\begin{array}{ccc:ccc:ccc} -s & -c & 0 & 0 & 0 & 0 & s & c & 0 \\ 0 & 0 & 0 & s & -c & 0 & -s & c & 0 \\ -c/r & s/r & 0 & c/r & s/r & 0 & 0 & -2s/r & 0 \end{array}\right] \begin{bmatrix} \Delta x_1 \\ \Delta y_1 \\ \Delta z_1 \\ \hdashline \Delta x_2 \\ \Delta y_2 \\ \Delta z_2 \\ \hdashline \Delta x_3 \\ \Delta y_3 \\ \Delta z_3 \end{bmatrix} \tag{11.9}$$

where $s = \sin(\alpha/2)$, $c = \cos(\alpha/2)$, and $r$ is the equilibrium distance between X and Y. (See Fig. I-12.)

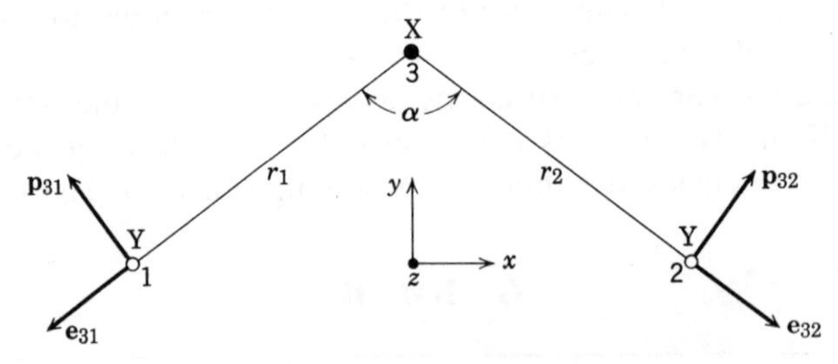


**Fig. I-12.** Unit vectors in a bent $XY_2$ molecule.

If unit vectors such as those in Fig. I-12 are considered, Eq. 11.9 can be written in a more compact form using vector notation:

$$\begin{bmatrix}\Delta r_1\\ \Delta r_2\\ \Delta\alpha\end{bmatrix}=\begin{bmatrix}\mathbf{e}_{31} & 0 & -\mathbf{e}_{31}\\ 0 & \mathbf{e}_{32} & -\mathbf{e}_{32}\\ \mathbf{p}_{31}/r & \mathbf{p}_{32}/r & -(\mathbf{p}_{31}+\mathbf{p}_{32})/r\end{bmatrix}\begin{bmatrix}\boldsymbol{\rho}_1\\ \boldsymbol{\rho}_2\\ \boldsymbol{\rho}_3\end{bmatrix} \tag{11.10}$$

Here $\boldsymbol{\rho}_1$, $\boldsymbol{\rho}_2$, and $\boldsymbol{\rho}_3$ are the displacement vectors of atoms 1, 2, and 3, respectively. Thus Eq. 11.10 can be written simply as

$$\mathbf{R}=\mathbf{S}\cdot\boldsymbol{\rho} \tag{11.11}$$

where the dot represents the scalar product of the two vectors. Here $\mathbf{S}$ is called the $\mathbf{S}$ matrix, and its components ($\mathbf{S}$ vector) can be written according to the following formulas: (1) bond stretching,

$$\Delta r_1=\Delta r_{31}=\mathbf{e}_{31}\cdot\boldsymbol{\rho}_1-\mathbf{e}_{31}\cdot\boldsymbol{\rho}_3 \tag{11.12}$$

and (2) angle bending,

$$\Delta\alpha=\Delta\alpha_{132}=[\mathbf{p}_{31}\cdot\boldsymbol{\rho}_1+\mathbf{p}_{32}\cdot\boldsymbol{\rho}_2-(\mathbf{p}_{31}+\mathbf{p}_{32})\cdot\boldsymbol{\rho}_3]/r \tag{11.13}$$

It is seen that the direction of the $\mathbf{S}$ vector is the direction in which a given displacement of the $i$th atom will produce the greatest increase in $\Delta r$ or $\Delta\alpha$. Formulas for obtaining the $\mathbf{S}$ vectors of other internal coordinates such as those of out-of-plane ($\Delta\theta$) and torsional ($\Delta\tau$) vibrations are also available.[3]

By using the $\mathbf{S}$ matrix, Eq. 11.7 is written as

$$\mathbf{G}=\mathbf{S}\mathbf{m}^{-1}\tilde{\mathbf{S}} \tag{11.14}$$

For a bent $XY_2$ molecule, this becomes

$$\mathbf{G}=\begin{bmatrix}\mathbf{e}_{31} & 0 & -\mathbf{e}_{31}\\ 0 & \mathbf{e}_{32} & -\mathbf{e}_{32}\\ \mathbf{p}_{31}/r & \mathbf{p}_{32}/r & -(\mathbf{p}_{31}+\mathbf{p}_{32})/r\end{bmatrix}\begin{bmatrix}\mu_1 & 0 & 0\\ 0 & \mu_1 & 0\\ 0 & 0 & \mu_3\end{bmatrix}$$

$$\times\begin{bmatrix}\mathbf{e}_{31} & 0 & \mathbf{p}_{31}/r\\ 0 & \mathbf{e}_{32} & \mathbf{p}_{32}/r\\ -\mathbf{e}_{31} & -\mathbf{e}_{32} & -(\mathbf{p}_{31}+\mathbf{p}_{32})/r\end{bmatrix}$$

$$=\begin{bmatrix}(\mu_3+\mu_1)\mathbf{e}_{31}^2 & \mu_3\mathbf{e}_{31}\cdot\mathbf{e}_{32} & \frac{\mu_1}{r}\mathbf{e}_{31}\cdot\mathbf{p}_{31}+\frac{\mu_3}{r}\mathbf{e}_{31}\cdot(\mathbf{p}_{31}+\mathbf{p}_{32})\\ & (\mu_3+\mu_1)\mathbf{e}_{32}^2 & \frac{\mu_1}{r}\mathbf{e}_{32}\cdot\mathbf{p}_{32}+\frac{\mu_3}{r}\mathbf{e}_{32}\cdot(\mathbf{p}_{31}+\mathbf{p}_{32})\\ & & \frac{\mu_1}{r^2}\mathbf{p}_{31}^2+\frac{\mu_1}{r^2}\mathbf{p}_{32}^2+\frac{\mu_3}{r}(\mathbf{p}_{31}+\mathbf{p}_{32})^2\end{bmatrix}$$

Considering

$$\mathbf{e}_{31}\cdot\mathbf{e}_{31}=\mathbf{e}_{32}\cdot\mathbf{e}_{32}=\mathbf{p}_{31}\cdot\mathbf{p}_{31}=\mathbf{p}_{32}\cdot\mathbf{p}_{32}=1, \qquad \mathbf{e}_{31}\cdot\mathbf{p}_{31}=\mathbf{e}_{32}\cdot\mathbf{p}_{32}=0$$

$$\mathbf{e}_{31}\cdot\mathbf{e}_{32}=\cos\alpha, \qquad \mathbf{e}_{31}\cdot\mathbf{p}_{32}=\mathbf{e}_{32}\cdot\mathbf{p}_{31}=-\sin\alpha$$

and

$$(\mathbf{p}_{31}+\mathbf{p}_{32})^2=2(1-\cos\alpha)$$

we find that the **G** matrix is calculated as

$$\mathbf{G}=\begin{bmatrix} \mu_3+\mu_1 & \mu_3\cos\alpha & -\dfrac{\mu_3}{r}\sin\alpha \\ & \mu_3+\mu_1 & -\dfrac{\mu_3}{r}\sin\alpha \\ & & \dfrac{2\mu_1}{r^2}+\dfrac{2\mu_3}{r^2}(1-\cos\alpha) \end{bmatrix} \tag{11.15}$$

If the **G** matrix elements obtained are written for each combination of internal coordinates, there results

$$\begin{aligned} G(\Delta r_1, \Delta r_1) &= \mu_3+\mu_1 \\ G(\Delta r_2, \Delta r_2) &= \mu_3+\mu_1 \\ G(\Delta r_1, \Delta r_2) &= \mu_3\cos\alpha \\ G(\Delta\alpha, \Delta\alpha) &= \frac{2\mu_1}{r^2}+\frac{2\mu_3}{r^2}(1-\cos\alpha) \\ G(\Delta r_1, \Delta\alpha) &= -\frac{\mu_3}{r}\sin\alpha \\ G(\Delta r_2, \Delta\alpha) &= -\frac{\mu_3}{r}\sin\alpha \end{aligned} \tag{11.16}$$

If such calculations are made for several types of molecules, it is immediately seen that the **G** matrix elements themselves have many regularities. Decius[39] developed general formulas for writing **G** matrix elements.* Some of them are as follows:

$$G_{rr}^2=\mu_1+\mu_2$$

$$G_{rr}^1=\mu_1\cos\phi$$

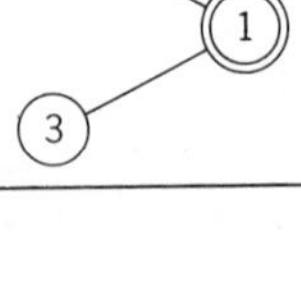


* See also Refs. 3 and 40.

$$G_{r\phi}^{2} = -\rho_{23}\mu_2 \sin\phi$$

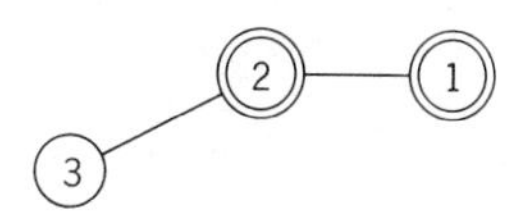


$$G_{r\phi}^{1}\begin{pmatrix}1\\1\end{pmatrix} = -(\rho_{13}\sin\phi_{213}\cos\psi_{234} + \rho_{14}\sin\phi_{214}\cos\psi_{243})\mu_1$$

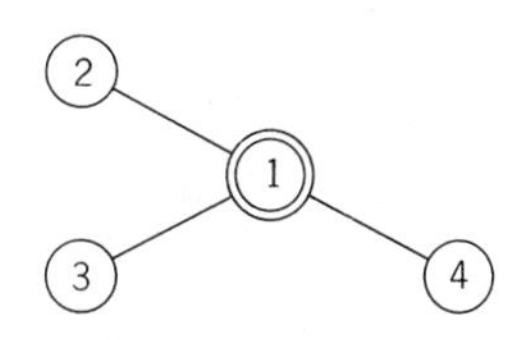


$$G_{\phi\phi}^{3} = \rho_{12}^{2}\mu_1 + \rho_{23}^{2}\mu_3 + (\rho_{12}^{2} + \rho_{23}^{2} - 2\rho_{12}\rho_{23}\cos\phi)\mu_2$$



$$G_{\phi\phi}^{2}\begin{pmatrix}1\\1\end{pmatrix} = (\rho_{12}^{2}\cos\psi_{314})\mu_1 + [(\rho_{12} - \rho_{23}\cos\phi_{123} - \rho_{24}\cos\phi_{124})\rho_{12}\cos\psi_{314} + (\sin\phi_{123}\sin\phi_{124}\sin^2\psi_{314} + \cos\phi_{324}\cos\psi_{314})\rho_{23}\rho_{24}]\mu_2$$

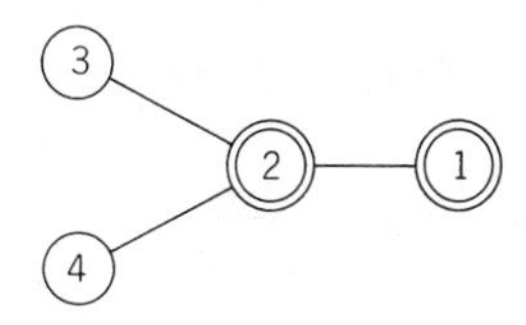


Here the atoms surrounded by a double circle are those common to both coordinates. The symbols $\mu$ and $\rho$ denote the reciprocals of mass and bond distance, respectively. The solid angle, $\psi_{\alpha\beta\gamma}$, in Fig. I-13 is defined as

$$\cos\psi_{\alpha\beta\gamma} = \frac{\cos\phi_{\alpha\delta\gamma} - \cos\phi_{\alpha\delta\beta}\cos\phi_{\beta\delta\gamma}}{\sin\phi_{\alpha\delta\beta}\sin\phi_{\beta\delta\gamma}} \tag{11.17}$$

The correspondence between the Decius formulas and the results obtained in Eq. 11.16 is evident.

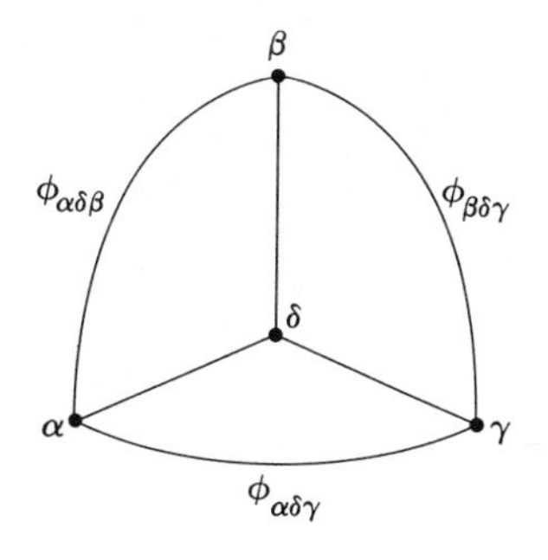


**Fig. I-13.**

With the Decius formulas, the **G** matrix elements of a pyramidal $XY_3$ molecule have been calculated and are shown in Table I-10.

TABLE I-10

| | $\Delta r_1$ | $\Delta r_2$ | $\Delta r_3$ | $\Delta\alpha_{23}$ | $\Delta\alpha_{31}$ | $\Delta\alpha_{12}$ |
|---|---|---|---|---|---|---|
| $\Delta r_1$ | A | B | B | C | D | D |
| $\Delta r_2$ | — | A | B | D | C | D |
| $\Delta r_3$ | — | — | A | D | D | C |
| $\Delta\alpha_{23}$ | — | — | — | E | F | F |
| $\Delta\alpha_{31}$ | — | — | — | — | E | F |
| $\Delta\alpha_{12}$ | — | — | — | — | — | E |

$$A = G_{rr}^2 = \mu_X + \mu_Y$$

$$B = G_{rr}^1 = \mu_X \cos\alpha$$

$$C = G_{r\phi}^1\binom{1}{1} = -\frac{2}{r}\frac{\cos\alpha(1-\cos\alpha)\mu_X}{\sin\alpha}$$

$$D = G_{r\phi}^2 = -\frac{\mu_X}{r}\sin\alpha$$

$$E = G_{\phi\phi}^3 = \frac{2}{r^2}[\mu_Y + \mu_X(1-\cos\alpha)]$$

$$F = G_{\phi\phi}^2\binom{1}{1} = \frac{\mu_Y}{r^2}\frac{\cos\alpha}{1+\cos\alpha} + \frac{\mu_X}{r^2}\frac{(1+3\cos\alpha)(1-\cos\alpha)}{1+\cos\alpha}$$

## I-12. UTILIZATION OF SYMMETRY PROPERTIES

In view of the equivalence of the two X–Y bonds of a bent $XY_2$ molecule, the **F** and **G** matrices obtained in Eqs. 11.2 and 11.15 are written as

$$\mathbf{F} = \begin{bmatrix} f_{11} & f_{12} & rf_{13} \\ f_{12} & f_{11} & rf_{13} \\ rf_{13} & rf_{13} & r^2 f_{33} \end{bmatrix} \tag{12.1}$$

$$\mathbf{G} = \begin{bmatrix} \mu_3+\mu_1 & \mu_3\cos\alpha & -\frac{\mu_3}{r}\sin\alpha \\ \mu_3\cos\alpha & \mu_3+\mu_1 & -\frac{\mu_3}{r}\sin\alpha \\ -\frac{\mu_3}{r}\sin\alpha & -\frac{\mu_3}{r}\sin\alpha & \frac{2\mu_1}{r^2}+\frac{2\mu_3}{r^2}(1-\cos\alpha) \end{bmatrix} \tag{12.2}$$

Both of these matrices are of the form

$$\begin{bmatrix} A & C & D \\ C & A & D \\ D & D & B \end{bmatrix} \tag{12.3}$$

The appearance of the same elements is evidently due to the equivalence of the two internal coordinates, $\Delta r_1$ and $\Delta r_2$. Such symmetrically equivalent sets of internal coordinates are seen in many other molecules, such as those in Fig. I-11. In these cases, it is possible to reduce the order of the **F** and **G** matrices (and hence the order of the secular equation resulting from them) by a coordinate transformation.

Let the internal coordinates $R$ be transformed by

$$\mathbf{R}^s = \mathbf{U}\mathbf{R} \tag{12.4}$$

Then

$$\begin{aligned} 2T &= \tilde{\dot{\mathbf{R}}}\mathbf{G}^{-1}\dot{\mathbf{R}} = \tilde{\dot{\mathbf{R}}}^s\tilde{\mathbf{U}}^{-1}\mathbf{G}^{-1}\mathbf{U}^{-1}\dot{\mathbf{R}}^s \\ &= \tilde{\dot{\mathbf{R}}}^s\mathbf{G}_s^{-1}\dot{\mathbf{R}}^s \\ 2V &= \tilde{\mathbf{R}}\mathbf{F}\mathbf{R} = \tilde{\mathbf{R}}^s\tilde{\mathbf{U}}^{-1}\mathbf{F}\mathbf{U}^{-1}\mathbf{R}^s \\ &= \tilde{\mathbf{R}}^s\mathbf{F}_s\mathbf{R}^s \end{aligned}$$

where

$$\begin{aligned} \mathbf{G}_s^{-1} &= \tilde{\mathbf{U}}^{-1}\mathbf{G}^{-1}\mathbf{U}^{-1} \quad \text{or} \quad \mathbf{G}_s = \mathbf{U}\mathbf{G}\tilde{\mathbf{U}} \\ \mathbf{F}_s &= \tilde{\mathbf{U}}^{-1}\mathbf{F}\mathbf{U}^{-1} \end{aligned} \tag{12.5}$$

If **U** is an orthogonal matrix ($\mathbf{U}^{-1} = \tilde{\mathbf{U}}$), Eq. 12.5 is written as

$$\mathbf{F}_s = \mathbf{U}\mathbf{F}\tilde{\mathbf{U}} \quad \text{and} \quad \mathbf{G}_s = \mathbf{U}\mathbf{G}\tilde{\mathbf{U}} \tag{12.6}$$

Both **GF** and $\mathbf{G}_s\mathbf{F}_s$ give the same roots, since

$$\begin{aligned} |\mathbf{G}_s\mathbf{F}_s - \mathbf{E}\lambda| &= |\mathbf{U}\mathbf{G}\tilde{\mathbf{U}}\tilde{\mathbf{U}}^{-1}\mathbf{F}\mathbf{U}^{-1} - \mathbf{E}\lambda| \\ &= |\mathbf{U}\mathbf{G}\mathbf{F}\mathbf{U}^{-1} - \mathbf{E}\lambda| \\ &= |\mathbf{U}|\,|\mathbf{G}\mathbf{F} - \mathbf{E}\lambda|\,|\mathbf{U}^{-1}| \\ &= |\mathbf{G}\mathbf{F} - \mathbf{E}\lambda| \end{aligned} \tag{12.7}$$

If we choose a proper **U** matrix from symmetry consideration, it is possible to factor the original **G** and **F** matrices into smaller ones. This, in turn, reduces the order of the secular equation to be solved, thus facilitating their solution. Their new coordinates $\mathbf{R}^s$ are called *symmetry coordinates.*

The **U** matrix is constructed by using the equation

$$\mathbf{R}^s = N \sum_K \chi_i(K) \cdot K(\Delta r_1) \tag{12.8}$$

Here $K$ is a symmetry operation, and the summation is made over all symmetry operations. Also, $\chi_i(K)$ is the character of the representation to which $R^s$ belongs. Called a generator, $\Delta r_1$ is, by symmetry operation $K$, transformed into $K(\Delta r_1)$, which is another coordinate of the same symmetrically equivalent set. Finally, $N$ is a normalizing factor.

As an example, consider a bent $XY_2$ molecule in which $\Delta r_1$ and $\Delta r_2$ are equivalent. Using $\Delta r_1$ as a generator, we obtain

| | $I$ | $C_2(z)$ | $\sigma(xz)$ | $\sigma(yz)$ |
|---|---|---|---|---|
| $K(\Delta r_1)$ | $\Delta r_1$ | $\Delta r_2$ | $\Delta r_2$ | $\Delta r_1$ |
| $\chi_{A_1}(K)$ | 1 | 1 | 1 | 1 |
| $\chi_{B_2}(K)$ | 1 | −1 | −1 | 1 |

Thus

$$R^s_{A_1} = N \sum \chi_{A_1}(K) \cdot K(\Delta r_1) = 2N(\Delta r_1 + \Delta r_2)$$

$$N = \frac{1}{2\sqrt{2}} \qquad \text{since } (2N)^2 + (2N)^2 = 1$$

Then

$$R^s_{A_1} = \frac{1}{\sqrt{2}}(\Delta r_1 + \Delta r_2) \tag{12.9}$$

Similarly,

$$R^s_{B_2} = \frac{1}{\sqrt{2}}(\Delta r_1 - \Delta r_2) \tag{12.10}$$

The remaining internal coordinate, $\Delta\alpha$, belongs to the $A_1$ species. Thus the complete **U** matrix is written as

$$\begin{bmatrix} R_1^s(A_1) \\ R_2^s(A_1) \\ R_3^s(B_2) \end{bmatrix} = \begin{bmatrix} \frac{1}{\sqrt{2}} & \frac{1}{\sqrt{2}} & 0 \\ 0 & 0 & 1 \\ \frac{1}{\sqrt{2}} & \frac{-1}{\sqrt{2}} & 0 \end{bmatrix} \begin{bmatrix} \Delta r_1 \\ \Delta r_2 \\ \Delta\alpha \end{bmatrix} \tag{12.11}$$

If the **G** and **F** matrices of type 12.3 are transformed by relations 12.6, where **U** is given by the matrix of Eq. 12.11, they become

$$\mathbf{F}_s, \mathbf{G}_s = \left[\begin{array}{cc|c} A+C & \sqrt{2}D & 0 \\ \sqrt{2}D & B & 0 \\ \hline 0 & 0 & A-C \end{array}\right] \tag{12.12}$$

or, more explicitly,

$$\mathbf{F}_s = \left[\begin{array}{cc|c} f_{11}+f_{12} & r\sqrt{2}f_{13} & 0 \\ r\sqrt{2}f_{13} & r^2 f_{33} & 0 \\ \hline 0 & 0 & f_{11}-f_{12} \end{array}\right] \tag{12.13}$$

$$\mathbf{G}_s = \left[\begin{array}{cc|c} \mu_3(1+\cos\alpha)+\mu_1 & -\frac{\sqrt{2}}{r}\mu_3 \sin\alpha & 0 \\ -\frac{\sqrt{2}}{r}\mu_3 \sin\alpha & \frac{2\mu_1}{r^2}+\frac{2\mu_3}{r^2}(1-\cos\alpha) & 0 \\ \hline 0 & 0 & \mu_3(1-\cos\alpha)+\mu_1 \end{array}\right] \tag{12.14}$$

In a pyramidal $XY_3$ molecule (Fig. I-11*c*), $\Delta r_1$, $\Delta r_2$, and $\Delta r_3$ are the equivalent set; so are $\Delta\alpha_{23}$, $\Delta\alpha_{31}$, and $\Delta\alpha_{12}$. It is already known from Eq. 8.7 that one $A_1$ and one $E$ vibration are involved both in the stretching and in the bending vibrations. Using $\Delta r_1$ as a generator, we obtain from Eq. 12.8

| | $I$ | $C_3^+$ | $C_3^-$ | $\sigma_1$ | $\sigma_2$ | $\sigma_3$ |
|---|---|---|---|---|---|---|
| $K(\Delta r_1)$ | $\Delta r_1$ | $\Delta r_2$ | $\Delta r_3$ | $\Delta r_1$ | $\Delta r_3$ | $\Delta r_2$ |
| $\chi_{A_1}(K)$ | 1 | 1 | 1 | 1 | 1 | 1 |
| $\chi_E(K)$ | 2 | −1 | −1 | 0 | 0 | 0 |

Then

$$R_{A_1}^s = \frac{1}{\sqrt{3}}(\Delta r_1 + \Delta r_2 + \Delta r_3) \tag{12.15}$$

$$R_{E_1}^s = \frac{1}{\sqrt{6}}(2\Delta r_1 - \Delta r_2 - \Delta r_3) \tag{12.16}$$

To find a coordinate that forms a degenerate pair with Eq. 12.16, we repeat the same procedure, using $\Delta r_2$ and $\Delta r_3$ as the generators. The results are

$$R^s_{E_2} = N(2\Delta r_2 - \Delta r_3 - \Delta r_1)$$
$$R^s_{E_3} = N(2\Delta r_3 - \Delta r_1 - \Delta r_2)$$

If we take a linear combination, $R^s_{E_2} + R^s_{E_3}$, we obtain Eq. 12.16. If we take $R^s_{E_2} - R^s_{E_3}$, we obtain

$$R^s_{E_4} = \frac{1}{\sqrt{2}}(\Delta r_2 - \Delta r_3) \qquad (12.17)$$

Since Eqs. 12.16 and 12.17 are mutually orthogonal, these two coordinates are taken as a degenerate pair. Similar results are obtained for three angle bending coordinates. Thus the complete **U** matrix is written as

$$\begin{bmatrix} R^s_1(A_1) \\ R^s_2(A_1) \\ R^s_{3a}(E) \\ R^s_{4a}(E) \\ R^s_{3b}(E) \\ R^s_{4b}(E) \end{bmatrix} = \begin{bmatrix} 1/\sqrt{3} & 1/\sqrt{3} & 1/\sqrt{3} & 0 & 0 & 0 \\ 0 & 0 & 0 & 1/\sqrt{3} & 1/\sqrt{3} & 1/\sqrt{3} \\ 2/\sqrt{6} & -1/\sqrt{6} & -1/\sqrt{6} & 0 & 0 & 0 \\ 0 & 0 & 0 & 2/\sqrt{6} & -1/\sqrt{6} & -1/\sqrt{6} \\ 0 & 1/\sqrt{2} & -1/\sqrt{2} & 0 & 0 & 0 \\ 0 & 0 & 0 & 0 & 1/\sqrt{2} & -1/\sqrt{2} \end{bmatrix} \begin{bmatrix} \Delta r_1 \\ \Delta r_2 \\ \Delta r_3 \\ \Delta \alpha_{23} \\ \Delta \alpha_{31} \\ \Delta \alpha_{12} \end{bmatrix}$$
$$(12.18)$$

The **G** matrix of a pyramidal $XY_3$ molecule has already been calculated (see Table I-10). By using Eq. 12.6 the new $\mathbf{G_s}$ matrix becomes

$$\mathbf{G_s} = \left[\begin{array}{cc|cc|cc} A+2B & C+2D & & & & \\ C+2D & E+2F & \multicolumn{2}{c|}{0} & \multicolumn{2}{c}{0} \\ \hline & & A-B & C-D & & \\ \multicolumn{2}{c|}{0} & C-D & E-F & \multicolumn{2}{c}{0} \\ \hline \multicolumn{2}{c|}{0} & \multicolumn{2}{c|}{0} & A-B & C-D \\ & & & & C-D & E-F \end{array}\right] \qquad (12.19)$$

Here $A$, $B$, and so forth, denote the elements in Table I-10. The **F** matrix transforms similarly. Therefore it is necessary only to solve two quadratic equations for the $A_1$ and $E$ species.

For the tetrahedral $XY_4$ molecule shown in Fig. I-11*e*, group theory (Secs. I-7 and I-8) predicts one $A_1$ and one $F_2$ stretching, and one $E$ and one $F_2$ bending, vibration. The **U** matrix for the four stretching coordinates is

$$\begin{bmatrix} R_1^s(A_1) \\ R_{2a}^s(F_2) \\ R_{2b}^s(F_2) \\ R_{2c}^s(F_2) \end{bmatrix} = \begin{bmatrix} 1/2 & 1/2 & 1/2 & 1/2 \\ 1/\sqrt{6} & 1/\sqrt{6} & -2/\sqrt{6} & 0 \\ 1/\sqrt{12} & 1/\sqrt{12} & 1/\sqrt{12} & -3/\sqrt{12} \\ -1/\sqrt{2} & 1/\sqrt{2} & 0 & 0 \end{bmatrix} \begin{bmatrix} \Delta r_1 \\ \Delta r_2 \\ \Delta r_3 \\ \Delta r_4 \end{bmatrix} \quad (12.20)$$

whereas the **U** matrix for the six bending coordinates becomes

$$\begin{bmatrix} R_1^s(A_1) \\ R_{2a}^s(E) \\ R_{2b}^s(E) \\ R_{3a}^s(F_2) \\ R_{3b}^s(F_2) \\ R_{3c}^s(F_2) \end{bmatrix}$$

$$= \begin{bmatrix} 1/\sqrt{6} & 1/\sqrt{6} & 1/\sqrt{6} & 1/\sqrt{6} & 1/\sqrt{6} & 1/\sqrt{6} \\ 2/\sqrt{12} & -1/\sqrt{12} & -1/\sqrt{12} & -1/\sqrt{12} & -1/\sqrt{12} & 2/\sqrt{12} \\ 0 & 1/2 & -1/2 & 1/2 & -1/2 & 0 \\ 2/\sqrt{12} & -1/\sqrt{12} & -1/\sqrt{12} & 1/\sqrt{12} & 1/\sqrt{12} & -2/\sqrt{12} \\ 1/\sqrt{6} & 1/\sqrt{6} & 1/\sqrt{6} & -1/\sqrt{6} & -1/\sqrt{6} & -1/\sqrt{6} \\ 0 & 1/2 & -1/2 & -1/2 & 1/2 & 0 \end{bmatrix} \begin{bmatrix} \Delta \alpha_{12} \\ \Delta \alpha_{23} \\ \Delta \alpha_{31} \\ \Delta \alpha_{14} \\ \Delta \alpha_{24} \\ \Delta \alpha_{34} \end{bmatrix} \quad (12.21)$$

The symmetry coordinate $R_1^s(A_1)$ in Eq. 12.21 represents a *redundant coordinate* (see Eq. 8.1). In such a case, a coordinate transformation reduces the order of the matrix by one, since all the **G** matrix elements related to this coordinate become zero. Conversely, this result provides a general method of finding redundant coordinates. Suppose that the elements of the **G** matrix are calculated in terms of internal coordinates such as those in Table I-10. If a suitable combination of internal coordinates is made so that $\sum_j G_{ij} = 0$ (where $j$ refers to all the equivalent internal coordinates), such a combination is a redundant coordinate. By using the **U** matrices in Eqs. 12.20 and 12.21, the problem of solving a tenth order secular equation for a tetrahedral $XY_4$ molecule is reduced to that of solving two first order ($A_1$ and $E$) and one quadratic ($F_2$) equation.

## I-13. POTENTIAL FIELDS AND FORCE CONSTANTS

Using Eqs. 11.1 and 12.1, we write the potential energy of a bent $XY_2$ molecule as

$$2V = f_{11}(\Delta r_1)^2 + f_{11}(\Delta r_2)^2 + f_{33}r^2(\Delta\alpha)^2 + 2f_{12}(\Delta r_1)(\Delta r_2) + 2f_{13}r(\Delta r_1)(\Delta\alpha) + 2f_{13}r(\Delta r_2)(\Delta\alpha) \quad (13.1)$$

This type of potential field is called a *generalized valence force* (GVF) field.* It consists of stretching and bending force constants, as well as the interaction force constants between them. When using such a potential field, four force constants are needed to describe the potential energy of a bent $XY_2$ molecule. Since only three vibrations are observed in practice, it is impossible to determine all four force constants simultaneously. One method used to circumvent this difficulty is to calculate the vibrational frequencies of isotopic molecules (e.g., $D_2O$ and HDO for $H_2O$), assuming the same set of force constants.† This method is satisfactory, however, only for simple molecules. As molecules become more complex, the number of interaction force constants in the GVF field becomes too large to allow any reliable evaluation.

In another approach, Shimanouchi[41] introduced the *Urey–Bradley force* (UBF) field, which consists of stretching and bending force constants, as well as repulsive force constants between nonbonded atoms. The general form of the potential field is given by

$$V = \sum_i [\tfrac{1}{2}K_i(\Delta r_i)^2 + K_i' r_i(\Delta r_i)] + \sum_i [\tfrac{1}{2}H_i r_{i\alpha}^2(\Delta\alpha_i)^2 + H_i' r_{i\alpha}^2(\Delta\alpha_i)] + \sum_i [\tfrac{1}{2}F_i(\Delta q_i)^2 + F_i' q_i(\Delta q_i)] \quad (13.2)$$

Here $\Delta r_i$, $\Delta\alpha_i$, and $\Delta q_i$ are the changes in the bond lengths, bond angles, and distances between nonbonded atoms, respectively. The symbols $K_i$, $K_i'$, $H_i$, $H_i'$, and $F_i$, $F_i'$ represent the stretching, bending, and repulsive force constants, respectively. Furthermore, $r_i$, $r_{i\alpha}$, and $q_i$ are the values of the distances at the equilibrium positions and are inserted to make the force constants dimensionally similar.

Using the relation

$$q_{ij}^2 = r_i^2 + r_j^2 - 2r_i r_j \cos\alpha_{ij} \quad (13.3)$$

---

* A potential field consisting of stretching and bending force constants only is called a *simple valence force* field.

† In addition to isotope frequency shifts, mean amplitudes of vibration, Coriolis coupling constants, centrifugal distortion constants, and so forth may be used to refine the force constants of small molecules (see Ref. 41a).

and considering that the first derivatives can be equated to zero in the equilibrium case, we can write the final form of the potential field as

$$\begin{aligned} V = \tfrac{1}{2}\sum_{i}\left[K_i + \sum_{j(\neq i)}(t_{ij}^2F'_{ij} + s_{ij}^2F_{ij})\right](\Delta r_i)^2 \\ + \tfrac{1}{2}\sum_{i<j}(H_{ij} - s_{ij}s_{ji}F'_{ij} + t_{ij}t_{ji}F_{ij})(\sqrt{r_ir_j}\,\Delta\alpha_{ij})^2 \\ + \sum_{i<j}(-t_{ij}t_{ji}F'_{ij} + s_{ij}s_{ji}F_{ij})(\Delta r_i)(\Delta r_j) \\ + \sum_{i\neq j}(t_{ij}s_{ji}F'_{ij} + t_{ji}s_{ij}F_{ij})\left(\frac{r_j}{r_i}\right)^{1/2}(\Delta r_i)(\sqrt{r_ir_j}\,\Delta\alpha_{ij}) \end{aligned} \tag{13.4*}$$

Here

$$\begin{aligned} s_{ij} &= \frac{r_i - r_j\cos\alpha_{ij}}{q_{ij}} \\ s_{ji} &= \frac{r_j - r_i\cos\alpha_{ij}}{q_{ij}} \\ t_{ij} &= \frac{r_j\sin\alpha_{ij}}{q_{ij}} \\ t_{ji} &= \frac{r_i\sin\alpha_{ij}}{q_{ij}} \end{aligned} \tag{13.5}$$

In a bent $XY_2$ molecule, Eq. 13.4 becomes

$$\begin{aligned} V = \tfrac{1}{2}(K + t^2F' + s^2F)[(\Delta r_1)^2 + (\Delta r_2)^2] \\ + \tfrac{1}{2}(H - s^2F' + t^2F)(r\,\Delta\alpha)^2 \\ + (-t^2F' + s^2F)(\Delta r_1)(\Delta r_2) \\ + ts(F' + F)(\Delta r_1)(r\,\Delta\alpha) \\ + ts(F' + F)(\Delta r_2)(r\,\Delta\alpha) \end{aligned} \tag{13.6}$$

where

$$s = \frac{r(1 - \cos\alpha)}{q}$$

$$t = \frac{r\sin\alpha}{q}$$

---

* In the case of tetrahedral molecules, a term

$$\sum_{i\neq j\neq k}\left(\frac{\kappa}{\sqrt{2}}\right)r_{ij}r_{ik}(r_{ij}\,\Delta\alpha_{ij})(r_{ik}\,\Delta\alpha_{ik})$$

must be added, where $\kappa$ is called the internal tension.

Comparing Eqs. 13.6 and 13.1, we obtain the following relations between the force constants of the generalized valence force field and those of the Urey–Bradley force field:

$$\begin{aligned} f_{11} &= K + t^2F' + s^2F \\ r^2f_{33} &= (H - s^2F' + t^2F)r^2 \\ f_{12} &= -t^2F' + s^2F \\ rf_{13} &= ts(F' + F)r \end{aligned} \tag{13.7}$$

Although the Urey–Bradley field has four force constants, $F'$ is usually taken as $-\frac{1}{10}F$, on the assumption that the repulsive energy between nonbonded atoms is proportional to $1/r$.[9]* Thus only three force constants, $K$, $H$, and $F$, are needed to construct the **F** matrix. The *orbital valence force* (OVF) field developed by Heath and Linnett[42] is similar to the UBF field. The OVF field uses the angle ($\Delta\beta$) which represents the distortion of the bond from the axis of the bonding orbital instead of the angle between two bonds ($\Delta\alpha$).

The number of force constants in the Urey–Bradley field is, in general, much smaller than that in the generalized valence force field. In addition, the UBF field has the advantages that (1) the force constants have clearer physical meanings than those of the GVF field, and (2) they are often transferable from molecule to molecule. For example, the force constants obtained for $SiCl_4$ and $SiBr_4$ can be used for $SiCl_3Br$, and $SiCl_2Br_2$, and $SiClBr_3$. Mizushima, Shimanouchi, and their co-workers[41a] and Overend and Scherer[43] have given many examples that demonstrate the transferability of the force constants in the UBF field. This property of the Urey–Bradley force constants is highly useful in calculations for complex molecules. It should be mentioned, however, that ignorance of the interactions between nonneighboring stretching vibrations and between bending vibrations in the Urey–Bradley field sometimes causes difficulties in adjusting the force constants to fit the observed frequencies. In such a case, it is possible to improve the results by introducing more force constants.[43,44]

Evidently, the values of force constants depend on the force field initially assumed. Thus a comparison of force constants between molecules should not be made unless they are obtained by using the same force field. The normal coordinate analysis developed in Secs. I-11 to I-13 has already been applied to a number of molecules of various structures. In Parts II, III, and IV, references are cited for each type. Appendix V lists the **G** and **F** matrix elements for typical molecules.

* This assumption does not cause serious error in final results, since $F'$ is small in most cases.

## I-14. SOLUTION OF THE SECULAR EQUATION

Once the **G** and **F** matrices are obtained, the next step is to solve the matrix secular equation:

$$|\mathbf{GF}-\mathbf{E}\lambda|=0 \tag{11.6}$$

In diatomic molecules, $\mathbf{G}=G_{11}=1/\mu$ and $\mathbf{F}=F_{11}=K$. Then $\lambda=G_{11}F_{11}$ and $\tilde{\nu}=\sqrt{\lambda}/2\pi c=\sqrt{K/\mu}/2\pi c$ (Eq. 2.6). If the units of mass and force constant are atomic weight and mdyn/Å (or $10^5$ dyn/cm), respectively,* $\lambda$ is related to $\tilde{\nu}(\text{cm}^{-1})$ by

$$\tilde{\nu}=1302.83\sqrt{\lambda}$$

or

$$\lambda=0.58915\left(\frac{\tilde{\nu}}{1000}\right)^2 \tag{14.1}$$

As an example, for the HF molecule $\mu=0.9573$ and $K=9.65$ in these units. Then, from Eqs. 2.6 and 14.1, $\tilde{\nu}$ is 4139 cm$^{-1}$.

The **F** and **G** matrix elements of a bent $XY_2$ molecule are given in Eqs. 12.15 and 12.16, respectively. The secular equation for the $A_1$ species is quadratic:

$$|\mathbf{GF}-\mathbf{E}\lambda|=\begin{vmatrix} G_{11}F_{11}+G_{12}F_{21}-\lambda & G_{11}F_{12}+G_{12}F_{22} \\ G_{21}F_{11}+G_{22}F_{21} & G_{21}F_{12}+G_{22}F_{22}-\lambda \end{vmatrix}=0 \tag{14.2}$$

If this is expanded into an algebraic equation, the following result is obtained:

$$\lambda^2-(G_{11}F_{11}+G_{22}F_{22}+2G_{12}F_{12})\lambda+(G_{11}G_{22}-G_{12}^2)(F_{11}F_{22}-F_{12}^2)=0 \tag{14.3}$$

For the $H_2O$ molecule,

$$\mu_1=\mu_H=\frac{1}{1.008}=0.99206$$

$$\mu_3=\mu_O=\frac{1}{15.995}=0.06252$$

$$r=0.96\ (\text{Å}),\qquad \alpha=105°$$

$$\sin\alpha=\sin 105°=0.96593$$

$$\cos\alpha=\cos 105°=-0.25882$$

* Although the bond distance is involved in both the **G** and **F** matrices, it is canceled during multiplication of the **G** and **F** matrix elements. Therefore any unit can be used for the bond distance.

Then the **G** matrix elements of Eq. 12.16 are

$$G_{11} = \mu_1 + \mu_3(1 + \cos\alpha) = 1.03840$$

$$G_{12} = -\frac{\sqrt{2}}{r}\mu_3 \sin\alpha = -0.08896$$

$$G_{22} = \frac{1}{r^2}[2\mu_1 + 2\mu_3(1 - \cos\alpha)] = 2.32370$$

If the force constants in terms of the generalized valence force field are selected as

$$f_{11} = 8.428, \qquad f_{12} = -0.105$$
$$f_{13} = 0.252, \qquad f_{33} = 0.768$$

the **F** matrix elements of Eq. 12.15 are

$$F_{11} = f_{11} + f_{12} = 8.32300$$

$$F_{12} = \sqrt{2}rf_{13} = 0.35638$$

$$F_{22} = r^2 f_{33} = 0.70779$$

Using these values, we find that Eq. 14.3 becomes

$$\lambda^2 - 10.22389\lambda + 13.86234 = 0$$

The solution of this equation gives

$$\lambda_1 = 8.61475, \qquad \lambda_2 = 1.60914$$

If these values are converted to $\tilde{\nu}$ through Eq. 14.1, we obtain

$$\tilde{\nu}_1 = 3824\ \text{cm}^{-1}, \qquad \tilde{\nu}_2 = 1653\ \text{cm}^{-1}$$

With the same set of force constants, the frequency of the $B_2$ vibration is calculated as

$$\lambda_3 = G_{33}F_{33} = [\mu_1 + \mu_3(1 - \cos\alpha)](f_{11} - f_{12})$$
$$= 9.13681$$
$$\tilde{\nu}_3 = 3938\ \text{cm}^{-1}$$

The observed frequencies corrected for anharmonicity are as follows: $\omega_1 = 3825\ \text{cm}^{-1}$, $\omega_2 = 1654\ \text{cm}^{-1}$, and $\omega_3 = 3936\ \text{cm}^{-1}$.

If the secular equation is third order, it gives rise to a cubic equation:

$$\lambda^3-(G_{11}F_{11}+G_{22}F_{22}+G_{33}F_{33}+2G_{12}F_{12}+2G_{13}F_{13}+2G_{23}F_{23})\lambda^2$$
$$+\left\{\begin{vmatrix}G_{11} & G_{12}\\ G_{21} & G_{22}\end{vmatrix}\begin{vmatrix}F_{11} & F_{12}\\ F_{21} & F_{22}\end{vmatrix}+\begin{vmatrix}G_{12} & G_{13}\\ G_{22} & G_{23}\end{vmatrix}\begin{vmatrix}F_{12} & F_{13}\\ F_{22} & F_{23}\end{vmatrix}\right.$$
$$+\begin{vmatrix}G_{11} & G_{13}\\ G_{21} & G_{23}\end{vmatrix}\begin{vmatrix}F_{11} & F_{13}\\ F_{21} & F_{23}\end{vmatrix}+\begin{vmatrix}G_{11} & G_{12}\\ G_{31} & G_{32}\end{vmatrix}\begin{vmatrix}F_{11} & F_{12}\\ F_{31} & F_{32}\end{vmatrix}$$
$$+\begin{vmatrix}G_{12} & G_{13}\\ G_{32} & G_{33}\end{vmatrix}\begin{vmatrix}F_{12} & F_{13}\\ F_{32} & F_{33}\end{vmatrix}+\begin{vmatrix}G_{11} & G_{13}\\ G_{31} & G_{33}\end{vmatrix}\begin{vmatrix}F_{11} & F_{13}\\ F_{31} & F_{33}\end{vmatrix}$$
$$+\begin{vmatrix}G_{21} & G_{22}\\ G_{31} & G_{32}\end{vmatrix}\begin{vmatrix}F_{21} & F_{22}\\ F_{31} & F_{32}\end{vmatrix}+\begin{vmatrix}G_{22} & G_{23}\\ G_{32} & G_{33}\end{vmatrix}\begin{vmatrix}F_{22} & F_{23}\\ F_{32} & F_{33}\end{vmatrix}$$
$$\left.+\begin{vmatrix}G_{21} & G_{23}\\ G_{31} & G_{33}\end{vmatrix}\begin{vmatrix}F_{21} & F_{23}\\ F_{31} & F_{33}\end{vmatrix}\right\}\lambda-\begin{vmatrix}G_{11} & G_{12} & G_{13}\\ G_{21} & G_{22} & G_{23}\\ G_{31} & G_{32} & G_{33}\end{vmatrix}\begin{vmatrix}F_{11} & F_{12} & F_{13}\\ F_{21} & F_{22} & F_{23}\\ F_{31} & F_{32} & F_{33}\end{vmatrix}=0 \tag{14.4}$$

Thus it is possible to solve the secular equation by expanding it into an algebraic equation. If the order of the secular equation is higher than three, however, direct expansion such as that just shown becomes too cumbersome. There are several methods of calculating the coefficients of an algebraic equation using indirect expansion.[3] The use of an electronic computer greatly reduces the burden of calculation. Excellent programs written by Schachtschneider[45] are available for the vibrational analysis of polyatomic molecules.

## I-15. VIBRATIONAL FREQUENCIES OF ISOTOPIC MOLECULES

As stated in Sec. I-13, the vibrational frequencies of isotopic molecules are very useful in refining a set of force constants in vibrational analysis. For large molecules, isotopic substitution is indispensable in making band assignments, since only vibrations involving the motion of the isotopic atom will be shifted by isotopic substitution.

Two important rules hold for the vibrational frequencies of isotopic molecules. The first, called the *product rule*, can be derived as follows.

Let $\lambda_1, \lambda_2, \ldots, \lambda_n$ be the roots of the secular equation $|\mathbf{GF}-\mathbf{E}\lambda|=0$. Then

$$\lambda_1\lambda_2\cdots\lambda_n=|\mathbf{G}|\,|\mathbf{F}| \tag{15.1}$$

holds for a given molecule. Since the isotopic molecule has exactly the same $|\mathbf{F}|$ as that in Eq. 15.1, a similar relation

$$\lambda'_1\lambda'_2 \cdots \lambda'_n = |\mathbf{G}'|\,|\mathbf{F}|$$

holds for this molecule. It follows that

$$\frac{\lambda_1\lambda_2 \cdots \lambda_n}{\lambda'_1\lambda'_2 \cdots \lambda'_n} = \frac{|\mathbf{G}|}{|\mathbf{G}'|} \tag{15.2}$$

Since

$$\tilde{\nu} = \frac{1}{2\pi c}\sqrt{\lambda}$$

Eq. 15.2 can be written as

$$\frac{\tilde{\nu}_1\tilde{\nu}_2 \cdots \tilde{\nu}_n}{\tilde{\nu}'_1\tilde{\nu}'_2 \cdots \tilde{\nu}'_n} = \sqrt{\frac{|\mathbf{G}|}{|\mathbf{G}'|}} \tag{15.3}$$

This rule has been confirmed by using pairs of molecules such as $H_2O$ and $D_2O$, $CH_4$ and $CD_4$. The rule is also applicable to the product of vibrational frequencies belonging to a single symmetry species.

A more general form of Eq. 15.3 is given by the *Redlich–Teller product rule*[1]:

$$\frac{\tilde{\nu}_1\tilde{\nu}_2 \cdots \tilde{\nu}_n}{\tilde{\nu}'_1\tilde{\nu}'_2 \cdots \tilde{\nu}'_n} = \sqrt{\left(\frac{m'_1}{m_1}\right)^{\alpha}\left(\frac{m'_2}{m_2}\right)^{\beta} \cdots \left(\frac{M}{M'}\right)^{t}\left(\frac{I_x}{I'_x}\right)^{\delta_x}\left(\frac{I_y}{I'_y}\right)^{\delta_y}\left(\frac{I_z}{I'_z}\right)^{\delta_z}} \tag{15.4}$$

Here $m_1$, $m_2, \ldots$ are the masses of the representative atoms of the various sets of equivalent nuclei (atoms represented by $m$, $m_0$, $m_{xy}, \ldots$ in the tables given in Appendix II); $\alpha$, $\beta, \ldots$ are the coefficients of $m$, $m_0$, $m_{xy}, \ldots$; $M$ is the total mass of the molecule; $t$ is the number of $T_x$, $T_y$, $T_z$ in the symmetry type considered; $I_x$, $I_y$, $I_z$ are the moments of inertia about the $x$, $y$, $z$ axes, respectively, which go through the center of the mass; and $\delta_x$, $\delta_y$, $\delta_z$ are 1 or 0, depending on whether or not $R_x$, $R_y$, $R_z$ belong to the symmetry type considered. On both the left and the right-hand sides, $\alpha$, $\beta, \ldots, t$, $\delta_x$, $\delta_y$, $\delta_z$ are counted only once for a degenerate vibration.

Another useful rule in regard to the vibrational frequencies of isotopic molecules, called the *sum rule*, can be derived as follows. It is obvious from Eqs. 14.3 and 14.4 that

$$\lambda_1 + \lambda_2 + \cdots + \lambda_n = \sum_n \lambda = \sum_{i,j} G_{ij}F_{ij} \tag{15.5}$$

Let $\sigma_k$ denote $\sum_{ij} G_{ij}F_{ij}$ for $k$ different isotopic molecules, all of which have the same **F** matrix. If a suitable combination of molecules is taken, so that

$$\sigma_1+\sigma_2+\cdots+\sigma_k=\left(\sum G_{ij}F_{ij}\right)_1+\left(\sum G_{ij}F_{ij}\right)_2+\cdots+\left(\sum G_{ij}F_{ij}\right)_k$$

$$=\left[\left(\sum G_{ij}\right)_1+\left(\sum G_{ij}\right)_2+\cdots+\left(\sum G_{ij}\right)_k\right]\left(\sum F_{ij}\right)$$

$$=0$$

then it follows that

$$\left(\sum \lambda\right)_1+\left(\sum \lambda\right)_2+\cdots+\left(\sum \lambda\right)_k=0 \qquad (15.6)$$

This rule has been verified for such combinations as $H_2O$, $D_2O$, and HDO, where

$$2\sigma(\mathrm{HDO})-\sigma(\mathrm{H_2O})-\sigma(\mathrm{D_2O})=0$$

Such relations between the frequencies of isotopic molecules are highly useful in making band assignments.

In general, the relative magnitude of isotopic shifts ($\Delta\nu/\nu$) becomes smaller as the relative mass difference between isotopes ($\Delta m/m$) becomes smaller. Thus the isotope shift decreases in the following order of pairs: ($^1$H, $^2$D) > ($^6$Li, $^7$Li) > ($^{10}$B, $^{11}$B) > ($^{14}$N, $^{15}$N) > . . . . It was thought initially that the isotope shifts of the metal–ligand vibrations involving heavy metal isotopes such as ($^{58}$Ni, $^{62}$Ni) and ($^{116}$Sn, $^{124}$Sn) would be too small to be detected under ordinary experimental conditions. It was shown, however, that the metal–ligand stretching vibrations involving these and many other metal isotopes exhibit shifts ranging from 10 to 2 cm$^{-1}$, values beyond the experimental error of common infrared instruments (±0.5 cm$^{-1}$). Thus the *metal isotope technique* provides a definitive method of assigning the metal–ligand stretching vibrations. Studies of metal–ligand bending vibrations are more complicated, since their shifts range from 2 to 0.5 cm$^{-1}$, values close to instrumental errors. The metal isotope technique has been applied to many coordination compounds to assign metal–ligand vibrations, as well as to refine metal–ligand force constants. For details, see Refs. 46 and 47.

## I-16. GROUP FREQUENCIES AND BAND ASSIGNMENTS

From observation of the infrared spectra of a number of compounds having a common group of atoms, it is found that, regardless of the rest of the molecule, this common group absorbs over a narrow range of frequencies, called the *group frequency*. For example, the group frequencies of the methyl group are 3000–2860, 1470–1400, 1380–1200, and 1200–800 $cm^{-1}$. Group frequencies have been found for a number of organic and inorganic groups, and they have been summarized as *group frequency charts*,[35,36] which are highly useful in identifying the atomic groups from infrared spectra. Group frequency charts for inorganic and coordination compounds are given in Appendix VI as well as in Figs. II-21 and II-22.

The concept of group frequency rests on the assumption that the vibrations of a particular group are relatively independent of those of the rest of the molecule. As stated in Sec. I-3, however, all the nuclei of the molecule perform their harmonic oscillations in a normal vibration. Thus an *isolated vibration*, which the group frequency would have to be, cannot be expected in polyatomic molecules. If, however, a group includes relatively light atoms such as hydrogen (OH, NH, $NH_2$, CH, $CH_2$, $CH_3$, etc.) or relatively heavy atoms such as the halogens (CCl, CBr, CI, etc.), as compared to other atoms in the molecule, the idea of an isolated vibration may be justified, since the amplitudes (or velocities) of the harmonic oscillation of these atoms are relatively larger or smaller than those of the other atoms in the same molecule. Vibrations of groups having multiple bonds (C≡C, C≡N, C=C, C=N, C=O, etc.) may also be relatively independent of the rest of the molecule if the groups do not belong to a conjugated system.

If atoms of similar mass are connected by bonds of similar strength (force constant), the amplitude of oscillation is similar for each atom of the whole system. Therefore it is not possible to isolate the group frequencies in a system like the following:

$$\text{—O—}\overset{|}{\underset{|}{\text{C}}}\text{—}\overset{|}{\underset{|}{\text{C}}}\text{—N}\Big\langle$$

A similar situation may occur in a system in which resonance effects average out the single and multiple bonds by conjugation. Examples of this effect are seen in the metal chelate compounds of $\beta$-diketones, $\alpha$-diimines, and oxalic acid (discussed in Part III). When the group frequency approximation is permissible, the mode of vibration corresponding to this frequency can be inferred empirically from the band assignments obtained theoretically for simple molecules. If *coupling* between various group vibrations is serious, it is necessary to make a theoretical analysis for each individual compound, using a method like the following one.

As stated in Sec. I-3, the generalized coordinates are related to the normal coordinates by

$$q_k = \sum_i B_{ki} Q_i \tag{3.10}$$

In matrix form, this is written as

$$\mathbf{q} = \mathbf{B}_q \mathbf{Q} \tag{16.1}$$

It can be shown[3] that the internal coordinates are also related to the normal coordinates by

$$\mathbf{R} = \mathbf{L}\mathbf{Q} \tag{16.2}$$

This is written more explicitly as

$$\begin{aligned} R_1 &= l_{11}Q_1 + l_{12}Q_2 + \cdots + l_{1N}Q_N \\ R_2 &= l_{21}Q_1 + l_{22}Q_2 + \cdots + l_{2N}Q_N \\ &\ \ \vdots \\ R_i &= l_{i1}Q_1 + l_{i2}Q_2 + \cdots + l_{iN}Q_N \end{aligned} \tag{16.3}$$

In a normal vibration in which the normal coordinate $Q_N$ changes with frequency $\nu_N$, all the internal coordinates, $R_1, R_2, \ldots, R_i$, change with the same frequency. The amplitude of oscillation is, however, different for each internal coordinate. The relative ratio of the amplitudes of the internal coordinates in a normal vibration associated with $Q_N$ is given by

$$l_{1N} : l_{2N} : \cdots : l_{iN} \tag{16.4}$$

If one of these elements is relatively large compared to the others, the normal vibration is said to be predominantly due to the vibration caused by the change of this coordinate.

The ratio of $l$'s given by Eq. 16.4 can be obtained as a column matrix (or eigenvector) $\boldsymbol{l}_N$, which satisfies the relation[3,4]

$$\mathbf{GF}\boldsymbol{l}_N = \boldsymbol{l}_N \lambda_N \tag{16.5}$$

It consists of $i$ elements, $l_{1N}, l_{2N}, \ldots, l_{iN}$, $i$ being the number of internal coordinates, and can be calculated if the **G** and **F** matrices are known. An assembly by columns of the $l$ elements obtained for each $\lambda$ gives the relation

$$\mathbf{GFL} = \mathbf{L}\boldsymbol{\Lambda} \tag{16.6}$$

where $\boldsymbol{\Lambda}$ is a diagonal matrix whose elements consist of $\lambda$ values.

As an example, calculate the **L** matrix of the $H_2O$ molecule, using the results obtained in Sec. I-14. The **G** and **F** matrices for the $A_1$ species are as follows

$$\mathbf{G} = \begin{bmatrix} 1.03840 & -0.08896 \\ -0.08896 & 2.32370 \end{bmatrix} \qquad \mathbf{F} = \begin{bmatrix} 8.32300 & 0.35638 \\ 0.35638 & 0.70779 \end{bmatrix}$$

with $\lambda_1 = 8.61475$ and $\lambda_2 = 1.60914$. The **GF** product becomes

$$\mathbf{GF} = \begin{bmatrix} 8.61090 & 0.30710 \\ 0.08771 & 1.61299 \end{bmatrix}$$

The **L** matrix can be calculated from Eq. 16.6:

$$\begin{bmatrix} 8.61090 & 0.30710 \\ 0.08771 & 1.61299 \end{bmatrix} \begin{bmatrix} l_{11} & l_{12} \\ l_{21} & l_{22} \end{bmatrix} = \begin{bmatrix} l_{11} & l_{12} \\ l_{21} & l_{22} \end{bmatrix} \begin{bmatrix} 8.61475 & 0 \\ 0 & 1.60914 \end{bmatrix}$$

However, this equation gives only the ratios $l_{11}:l_{21}$ and $l_{12}:l_{22}$. To determine their values, it is necessary to use the following normalization condition:

$$\mathbf{L}\tilde{\mathbf{L}} = \mathbf{G} \qquad (16.7)^*$$

Then the final result is

$$\begin{bmatrix} l_{11} & l_{12} \\ l_{21} & l_{22} \end{bmatrix} = \begin{bmatrix} 1.01683 & -0.06686 \\ 0.01274 & 1.52432 \end{bmatrix}$$

This result indicates that, in the normal vibration $Q_1$, the relative ratio of amplitudes of two internal coordinates, $R_1$ (symmetric OH stretching) and $R_2$ (HOH bending), is 1.0168:0.0127. Therefore this vibration (3824 cm$^{-1}$) is assigned to an almost pure OH stretching mode. The relative ratio of amplitudes for the $Q_2$ vibration is −0.0669:1.5243. Thus this vibration is assigned to an almost pure HOH bending mode.

In other cases, the $l$ values do not provide the band assignments that are expected empirically. This occurs because the dimension of $l$ for a stretching coordinate is different from that for a bending coordinate. Morino and Kuchitsu[48] proposed that the potential energy distribution of

---

* This equation can be derived as follows. According to Eq. 11.3, $2T = \tilde{\dot{\mathbf{R}}}\mathbf{G}^{-1}\dot{\mathbf{R}}$. On the other hand, Eq. 16.2 gives $\dot{\mathbf{R}} = \mathbf{L}\dot{\mathbf{Q}}$ and $\tilde{\dot{\mathbf{R}}} = \tilde{\dot{\mathbf{Q}}}\tilde{\mathbf{L}}$. Thus $2T = \tilde{\dot{\mathbf{Q}}}\tilde{\mathbf{L}}\mathbf{G}^{-1}\mathbf{L}\dot{\mathbf{Q}}$. Comparing this with $2T = \tilde{\dot{\mathbf{Q}}}\mathbf{E}\dot{\mathbf{Q}}$ (matrix form of Eq. 3.11), we obtain $\tilde{\mathbf{L}}\mathbf{G}^{-1}\mathbf{L} = \mathbf{E}$ or $\mathbf{L}\tilde{\mathbf{L}} = \mathbf{G}$.

a normal vibration $Q_N$, defined by

$$V(Q_N) = \tfrac{1}{2} Q_N^2 \sum_{ij} F_{ij} l_{iN} l_{jN} \qquad (16.8)^*$$

gives a better measure for making band assignments. In general, the value of $F_{ij} l_{iN} l_{jN}$ is large when $i = j$. Therefore the $F_{ii} l_{iN}^2$ terms are most important in determining the distribution of the potential energy. Thus the ratios of the $F_{ii} l_{iN}^2$ terms provide a measure of the relative contribution of each internal coordinate $R_i$ to the normal coordinate $Q_N$. If any $F_{ii} l_{iN}^2$ term is exceedingly large compared with the others, the vibration is assigned to the mode associated with $R_i$. If $F_{ii} l_{iN}^2$ and $F_{jj} l_{jN}^2$ are relatively large compared with the others, the vibration is assigned to a mode associated with both $R_i$ and $R_j$ (coupled vibration).

As an example, let us calculate the potential energy distribution for the $H_2O$ molecule. Using the **F** and **L** matrices obtained previously, we find that the $\tilde{\mathbf{L}}\mathbf{F}\mathbf{L}$ matrix is calculated to be

$$\begin{bmatrix} \begin{pmatrix} l_{11}^2 F_{11} & + & l_{21}^2 F_{22} & + & 2 l_{21} l_{11} F_{12} \\ 8.60551 & & 0.00011 & & 0.00923 \end{pmatrix} & 0 \\ 0 & \begin{pmatrix} l_{12}^2 F_{11} & + & l_{22}^2 F_{22} & + & 2 l_{12} l_{22} F_{12} \\ 0.03721 & & 1.64459 & & -0.07264 \end{pmatrix} \end{bmatrix}$$

Then the potential energy distribution in each normal vibration ($F_{ii} l_{iN}^2$) is given by

$$\begin{array}{c} \\ R_1 \\ R_2 \end{array} \begin{array}{c} \lambda_1 \qquad\quad \lambda_2 \\ \begin{bmatrix} 8.60551 & 0.03721 \\ 0.00011 & 1.64459 \end{bmatrix} \end{array}$$

More conveniently, the result is expressed by calculating $(F_{ii} l_{iN}^2 / \sum F_{ii} l_{iN}^2) \times 100$ for each coordinate:

$$\begin{array}{c} \\ R_1 \\ R_2 \end{array} \begin{array}{c} \lambda_1 \qquad\quad \lambda_2 \\ \begin{bmatrix} 99.99 & 2.21 \\ 0.01 & 97.79 \end{bmatrix} \end{array}$$

---

* According to Eq. 11.1, the potential energy is written as $2V = \tilde{\mathbf{R}}\mathbf{F}\mathbf{R}$. Using Eq. 16.2, we can write this as $2V = \tilde{\mathbf{Q}}\tilde{\mathbf{L}}\mathbf{F}\mathbf{L}\mathbf{Q}$. On the other hand, Eq. 3.12 can be written as $2V = \tilde{\mathbf{Q}}\mathbf{\Lambda}\mathbf{Q}$. A comparison of these two expression gives $\mathbf{\Lambda} = \tilde{\mathbf{L}}\mathbf{F}\mathbf{L}$. If this is written for one normal vibration whose frequency is $\lambda_N$, we have

$$\lambda_N = \sum_{ij} \tilde{l}_{Ni} F_{ij} l_{jN} = \sum_{ij} F_{ij} l_{iN} l_{jN}$$

Then the potential energy due to this vibration is expressed by Eq. 16.8.

In this case, the final results are the same whether the band assignments are based on the **L** matrix or on the potential energy distribution: $Q_1$ is the symmetric OH stretching and $Q_2$ is the HOH bending. In other cases, different results may be obtained, depending on which criterion is used for band assignments.

A more rigorous method of determining the vibrational mode is to draw the displacements of individual atoms in terms of rectangular coordinates. As in Eq. 16.2, the relationship between the rectangular and normal coordinates is given by

$$\mathbf{X} = \mathbf{L}_x \mathbf{Q} \tag{16.9}$$

The $\mathbf{L}_x$ matrix can be obtained from the relationship[49]

$$\mathbf{L}_x = \mathbf{M}^{-1}\tilde{\mathbf{B}}\mathbf{G}^{-1}\mathbf{L} \tag{16.10)*$$

The matrices on the right have already been defined.

Three-dimensional drawings of normal modes such as those shown in Part II can be made from the cartesian displacement calculations obtained above. However, hand plotting of these data is laborious and complicated. Use of computer plotting programs greatly facilitates this process.[50]

## I-17. INTENSITY OF INFRARED ABSORPTION[3,51,52]

The absorption of strictly monochromatic light ($\nu$) is expressed by the Lambert–Beer law:

$$I_\nu = I_{0,\nu} e^{-\alpha_\nu p l} \tag{17.1}$$

where $I_\nu$ is the intensity of the light transmitted by a cell of length $l$ containing a gas at pressure $p$, $I_{0,\nu}$ is the intensity of the incident light, and $\alpha_\nu$ is the absorption coefficient for unit pressure. The true integrated absorption coefficient $A$ is defined by

$$A = \int_{\text{band}} \alpha_\nu \, d\nu = \frac{1}{pl} \int_{\text{band}} \ln\left(\frac{I_{0,\nu}}{I_\nu}\right) d\nu \tag{17.2}$$

where the integration is carried over the entire frequency region of a band.

In practice, $I_\nu$ and $I_{0,\nu}$ cannot be measured accurately, since no spectrophotometers have infinite resolving power. Therefore we measure

---

* By combining Eqs. 11.8 and 16.9, we have $\mathbf{R} = \mathbf{BX} = \mathbf{BL}_x\mathbf{Q}$. Since $\mathbf{R} = \mathbf{LQ}$ (Eq. 16.2), it follows that $\mathbf{LQ} = \mathbf{BL}_x\mathbf{Q}$ or $\mathbf{L} = \mathbf{BL}_x$. The kinetic energy is written as $2T = \tilde{\dot{\mathbf{X}}}\mathbf{M}\dot{\mathbf{X}}$. In terms of internal coordinates, it is written as $2T = \tilde{\dot{\mathbf{R}}}\mathbf{G}^{-1}\dot{\mathbf{R}} = \tilde{\dot{\mathbf{X}}}\tilde{\mathbf{B}}\mathbf{G}^{-1}\mathbf{B}\dot{\mathbf{X}}$. By comparing these two expressions, we have $\mathbf{M} = \tilde{\mathbf{B}}\mathbf{G}^{-1}\mathbf{B}$. Then we can write $\mathbf{L}_x = \mathbf{M}^{-1}\mathbf{ML}_x = \mathbf{M}^{-1}\tilde{\mathbf{B}}\mathbf{G}^{-1}\mathbf{BL}_x = \mathbf{M}^{-1}\tilde{\mathbf{B}}\mathbf{G}^{-1}\mathbf{L}$.

instead the apparent intensity $T_\nu$:

$$T_\nu = \int_{\text{slit}} I(\nu) g(\nu, \nu')\, d\nu \tag{17.3}$$

where $g(\nu, \nu')$ is a function indicating the amount of light of frequency $\nu$ when the spectrophotometer reading is set at $\nu'$. Then the apparent integrated absorption coefficient $B$ is defined by

$$B = \frac{1}{pl} \int_{\text{band}} \ln \frac{\int_{\text{slit}} I_0(\nu) g(\nu, \nu')\, d\nu}{\int_{\text{slit}} I(\nu) g(\nu, \nu')\, d\nu}\, d\nu' \tag{17.4}$$

It can be shown that

$$\lim_{pl \to 0} (A - B) = 0 \tag{17.5}$$

if $I_0$ and $\alpha_\nu$ are constant within the slit width used. (This condition is approximated by using a narrow slit.) In practice, we plot $B/pl$ against $pl$, and extrapolate the curve to $pl \to 0$. To apply this method to gaseous molecules, it is necessary to broaden the vibrational–rotational bands by adding a high pressure inert gas (pressure broadening).

For liquids and solutions, $p$ and $\alpha$ in preceding equations are replaced by $M$ (molar concentration) and $\varepsilon$ (molar absorption coefficient), respectively. However, the extrapolation method just described is not applicable, since experimental errors in determining $B$ values become too large at low concentration or at small cell length. The true integrated absorption coefficient of a liquid can be calculated if we assume that the shape of an absorption band is represented by the Lorentz equation and that the slit function is triangular.[53]

Theoretically, the true integrated absorption coefficient $A_N$ of the $N$th normal vibration is given by[3]

$$A_N = \frac{n\pi}{3c} \left[ \left( \frac{\partial \mu_x}{\partial Q_N} \right)_0^2 + \left( \frac{\partial \mu_y}{\partial Q_N} \right)_0^2 + \left( \frac{\partial \mu_z}{\partial Q_N} \right)_0^2 \right] \tag{17.6}$$

where $n$ is the number of molecules per cubic centimeter, and $c$ is the velocity of light. As shown by Eq. 16.2, an internal coordinate $R_i$ is related to a set of normal coordinates by

$$R_i = \sum_N L_{iN} Q_N \tag{17.7}$$

If the additivity of the bond dipole moment is assumed, it is possible to write

$$\begin{aligned}\frac{\partial \mu}{\partial Q_N} &= \sum_i \left(\frac{\partial \mu}{\partial R_i}\right)\left(\frac{\partial R_i}{\partial Q_N}\right) \\ &= \sum_i \left(\frac{\partial \mu}{\partial R_i}\right) L_{iN}\end{aligned} \tag{17.8}$$

Then Eq. 17.6 is written as

$$\begin{aligned}A_N &= \frac{n\pi}{3c}\left[\left(\sum_i \frac{\partial \mu_x}{\partial R_i} L_{iN}\right)_0^2 + \left(\sum_i \frac{\partial \mu_y}{\partial R_i} L_{iN}\right)_0^2 + \left(\sum \frac{\partial \mu_z}{\partial R_i} L_{iN}\right)_0^2\right] \\ &= \frac{n\pi}{3c}\sum_i \left[\left(\frac{\partial \mu_x}{\partial R_i}\right)_0^2 + \left(\frac{\partial \mu_y}{\partial R_i}\right)_0^2 + \left(\frac{\partial \mu_z}{\partial R_i}\right)_0^2\right](L_{iN})^2\end{aligned} \tag{17.9}$$

This equation shows that the intensity of an infrared band depends on the values of the $\partial\mu/\partial R$ terms as well as of the $L$ matrix elements.

Equation 17.9 has been applied to relatively small molecules to calculate the $\partial\mu/\partial R$ terms from the observed intensity and known $L_{iN}$ values.[7] However, the additivity of the bond dipole moment does not strictly hold, and the results obtained are often inconsistent and conflicting. Thus far, very few studies have been made on infrared intensities of large molecules because of these difficulties.

## I-18. DEPOLARIZATION OF RAMAN LINES[16]

As stated in Secs. I-7 and I-8, it is possible, by using group theory, to classify the normal vibration into various symmetry species. Experimentally, measurements of the infrared dichroism and polarization properties of Raman lines of an orientated crystal provide valuable information about the symmetry of normal vibrations (Sec. I-22). Here we consider the polarization properties of Raman lines in liquids and solutions in which molecules or ions take completely random orientations.

Suppose that we irradiate a molecule fixed at the origin of a space-fixed coordinate system with natural light from the positive-$y$ direction, and observe the Raman scattering in the $x$ direction as shown in Fig. I-14. The incident light vector $E$ may be resolved into two components, $E_x$ and $E_z$, of equal magnitude ($E_y = 0$). Both components give induced dipole moments, $P_x$, $P_y$, and $P_z$. However, only $P_y$ and $P_z$ contribute to the scattering along the $x$ axis, since an oscillating dipole cannot radiate in its

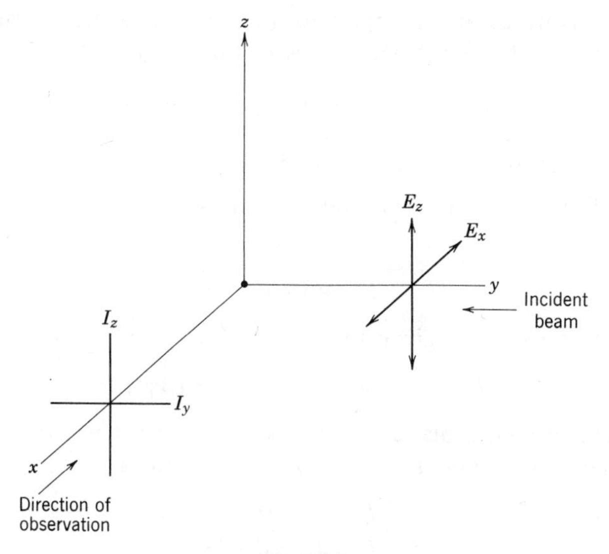


**Fig. I-14.**

own direction. Then, from Eq. 5.7, we have

$$P_y = \alpha_{yx}E_x + \alpha_{yz}E_z \tag{18.1*}$$

$$P_z = \alpha_{zx}E_x + \alpha_{zz}E_z \tag{18.2}$$

The intensity of the scattered light is proportional to the sum of squares of the individual $\alpha_{ij}E_j$ terms. Thus the ratio of the intensities in the $y$ and $z$ directions is

$$\rho_n = \frac{I_y}{I_z} = \frac{\alpha_{yx}^2 E_x^2 + \alpha_{yz}^2 E_z^2}{\alpha_{zx}^2 E_x^2 + \alpha_{zz}^2 E_z^2} \tag{18.3}$$

where $\rho_n$ is called the *depolarization ratio for natural light* ($n$).

In a homogeneous liquid or gas, the molecules are randomly orientated, and we must consider the polarizability components averaged over all molecular orientations. The results are expressed in terms of two quantities: $\bar{\alpha}$ (*mean value*) and $\gamma$ (*anisotropy*):

$$\bar{\alpha} = \tfrac{1}{3}(\alpha_{xx} + \alpha_{yy} + \alpha_{zz}) \tag{18.4}$$

$$\gamma^2 = \tfrac{1}{2}[(\alpha_{xx} - \alpha_{yy})^2 + (\alpha_{yy} - \alpha_{zz})^2 + (\alpha_{zz} - \alpha_{xx})^2 + 6(\alpha_{xy}^2 + \alpha_{yz}^2 + \alpha_{zx}^2)] \tag{18.5}$$

* In the case of Raman scattering, it is the $\partial\alpha/\partial Q$ term that should be used in Eq. 18.1 and the following equations.

These two quantities are invariant to any coordinate transformation. It can be shown[3] that the average values of the squares of $\alpha_{ij}$ are

$$\overline{(\alpha_{xx})^2} = \overline{(\alpha_{yy})^2} = \overline{(\alpha_{zz})^2} = \tfrac{1}{45}[45(\bar{\alpha})^2 + 4\gamma^2] \qquad (18.6)$$

$$\overline{(\alpha_{xy})^2} = \overline{(\alpha_{yz})^2} = \overline{(\alpha_{zx})^2} = \tfrac{1}{15}\gamma^2 \qquad (18.7)$$

Since $E_x = E_z = E$, Eq. 18.3 can be written as

$$\rho_n = \frac{I_y}{I_z} = \frac{6\gamma^2}{45(\bar{\alpha})^2 + 7\gamma^2} \qquad (18.8)$$

The total intensity, $I_n$, is given by

$$I_n = I_y + I_z = \text{const}\,[\tfrac{1}{45}\{45(\bar{\alpha})^2 + 13\gamma^2\}]E^2 \qquad (18.9)$$

If the incident light is plane polarized (e.g., laser beam), with its electric vector in the $z$ direction ($E_x = 0$), Eq. 18.8 becomes

$$\rho_p = \frac{I_y}{I_z} = \frac{3\gamma^2}{45(\bar{\alpha})^2 + 4\gamma^2} \qquad (18.10)$$

where $\rho_p$ is the *depolarization ratio for polarized light* ($p$). In this case, the total intensity is given by

$$I_p = I_y + I_z = \text{const}\,[\tfrac{1}{45}\{45(\bar{\alpha})^2 + 7\gamma^2\}]E^2 \qquad (18.11)$$

The symmetry property of a normal vibration can be determined by measuring the depolarization ratio. From an inspection of character tables (Appendix I), it is obvious that $\bar{\alpha}$ is nonzero only for totally symmetric vibrations. Then Eq. 18.8 gives $0 \leqslant \rho_n < \frac{6}{7}$, and the Raman lines are said to be *polarized.* For all nontotally symmetric vibrations, $\bar{\alpha}$ is zero, and $\rho_n = \frac{6}{7}$. Then the Raman lines are said to be *depolarized.* If the exciting line is plane polarized, these criteria must be changed according to Eq. 18.10. Thus $0 \leqslant \rho_p < \frac{3}{4}$ for totally symmetric vibrations, and $\rho_p = \frac{3}{4}$ for nontotally symmetric vibrations. Figure I-15 shows the Raman spectra of $CCl_4$ (500–150 cm$^{-1}$) in two directions of polarization obtained with the 488 nm excitation. The three bands at 459, 314, and 218 cm$^{-1}$ give $\rho_p$ values of approximately 0.02, 0.75, and 0.75, respectively. Thus it is concluded that the 459 cm$^{-1}$ band is polarized ($A_1$), whereas the two bands at 314 ($F_2$) and 218 ($E$) cm$^{-1}$ are depolarized.

As stated in Sec. I-5, the polarizability tensors are symmetric in normal Raman scattering. If the exciting frequency approaches that of an electronic absorption, some scattering tensors become antisymmetric,* and resonance Raman scattering can occur (Sec. I-20). In this case, Eq. 18.10

---

* A tensor is called antisymmetric if $\alpha_{xx} = \alpha_{yy} = \alpha_{zz} = 0$ and $\alpha_{xy} = -\alpha_{yx}$, $\alpha_{yz} = -\alpha_{zy}$, and $\alpha_{zx} = -\alpha_{xz}$.

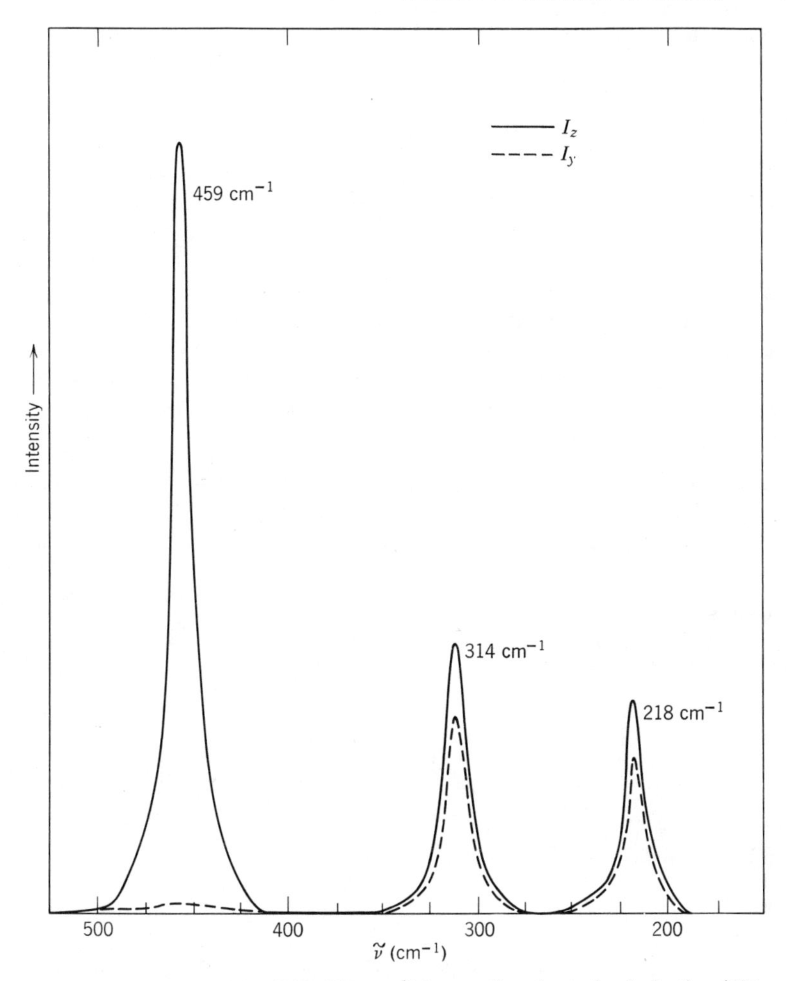


**Fig. I-15.** Raman spectra of $CCl_4$ (500–150 $cm^{-1}$) in two directions of polarization (488 nm excitation).

must be written in a more general form:[54]

$$\rho_p = \frac{3g^s + 5g^a}{10g^o + 4g^s} \tag{18.12}$$

where

$$\begin{aligned} g^o &= 3(\bar{\alpha})^2 \\ g^s &= \tfrac{1}{3}[(\alpha_{xx} - \alpha_{yy})^2 + (\alpha_{xx} - \alpha_{zz})^2 + (\alpha_{yy} - \alpha_{zz})^2] \\ &\quad + \tfrac{1}{2}[(\alpha_{xy} + \alpha_{yx})^2 + (\alpha_{xz} + \alpha_{zx})^2 + (\alpha_{yz} + \alpha_{zy})^2] \\ g^a &= \tfrac{1}{2}[(\alpha_{xy} - \alpha_{yx})^2 + (\alpha_{xz} - \alpha_{zx})^2 + (\alpha_{yz} - \alpha_{zy})^2] \end{aligned} \tag{18.13}$$

If we define

$$\gamma_s^2 = \tfrac{3}{2}g^s \quad \text{and} \quad \gamma_{as}^2 = \tfrac{3}{2}g^a \tag{18.14}$$

Eq. 18.12 can be written as

$$\rho_p = \frac{3\gamma_s^2 + 5\gamma_{as}^2}{45(\bar{\alpha})^2 + 4\gamma_s^2} \tag{18.15}$$

In normal Raman scattering, $\gamma_s^2 = \gamma^2$ and $\gamma_{as}^2 = 0$. Then Eq. 18.15 is reduced to Eq. 18.10.

The symmetry properties of resonance Raman lines can be predicted on the basis of Eq. 18.15. For totally symmetric vibrations, $\bar{\alpha} \neq 0$ and $\gamma_{as} = 0$. Then Eq. 18.15 gives $0 \leqslant \rho_p < \frac{3}{4}$. Nontotally symmetric vibrations ($\bar{\alpha} = 0$) are classified into two types; those which have symmetric scattering tensors, and those which have antisymmetric scattering tensors. If the tensor is symmetric, $\gamma_{as} = 0$ and $\gamma_s \neq 0$. Then Eq. 18.15 gives $\rho_p = \frac{3}{4}$ (depolarized). If the tensor is antisymmetric, $\gamma_{as} \neq 0$ and $\gamma_s = 0$. Then Eq. 18.15 gives $\rho_p = \infty$ (*inverse polarization*). In the case of the $\mathbf{D}_{4h}$ point group, the $B_{1g}$ and $B_{2g}$ representations belong to the former type, whereas the $A_{2g}$ representation belongs to the latter.[54a] As will be shown in Sec. I-20, Spiro and Strekas[54] observed for the first time inversely polarized bands in the resonance Raman spectra of heme proteins.

## I-19. INTENSITY OF RAMAN SCATTERING

According to the quantum mechanical theory of light scattering[7,16,20] the intensity per unit solid angle of scattered light arising from a transition between states $m$ and $n$ is given by

$$I_{n \leftarrow m} = \text{const}\,(\nu_0 + \nu_{mn})^4 \sum_{\rho\sigma} |(P_{\rho\sigma})_{mn}|^2 \tag{19.1}$$

where

$$(P_{\rho\sigma})_{mn} = (\alpha_{\rho\sigma})_{mn}E = \frac{1}{h}\sum_r \left[\frac{(M_\rho)_{rn}(M_\sigma)_{mr}}{\nu_{rm} - \nu_0} + \frac{(M_\rho)_{mr}(M_\sigma)_{rn}}{\nu_{rn} + \nu_0}\right]E \tag{19.2}$$

Here $\nu_0$ is the frequency of the incident light; $\nu_{rm}$, $\nu_{rn}$, and $\nu_{mn}$ are the frequencies corresponding to the energy differences between subscripted states; terms of the type $(M_\sigma)_{mr}$ are the cartesian components of transition moments such as $\int \Psi_r^* \mu_\sigma \Psi_m d\tau$; and $E$ is the electric vector of the incident light. It should be noted here that the states denoted by $m$, $n$, and $r$ represent vibronic states $\psi_g(\xi, Q)\phi_i^g(Q)$, $\psi_g(\xi, Q)\phi_j^g(Q)$, and $\psi_e(\xi, Q)\phi_v^e(Q)$, respectively, where $\psi_g$ and $\psi_e$ are electronic ground and excited state wave functions, respectively, and $\phi_i^g$, $\phi_j^g$, and $\phi_v^e$ are vibrational functions. Finally, $\sigma$ and $\rho$ denote $x$, $y$, and $z$ components.

Since the electric dipole operator acts only on the electronic wave functions, the $(\alpha_{\rho\sigma})_{mn}$ term in Eq. 19.2 can be written in the form[20]

$$(\alpha_{\rho\sigma})_{mn} = \frac{1}{h}\int \phi_j^g (\alpha_{\rho\sigma})_{gg} \phi_i^g \, dQ \tag{19.3}$$

where

$$(\alpha_{\rho\sigma})_{gg} = \sum_e \left( \frac{\int \psi_g^* \mu_\sigma \psi_e \, d\tau \cdot \int \psi_e^* \mu_\rho \psi_g \, d\tau}{\bar{\nu}_{eg} - \nu_0} + \frac{\int \psi_g^* \mu_\rho \psi_e \, d\tau \cdot \int \psi_e^* \mu_\sigma \psi_g \, d\tau}{\bar{\nu}_{eg} + \nu_0} \right)$$

Here $\bar{\nu}_{eg}$ corresponds to the energy of a pure electronic transition between the ground and excited states.

To discuss the Raman scattering, we expand the $(\alpha_{\rho\sigma})_{gg}$ term as a Taylor series with respect to the normal coordinate $Q$:

$$(\alpha_{\rho\sigma})_{gg} = (\alpha_{\rho\sigma})_{gg}^0 + \left[\frac{\partial(\alpha_{\rho\sigma})_{gg}}{\partial Q}\right]_0 Q + \cdots \tag{19.4}$$

Then, we write Eq. 19.3 as

$$(\alpha_{\rho\sigma})_{mn} = \frac{1}{h}(\alpha_{\rho\sigma})_{gg}^0 \int \phi_j^g \phi_i^g \, dQ + \frac{1}{h}\left[\frac{\partial(\alpha_{\rho\sigma})_{gg}}{\partial Q}\right]_0 \int \phi_j^g Q \phi_i^g \, dQ \tag{19.5}$$

The first term on the right is zero unless $i = j$. This term is responsible for Rayleigh scattering. The second term determines the activity of fundamental vibrations in Raman scattering; it vanishes for a harmonic oscillator unless $j = i \pm 1$.

If we consider a Stokes transition, $\upsilon \rightarrow \upsilon + 1$, Eq. 19.5 is written as[3]

$$(\alpha_{\rho\sigma})_{\upsilon,\upsilon+1} = \frac{1}{h}\left[\frac{\partial(\alpha_{\rho\sigma})_{gg}}{\partial Q}\right]_0 \sqrt{\frac{(\upsilon+1)h}{8\pi^2\mu\nu}} \tag{19.6}$$

where $\mu$ and $\nu$ are the reduced mass and the Stokes frequency. Then Eq. 19.1 is written as

$$I = \text{const}\,(\nu_0 - \nu)^4 \frac{E^2}{h^2}\left[\frac{\partial(\alpha_{\rho\sigma})_{gg}}{\partial Q}\right]_0^2 \frac{(\upsilon+1)h}{8\pi^2\mu\nu} \tag{19.7}$$

In Sec. I-18, we derived a classical equation for Raman intensity:

$$\begin{aligned} I_n &= \text{const}\left(\frac{\partial \alpha}{\partial Q}\right)^2 E^2 \\ &= \text{const}\,[\tfrac{1}{45}\{45(\bar{\alpha})^2 + 13\gamma^2\}]E^2 \end{aligned} \tag{18.9}$$

By replacing the $\partial\alpha/\partial Q$ term of Eq. 19.7 with the square bracket term of

Eq. 18.9, we obtain

$$I_n = \text{const}\,(\nu_0 - \nu)^4 \frac{(\upsilon+1)}{8\pi^2\mu\nu} \frac{E^2}{h^2} [\tfrac{1}{45}\{45(\bar{\alpha})^2 + 13\gamma^2\}] \tag{19.8}$$

At room temperature, most of the scattering molecules are in the $\upsilon = 0$ state, but some are in higher vibrational states. Using the Maxwell–Boltzmann distribution law, we find that the fraction of molecules $f_\upsilon$ with vibrational quantum number $\upsilon$ is given by

$$f_\nu = \frac{e^{-[\upsilon+(1/2)]h\upsilon/kT}}{\sum\limits_{\nu} e^{-[\upsilon+(1/2)]h\upsilon/kT}} \tag{19.9}$$

Then the total intensity is proportional to $\sum\limits_{\upsilon} f_\upsilon(\upsilon+1)$, which is equal to $(1 - e^{-h\nu/kT})^{-1}$ (see Ref. 16). Hence we can rewrite Eq. 19.8 in the form

$$I_n = KI_0 \frac{(\nu_0 - \nu)^4}{\mu\nu(1 - e^{-h\upsilon/kT})} [45(\bar{\alpha})^2 + 13\gamma^2] \tag{19.10}$$

Here $I_0$ is the incident light intensity which is proportional to $E^2$, and $K$ summarizes all other constant terms.

If the incident light is polarized, the form of Eq. 19.10 is slightly modified:

$$I_\rho = KI_0 \frac{(\nu_0 - \nu)^4}{\mu\nu(1 - e^{-h\upsilon/kT})} [45(\bar{\alpha})^2 + 7\gamma^2] \tag{19.11}$$

As shown in Sec. I-18, the degree of depolarization $\rho_p$ is

$$\rho_p = \frac{3\gamma^2}{45(\bar{\alpha})^2 + 4\gamma^2} \quad \text{or} \quad \gamma^2 = \frac{45(\bar{\alpha})^2\rho_p}{3 - 4\rho_p} \tag{19.12}$$

Since $\rho_p = \frac{3}{4}$ for nontotally symmetric vibrations, Eq. 19.12 holds only for totally symmetric vibrations. Then Eq. 19.11 is written as

$$I_p = K'I_0 \frac{(\nu_0 - \nu)^4}{\mu\nu(1 - e^{-h\upsilon/kT})} \left(\frac{1 + \rho_p}{3 - 4\rho_p}\right) (\bar{\alpha})^2 \tag{19.13}$$

In the case of a solution, the intensity is proportional to the molar concentration, $C$. Then Eq. 19.13 is written as

$$I_p = K''I_0 \frac{C(\nu_0 - \nu)^4}{\mu\nu(1 - e^{-h\upsilon/kT})} \left(\frac{1 + \rho_p}{3 - 4\rho_p}\right) (\bar{\alpha})^2 \tag{19.14}$$

If we compare the intensities of totally symmetric vibrations ($A_1$ mode) of two tetrahedral $XY_4$ type molecules, the intensity ratio is given by

$$\frac{I_1}{I_2} = \frac{C_1}{C_2}\left(\frac{\tilde{\nu}_0 - \tilde{\nu}_1}{\tilde{\nu}_0 - \tilde{\nu}_2}\right)^4 \frac{\tilde{\nu}_2\mu_2}{\tilde{\nu}_1\mu_1} \frac{(1 - e^{-hc\tilde{\nu}_2/kT})(\bar{\alpha}_1)^2}{(1 - e^{-hc\tilde{\nu}_1/kT})(\bar{\alpha}_2)^2} \tag{19.15}$$

In this case, the $\rho_p$ term drops out, since $\gamma^2 = 0$ for isotropic molecules such as tetrahedral $XY_4$ and octahedral $XY_6$ types. By using $CCl_4$ as the standard, it is possible to determine the relative value of the $\partial\alpha/\partial Q$ term, which provides information about the degree of covalency and the bond order.[16]

## I-20. RESONANCE RAMAN SPECTRA[16,19,20,55]

In normal Raman spectroscopy, the exciting frequency lies in the region where the compound has no electronic absorption band (Sec. I-1). In resonance Raman spectroscopy, the exciting frequency falls within the electronic band. In the gaseous phase, this tends to cause resonance fluorescence since the rotational–vibrational levels are discrete. In the liquid and solid states, however, these levels are no longer discrete because of molecular collisions and/or intermolecular interactions. If such a broad vibronic band is excited, it tends to give resonance Raman rather than resonance fluorescence spectra.[55,56]

As an example, Fig. I-16 shows the resonance Raman spectra of $TiI_4$ in cyclohexane obtained by Clark and Mitchell.[57] Figure I-17 shows the positions of the exciting frequencies relative to the electronic absorption spectrum. As the exciting wavelength is changed from 647.1 to 514.5 nm, the intensity of the $\nu_1(A_1)$ band is enhanced relative to the 806 $cm^{-1}$ band of cyclohexane (the internal standard). It is seen that the intensity of the 161 $cm^{-1}$ band ($\nu_1$) is maximized when the exciting frequency is near the absorption maximum (515 nm). In one experiment, a series of overtones up to $13\nu_1$ was observed, although Fig. I-16 shows the series up to $5\nu_1$. These overtone frequencies have been used to calculate the anharmonicity constant with great accuracy.[57]

Resonance Raman spectroscopy is particularly suited to the study of biological macromolecules such as heme proteins because only a dilute solution (biological condition) is needed to observe the spectrum and only vibrations localized within the chromophoric group are enhanced when the exciting frequency approaches that of the relevant chromophore. This *selectivity* is highly important in studying the theoretical relationship between the electronic transition and the vibrations to be resonance-enhanced.

The origin of resonance Raman enhancement is explained in terms of Eq. 19.2. In normal Raman spectroscopy, $\nu_0$ is chosen in the region that

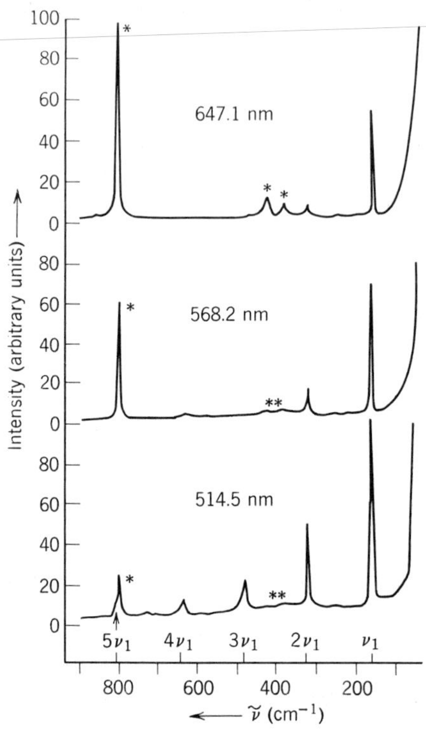


**Fig. I-16.** Raman spectra of titanium tetraiodide in cyclohexane obtained with 647.1, 568.2, and 514.5 nm excitation. Solvent peaks are marked with an asterisk, and overtones as $n\nu_1$.

is far from the electronic absorption. Then $\nu_{rm} \gg \nu_0$, and $\alpha_{\rho\sigma}$ is independent of the exciting frequency $\nu_0$. In resonance Raman spectroscopy, the denominator, $\nu_{rm} - \nu_0$, becomes very small as $\nu_0$ approaches $\nu_{rm}$. Thus the first term in the square brackets of Eq. 19.2 dominates all other terms and results in striking enhancement of Raman lines. However, Eq. 19.2 cannot account for the selectivity of resonance Raman enhancement since it is not specific about the states of the molecule. Albrecht[58] derived a

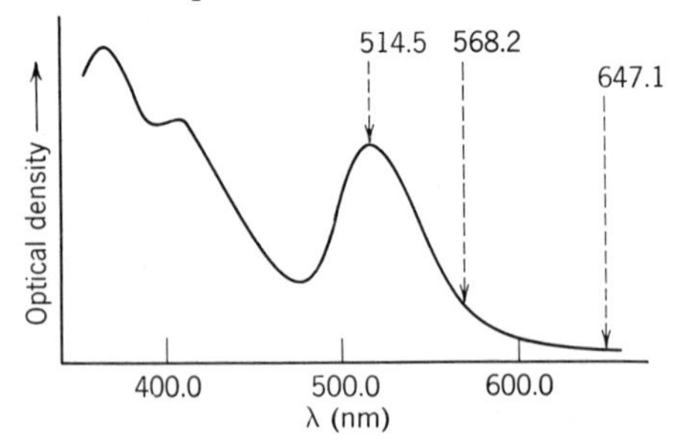


**Fig. I-17.** Electronic spectrum of titanium tetraiodide.

more specific equation for the initial and final states of resonance Raman scattering by introducing the Herzberg–Teller expansion of electronic wave functions into the Kramers–Heisenberg dispersion formula. The results are as follows:

$$(\alpha_{\rho\sigma})_{gi,gj} = A + B + C \tag{20.1}$$

$$A = \sum_{e\neq g}{}' \sum_{\upsilon} \left[ \frac{(g^0|\, R_\sigma \,|e^0)(e^0|\, R_\rho \,|g^0)}{E_{e\upsilon} - E_{gi} - E_0} + (\text{nonresonance term}) \right] \langle i \mid \upsilon \rangle \langle \upsilon \mid j \rangle \tag{20.2}$$

$$B = \sum_{e\neq g}{}' \sum_{\upsilon} \sum_{s\neq e}{}' \sum_{a} \left[ \left\{ \frac{(g^0|\, R_\sigma \,|e^0)(e^0|\, h_a \,|s^0)(s^0|\, R_\rho \,|g^0)}{E_{e\upsilon} - E_{gi} - E_0} + (\text{nonresonance term}) \right\} \times \frac{\langle i \mid \upsilon \rangle \langle \upsilon |\, Q_a \,| j \rangle}{E_e^0 - E_s^0} + \left\{ \frac{(g^0|\, R_\sigma \,|s^0)(s^0|\, h_a \,|e^0)(e^0|\, R_\rho \,|g^0)}{E_{e\upsilon} - E_{gi} - E_0} + (\text{nonresonance term}) \right\} \times \frac{\langle i |\, Q_a \,| \upsilon \rangle \langle \upsilon \mid j \rangle}{E_e^0 - E_s^0} \right] \tag{20.3}$$

$$C = \sum_{e\neq g}{}' \sum_{t\neq g}{}' \sum_{\upsilon} \sum_{a} \left[ \left\{ \frac{(g^0|\, h_a \,|t^0)(t^0|\, R_\sigma \,|e^0)(e^0|\, R_\rho \,|g^0)}{E_{e\upsilon} - E_{gi} - E_0} + (\text{nonresonance term}) \right\} \times \frac{\langle i \mid \upsilon \rangle \langle \upsilon |\, Q_a \,| j \rangle}{E_g^0 - E_t^0} + \left\{ \frac{(g^0|\, R_\sigma \,|e^0)(e^0|\, R_\rho \,|t^0)(t^0|\, h_a \,|g^0)}{E_{e\upsilon} - E_{gi} - E_0} + (\text{nonresonance term}) \right\} \times \frac{\langle i |\, Q_a \,| \upsilon \rangle \langle \upsilon \mid j \rangle}{E_g^0 - E_t^0} \right] \tag{20.4}$$

The notations $g$, $i$, $j$, $e$, and $\upsilon$ were explained in Sec. I-19. Other notations are as follows: $s$, another excited electronic state; $h_a$, the vibronic coupling operator $\partial\mathcal{H}/\partial Q_a$, $\mathcal{H}$ and $Q_a$ being the electronic Hamiltonian and the $a$th normal coordinate of the electronic ground state, respectively; $E_{gi}$, $E_{gj}$, and $E_{e\upsilon}$, the energies of states $gi$, $gj$, and $e\upsilon$, respectively; $|g^0)$, $|e^0)$, and $|s^0)$, the electronic wavefunctions for the equilibrium nuclear positions of the ground and excited states; $E_e^0$ and $E_s^0$, the corresponding energies of the electronic states, $e^0$ and $s^0$, respectively; and $E_0$, the energy of the exciting light. The nonresonance terms are similar to the preceding terms except that the denominator is $(E_{e\upsilon} - E_{gj} + E_0)$ instead of $(E_{e\nu} - E_{gi} - E_0)$ and that $R_\sigma$ and $R_\rho$ in the numerator are interchanged. These terms can be neglected under the strict resonance condition since the resonance terms become very large. The $C$ term is usually neglected because its components are denominated by $E_g^0 - E_t^0$, where $t$ refers to an excited state which is much higher in energy than the first excited state.

The relative importance of the $A$ and $B$ terms has been discussed by several investigators. For example, Spiro and his co-workers[59] carried out

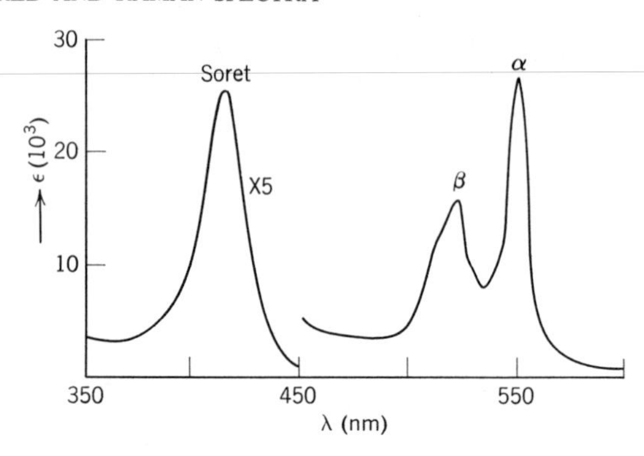


**Fig. I-18.** Electronic spectrum of ferrocytochrome *c*.[59]

an extensive study on resonance Raman spectra of various heme proteins under strict resonance conditions, and were able to account for the observed spectra with only the $B$ term. As is shown in Fig. I-18, ferrocytochrome $c$ exhibits two electronic bands refered to as the $\alpha$-band and Soret band, along with a vibronic side band ($\beta$-band) in the 400–600 nm region. According to molecular orbital calculations, the $\alpha$-band and Soret band are due to the $a_{2u} \rightarrow e_g$ and $a_{1u} \rightarrow e_g$ electronic transitions of the Fe-porphin core chromophore ($\mathbf{D}_{4h}$ symmetry), respectively. Since both transitions are of the $E_u$ type, they are subject to strong interaction. Now, the integral, $(e^0| h_a |s^0)$, of the $B$ term in Eq. 20.3 determines the symmetry of the normal modes, which may be resonance enhanced through vibrational coupling of the two states, $e$ and $s$. In this case, vibrations belonging to $E_u \times E_u = A_{1g} + B_{1g} + B_{2g} + A_{2g}$ species may be resonance enhanced. In fact, Spiro and his co-workers[59] measured the resonance Raman spectra of ferrocytochrome $c$ in the region of the $\beta$-band and observed a number of vibrations belonging to these species. Figure I-19 shows the resonance Raman spectra of cytochrome $c$ (514.5 nm excitation) obtained at two directions of polarization. As discussed in Sec. I-18, the $A_{1g}$, $B_{1g}$ and $B_{2g}$, and $A_{2g}$ vibrations are expected to be polarized ($p$), depolarized ($dp$), and inversely polarized ($ip$),* respectively. These polarization properties, together with their vibrational frequencies, were used by Spiro and his co-workers to make complete assignments of vibrational spectra of the Fe-porphin skeletons of a series of heme proteins. They showed that the resonance Raman spectrum may be used to predict the oxidation and spin states of the Fe

* For the $A_{2g}$ modes, $\frac{3}{4} < \rho_p < \infty$ was observed rather than $\rho_p = \infty$, as predicted by the theory. This discrepancy was attributed to the lowering of the $\mathbf{D}_{4h}$ symmetry of the complex due to the presence of the peripheral groups in cytochrome $c$ (Ref. 54).

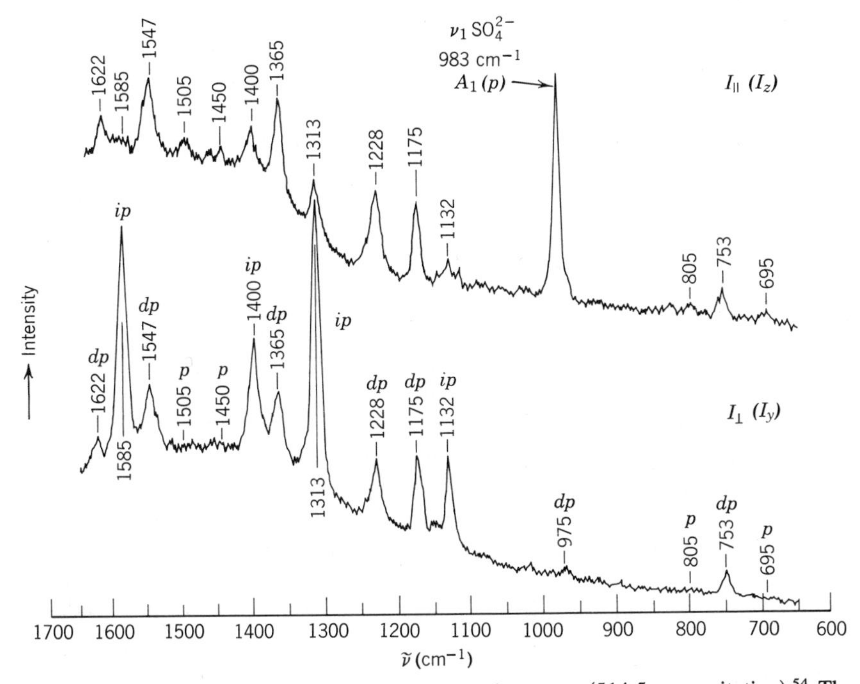


**Fig. I-19.** Resonance Raman spectra of ferrocytochrome *c*. (514.5 nm excitation).[54] The scattering geometry is shown in Fig. I-14.

atom in heme proteins. For example, the Fe atom in oxyhemoglobin has been shown to be low spin Fe(III). It should be noted that the $A_{2g}$ mode, which is normally Raman inactive, is observed under the resonance condition.

A more complete study of resonance Raman spectra involves the observation of *excitation profiles* (Raman intensity plotted as a function of the exciting frequency for each mode), and the simulation of observed excitation profiles based on various theories of resonance Raman spectroscopy. For example, Inagaki et al.[60] showed from such studies that the $A$ term (Eq. 20.2) dominates the resonance Raman scattering of $\beta$-carotene.

## I-21. VIBRATIONAL SPECTRA IN GASEOUS PHASE[1,2] AND INERT GAS MATRICES[34]

As distinct from molecules in condensed phases, those in the gaseous phase are free from intermolecular interactions. If the molecules are relatively small, vibrational spectra of gases exhibit rotational fine structure (see Fig. I-4) from which moments of inertia and hence internuclear distances and bond angles can be calculated.[1,2] Furthermore, detailed analysis of rotational fine structure provides information about the magnitude of rotation–vibration interaction (Coriolis coupling), centrifugal distortion, anharmonicity, and even nuclear spin statistics in some cases. In the past, infrared spectroscopy was the main tool in measuring gas-phase vibrational spectra. Recently, Raman spectroscopy has been playing a significant role because of the development of powerful laser sources and high resolution spectrophotometers. For example, Clark and Rippon[61a] measured gas-phase Raman spectra of Group IV tetrahalides, and calculated the Coriolis coupling constants from the observed band contours. For gas-phase Raman spectra of other inorganic compounds, see Ref. 61. Unfortunately, the majority of inorganic and coordination compounds exist as solids at room temperature. Although some of these compounds can be vaporized at high temperatures without decomposition, it is rather difficult to measure their spectra by the conventional method. Furthermore, high temperature spectra are difficult to interpret because of the increased importance of rotational and vibrational hot bands.

In 1954, Pimentel and his co-workers[62] developed the *matrix isolation technique* to study the infrared spectra of unstable and stable species. In this method, solute molecules and inert gas molecules such as Ar and $N_2$ are mixed at a ratio of 1:500 or greater and deposited on an IR window such as a CsI crystal cooled to 10–15°K. Since the solute molecules trapped in an inert gas matrix are completely isolated from each other, the matrix isolation spectrum is similar to the gas-phase spectrum; no crystal field splittings and no lattice modes are observed. However, the former spectrum is simpler than the latter because, except for a few small hydride molecules, no rotational transitions are observed because of the rigidity of the matrix environment at low temperatures. The lack of rotational structure and intermolecular interactions results in a sharpening of the solute band so that even very closely located metal isotope peaks can be resolved in a matrix environment. This technique is also applicable to a compound which is not volatile at room temperature. For example, matrix isolation spectra of metal halides can be measured by vaporizing these compounds at high temperatures in a Knudsen cell and cocondensing their vapors with inert gas molecules on a cold window.[63] The recent development of closed-cycle helium refrigerators greatly facilitated the experimental technique. The matrix isolation technique has now been applied to a number of inorganic and coordination compounds to obtain structural and bonding information. Some important applications of this technique are described below.

### (A) Vibrational Spectra of Radicals

Highly reactive radicals can be produced *in situ* in inert gas matrices by photolysis and other techniques. Since these radicals are stabilized in matrix environments, their spectra can be measured by routine spectroscopic techniques. For example, the spectrum of the HOO radical[64] was obtained by measuring the spectrum of the photolysis product of a mixture of HI and $O_2$ in an Ar matrix at ~4°K. Part II lists the vibrational frequencies of many other radicals, such as NH, OF, and FCO, obtained by similar methods.

### (B) Vibrational Spectra of High-Temperature Species

Alkali halide vapors produced at high temperatures consist mainly of monomers and dimers. The vibrational spectra of these salts at high temperatures are difficult to measure and difficult to interpret because of the presence of hot bands. The matrix isolation technique utilizing a Knudsen cell has solved this problem. The vibrational frequencies of some of these high temperature species are listed in Part II.

### (C) Isotope Splittings

As stated before, it is possible to observe individual peaks due to heavy metal isotopes in inert gas matrices since the bands are extremely sharp (half-band width, 1.5–1.0 $cm^{-1}$) under these conditions. Figure I-20*a* shows the infrared spectrum of the $\nu_7$ band (coupled vibration between CrC stretching and CrCO bending modes) of $Cr(CO)_6$ in a $N_2$ matrix.[65] The bottom curve shows a computer simulation using the measured isotope shift of 2.5 $cm^{-1}$ per atomic mass unit, a 1.2 $cm^{-1}$ half-band width, and the percentages of natural abundance of Cr isotopes: $^{50}Cr$ (4.31%), $^{52}Cr$ (83.76%), $^{53}Cr$ (9.55%), and $^{54}Cr$ (2.38%). These isotope frequencies are highly important in calculating force constants and anharmonicity corrections. In Parts II and III, isotope frequencies will be given for some representative compounds.

### (D) Chemical Synthesis

Recently, the matrix isolation technique has been used to synthesize a number of unstable and transient coordination compounds. For example, a series of nickel carbonyls of the type $Ni(CO)_x$, where $x = 1, 2, 3$, and 4, have been synthesized by allowing metal vapor to react with CO diluted in Ar on a cold window and warming the matrix carefully. Figure I-21 shows the result obtained by DeKock.[66] The structures of $Ni(CO)_2$ and $Ni(CO)_3$ were concluded to be linear and trigonal–planar, respectively, since these compounds exhibit only one CO stretching band in the infrared. Similar methods have been applied to the synthesis of a number

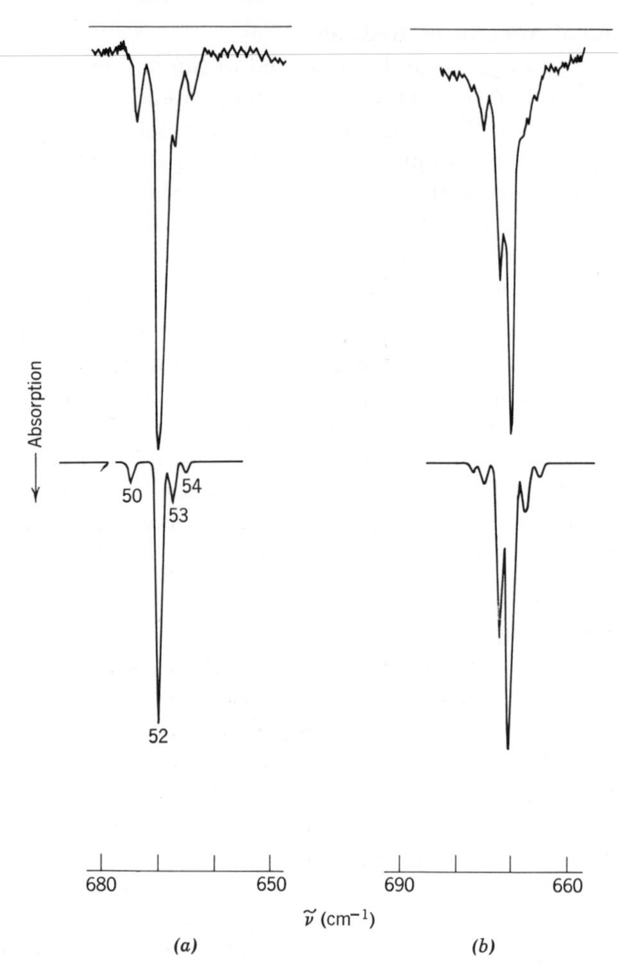


**Fig. I-20.** Matrix isolation infrared spectra of $Cr(CO)_6$: (*a*), $N_2$ matrix, (*b*) Ar matrix.

of coordination compounds $ML_x$, where M is Pt, Pd, Ni, etc., and L is CO, $N_2$, $O_2$, $PF_3$, etc. More detailed discussions of individual compounds will be given in Part III.

### (E) Matrix Effect

The vibrational frequencies of matrix-isolated molecules give slight shifts when the matrix gas is changed. This result suggests the presence of weak interaction between the solute and matrix molecules. In some cases, the spectra are complicated by the presence of more than one trapping site. For example, the infrared spectrum of $Cr(CO)_6$ in an Ar matrix (Fig.

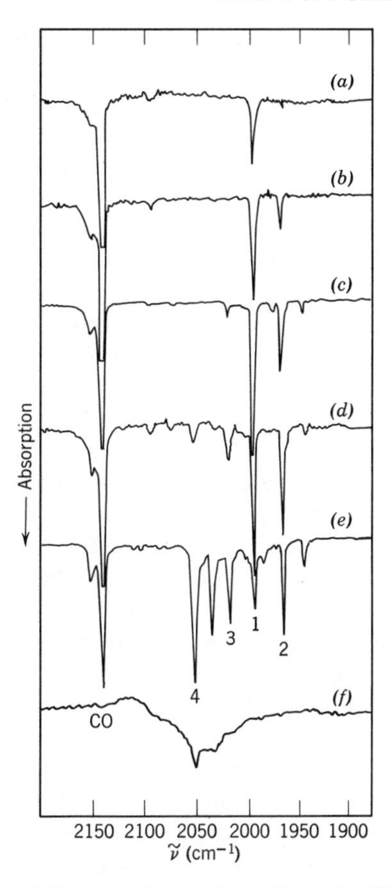


**Fig. I-21.** Infrared spectra of Ni atoms deposited in a 500:1 Ar–CO matrix and subsequent annealing: (*a*) original, (*b*) 17°K, (*c*) 18°K, (*d*) 19°K, (*e*) 26°K, (*f*) 35°K (temperatures are relative). The arabic numerals refer to the relative rate of growth and disappearance of the bands and hence to *n* in $Ni(CO)_n$.

I-20*b*) is markedly different from that in a $N_2$ matrix (Fig. I-20*a*).[65] The former spectrum can be interpreted by assuming two different sites in an Ar matrix. The computer-simulation spectrum (bottom curve) was obtained by assuming that these two sites are populated in a 2:1 ratio, the frequency separation of the corresponding peaks being 2 cm$^{-1}$. Thus it is always desirable to obtain the matrix isolation spectra in several different environments.

## I-22. VIBRATIONAL SPECTRA OF CRYSTALS

Because of intermolecular interactions, the symmetry of a molecule is generally lower in the crystalline state than in the gaseous (isolated) state.* This change in symmetry may split the degenerate vibrations and activate infrared (or Raman) inactive vibrations. In addition, the spectra obtained in the crystalline state exhibit *lattice modes*—vibrations due to translatory and rotatory motions of a molecule in the crystalline lattice. Although their frequencies are usually lower than 300 $cm^{-1}$, they may appear in the high frequency region as the combination bands with internal modes (see Fig. II-10, for example). Thus the vibrational spectra of crystals must be interpreted with caution, especially in the low frequency region.

To analyze the spectra of crystals, it is necessary to carry out the site group or factor group analysis described in the following subsection.

**(A) Site Group Analysis**

According to Halford,[67] the vibrations of a molecule in the crystalline state are governed by a new selection rule derived from *site symmetry*—a local symmetry around the center of gravity of a molecule in a unit cell. The site symmetry can be found by using the following two conditions: (1) the site group must be a subgroup of both the space group of the crystal and the molecular point group of the isolated molecule, and (2) the number of equivalent sites must be equal to the number of molecules in the unit cell. Halford derived a complete table that lists possible site symmetries and the number of equivalent sites for 230 space groups.† Suppose that the space group of the crystal, the number of molecules in the unit cell (Z), and the point group of the isolated molecule are known. Then the site symmetry can be found from Halford's table. In general, the site symmetry is lower than the molecular symmetry in an isolated state.

The vibrational spectra of calcite and aragonite crystals are markedly different, although both have the same composition (Sec. II-4). This result can be explained if we consider the difference in site symmetry of the $CO_3^{2-}$ ion between these crystals. According to X-ray analysis, the space group of calcite is $\mathbf{D}_{3d}^6$ and Z is two. Halford's table gives

$$\mathbf{D}_3(2), \quad \mathbf{C}_{3i}(2), \quad \infty\mathbf{C}_3(4), \quad \mathbf{C}_i(6), \quad \infty\mathbf{C}_2(6)$$

as possible site symmetries for space group $\mathbf{D}_{3d}^6$ (the number in front of point group notation indicates the number of distinct sets of sites, and that in parenthesis denotes the number of equivalent sites for each distinct set). Rule 2 eliminates all but $\mathbf{D}_3(2)$ and $\mathbf{C}_{3i}(2)$. Rule 1 eliminates the latter since $\mathbf{C}_{3i}$ is not a subgroup of $\mathbf{D}_{3h}$. Thus the site symmetry of

* The symmetry of a molecule may be the lowest in solution (or liquid) because it interacts with randomly oriented molecules.

† For a more complete table, see Ref. 67a.

the $CO_3^{2-}$ ion in calcite must be $\mathbf{D}_3$. On the other hand, the space group of aragonite is $\mathbf{D}_{2h}^{16}$ and Z is four. Halford's table gives

$$2\mathbf{C}_i(4), \qquad \infty\mathbf{C}_s(4)$$

Since $\mathbf{C}_i$ is not a subgroup of $\mathbf{D}_{3h}$, the site symmetry of the $CO_3^{2-}$ ion in aragonite must be $\mathbf{C}_s$. Thus the $\mathbf{D}_{3h}$ symmetry of the $CO_3^{2-}$ ion in an isolated state is lowered to $\mathbf{D}_3$ in calcite and to $\mathbf{C}_s$ in aragonite. Then the selection rules are changed as shown in Table I-11.

TABLE I-11. CORRELATION TABLE FOR $\mathbf{D}_{3h}$, $\mathbf{D}_3$, $\mathbf{C}_{2v}$, AND $\mathbf{C}_S$

| Point Group | $\nu_1$ | $\nu_2$ | $\nu_3$ | $\nu_4$ |
|---|---|---|---|---|
| $\mathbf{D}_{3h}$ | $A_1'(R)$ | $A_2''$(I) | $E'$(I, R) | $E'$(I, R) |
| $\mathbf{D}_3$ | $A_1$(R) | $A_2$(I) | $E$(I, R) | $E$(I, R) |
| $\mathbf{C}_{2v}$ | $A_1$(I, R) | $B_1$(I, R) | $A_1$(I, R) + $B_2$(I, R) | $A_1$(I, R) + $B_2$(I, R) |
| $\mathbf{C}_S$ | $A'$(I, R) | $A''$(I, R) | $A'$(I, R) + $A'$(I, R) | $A'$(I, R) + $A'$(I, R) |

There is no change in the selection rule in going from the free $CO_3^{2-}$ ion to calcite. In aragonite, however, $\nu_1$ becomes infrared active, and $\nu_3$ and $\nu_4$ each split into two bands. The observed spectra of calcite and aragonite are in good agreement with these predictions (see Table II-4*b*).

### (B) Factor Group Analysis[33]

A more complete analysis including lattice modes can be made by the method of factor group analysis developed by Bhagavantam and Venkatarayudu.[68] In this method, we consider all the normal vibrations for an entire Bravais primitive cell.* Figure I-22 illustrates the Bravais cell of calcite, which consists of the following symmetry elements: $I$, $2S_6$, $2S_6^2 \equiv 2C_3$, $S_6^3 \equiv i$, $3C_2$, and $3\sigma_v$ (glide plane). These elements are exactly the same as those of the point group $\mathbf{D}_{3d}$, although the last element is a glide plane rather than a plane of symmetry in a single molecule.

It is possible to derive the 230 space groups by combining operations possessed by the 32 crystallographic point groups† with operations such as pure translation, screw rotation (translation + rotation), and glide plane reflection (translation + reflection). If we regard the translations that carry

* Every molecule (or ion) in a Bravais primitive cell can be superimposed on that of the neighboring cell by simple translation. In the case of calcite, the Bravais primitive cell is the same as the crystallographic unit cell. However, this is not always the case.

† In crystals, the number of point groups is limited to 32 since only $C_1$, $C_2$, $C_3$, $C_4$, and $C_6$ are possible due to the space-filling requirements of the crystal lattice.

a point in a unit cell into the equivalent point in another cell as identity, we define the 230 factor groups that are the subgroups of the corresponding space groups. In the case of calcite, the factor group consists of the symmetry elements described above, and is denoted by the same notation as that used for the space group ($\mathbf{D}_{3d}^6$). The site group discussed previously is a subgroup of a factor group.

Since the Bravais cell contains 10 atoms, it has $3 \times 10 - 3 = 27$ normal vibrations, excluding three translational motions of the cell as a whole.* These 27 vibrations can be classified into various symmetry species of the factor group $\mathbf{D}_{3d}^6$, using a procedure similar to that described in Sec. I-7 for internal vibrations. First, we calculate the characters of representations corresponding to the entire freedom possessed by the Bravais primitive cell ($\chi_R(N)$), translational motions of the whole cell ($\chi_R(T)$),

TABLE I-12. FACTOR GROUP ANALYSIS OF CALCITE CRYSTAL[a]

| $\mathbf{D}_{3d}^6$ | $I$ | $2S_6$ | $2C_3 \equiv 2S_6^2$ | $i \equiv S_6^3$ | $3C_2$ | $3\sigma_v$ | Number of Vibrations $N$ | $T$ | $T'$ | $R'$ | $n$ | | |
|---|---|---|---|---|---|---|---|---|---|---|---|---|---|
| $A_{1g}$ | 1 | 1 | 1 | 1 | 1 | 1 | 1 | 0 | 0 | 0 | 1 | | $\alpha_{xx}+\alpha_{yy}, \alpha_{zz}$ |
| $A_{1u}$ | 1 | −1 | 1 | −1 | 1 | −1 | 2 | 0 | 1 | 0 | 1 | | |
| $A_{2g}$ | 1 | 1 | 1 | 1 | −1 | −1 | 3 | 0 | 1 | 1 | 1 | | |
| $A_{2u}$ | 1 | −1 | 1 | −1 | −1 | 1 | 4 | 1 | 1 | 1 | 1 | $T_z$ | |
| $E_g$ | 2 | −1 | −1 | 2 | 0 | 0 | 4 | 0 | 1 | 1 | 2 | | $(\alpha_{xx}-\alpha_{yy}, \alpha_{xy}), (\alpha_{xz}\alpha_{yz})$ |
| $E_u$ | 2 | 1 | −1 | −2 | 0 | 0 | 6 | 1 | 2 | 1 | 2 | $(T_x, T_y)$ | |
| $N_R(p)$ | 10 | 2 | 4 | 2 | 4 | 0 | | | | | | | |
| $N_R(s)$ | 4 | 2 | 4 | 2 | 2 | 0 | | | | | | | |
| $N_R(s-\upsilon)$ | 2 | 0 | 2 | 0 | 2 | 0 | | | | | | | |
| $\chi_R(N)$ | 30 | 0 | 0 | −6 | −4 | 0 | ← $N$ | | | | | | |
| $\chi_R(T)$ | 3 | 0 | 0 | −3 | −1 | 1 | ← $T$ | | | | | | |
| $\chi_R(T')$ | 9 | 0 | 0 | −3 | −1 | −1 | ← $T'$ | | | | | | |
| $\chi_R(R')$ | 6 | 0 | 0 | 0 | −2 | 0 | ← $R'$ | | | | | | |
| $\chi_R(n)$ | 12 | 0 | 0 | 0 | 0 | 0 | ← $n$ | | | | | | |

[a] $p$, total number of atoms in the Bravais primitive cell.
$s$, total number of molecules (ions) in the primitive cell.
$\upsilon$, total number of monoatomic molecules (ions) in the primitive cell.
$N_R(p)$, number of atoms unchanged by symmetry operation, $R$.
$N_R(s)$, number of molecules (ions) whose center of gravity is unchanged by symmetry operation, $R$.
$N_R(s-\upsilon)$, $N_R(s)$ minus number of monoatomic molecules (ions) unchanged by symmetry operation, $R$.
$\chi_R(N) = N_R(p)[\pm(1+2\cos\theta)]$, character of representation for entire freedom possessed by the primitive cell.
$\chi_R(T) = \pm(1+2\cos\theta)$, character of representation for translational motions of the whole primitive cell.
$\chi_R(T') = \{N_R(s)-1\}\{\pm(1+2\cos\theta)$, character of representation for translatory lattice modes.
$\chi_R(R') = N_R(s-\upsilon)\{\pm(1+2\cos\theta)$, character of representation for rotary lattice modes.
$\chi_R(n) = \chi_R(N) - \chi_R(T) - \chi_R(T\,) - \chi_R(R')$, character of representation for internal modes.
Note that + and − signs are for proper and improper rotations, respectively. The symbol $\theta$ should be taken as defined in Sec. I-7.

* These three motions give "acoustic modes" that propagate sound waves through the crystal.

translatory lattice modes ($\chi_R(T')$), rotatory lattice modes ($\chi_R(R')$), and internal modes ($\chi_R(n)$), using the equations given in Table I-12. Then each of these characters is resolved into the symmetry species of the point group, $\mathbf{D}_{3d}$. The final results show that three internal modes ($A_{2u}$ and two $E_u$), three translatory modes ($A_{2u}$ and two $E_u$), and two rotatory modes ($A_{2u}$ and $E_u$) are infrared active, and three internal modes ($A_{1g}$ and two $E_g$), one translatory mode ($E_g$), and one rotatory mode ($E_g$) are Raman active. As will be shown in Sec. I-22(c), these predictions are in perfect agreement with the observed spectra.

The correlation method developed by Fateley et al.[14] is simpler than factor group analysis and gives the same results. A complete normal coordinate analysis on the whole Bravais cell of calcite type crystals was carried out by Nakagawa and Walter.[69] Similar treatments have been extended to nitro and aquo complexes.[70]

### (C) Infrared Dichroism

Suppose that we irradiate a single crystal of calcite with polarized infrared radiation whose electric vector vibrates along the $c$ axis ($z$ direction) in Fig. I-22. Then the infrared spectrum shown by the solid

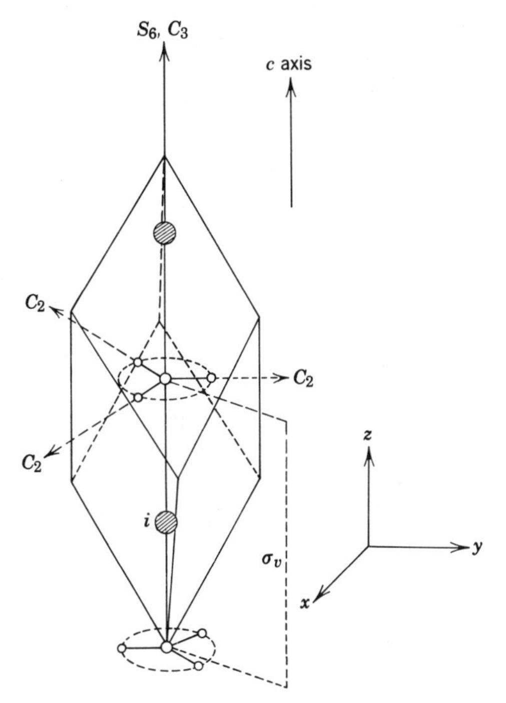


**Fig. I-22.** The Bravais primitive cell of calcite.

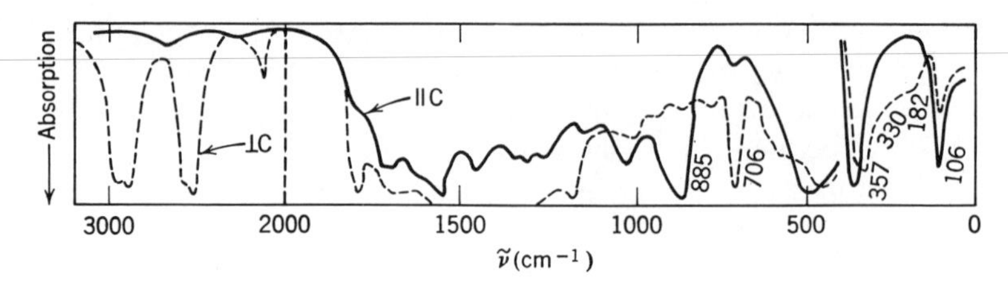


**Fig. I-23.** Infrared dichroism of calcite.[71]

curve of Fig. I-23 is obtained.[71] According to Table I-12, only the $A_{2u}$ vibrations are activated under such conditions. Thus the three bands observed at 885($v$), 357($t$) and 106($r$) $cm^{-1}$ are assigned to the $A_{2u}$ species. The spectrum shown by the dotted curve is obtained if the direction of polarization is perpendicular to the $c$ axis ($x$, $y$ plane). In this case, only the $E_u$ vibrations should be infrared active. Therefore the five bands observed at 1484($v$), 706($v$), 330($t$), 182($t$), and 106($r$) $cm^{-1}$ are assigned to the $E_u$ species. As shown above, polarized infrared studies of single crystals provide valuable information about the symmetry properties of normal vibrations if the crystal structure is known from other sources.

The ratio of the absorption intensities in the directions parallel and perpendicular to a crystal axis, called the *dichroic ratio*, is defined by

$$R = \frac{\int_{\text{band}} \varepsilon_{\parallel}(\tilde{\nu})\, d\tilde{\nu}}{\int_{\text{band}} \varepsilon_{\perp}(\tilde{\nu})\, d\tilde{\nu}} \tag{22.1}$$

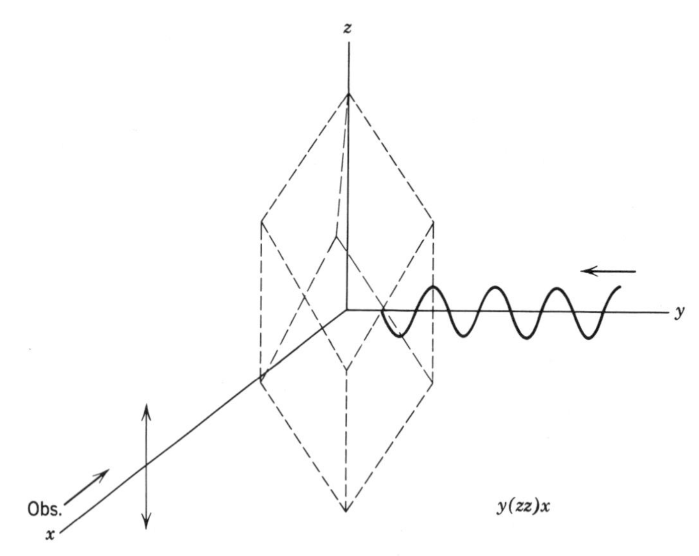


**Fig. I-24.**

Here $\varepsilon_{\parallel}$ and $\varepsilon_{\perp}$ denote the absorption coefficients for radiation whose electric vector vibrates in the directions parallel and perpendicular, respectively, to the crystal axis. In the case of calcite, the maximum dichroic ratio is expected, since the carbonate ions are oriented parallel to each other. The dichroic ratio will be smaller if the molecules are not parallel to each other in a crystal lattice.

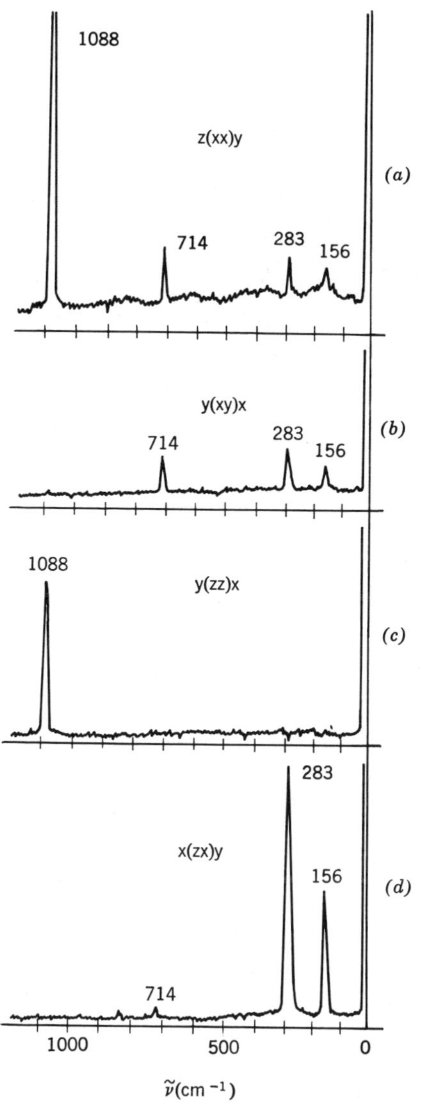


**Fig. I-25.** Polarized Raman spectra of calcite.[73]

## (D) Polarized Raman Spectra

Polarized Raman spectra provide more information about the symmetry properties of normal vibrations than do polarized infrared spectra. Again consider a single crystal of calcite. According to Table I-12, the $A_{1g}$ vibrations become Raman active if any one of the polarizability components, $\alpha_{xx}$, $\alpha_{yy}$, and $\alpha_{zz}$, is changed. Suppose that we irradiate a calcite crystal from the $y$ direction, using polarized radiation whose electric vector vibrates parallel to the $z$ axis (see Fig. I-24), and observe the Raman scattering in the $x$ direction with its polarization in the $z$ direction. This condition is abbreviated as $y(zz)x$. In this case, Eq. 5.7 is simplified to $P_z = \alpha_{zz}E_z$ because $E_x = E_y = 0$ and $P_x = P_y = 0$. Since $\alpha_{zz}$ belongs to the $A_{1g}$ species, only the $A_{1g}$ vibrations are observed under this condition. Figure I-25*c* illustrates the Raman spectrum obtained with this condition. Thus the strong Raman line at 1088 $cm^{-1}$($v$) is assigned to the $A_{1g}$ species. Both the $A_{1g}$ and $E_g$ vibrations are observed if the $z(xx)y$ condition is used. The Raman spectrum (Fig. I-25*a*) shows that five Raman lines [1088($v$), 714($v$), 283($r$), 156($t$), and 1434($v$) $cm^{-1}$ (not shown)] are observed under this condition. Since the 1088 $cm^{-1}$ line belongs to the $A_{1g}$ species, the remaining four must belong to the $E_g$ species. These assignments can also be confirmed by measuring Raman spectra using the $y(xy)x$ and $x(zx)y$ conditions (Fig. I-25*b* and *d*). These experiments were originally performed by Bhagavantam,[72] using a mercury line as a Raman source. However, recently developed gas lasers are ideal for such experiments, since they provide strong and completely polarized radiation.

For a review of single-crystal Raman spectroscopy, see Ref. 74.

## General References

### Theory of Molecular Vibrations


1. G. Herzberg, *Molecular Spectra and Molecular Structure.* Vol. II: *Infrared and Raman Spectra of Polyatomic Molecules,* Van Nostrand, Princeton, N.J., 1945.
2. G. Herzberg, *Molecular Spectra and Molecular Structure.* Vol. I: *Spectra of Diatomic Molecules,* Van Nostrand, Princeton, N.J., 1950.
3. E. B. Wilson, J. C. Decius, and P. C. Cross, *Molecular Vibrations,* McGraw-Hill, New York, 1955.
4. G. W. King, *Spectroscopy and Molecular Structure,* Holt, Rinehart, and Winston, New York, 1964.
5. C. J. H. Schutte, *The Theory of Molecular Spectroscopy.* Vol. I: *The Quantum Mechanics and Group Theory of Vibrating and Rotating Molecules,* North Holland, Amsterdam, 1976.
6. P. Gans, *Vibrating Molecules,* Chapman and Hall, London, 1971.
7. D. Steele, *Theory of Vibrational Spectroscopy,* Saunders, London, 1971.

8. L. A. Woodward, *Introduction to the Theory of Molecular Vibrations and Vibrational Spectroscopy*, Oxford University Press, London, 1972.
9. C. N. Banwell, *Fundamentals of Molecular Spectroscopy*, 2nd ed., McGraw-Hill, London, 1972.
10. S. J. Cyvin, *Molecular Structures and Vibrations*, Elsevier, Amsterdam, 1972.
10a. A. Fadini, *Molekülkraftkonstanten*, Steinkopff-Verlag, Darmstadt, 1975.

**Symmetry and Group Theory**

11. L. H. Hall, *Group Theory and Symmetry in Chemistry*, McGraw-Hill, New York, 1969.
12. F. A. Cotton, *Chemical Application of Group Theory*, 2nd ed., Wiley-Interscience, New York, 1971.
13. M. Orchin and H. H. Jaffé, *Symmetry, Orbitals and Spectra*, Wiley, New York, 1971.
14. W. G. Fateley, F. R. Dollish, N. T. McDevitt, and F. F. Bentley, *Infrared and Raman Selection Rules for Molecular and Lattice Vibrations: The Correlation Method*, Wiley-Interscience, New York, 1972.
15. J. R. Ferraro and J. S. Ziomek, *Introductory Group Theory and Its Application to Molecular Structure*, 2nd ed., Plenum Press, New York, 1975.

**Raman Spectroscopy**

16. H. A. Szymanski, ed., *Raman Spectroscopy: Theory and Practice*, Vol. 1, 1967, and Vol. 2, 1970, Plenum Press, New York.
17. T. R. Gilson and P. J. Hendra, *Laser Raman Spectroscopy*, Wiley, New York, 1970.
18. M. C. Tobin, *Laser Raman Spectroscopy*, Wiley-Interscience, New York, 1971.
19. J. P. Mathieu, ed., *Advances in Raman Spectroscopy*, Vol. 1, Heyden, London, 1973.
20. J. A. Koningstein, *Introduction to the Theory of the Raman Effect*, D. Reidel, Dordrecht (Holland), 1973.

**Vibrational Spectra of Inorganic, Coordination, and Organometallic Compounds**

21. H. Siebert, *Anwendungen der Schwingung-spektroskopie in der Anorganischen Chemie*, Springer-Verlag, Berlin, 1966.
22. D. M. Adams, *Metal-Ligand and Related Vibrations*, Edward Arnold, London, 1967.
23. J. R. Ferraro, *Low-frequency Vibrations of Inorganic and Coordination Compounds*, Plenum Press, New York, 1971.
24. L. H. Jones, *Inorganic Vibrational Spectroscopy*, Vol. 1, Marcel Dekker, New York, 1971.
25. R. A. Nyquist and R. O. Kagel, *Infrared Spectra of Inorganic Compounds*, Academic Press, New York, 1971.
26. S. D. Ross, *Inorganic Infrared and Raman Spectra*, McGraw-Hill, New York, 1972.
27. N. N. Greenwood, E. J. F. Ross, and B. P. Straughan, *Index of Vibrational Spectra of Inorganic and Organometallic Compounds*, Vol. 1, Butterworths, London, 1972.
28. *Spectroscopic Properties of Inorganic and Organometallic Compounds*, Vols. 1–9, The Chemical Society, London, 1968–1975.
28a. E. Maslowsky, Jr., *Vibrational Spectra of Organometallic Compounds*, Wiley, New York, 1976.

**Miscellaneous**

29. L. H. Little, *Infrared Spectroscopy of Adsorbed Species*, Academic Press, New York, 1966.
30. M. L. Hair, *Infrared Spectroscopy in Surface Chemistry*, Marcel Dekker, New York, 1967.

31. B. Meyer, *Low-Temperature Spectroscopy*, Elsevier, Amsterdam, 1971.
32. F. S. Parker, *Applications of Infrared Spectroscopy in Biochemistry, Biology and Medicine*, Plenum Press, New York, 1971.
33. G. Turrell, *Infrared and Raman Spectra of Crystals*, Academic Press, New York, 1972.
34. H. E. Hallam, ed., *Vibrational Spectroscopy of Trapped Species*, Wiley, New York, 1973.
35. L. J. Bellamy, *Infrared Spectra of Complex Molecules*, 3rd ed., Wiley, New York, 1975.
36. N. B. Colthup, L. H. Daly, and S. E. Wiberley, *Introduction to Infrared and Raman Spectroscopy*, 2nd ed., Academic Press, New York, 1975.
36a. C. Karr, ed., *Infrared and Raman Spectroscopy of Lunar and Terrestrial Minerals*, Academic Press, New York, 1975.
36b. M. Moskovits and G. A. Ozin, eds., *Cryochemistry*, Wiley, New York, 1976.

## References

37. P. Pulay, *Mol. Phys.*, **17,** 197 (1969); **18,** 473 (1970); P. Pulay and W. Meyer, *J. Mol. Spectrosc.*, **40,** 59 (1971).
38. E. B. Wilson, *J. Chem. Phys.*, **7,** 1047 (1939); **9,** 76 (1941).
39. J. C. Decius, *J. Chem. Phys.*, **16,** 1025 (1948).
40. T. Shimanouchi, *J. Chem. Phys.*, **25,** 660 (1956).
41. T. Shimanouchi, *J. Chem. Phys.*, **17,** 245, 734, 848 (1949).
41a. T. Shimanouchi, "The Molecular Force Field," in *Physical Chemistry: An Advanced Treatise*, Vol. 4, D. Henderson, ed., Academic Press, New York, 1970.
42. D. F. Heath and J. W. Linnett, *Trans. Faraday Soc.*, **44,** 556, 873, 878, 884 (1948); **45,** 264 (1949).
43. J. Overend and J. R. Scherer, *J. Chem. Phys.*, **32,** 1289, 1296, 1720 (1960); **33,** 446 (1960); **34,** 547 (1961); **36,** 3308 (1962).
44. T. Shimanouchi, *Pure Appl. Chem.*, **7,** 131 (1963).
45. J. H. Schachtschneider, "Vibrational Analysis of Polyatomic Molecules," Pts. V and VI, Tech. Rep. 231-64 and 53-65, Shell Development Co., Emeryville, Calif., 1964 and 1965.
46. K. Nakamoto, *Angew. Chem.*, **11,** 666 (1972).
47. N. Mohan, K. Nakamoto, and A. Müller, "The Metal Isotope Effect on Molecular Vibrations," in *Advances in Infrared and Raman Spectroscopy*, Vol. 1, R. J. H. Clark and R. E. Hester, eds., Heyden, London, 1976.
48. Y. Morino and K. Kuchitsu, *J. Chem. Phys.*, **20,** 1809 (1952).
49. B. L. Crawford and W. H. Fletcher, *J. Chem. Phys.*, **19,** 141 (1951).
50. P. LaBonville and J. M. Williams, *Appl. Spectrosc.*, **25,** 672 (1971).
51. E. B. Wilson and A. J. Wells, *J. Chem. Phys.*, **14,** 578 (1946).
52. A. M. Thorndike, E. B. Wilson, and A. J. Wells, *J. Chem. Phys.*, **15,** 157 (1947).
53. D. A. Ramsay, *J. Am. Chem. Soc.*, **74,** 72 (1952).
54. T. G. Spiro and T. C. Strekas, *Proc. Nat. Acad. Sci.*, **69,** 2622 (1972).
54a. W. M. McClain, *J. Chem. Phys.*, **55,** 2789 (1971).
55. W. Kiefer, *Appl. Spectrosc.*, **28,** 115 (1974).
56. M. Mingardi and W. Siebrand, *J. Chem. Phys.*, **62,** 1074 (1975).
57. R. J. H. Clark and P. D. Mitchell, *J. Am. Chem. Soc.*, **95,** 8300 (1973).
58. A. C. Albrecht, *J. Chem. Phys.*, **34,** 1476 (1961).
59. T. G. Spiro, *Acc. Chem. Res.*, **7,** 339 (1974).

60. F. Inagaki, M. Tasumi, and T. Miyazawa, *J. Mol. Spectrosc.*, **50,** 286 (1974).
61. G. A. Ozin, "Single Crystal and Gas Phase Raman Spectroscopy in Inorganic Chemistry," in *Progress in Inorganic Chemistry*, S. J. Lippard, ed., Vol. 14, Wiley-Interscience, New York, 1971.
61a. R. J. H. Clark and D. M. Rippon, *J. Mol. Spectrosc.*, **44,** 479 (1972).
62. E. Whittle, D. A. Dows, and G. C. Pimentel, *J. Chem. Phys.*, **22,** 1943 (1954).
63. M. J. Linevsky, *J. Chem. Phys.*, **34,** 587 (1961).
64. D. E. Milligan and M. E. Jacox, *J. Chem. Phys.*, **38,** 2627 (1963).
65. D. Tevault and K. Nakamoto, *Inorg. Chem.*, **14,** 2371 (1975).
66. R. L. DeKock, *Inorg. Chem.*, **10,** 1205 (1971).
67. R. S. Halford, *J. Chem. Phys.*, **14,** 8 (1946).
67a. W. G. Fateley, *Appl. Spectrosc.*, **27,** 395 (1973); W. G. Fateley, N. J. McDevitt, and F. F. Bentley, *ibid.*, **25,** 155 (1971).
68. S. Bhagavantam and T. Venkatarayudu, *Proc. Indian Acad. Sci.*, **9A,** 224 (1939); "Theory of Groups and Its Application to Physical Problems," Andhra University, Waltair, 1951.
69. I. Nakagawa and J. L. Walter, *J. Chem. Phys.*, **51,** 1389 (1969).
70. I. Nakagawa, *Coord. Chem. Rev.*, **4,** 423 (1969).
71. M. Tsuboi, *Infrared Absorption Spectra*, Vol. 6, Nankodo, Tokyo, 1958, p. 41.
72. S. Bhagavantam, *Proc. Indian Acad. Sci.*, **11A,** 62 (1940).
73. S. P. S. Porto, J. A. Giordmaine, and T. C. Damen, *Phys. Rev.*, **147,** 608 (1966).
74. I. R. Beattie and T. R. Gilson, *Proc. Roy. Soc. London*, **A307,** 407 (1968).

# Inorganic Compounds

## *Part II*

## II-1. DIATOMIC MOLECULES

As shown in Sec. I-2, diatomic molecules have only one vibration along the chemical bond; its frequency is given by

$$\tilde{\nu} = \frac{1}{2\pi c}\sqrt{\frac{K}{\mu}}$$

where $K$ is the force constant, $\mu$ the reduced mass, and $c$ the velocity of light. In homopolar XX molecules ($\mathbf{D}_{\infty h}$), the vibration is not infrared active but is Raman active, whereas it is both infrared and Raman active in heteropolar XY molecules ($\mathbf{C}_{\infty v}$). Table II-1*a* lists the frequencies corrected for anharmonicity ($\omega_e$) and anharmonicity constants. The force constants can be calculated directly from these $\omega_e$ values.

Table II-1*b* lists the observed frequencies of a number of diatomic molecules, ions, and radicals as reported recently. The matrix isolation technique was employed extensively to observe the spectra of unstable and reactive species (Sec. I-21). Resonance Raman spectra of colored diatomic species exhibit a series of overtone frequencies which can be used to calculate anharmonicity constants (Sec. I-20).

Tables II-1*a* and II-1*b* show several interesting trends in frequencies and force constants. In the same family of the periodic table, we find the following:

| | $F_2$ | | $Cl_2$ | | $Br_2$ | | $I_2$ |
|---|---|---|---|---|---|---|---|
| $\tilde{\nu}$ (cm$^{-1}$) | 892 | > | 546 | > | 319 | > | 215 |
| $K$ (mdyn/Å) | 4.45 | > | 3.19 | > | 2.46 | > | 1.76 |

A similar trend is seen for $Li_2 > Na_2 > K_2 > Rb_2$. For a series of hydrogen halides, we find:

| | HF | | HCl | | HBr | | HI |
|---|---|---|---|---|---|---|---|
| $\tilde{\nu}$ (cm$^{-1}$) | 4138.5 | > | 2991 | > | 2649.7 | > | 2309.5 |
| $K$ (mdyn/Å) | 9.66 | > | 5.16 | > | 4.12 | > | 3.12 |

The same trend is seen for HBe > HMg > HCa > HSr. Across the periodic table, a series such as the following:

| | HC | | HN | | HO | | HF |
|---|---|---|---|---|---|---|---|
| $\tilde{\nu}$ (cm$^{-1}$) | 2861.6 | < | (3300) | < | 3735.2 | < | 4138.5 |
| $K$ (mdyn/Å) | 4.49 | < | 6.03 | < | 7.79 | < | 9.66 |

TABLE II-1*a*. HARMONIC FREQUENCIES AND ANHARMONICITY CONSTANTS OF SOME DIATOMIC MOLECULES AND IONS ($cm^{-1}$)[a,b]

| Molecule | $\omega_e$ | $x_e\omega_e$ | Molecule | $\omega_e$ | $x_e\omega_e$ |
|---|---|---|---|---|---|
| HH | 4395.2 | 117.91 | HCo | (1890) | — |
| $[HH]^+$ | 2297 | 62 | HNi | [1926.6] | — |
| HD | 3817.1 | 94.96 | $^7Li^7Li$ | 351.4 | 2.59 |
| DD | 3118.5 | 64.10 | $^7LiF$ (4) | 906.2 | 7.90 |
| $H^7Li$ | 1405.7 | 23.20 | $^7LiCl$ (4) | 641 | 4.2 |
| $H^{23}Na$ | 1172.2 | 19.72 | $^7LiBr$ (4) | 563 | 3.53 |
| $H^{39}K$ | 985.0 | 14.65 | $^7LiI$ (4) | 498 | 3.39 |
| HRb | 936.8 | 14.15 | $^{23}Na^{23}Na$ | 159.2 | 0.73 |
| $H^{138}Cs$ | 890.7 | 12.6 | $^{23}NaK$ | 123.3 | 0.40 |
| $H^9Be$ | 2058.6 | 35.5 | $^{23}NaRb$ | 106.6 | 0.46 |
| $H^{24}Mg$ | 1495.7 | 31.5 | $^{23}Na^{19}F$ (5) | 536.1 | 3.83 |
| $H^{40}Ca$ | 1299 | 19.5 | $^{23}NaCl$ (6) | 366 | 2.05 |
| HSr | 1206.2 | 17.0 | $^{23}NaBr$ (6) | 302 | 1.50 |
| HBa | 1172 | 16 | $^{23}NaI$ (6) | 258 | 1.08 |
| $H^{11}B$ | (2366) | (49) | $^{39}K^{39}K$ | 92.6 | 0.35 |
| $H^{27}Al$ | 1682.6 | 29.15 | KF (7) | 426 | 2.4 |
| $H^{115}In$ | 1474.7 | 24.7 | KCl (6) | 281 | 1.36 |
| HTl | 1390.7 | 22.7 | KBr (6) | 213 | 0.80 |
| $H^{12}C$ | 2861.6 | 64.3 | KI | 212 | 0.70 |
| $D^{12}C$ | 2101.0 | 34.7 | $^{85}Rb^{85}Rb$ | 57.3 | 0.96 |
| $H^{28}Si$ (2) | 2042.5 | 35.67 | $Rb^{133}Cs$ | 49.4 | — |
| HSn | (1580) | — | RbF (7) | 376 | 1.9 |
| HPb | 1564.1 | 29.75 | RbCl (6) | 228 | 0.92 |
| $H^{14}N$ | (3300) | — | $^{133}Cs^{133}Cs$ | 42.0 | 0.08 |
| $H^{31}P$ | (2380) | — | CsF (7) | 353 | 1.7 |
| HBi | 1698.9 | 31.6 | CsCl (6) | 209 | 0.75 |
| $H^{16}O$ | 3735.2 | 82.81 | $^{133}CsBr$ | (194) | (2.0) |
| $D^{16}O$ | 2720.9 | 44.2 | $^{133}Cs^{127}I$ | 142 | (1.2) |
| $H^{19}F$ | 4138.5 | 90.07 | $^9Be^{19}F$ | 1265.6 | 9.12 |
| $D^{19}F$ | 2998.3 | 45.71 | $^9Be^{35}Cl$ | 846.6 | 5.11 |
| $H^{35}Cl$ (3) | 2991.0 | 52.85 | $^9Be^{16}O$ | 1487.3 | 11.83 |
| $D^{35}Cl$ (3) | 2145.2 | 27.18 | $^{24}Mg^{19}F$ | 717.6 | 3.84 |
| HBr | 2649.7 | 45.21 | $^{24}Mg^{35}Cl$ | 465.4 | 2.05 |
| HI | 2309.5 | 39.73 | $^{24}Mg^{79}Br$ | 373.8 | 1.34 |
| $H^{63}Cu$ | 1940.4 | 37.0 | $Mg^{127}I$ | [312] | — |
| HAg | 1760.0 | 34.05 | $^{24}Mg^{16}O$ | 785.1 | 5.18 |
| $H^{197}Au$ | 2305.0 | 43.12 | MgS | 525.2 | 2.93 |
| HZn | 1607.6 | 55.14 | $^{40}Ca^{19}F$ | 587.1 | 2.74 |
| HCd | 1430.7 | 46.3 | $Ca^{35}Cl$ | 369.8 | 1.31 |
| HHg | 1387.1 | 83.01 | $Ca^{79}Br$ | 285.3 | 0.86 |
| $H^{55}Mn$ | [1490.6] | — | $Ca^{127}I$ | 242.0 | 0.64 |

TABLE II-1*a* (*Continued*)

| Molecule | $\omega_e$ | $x_e\omega_e$ | Molecule | $\omega_e$ | $x_e\omega_e$ |
|---|---|---|---|---|---|
| $Ca^{16}O$ | 650 | 6.6 | SiSi | (750) | — |
| $Sr^{19}F$ | 500.1 | 2.21 | $^{28}Si^{19}F$ | 856.7 | 4.7 |
| $Sr^{35}Cl$ | 302.3 | 0.95 | $^{28}Si^{35}Cl$ | 535.4 | 2.20 |
| $Sr^{79}Br$ | 216.5 | 0.51 | SiBr | 425.4 | 1.5 |
| $Sr^{127}I$ | 173.9 | 0.42 | $^{28}Si^{14}N$ | 1151.7 | 6.56 |
| $Sr^{16}O$ | 653.5 | 4.0 | $^{28}Si^{16}O$ | 1242.0 | 6.05 |
| $Ba^{19}F$ | 468.9 | 1.79 | $^{28}Si^{32}S$ | 749.5 | 2.56 |
| $^{138}Ba^{35}Cl$ | 279.3 | 0.89 | $^{28}SiSe$ | 580.0 | 1.78 |
| $Ba^{79}Br$ | 193.8 | 0.42 | $^{28}SiTe$ | 481.2 | 1.30 |
| $Ba^{16}O$ | 669.8 | 2.05 | $Ge^{19}F$ | 665 | 2.79 |
| $^{11}B^{11}B$ | 1051.3 | 9.4 | $^{74}Ge^{35}Cl$ | 407.6 | 1.36 |
| $^{11}B^{19}F$ | 1399.8 | 11.3 | GeBr | 296.6 | 0.9 |
| $^{11}B^{35}Cl$ | 839.1 | 5.11 | $^{74}Ge^{16}O$ | 985.7 | 4.30 |
| $^{11}B^{79}Br$ | 684.3 | 3.52 | $^{74}Ge^{32}S$ | 575.8 | 1.80 |
| $^{11}B^{14}N$ | 1514.6 | 12.3 | $^{74}Ge^{80}Se$ | 406.8 | 1.2 |
| $^{11}B^{16}O$ | 1885.4 | 11.77 | $^{74}Ge^{130}Te$ | 323.4 | 1.0 |
| $^{27}Al^{19}F$ | 814.5 | 8.1 | $Sn^{19}F$ | 582.9 | 2.69 |
| $^{27}Al^{35}Cl$ | 481.3 | 1.95 | $Sn^{35}Cl$ | 352.5 | 1.06 |
| $^{27}Al^{79}Br$ | 378.0 | 1.28 | SnBr | 247.7 | 0.62 |
| $^{27}Al^{127}I$ | 316.1 | 1.0 | $Sn^{16}O$ | 822.4 | 3.73 |
| $^{27}Al^{16}O$ | 978.2 | 7.12 | SnS | 487.7 | 1.34 |
| $^{69}Ga^{35}Cl$ | 365.0 | 1.1 | SnSe | 331.2 | 0.74 |
| $^{69}Ga^{81}Br$ | 263.0 | 0.81 | SnTe | 259.5 | 0.50 |
| $^{69}Ga^{127}I$ | 216.4 | 0.5 | PbPb | 256.5 | 2.96 |
| $Ga^{16}O$ | 767.7 | 6.34 | $Pb^{19}F$ | 507.2 | 2.30 |
| $^{115}In^{35}Cl$ | 317.4 | 1.01 | $Pb^{35}Cl$ | 303.8 | 0.88 |
| $^{115}In^{81}Br$ | 221.0 | 0.65 | $Pb^{79}Br$ | 207.5 | 0.50 |
| $^{115}In^{127}I$ | 177.1 | 0.4 | $Pb^{127}I$ | 160.5 | 0.25 |
| $In^{16}O$ | 703.1 | 3.71 | $Pb^{16}O$ | 721.8 | 3.70 |
| $Tl^{19}F$ | 475.0 | 1.89 | $^{208}Pb^{32}S$ | 428.1 | 1.20 |
| $Tl^{35}Cl$ | 287.5 | 1.24 | PbSe | 277.6 | 0.51 |
| $Tl^{81}Br$ | 192.1 | 0.39 | PbTe | 211.8 | 0.12 |
| $Tl^{127}I$ | 150 | — | $^{14}N^{14}N$ | 2359.6 | 14.46 |
| $^{12}C^{12}C$ | 1641.4 | 11.67 | $[^{14}N^{14}N]^+$ | 2207.2 | 16.14 |
| $^{12}C^{35}Cl$ | 846 | 1.0 | $^{14}N^{16}O$ | 1903.9 | 13.97 |
| $^{12}C^{14}N$ | 2068.7 | 13.14 | $^{14}NS$ | 1220.0 | 7.75 |
| $^{12}C^{31}P$ | 1239.7 | 6.86 | $^{14}NBr$ | 693 | 5.0 |
| $^{12}C^{16}O$ | 2170.2 | 13.46 | $^{14}N^{31}P$ | 1337.2 | 6.98 |
| $[^{12}C^{16}O]^+$ | 2214.24 | 15.16 | $^{14}N^{75}As$ | 1068.0 | 5.36 |
| $^{12}C^{32}S$ | 1285.1 | 6.5 | $^{14}NSb$ | 942.0 | 5.6 |
| $^{12}CSe$ | 1036.0 | 4.8 | $^{31}P^{31}P$ | 780.4 | 2.80 |

TABLE II-1*a* (*Continued*)

| Molecule | $\omega_e$ | $x_e\omega_e$ | Molecule | $\omega_e$ | $x_e\omega_e$ |
|---|---|---|---|---|---|
| $^{31}P^{16}O$ | 1230.6 | 6.52 | $^{63}Cu^{127}I$ | 264.8 | 0.71 |
| $^{75}As^{75}As$ | 429.4 | 1.12 | $Cu^{16}O$ | 628 | 3 |
| $[^{75}As^{75}As]^+$ | (314.8) | (1.25) | $^{107}Ag^{35}Cl$ | 343.6 | 1.16 |
| $^{75}As^{16}O$ | 967.4 | 5.3 | $^{109}Ag^{81}Br$ | 247.7 | 0.68 |
| SbSb | 269.9 | 0.59 | $^{107}Ag^{127}I$ | 206.2 | 0.43 |
| $Sb^{209}Bi$ | 220.0 | 0.50 | $Ag^{16}O$ | 493.2 | 4.10 |
| $Sb^{19}F$ | 614.2 | 2.77 | $^{197}Au^{35}Cl$ | 382.8 | 1.30 |
| $Sb^{35}Cl$ | 369.0 | 0.92 | $Zn^{19}F$ | (630) | (3.5) |
| $Sb^{14}N$ | 942.0 | 5.6 | $Zn^{35}Cl$ | 390.5 | 1.55 |
| $Sb^{16}O$ | 817.2 | 5.30 | ZnBr | (220) | — |
| $^{209}Bi^{209}Bi$ | 172.7 | 0.32 | $^{64}Zn^{127}I$ | 223.4 | 0.75 |
| $^{209}Bi^{19}F$ | 510.8 | 2.05 | $Cd^{19}F$ | (535) | — |
| $^{209}Bi^{35}Cl$ | 308.0 | 0.96 | $Cd^{35}Cl$ | 330.5 | 1.2 |
| $^{209}Bi^{79}Br$ | 209.3 | 0.47 | CdBr | 230.0 | 0.50 |
| $^{209}Bi^{127}I$ | 163.9 | 0.31 | $Cd^{127}I$ | 178.5 | 0.63 |
| $^{209}Bi^{16}O$ | 702.1 | 5.20 | $Hg^{19}F$ | 490.8 | 4.05 |
| $^{16}O^{16}O$ | 1580.4 | 12.07 | $Hg^{35}Cl$ | 292.6 | 1.60 |
| $[OO]^+$ | 1876.4 | 16.53 | $^{202}Hg^{81}Br$ | 186.3 | 0.98 |
| $^{16}OCl$ | (780) | — | $Hg^{127}I$ | 125.6 | 1.09 |
| $^{16}OBr$ | 713 | 7 | HgTl | 26.9 | 0.69 |
| $^{16}OI$ | (687) | (5) | $^{45}Sc^{16}O$ | 971.6 | 3.95 |
| $^{16}O^{32}S$ | 1123.7 | 6.12 | $^{89}Y^{16}O$ | 852.5 | 2.45 |
| $^{32}S^{32}S$ | 725.7 | 2.85 | $^{139}La^{16}O$ | 811.6 | 2.23 |
| $Se^{16}O$ | 907.1 | 4.61 | $Ce^{16}O$ | 865.0 | 2.99 |
| $^{80}Se^{80}Se$ | 391.8 | 1.06 | $^{141}Pr^{16}O$ | 818.9 | 1.20 |
| $Te^{16}O$ | 796.0 | 3.50 | $Ge^{16}O$ | 841.0 | 3.70 |
| TeTe | 251 | 0.55 | $Lu^{16}O$ | 841.7 | 4.07 |
| $^{19}F^{19}F$ | [892.1] | — | YbCl | 293.6 | 1.23 |
| $^{19}F^{35}Cl$ | 793.2 | 9.9 | $^{48}Ti^{35}Cl$ | 456.4 | 6.3 |
| $^{19}F^{79}Br$ | 671 | 3 | $^{48}Ti^{16}O$ | 1008.4 | 4.61 |
| $^{35}Cl^{35}Cl$ | 564.9 | 4.0 | $^{90}Zr^{16}O$ | 936.6 | 3.45 |
| $[^{35}Cl^{35}Cl]^+$ | 645.3 | 2.90 | $V^{16}O$ | 1012.7 | 4.9 |
| ClBr (8) | 442.5 | 1.5 | $Cr^{16}O$ | 898.8 | 6.5 |
| $^{79}Br^{79}Br$ (9) | 325.4 | 1.098 | $^{55}MnF$ | 618.8 | 3.01 |
| $^{79}Br^{81}Br$ | 323.2 | 1.07 | $^{55}Mn^{35}Cl$ | 384.9 | 1.4 |
| $^{127}I^{35}Cl$ | 384.2 | 1.47 | $^{55}MnBr$ | 289.7 | 0.9 |
| $^{127}I^{79}Br$ | 268.4 | 0.78 | $^{55}Mn^{16}O$ | 840.7 | 4.89 |
| $^{127}I^{127}I$ | 214.6 | 0.61 | $Fe^{35}Cl$ | 406.6 | 1.2 |
| $^{63}Cu^{19}F$ | 622.7 | 3.95 | $Fe^{16}O$ | 880 | 5 |
| $^{63}Cu^{35}Cl$ | 416.9 | 1.57 | CoCl | 421.2 | 0.74 |
| $^{63}Cu^{79}Br$ | 314.1 | 0.87 | NiCl | 419.2 | 1.04 |

TABLE II-1*b*. OBSERVED FREQUENCIES OF DIATOMIC MOLECULES, IONS, AND RADICALS ($cm^{-1}$)

| Species | State[a] | $\tilde{\nu}$ | Refs. | Species | State[a] | $\tilde{\nu}$ | Refs. |
|---|---|---|---|---|---|---|---|
| HSi | mat | 1967 | 10 | NCl | mat | 818.5, 825 | 32 |
| $H^{14}N$ | mat | 3133 | 11 | NBr | mat | 691 | 32 |
| $H^{16}O$ | mat | 3452, 3428 | 12 | $[NO]^{2-}$ | mat | 897 | 33 |
| $[HO]^-$ | solid | 3637 | 13 | $[NO]^-$ | mat | 1350 | 34 |
| HS | mat | 2541 | 14 | $^{14}N^{16}O$ | mat | 1880 | 35 |
| LiF | mat | 867 | 15, 16 | $[NO]^+$ | solid | 2273 | 36 |
| LiCl | mat | 569 | 15, 16 | PP | gas | 775 | 37 |
| LiBr | mat | 512 | 16 | AsAs | gas | 421 | 37 |
| LiI | mat | 433 | 16 | $[OO]^-$ | mat | 1097 | 38 |
| LiO | mat | 745 | 17 | $^{16}O^{16}O$ | gas | 1580 | 39 |
| NaF | mat | 515 | 15, 16 | $[OO]^+$ | solid | 1865 | 40 |
| MgF | mat | 740 | 18 | OF | mat | 1029 | 41 |
| AlF | mat | 785 | 19 | $[SS]^-$ | solid | $594^b$ | 42 |
| TlF | mat | 441 | 20 | SS | mat | 716 | 43 |
| TlCl | mat | 261 | 20 | $[SSe]^-$ | solid | $464^b$ | 42 |
| TlBr | mat | 179 | 20 | $[SeSe]^-$ | solid | $325^b$ | 42 |
| TlI | mat | 143 | 20 | $[FF]^-$ | mat | 475 | 44 |
| $^{12}CF$ | mat | 1279 | 21 | $[FCl]^+$ | sol'n | 819 | 45 |
| $[CN]^-$ | sol'n | 2080 | 22 | $[^{35}Cl^{35}Cl]^-$ | mat | $249^b$ | 46 |
| CN | mat | 2046 | 23 | $^{35}Cl^{35}Cl$ | mat | 546 | 47 |
| CO | mat | 2138 | 24 | ClO | mat | ~970 | 48, 41 |
| $[CO]^+$ | gas | 2214 | 25 | BrBr | gas | $319^b$ | 49 |
| $^{12}C^{32}S$ | gas | 1274 | 26 | $[BrBr]^+$ | sol'n | 362 | 50 |
| $Si^{16}O$ | mat | 1224 | 27 | $Br^{35}Cl$ | gas | $440^b$ | 49 |
| $^{74}Ge^{16}O$ | mat | 973 | 28 | II | gas | $215^b$ | 49 |
| $^{74}Ge^{32}S$ | mat | 566.6 | 29 | $[II]^+$ | sol'n | $238^b$ | 51 |
| $^{74}Ge^{80}Se$ | mat | 397.2 | 29 | $I^{35}Cl$ | gas | $381^b$ | 49 |
| $^{74}GeTe$ | mat | 317.6 | 29 | IBr | gas | $265^b$ | 49 |
| SnO | mat | 816.1 | 30 | TiO | mat | 1005 | 52 |
| $^{120}Sn^{32}S$ | mat | 480.5 | 29 | ZrO | mat | 959 | 52 |
| $^{120}Sn^{80}Se$ | mat | 325.2 | 29 | TaO | mat | 1020 | 53 |
| PbO | mat | 718.4 | 31 | WO | mat | 1055 | 54 |
| $Pb^{32}S$ | mat | 423.1 | 29 | $Nb^{14}N$ | mat | 1003 | 55 |
| $Pb^{80}Se$ | mat | 275.1 | 29 | $Nb^{16}O$ | mat | 968 | 55 |
| $^{14}NF$ | mat | 1115 | 32 | $U^{16}O$ | mat | 776 | 56 |

[a] mat: inert gas matrix.
[b] From resonance Raman spectra

[a] The compounds are listed in the order of the periodic table: IA, IIA, . . . , VIIIA, IB, IIB, . . . , VIIIB.
[b] Most of these values (gas-phase spectra) were taken from Herzberg.[1] Data obtained from other sources are indicated by reference numbers in parentheses after chemical formulas. Values in parentheses are not certain, and those in square brackets are observed frequencies without anharmonicity correction.

is found. Other interesting series are these:

| | $N_2$ | | CO | | NO | | $O_2$ |
|---|---|---|---|---|---|---|---|
| $\tilde{\nu}$ ($cm^{-1}$) | 2359.6 | > | 2138 | > | 1880 | > | 1580 |
| $K$ (mdyn/Å) | 22.98 | > | 18.47 | > | 15.55 | > | 11.77 |

and these:

| | $O_2^+$ | | $O_2$ | | $O_2^-$ |
|---|---|---|---|---|---|
| $\tilde{\nu}$ ($cm^{-1}$) | 1865 | > | 1580 | > | 1097 |
| $K$ (mdyn/Å) | 16.39 | > | 11.77 | > | 5.22 |

As stated in Sec. I-2, the force constant is not directly related to the dissociation energy. For a series of similar molecules, however, there is an approximate linear relationship between the two. The last series above is of particular interest since it shows the effect of charge on the frequency and bond energy. Here $\nu(O_2^+)$ is higher than $\nu(O_2)$ because $O_2^+$ is formed by losing one electron from an antibonding orbital of $O_2$, while $\nu(O_2^-)$ is lower than $\nu(O_2)$ because one extra electron of $O_2^-$ enters an antibonding orbital of $O_2$. The $O_2^+$ ion was found in compounds such as $O_2^+[MF_6]^-$ (M = As, Sb, etc.) and $O_2^+[M_2F_{11}]^-$ (M = Nb, Ta, etc.),[57] whereas the $O_2^-$ ion was observed in a triangular $M^+O_2^-$ complex formed by the reaction of alkali metals with $O_2$ in an Argon matrix.[38]

The effect of changing the charge on the vibrational frequency is also seen in other series: $N_2 > N_2^+$, $NO^+ > NO > NO^- > NO^{2-}$, $CO^+ > CO$, $F_2 > F_2^-$, etc. As will be shown in Part III, these frequency trends provide valuable information about the nature of the metal–ligand bond involving these diatomic species.

Alkali halide vapor consists mainly of monomers and dimers. As stated in Sec. I-21, Linevsky[58] developed a technique to isolate these monomers and dimers, produced at high temperature, in inert gas matrices. This technique has been used extensively to study the infrared spectra and structures of a number of inorganic salts. Some references on lithium halide dimers are as follows: $(LiF)_2$ (59), $(LiCl)_2$ (60, 61), and $(LiBr)_2$ (60, 61). These dimers are known to be cyclic-planar ($\mathbf{D}_{2h}$). On the other hand, $(TlX)_2$(X = F and Cl) are linear and symmetrical ($\mathbf{D}_{\infty h}$, X–Tl–Tl–X).[62] Such structures have been well known for $(HgX)_2$ (X = Cl, Br, and I).[63] The dimer $(LiO)_2$ is also cyclic-planar.[64] However, $(NH)_2$ is *trans*-planar ($\mathbf{C}_{2h}$) in a $N_2$ matrix[65] but takes the *cis* structure upon complex formation with the $Cr(CO)_5$ group.[66]

Hydrogen halides polymerize in the condensed phases; hydrogen fluoride polymerizes even in the gaseous phase.[67] The HX stretching bands are shifted markedly to lower frequencies by polymerization. For example, the monomer frequencies of HF (3962 $cm^{-1}$), HCl (2886 $cm^{-1}$), HBr (2558 $cm^{-1}$), and HI (2230 $cm^{-1}$) in the gaseous phase are lowered to 3420–3060,[68] 2746–2704,[69] 2438–2404,[69] and 2120 $cm^{-1}$,[69], respectively, in the solid phase. The infrared spectra of monomeric and polymeric hydrogen halides in inert gas matrices have been reported.[70]

Halogens form molecular compounds with organic solvents. For example, the band at 213 $cm^{-1}$ of gaseous $I_2$ is shifted to 201 $cm^{-1}$ in benzene solution,[71] and the band at 381.5 $cm^{-1}$ of gaseous ICl is shifted to 275 $cm^{-1}$ in pyridine solution.[72] The Raman spectra of $I_2$, $Br_2$, and ICl have been studied in many solvents.[73] These frequencies are much lower than the corresponding gas-phase frequencies because of charge-transfer interaction with solvent molecules.

The hydroxyl ion $[OH]^-$ is characterized by a sharp band at 3700–3500 $cm^{-1}$. For example, $LiOH \cdot H_2O$ exhibits a sharp OH stretching band at 3574 $cm^{-1}$ and a broad $OH_2$ stretching band in the 3200–2800 $cm^{-1}$ region.[74] The cyanide ion $[CN]^-$ exhibits a relatively sharp band in the 2250–2050 $cm^{-1}$ region. The CN stretching bands are at 2080 $cm^{-1}$ for ionic cyanides such as Na[CN] and K[CN][75] and at 2170–2250 $cm^{-1}$ for covalent cyanides such as Cu[CN] and Au[CN], in which two metals are bridged by the CN groups.[76] The vibrational frequencies of cyanogen, N≡C—C≡N, have been determined and its harmonic frequencies and anharmonicity constants calculated using $^{13}C$ and $^{15}N$ isotope data.[77] The NO stretching frequencies of nitric oxide (NO) in an Ar matrix are 1883 (monomer), 1862 and 1768 (*cis*-dimer), and 1740 (*trans*-dimer) $cm^{-1}$.[78]

Finally, many compounds containing the M=O groups, such as V=O, Nb=O, Ta=O, Mo=O, W=O, Re=O, Ru=O, and Os=O, exhibit the M=O stretching bands in the 1050–800 $cm^{-1}$ region.

## II-2. TRIATOMIC MOLECULES

The three normal modes of linear $X_3$ ($\mathbf{D}_{\infty h}$) and YXY ($\mathbf{D}_{\infty h}$) type molecules were shown in Fig. I-7; $\nu_1$ is Raman active but not infrared active, whereas $\nu_2$ and $\nu_3$ are infrared active but not Raman active (mutual exclusion rule). However, all three vibrations become infrared as well as Raman active in linear XYZ type molecules ($\mathbf{C}_{\infty v}$), shown in Fig. II-1. The three normal modes of bent $X_3$ ($\mathbf{C}_{2v}$) and YXY ($\mathbf{C}_{2v}$) type molecules were also shown in Fig. I-7. In this case, all three vibrations are both infrared and Raman active. The same holds for bent XXY and XYZ ($\mathbf{C}_s$) type molecules. In the following, the vibrational frequencies of a number of triatomic molecules are listed for each class of compounds.

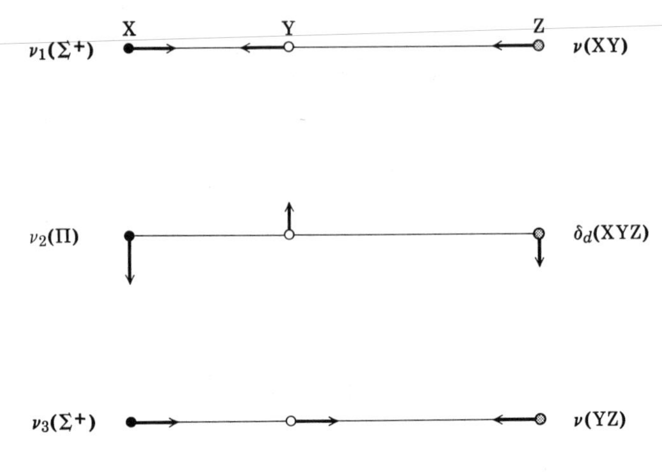


**Fig. II-1.** Normal modes of vibration of linear XYZ molecules.

Table II-2*a* lists the vibrational frequencies of $XY_2$ type metal halides. Most of these data were obtained in inert gas matrices. Although the structures of these halides are classified as either linear or bent, it should be noted that the bond angles in the latter type range from 95° to 170°.[79] Thus, some bent molecules are almost linear. These two types can be distinguished by the infrared activity of the $\nu_1$ mode; it is active for bent but not active for linear molecules. However, this simple criterion has led to conflicting results in some cases.[80] The bond angle of the YXY type molecule can be determined by the metal (X atom) isotope frequencies of the $\nu_3$ modes observed in inert gas matrices. If a pair of $\tilde{\nu}_3$ frequencies is determined, the bond angle ($2\alpha$) can be calculated from the equation[81]

$$\left(\frac{\tilde{\nu}'_3}{\tilde{\nu}_3}\right)^2 = \frac{M_X}{M_{X'}}\left(\frac{M_{X'}+2M_Y \sin^2 \alpha}{M_X+2M_Y \sin^2 \alpha}\right)$$

where $M$ denotes the mass of the atom subscripted, and the prime indicates an isotope. (For the derivation of this equation, see Part I.) Figure II-2 shows the $\nu_3$ spectra of $NiF_2$ in Ne and Ar matrices.[81] Using these isotope frequencies, the FNiF angle was calculated to be 154–167°. It should be noted in this table and others that the majority of compounds follow the trend $\nu_3 > \nu_1$ and that the exceptions to this rule occur in some bent molecules.

TABLE II-2a. VIBRATIONAL FREQUENCIES OF $XY_2$ TYPE METAL HALIDES ($cm^{-1}$)

| Compound[a] | Structure | $\nu_1$ | $\nu_2$ | $\nu_3$ | Refs. |
|---|---|---|---|---|---|
| $BeF_2$ | linear | (680) | 345 | 1555 | 82 |
| $BeCl_2$ | linear | (390) | 250 | 1135 | 82 |
| $BeBr_2$ | linear | (230) | 220 | 1010 | 82 |
| $BeI_2$ | linear | (160) | (175) | 873 | 82 |
| $MgF_2$ | bent | 477 | 240 | 840 | 83 |
| $^{24}Mg^{35}Cl_2$ | linear | — | 87.7 | 590.0 | 84 |
| $^{40}CaF_2$ | bent | 484.8 | 163.4 | 553.7 | 85 |
| $^{40}Ca^{35}Cl_2$ | linear | — | 63.6 | 402.3 | 84 |
| $^{86}SrF_2$ | bent | 441.5 | 82.0 | 443.4 | 85 |
| $^{88}Sr^{35}Cl_2$ | bent | 269.3 | 43.7 | 299.5 | 84 |
| $BaF_2$ | bent | 413.2 | (64) | 389.6 | 85 |
| $BaCl_2$ | bent | 255.2 | — | 260.0 | 84 |
| $CF_2$ | bent | 1102 | 668 | 1222 | 86 |
| $C^{35}Cl_2$ | bent | 719.5 | — | 745.7 | 87 |
| $CBr_2$ | bent | 595.0 | — | 640.5 | 88 |
| $SiF_2$ | bent | 851.5 | 345 | 864.6 | 89 |
| $^{28}Si^{35}Cl_2$ | bent | 502 | — | 513 | 90 |
| $^{28}SiBr_2$ | bent | 402.6 | — | 399.5 | 91 |
| $GeF_2$ | bent | 685 | 263 | 655 | 92 |
| $GeCl_2$ | bent | 398 | — | 373 | 93 |
| $SnF_2$ | bent | 592.7 | 197 | 570.9 | 94 |
| $SnCl_2$ | bent | 354 | 120 | 334 | 93 |
| $SnBr_2$ | bent | 237 | 84 | 223 | 95 |
| $PbF_2$ | bent | 531.2 | 165 | 507.2 | 94 |
| $PbCl_2$ | bent | 297 | — | 321 | 93 |
| $PbBr_2$ (g) | bent | 200 | 64 | — | 96 |
| $NF_2$ | bent | 1069.6 | 573.4 | 930.7 | 97 |
| $PF_2$ | bent | 834.0 | — | 843.5 | 98 |
| $PCl_2$ | bent | 452.0 | — | 524.9 | 99 |
| $PBr_2$ | bent | 369.0 | — | 410.0 | 99 |
| $O^{35}Cl_2$ | bent | 630.7 | 296.4 | 670.8 | 100 |
| $SF_2$ (g) | bent | 840 | 357 | 809 | 101 |
| $SCl_2$ (l) | bent | 514 | 208 | 535 | 102 |
| $SeCl_2$ (g) | bent | 415 | 153 | 377 | 103 |
| $TeCl_2$ (g) | bent | 377 | 125 | — | 103 |
| $KrF_2$ (g) | linear | 449 | 233 | 596,580 | 104 |
| $XeF_2$ | linear | 515 | 213.2 | 558 | 105 |
| $Xe^{35}Cl_2$ | linear | — | — | 314.1 | 106 |
| $^{63}CuF_2$ | bent | — | 183.0 | 743.9 | 81 |
| $[CuCl_2]^-$(s) | linear | 300 | 109 | 405 | 107 |
| $[CuBr_2]^-$(s) | linear | 193 | 81 | 322 | 107 |

TABLE II-2*a* (*Continued*)

| Compound[a] | Structure | $\nu_1$ | $\nu_2$ | $\nu_3$ | Refs. |
|---|---|---|---|---|---|
| $[CuI_2]^-$ (s) | linear | 148 | 65 | 279 | 107 |
| $[AgCl_2]^-$ (s) | linear | 268 | 88 | 333 | 107 |
| $[AgBr_2]^-$ (s) | linear | 170 | 61 | 253 | 107 |
| $[AgI_2]^-$ (s) | linear | 133 | 49 | 215 | 107 |
| $[AuCl_2]^-$ (s) | linear | 329 | 120,112 | 350 | 108 |
| $[AuBr_2]^-$ (s) | linear | 209 | 79,75 | 254 | 108 |
| $[AuI_2]^-$ (s) | linear | 158 | 67,59 | 210 | 108 |
| $ZnF_2$ | linear | 600 | 150 | 758 | 109 |
| $ZnCl_2$ | linear | 351 | 102 | 508 | 109 |
| $ZnBr_2$ | linear | 217 | 74 | 404 | 109 |
| $ZnI_2$ | linear | 155 | 62 | 346 | 109 |
| $CdF_2$ | linear | 572 | 123 | 662 | 109 |
| $CdCl_2$ | linear | 327 | 88 | 419 | 109 |
| $CdBr_2$ | linear | 205 | 62 | 319 | 109 |
| $CdI_2$ | linear | 149 | (50) | 269 | 109 |
| $HgF_2$ | linear | 588 | 171 | 641 | 109 |
| $HgCl_2$ | linear | 348 | 107 | 405 | 109 |
| $HgBr_2$ | linear | 219 | 73 | 294 | 109 |
| $HgI_2$ | linear | 158 | 63 | 237 | 109 |
| $TiF_2$ | bent | 665 | ~180 | 766 | 110 |
| $VF_2$ | bent | — | — | 733.2 | 79 |
| $CrF_2$ | linear | — | 155.4 | 654.5 | 79 |
| $CrCl_2$ | linear | — | — | 493.5 | 111 |
| $MnF_2$ | linear | — | 124.8 | 700.1 | 79 |
| $MnCl_2$ | linear | — | 83 | 476.8 | 111 |
| $FeF_2$ | linear | — | 141.0 | 731.3 | 79 |
| $FeCl_2$ | linear | — | 88 | 493.2 | 111 |
| $CoF_2$ | bent | — | 151.0 | 723.5 | 79 |
| $CoCl_2$ | linear | — | 94.5 | 493.4 | 111 |
| $NiF_2$ | bent | — | 139.7 | 779.6 | 79 |
| $NiCl_2$ | linear | (365) | 85 | 520.6 | 111 |
| $NiBr_2$ | linear | — | 69 | 414.2 | 111 |

[a] All data were obtained in inert gas matrices except those for which the physical state is indicated as g (gas), l (liquid) or s (solid).

The structure of the dimeric species $(MX_2)_2$ is known to be cyclic-planar:

$$X{-}M\langle{}^{X}_{X}\rangle M{-}X \qquad (\mathbf{D}_{2h})$$[112,113]

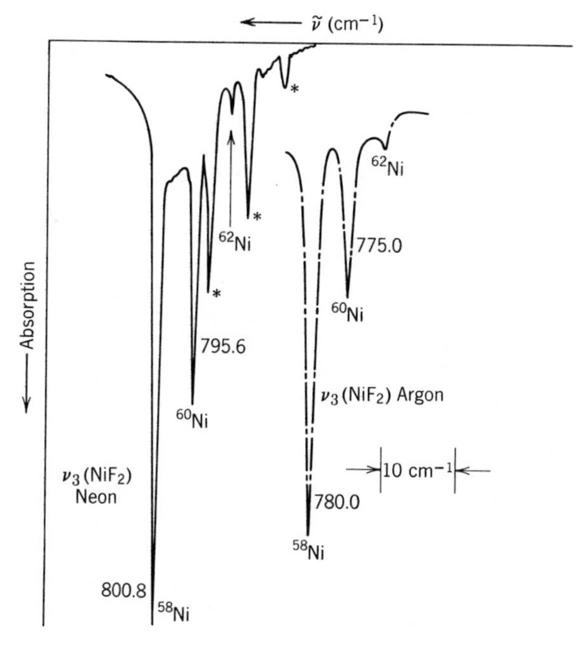


**Fig. II-2.** Infrared spectrum ($\nu_3$) of $NiF_2$ in Ne and Ar matrices. Matrix splitting, indicated by * is present in the Ne matrix spectrum.

although an exception is reported for $(GeF_2)_2$:[114]

F F
Ge Ge ($\mathbf{C}_{2h}$)
F F

(Nonplanar)

Table II-2*b* lists the vibrational frequencies of triatomic oxides, sulfides, and selenides. Most of these data were obtained in inert gas matrices. The dioxo groups, such as $VO_2^+$, $MoO_2^{2+}$, $WO_2^{2+}$, $ReO_2^+$, and $UO_2^{2+}$, exhibit strong bands in the 1000–850 cm$^{-1}$ region. *Trans*-Dioxo groups exhibit only one M=O stretching band, whereas *cis*-dioxo groups show two such bands.[151] For example, *cis*-$VO_2^+$ groups show one band at 907–876 (antisymmetric) and the other at 922–910 cm$^{-1}$ (symmetric).[152] Similar results are reported for *cis*-$MoO_2^{2+}$,[153] and *cis*-$WO_2^{2+}$,[154] groups. The $UO_2^{2+}$, $NpO_2^{2+}$, $PuO_2^{2+}$, and $AmO_2^{2+}$ groups are linear and exhibit only one band in the 940–850 cm$^{-1}$ region.[155] Jones[156] derived an equation which relates the U=O stretching force constant to the U=O distance. McGlynn et al.[157] noted that, in a series of $K_xUO_2L_y(NO_3)_2$ type compounds, the U=O stretching frequencies decrease as L is

Table II-2*b*. Vibrational Frequencies of YXY and XXY Type Oxides and Related Compounds ($cm^{-1}$)

| Compound[a] | Structure | $\nu_1$[b] | $\nu_2$ | $\nu_3$[b] | Refs. |
|---|---|---|---|---|---|
| $^{6}Li^{6}LiO$ | linear | — | 118 | 1028.5 | 17 |
| $O^{6}LiO$ | bent | — | 243.4 | 730 | 17 |
| ONaO | bent | 1080.0 | 390.7 | 332.8 | 115 |
| OBO | linear | — | — | 1276 | 116 |
| AlOAl | bent | 714 | (120) | 994 | 117, 118 |
| GaOGa | bent | 472 | — | 809.4 | 119 |
| $O^{12}CO$ (g) | linear | (1337)[c] | 667 | 2349 | 120 |
| $O^{13}CO$ | linear | — | (649) | 2284 | 121 |
| SCS (g) | linear | 658 | 397 | 1533 | 122 |
| SeCSe | linear | 364 | 313 | 1303 | 123 |
| $O^{120}SnO$ | bent | — | — | 863.1 | 124 |
| $[ONO]^-$ | bent | — | — | 1244 | 125 |
| $O^{14}NO$ | bent | (1318) | 749 | 1610 | 126 |
| $[ONO]^+$ (s) | linear | 1396 | 570 | 2360 | 127 |
| NNO | linear | 2223.8 | 588.7 | 1284.9 | 128 |
| $[O_3]^-$ | bent | 1016 | 600.9 | 802.3 | 129 |
| $O_3$ | bent | 1134.9 | 716.0 | 1089.2 | 130 |
| $S_3$(g) | bent | 585 | 490 | 651 | 131 |
| $[S_3]^-$ (1) | bent | 533 | 232 | 580 | 132 |
| OSO | bent | 1152.0 | 519.6 | 1359.3, 1353.8 | 133 |
| $^{18}OS^{18}O$ (g) | bent | 1114 | 505 | 1335 | 134 |
| $[O^{32}SO]^-$ | bent | 984.8 | 495.6 | 1042.0 | 135 |
| $O^{80}SeO$ | bent | 922.0 | 372.5 | 965.6 | 136 |
| OTeO | bent | 831.7 | 294 | 848.3 | 137 |
| FOO | bent | ~1500 | 586.4 | 376.0 | 138 |
| FOF | bent | 925, 915 | 461 | 821 | 139 |
| $^{35}ClO^{35}Cl$ (s) | bent | 630.7 | 296.4 | 670.8 | 140 |
| ClOO | bent | 407 | 373 | 1441 | 141 |
| $[OClO]^-$ (s) | bent | 790 | 400 | 840 | 142 |
| OClO (g) | bent | 943.2 | 445 | 1110.5 | 142a |
| $[OClO]^+$ (s) | bent | ~1040 | ~520 | ~1290 | 143 |
| BrOBr (s) | bent | 504 | 197 | 587 | 144 |
| $[OBrO]^-$ | bent | 775 | 400 | 800 | 142 |
| OCeO | bent | 757 | — | 736.8 | 145 |
| OTbO | bent | 758.7 | — | 718.8 | 145 |
| OThO | bent | 787.4 | — | 735.3 | 145 |
| OPrO | bent | (700) | — | 730.4 | 145 |
| $O^{46}TiO$ | bent | — | — | 940.7 | 146 |
| OTaO | bent | 971 | — | 912 | 147 |
| OWO | bent | 992 | — | 928 | 148 |
| OUO | linear | (765.4) | — | 776.1 | 149 |
| $[OUO]^{2+}$ (s) | linear | (856) | — | 931 | 150 |

changed in the order of the spectrochemical series

$$[CN]^- > en > NH_3 > [NCS]^- > [ONO]^- > py > H_2O > F^- > [NO_3]^-$$

The relationship between the $UO_2$ frequency and the U=O distance has been studied by several other investigators.[158–160]

Table II-2*c* lists the vibrational frequencies of triatomic interhalogeno compounds. The resonance Raman spectrum of the $I_3^-$ ion gives a series of overtones of the $\nu_1$ vibration.[183,184] The resonance Raman spectra of the $I_2Br^-$ and $IBr_2^-$ ions and their complexes with amylose have been studied.[185] The same table also lists the vibrational frequencies of XHY type (X, Y: halogens) compounds. All these species are linear except the $ClHCl^-$ ion, which was found to be bent in an inelastic neutron scattering (INS) and Raman spectral study.[178]

Table II-2*d* lists the vibrational frequencies of bent $XH_2$ type molecules. The XH stretching frequencies are lower and the $XH_2$ bending frequencies are higher in condensed phases than in the vapor phase because of hydrogen bonding in the former. This trend is also seen for $H_2O$ and $H_2S$ dissolved in organic solvents such as pyridine and dioxane.[200,201] According to Walrafen,[191] liquid $H_2O$ exhibits $\nu_1$ and $\nu_3$ at 3450 and 3615 $cm^{-1}$, respectively. On the other hand, Senior and Thompson[202] assign both modes at 3450 $cm^{-1}$ and interpret the 3615 $cm^{-1}$ band as a combination $\nu_1 + \nu_0$ band where $\nu_0$ is the low frequency stretching mode of the O—H······O system. The spectrum of ice is complicated because of its polymorphism. Hornig et al.[192] assigned the spectrum of ice I as shown in Table II-2*d*. They also assigned some librational and translational modes. Bertie and Whalley[203] studied the vibrational spectra of ice in other phases and found the spectra to be consistent with reported crystal structures in each phase. For crystal water and aquo complexes, see Sec. III-5.

Table II-2*e* lists the vibrational frequencies of linear XYZ type molecules. Some of these vibrations split into two because of Fermi resonance or the crystal field effect. Vibrational spectra of coordination compounds containing pseudohalide ions such as $NCS^-$, $NCO^-$, and $N_3^-$ are discussed in Sec. III-11.

Table II-2*f* lists the vibrational frequencies of bent XYZ type molecules. The spectra of most of these compounds were measured in inert gas matrices.

---

[a] All data were obtained in inert gas matrices except those for which the physical state is indicated as g (gas), l (liquid), or s (solid).

[b] For XXY molecules, $\nu_1$ and $\nu_3$ are XX and XY stretching modes, respectively.

[c] Fermi resonance with $2\nu_2$ (see Sec. I-10).

TABLE II-2c. VIBRATIONAL FREQUENCIES OF TRIATOMIC HALOGENO COMPOUNDS ($cm^{-1}$)

| Compound[a] | Structure[b] | $\nu_1$[c] | $\nu_2$ | $\nu_3$[c] | Refs. | Compound[a] | Structure[b] | $\nu_1$[c] | $\nu_2$ | $\nu_3$[c] | Refs. |
|---|---|---|---|---|---|---|---|---|---|---|---|
| $FClF^+$ (s) | b | 807 | 387 | 830 | 161 | $BrICl^-$ (s) | l | 203 | — | 180 | 174 |
| FClF (m) | b | 535 | 242 | 575 | 162 | $BrII^-$ (s) | l | 117 | 84 | 168 | 174 |
| $FClF^-$ (s) | l | 510, 478 | — | 636 | 163 | $BrII^+$ (s) | b | 258 | — | 198 | 172 |
| $FClCl^+$ (s) | b | 744 | 299, 293 | 535, 528 | 164 | $I_3^-$ (sl) | l | 114 | 52 | 145 | 170 |
| $FBrF^+$ (s) | b | 706 | 366 | 715 | 165, 166 | $HFH^+$ (s) | b | 2970 | 1680 | 3080 | 175 |
| $FBrF^-$ (s) | l | 442 | 198 | 596, 562 | 167 | $FHF^-$ (s) | l | 596 | 1233 | 1450 | 176 |
| $ClFCl^+$ (s) | b | 529 | 293 | 586 | 168 | $ClHF^-$ (s) | l | 275 | 863, 823 | 2710 | 177 |
| $Cl_3^+$ (s) | b | 493, 485 | 225 | 508 | 164 | $ClHCl^-$ (s) | b | 199 | 660, 602 | 1670 | 178 |
| $Cl_3^-$ (s) | b | 268 | (165) | 242 | 169 | $ClHBr^-$ (s) | l | — | 508 | 1705 | 179 |
| $ClBrCl^-$ (sl) | l | 278 | ~135 | 225 | 170 | $ClHI^-$ (s) | l | — | 485 | 2200 | 179 |
| $ClICl^+$ (s) | b | 371 | 147 | 364 | 171, 171a | $BrHF^-$ (s) | l | 220 | 740 | 2900 | 177 |
| $ClICl^-$ (sl) | l | 269 | 127 | 226 | 170 | $BrHBr^-$ (s) | l | 126 | 1038 | 1420 | 180 |
| $ClII^+$ (s) | b | 360 | 126 | 197 | 172 | $BrHBr^-$ (m) | l | 164 | — | 728 | 181 |
| $Br_3^-$ (sl) | l | 164 | 53 | 191 | 170 | $IHF^-$ (s) | l | 180 | 635 | 3145 | 177 |
| $Br_3$ (m) | l | 190 | — | — | 173 | IHI (m) | l | (120.7) | — | 682.1 | 182 |
| $BrIBr^+$ (s) | l | 256 | 124 | 256 | 172 | | | | | | |

[a] m: inert gas matrix; sl: solution; s: solid.

[b] b: bent; l: linear.

[c] For XYZ type, $\nu_1$ and $\nu_3$ correspond to $\nu(XY)$ and $\nu(YZ)$, respectively.

TABLE II-2*d*. VIBRATIONAL FREQUENCIES OF BENT $XH_2$ TYPE MOLECULES ($cm^{-1}$)

| Molecule | State | $\nu_1$ | $\nu_2$ | $\nu_3$ | Refs. |
|---|---|---|---|---|---|
| $GeH_2$ | matrix | 1887 | 928 | 1864 | 186 |
| $NH_2$ | matrix | — | 1499 | 3220 | 187 |
| | surface | 3290 | 1610 | 3380 | 188 |
| $[NH_2]^-$ | solid | 3270 | 1556 | 3323 | 189 |
| $^{16}OH_2$ | gas | 3657 | 1595 | 3756 | 190 |
| | liquid | 3450 | 1640 | 3615 | 191 |
| | solid | 3400 | 1620 | 3220 | 192 |
| $^{16}OHD^a$ | gas | 2727 | 1402 | 3707 | 190 |
| | solid | 2416 | 1490 | 3275 | 192 |
| $^{16}OD_2$ | gas | 2671 | 1178 | 2788 | 190 |
| | solid | 2520, 2336 | 1210 | 2432 | 192 |
| $^{18}OD_2$ | gas | 2657 | 1169 | 2764 | 193 |
| $OT_2{}^b$ | gas | — | 996 | 2370 | 194 |
| $SH_2$ | gas | 2615 | 1183 | 2627 | 195 |
| | matrix | 2619.5 | — | 2632.6 | 196 |
| | solid | 2532, 2523 | 1186, 1171 | 2544 | 197 |
| $SD_2$ | gas | 1892 | 934 | 2000 | 198 |
| | solid | 1843, 1835 | 857, 847 | 1854 | 197 |
| $SeH_2$ | gas | 2345 | 1034 | 2358 | 199 |
| $SeD_2$ | gas | 1687 | 741 | 1697 | 199 |

[a] Here $\nu_1$ and $\nu_3$ denote $\nu$(OD) and $\nu$(OH), respectively.
[b] T: $^3H$ (tritium).

## II-3. PYRAMIDAL FOUR-ATOM MOLECULES

### (1) $XY_3$ Molecules ($C_{3v}$)

The four normal modes of vibrations of a pyramidal $XY_3$ molecule are shown in Fig. II-3. These four vibrations are both infrared and Raman active. Table II-3*a* lists the fundamental frequencies of $XH_3$ type molecules. Several bands marked by an asterisk are split into two by *inversion doubling*. As is shown in Fig. II-4, two configurations of the $XH_3$ molecule are equally probable. If the potential barrier between them is small, the molecule may resonate between the two structures. As a result, each vibrational level splits into two levels (positive and negative).[247] Transitions between levels of different sign are allowed in the infrared spectrum, whereas those between levels of the same sign are allowed in the Raman spectrum. The transition between the two levels at $\upsilon = 0$ is also observed in the microwave region ($\tilde{\nu} = 0.79$ $cm^{-1}$). If the potential

TABLE II-2e. VIBRATIONAL FREQUENCIES OF LINEAR XYZ AND $X_3$ MOLECULES ($cm^{-1}$)

| XYZ | State | $\nu_1$(XY) | $\nu_2(\delta)$ | $\nu_3$(YZ) | Refs. |
|---|---|---|---|---|---|
| HCN | gas | 3311 | 712 | 2097 | 204 |
| | matrix | 3306 | 721 | — | 205 |
| DCN | gas | 2630 | 569 | 1925 | 204 |
| TCN | gas | 2460 | 513 | 1724 | 206 |
| FCN | gas | 1077 | 449 | 2290 | 207 |
| ClCN | gas | 714 | 380 | 2219 | 208 |
| | matrix | 718 | 384,387[a] | — | 209 |
| BrCN | gas | 574 | 342.5 | 2200 | 208 |
| | matrix | 575 | 349,351[a] | — | 209 |
| ICN | gas | 485 | 304 | 2188 | 210 |
| NaCN | matrix | 382 | 170 | 2044 | 211 |
| HNC | matrix | 3620 | 477 | 2029 | 212 |
| $^6$LiNC | matrix | 722.9 | 121.7 | 2080.5 | 213 |
| $H^{11}BO$ | matrix | (2849) | 754 | 1817 | 214 |
| CCO | matrix | 1074 | 381 | 1978 | 215 |
| SCO | gas | 859 | 520 | 2062 | 216 |
| SCSe | gas | 1435 | (355) | 506 | 217 |
| SCTe | sol'n | 1347 | (377) | 423 | 217 |
| $[CNO]^-$ | solid | 2096 | 471 | 1106 | 218 |
| $[NNF]^+$ | solid | 2371 | 391 | 1057 | 219 |
| NCN | matrix | — | 423 | 1475 | 220 |
| NNO | gas | 2277 | 596.5 | 1300.3 | 221 |
| NNC | matrix | 1241 | 393 | 2847 | 222 |
| NCO | matrix | 1275 | 487 | 1922 | 223 |
| $[NCO]^-$ | solid | 2155 | 630 | 1282,1202[b] | 224 |
| $[NNN]^-$ | solid | 1344 | 645 | 2041 | 225 |
| $[NC^{32}S]^-$ | solid | 2053 | 486,471 | 748 | 226 |
| $[NCSe]^-$ | solid | 2070 | 424,416 | 558 | 227 |
| $[NCTe]^-$ | solid | 2075 | 366 | 451 | 224 |

[a] Splitting due to matrix environment.
[b] Fermi resonance between $\nu_3$ and $2\nu_2$.

barrier is sufficiently high and if the three Y groups are not identical, optical isomers may be anticipated.

As is seen in Table II-3*a*, $\nu_1$ and $\nu_3$ overlap or are close in most compounds. The presence of the hydronium ($H_3O^+$) ion in hydrated acids has been confirmed by observing its characteristic frequencies. For example, it was shown from infrared spectra that $H_2PtCl_6 \cdot 2H_2O$ should be formulated as $(H_3O)_2[PtCl_6]$.[257] For normal coordinate analysis of pyramidal $XH_3$ molecules, see Refs. 258–260.

TABLE II-2$f$. VIBRATIONAL FREQUENCIES OF BENT XYZ TYPE MOLECULES ($CM^{-1}$)

| XYZ | State | $\nu_1$(XY) | $\nu_2$ | $\nu_3$(YZ) | Refs. |
|---|---|---|---|---|---|
| HCO | matrix | 2488 | 1090 | 1861 | 228 |
| $H^{12}CF$ | matrix | — | 1405 | 1181.5 | 21 |
| HNO | matrix | 3450 | 1110 | 1563 | 229 |
| $H^{14}NF$ | matrix | — | 1432 | 1000 | 32 |
| HOO | matrix | 3414 | 1389 | 1101 | 230 |
| HOF | matrix | 3537.1 | 1359.0 | 886.0 | 231 |
| HOCl | matrix | 3581 | 1239 | 729 | 232 |
| HOBr | matrix | 3590 | 1164 | 626 | 232 |
| $H^{72}GeBr$ | matrix | 1858 | 701 | 283 | 233 |
| NOF | matrix | 1886.6 | 734.9 | 492.2 | 234 |
| NSCl | gas | 1325 | 273 | 414 | 235 |
| ONF | gas | 1843.5 | 765.8 | 519.9 | 236 |
| $ON^{35}Cl$ | gas | 1799.7 | 603.2 | 331.9 | 237 |
| ONBr | gas | 1799.0 | 542.0 | 266.4 | 238 |
| OCF | matrix | 1855 | 626 | 1018 | 239 |
| OCCl | matrix | 1880 | 281 | 570 | 240 |
| OOF | matrix | 1500.9 | 586.8 | 376.6 | 241 |
| $O^{35}ClF$ | matrix | 1038.0 | 593.5 | 315.2 | 242 |
| SNF | gas | 1372 | 640 | 366 | 243 |
| $^{35}ClCF$ | matrix | 742 | — | 1146 | 244 |
| $^{35}Cl^{32}SN$ | matrix | 1327.3 | 403.8 | 267.4 | 245 |
| ClSnBr | matrix | 328 | — | 228 | 246 |
| ClPbBr | matrix | 295 | — | 200 | 246 |
| $Br^{32}SN$ | matrix | 1312.9 | 346.1 | 226.2 | 245 |

Table II-3$b$ lists the vibrational frequencies of pyramidal $XY_3$ halogeno compounds. Clark and Rippon[270] have measured the Raman spectra of a number of these compounds in the gaseous phase. The compounds show a $\nu_1 > \nu_3$ and $\nu_2 > \nu_4$ trend, whereas the opposite trend holds for the neutral $XH_3$ molecules listed in Table II-3$a$. In some cases, two stretching frequencies ($\nu_1$ and $\nu_3$) are too close to be distinguished empirically. This is also true for two bending bands ($\nu_2$ and $\nu_4$). The symmetry of the $SnX_3^-$ ion in $[As(Ph)_4][SnX_3]$ (X = Br and I) is lowered to $\mathbf{C}_s$ in the solid state, so that $\nu_3$ and $\nu_4$ each split into two bands.[267] Normal coordinate analyses on $[SnX_3]^-$ (X = F, Cl, Br, and I)[267] and Group VB trihalides[275] have been carried out.

Table II-3$c$ lists the vibrational frequencies of pyramidal $XO_3$ type compounds. Rocchiciolli[286] has measured the infrared spectra of a number of sulfites, selenites, chlorates, and bromates. Dasent and Waddington[287] also measured the infrared spectra of metal iodates and

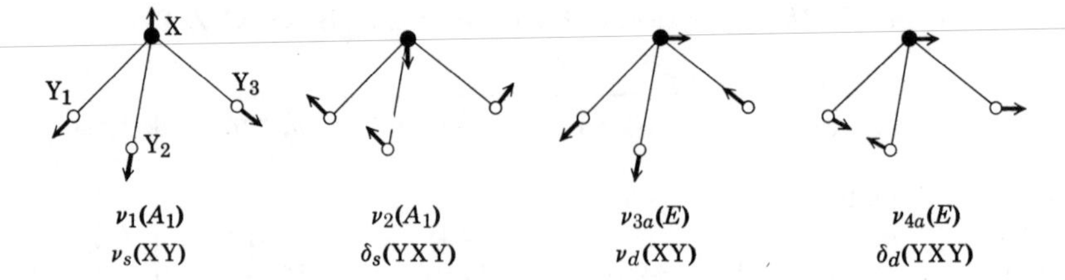


**Fig. II-3.** Normal modes of vibration of pyramidal $XY_3$ molecules.

suggested that extra bands at 480–420 cm$^{-1}$ may be due to the metal-oxygen vibrations. Again the $\nu_3$ and $\nu_4$ vibrations may split into two bands because of lowering of symmetry in the crystalline state. Although $\nu_2 > \nu_4$ holds in all cases, the order of two stretching frequencies ($\nu_1$ and $\nu_3$) depends on the nature of the central metal. Figure II-5 illustrates the infrared spectra of $KClO_3$ and $KIO_3$ obtained in the crystalline state.

### (2) $ZXY_2(C_s)$ and ZXYW ($C_1$) Molecules

Substitution of one of the Y atoms of a pyramidal $XY_3$ molecule by a Z atom lowers the symmetry from $\mathbf{C}_{3v}$ to $\mathbf{C}_s$. Then the degenerate vibrations split into two bands, and all six vibrations become infrared and Raman active. The relationship between $\mathbf{C}_{3v}$ and $\mathbf{C}_s$ is shown in Table II-3*d*. Table II-3*e* lists the vibrational frequencies of pyramidal $ZXY_2$

Table II-3*a*. Vibrational Frequencies of Pyramidal $XH_3$ Molecules (cm$^{-1}$)

| Molecule | State | $\nu_1(A_1)$ | $\nu_2(A_1)$ | $\nu_3(E)$ | $\nu_4(E)$ | Refs. |
|---|---|---|---|---|---|---|
| $NH_3$ | gas | 3335.9, 3337.5[a] | 931.6, 968.1[a] | 3414 | 1627.5 | 247 |
| | solid | 3223 | 1060 | 3378 | 1646 | 248 |
| $ND_3$ | gas | 2419 | 748.6, 749.0[a] | 2555 | 1191.0 | 247 |
| | solid | 2318 | 815 | 2500 | 1196 | 248 |
| $^{15}NH_3$ | gas | 3335 | 926, 961[a] | 3335 | 1625 | 249 |
| $NT_3$ | gas | 2016 | 647 | 2163 | 1000 | 250 |
| $PH_3$ | gas | 2327 | 990, 992[a] | 2421 | 1121 | 251 |
| $PD_3$ | gas | 1694 | 730 | (1698) | 806 | 251 |
| $AsH_3$ | gas | 2122 | 906 | 2185 | 1005 | 251 |
| $AsD_3$ | gas | 1534 | 660 | — | 714 | 251 |
| $SbH_3$ | gas | 1891 | 782 | 1894 | 831 | 252 |
| $SbD_3$ | gas | 1359 | 561 | 1362 | 593 | 252 |
| $[OH_3]^+SbCl_6^-$ | sol'n | 3560 | 1095 | 3510 | 1600 | 253 |
| $[OH_3]^+ClO_4^-$ | solid | 3285 | 1175 | 3100 | 1577 | 254 |
| $[OH_3]^+NO_3^-$ | solid | 2780 | 1135 | 2780 | 1680 | 255 |
| $[GeH_3]^-$ | liquid $NH_3$ | 1740 | 809 | — | 886 | 256 |

[a] Splitting due to Fermi resonance.

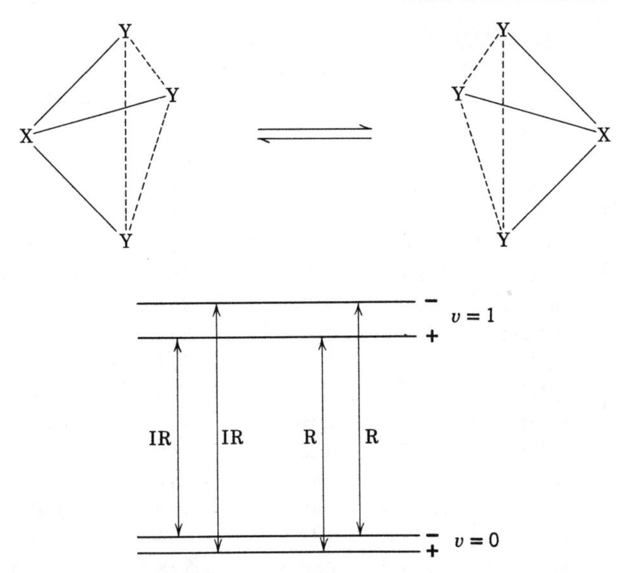


**Fig. II-4.** Inversion doubling of $XY_3$ type molecules.

molecules. Simon and Paetzold[307] made an extensive study of the vibrational spectra of selenium compounds. The ZXYW type molecule belongs to the $\mathbf{C}_1$ point group, and all six vibrations are infrared and Raman active. The vibrational spectra of OSClBr[308] and $[XSnYZ]^-$ (X, Y, Z: a halogen)[309] have been reported.

## II-4. PLANAR FOUR-ATOM MOLECULES

### (1) $XY_3$ Molecules ($D_{3h}$)

The four normal modes of vibration of planar $XY_3$ molecules are shown in Fig. II-6; $\nu_2$, $\nu_3$ and $\nu_4$ are infrared active, and $\nu_1$, $\nu_3$ and $\nu_4$ are Raman active. This case should be contrasted with pyramidal $XY_3$ molecules, for which all four vibrations are both infrared and Raman active.

Table II-4*a* lists the vibrational frequencies of planar $XY_3$ molecules. As stated above, pyramidal and planar structures can be distinguished easily on the basis of the difference in selection rules. In some cases, however, this approach leads to conflicting results. For example, rare-earth trifluorides in inert gas matrices exhibit infrared spectra that are consistent with planar structures except for $PrF_3$, which was thought to be pyramidal since it showed two stretching bands at 542 and 458 $cm^{-1}$.[324] Later, a

TABLE II-3b. VIBRATIONAL FREQUENCIES OF PYRAMIDAL $XY_3$ HALOGENO COMPOUNDS ($cm^{-1}$)

| Molecule | State | $\nu_1$ | $\nu_2$ | $\nu_3$ | $\nu_4$ | Refs. |
|---|---|---|---|---|---|---|
| $[InCl_3]^-$ | solid | 252 | 102 | 185 | 97 | 261 |
| $[InBr_3]^-$ | solid | 177 | 74 | 149 | 46 | 261 |
| $[InI_3]^-$ | solid | 136 | 78 | 110 | 40 | 261 |
| $CF_3$ | matrix | 1084 | 703 | 1250 | 600, 500 | 262 |
| $^{28}SiF_3$ | matrix | 832 | 406 | 954 | 290 | 263 |
| $^{28}SiCl_3$ | matrix | 470.2 | — | 582.0 | — | 264 |
| $GeCl_3$ | matrix | 388 | — | 362 | — | 265 |
| $[GeCl_3]^-$ | solid | 303 | — | 285 | — | 266 |
| $[SnF_3]^-$ | solid | 520 | 280 | 477 | 224 | 267 |
| $[SnCl_3]^-$ | sol'n | 297 | 128 | 256 | 103 | 267 |
| $[SnBr_3]^-$ | sol'n | 211 | 83 | 181 | 65 | 267 |
| $NF_3$ | gas | 1035 | 649 | 910 | 500 | 268 |
| $NCl_3$ | sol'n | 535 | 347 | 637 | 254 | 269 |
| $PF_3$ | gas | 893.2 | 486.5 | 858.4 | 345.6 | 270 |
| $PCl_3$ | gas | 515.0 | 258.3 | 504.0 | 186.0 | 270 |
| $PBr_3$ | gas | 390.0 | 159.9 | 384.4 | 112.8 | 270 |
| $PI_3$ | solid | 303 | 111 | 325 | 79 | 271 |
| $AsF_3$ | gas | 738.5 | 336.8 | 698.8 | 262.0 | 270 |
| $AsCl_3$ | gas | 416.5 | 192.5 | 391.0 | 150.2 | 270 |
| $AsBr_3$ | gas | 289.7 | 125.4 | 284.0 | 92.5 | 270 |
| $AsI_3$ | gas | 212.0 | 89.6 | (201) | 63.9 | 270 |
| $SbF_3$ | matrix | 654 | 259 | 624 | — | 272 |
| $SbCl_3$ | gas | 380.7 | 150.8 | 358.9 | 121.8 | 270 |
| $SbBr_3$ | gas | 256.0 | 101.2 | 248.9 | 76.2 | 270 |
| $SbI_3$ | gas | 186.5 | 74.0 | (147) | 54.3 | 270 |
| $BiCl_3$ | gas | 342 | 123 | 322 | 107 | 273 |
| $BiBr_3$ | gas | 220 | 77 | 214 | 63 | 274 |
| $BiI_3$ | solid | 145.5 | 90.2 | 115.2 | 71.0 | 275 |
| $[SF_3]^+$ | melt | 943 | 690 | 922 | 356 | 276 |
| $[SCl_3]^+$ | solid | 498 | 276 | 533, 521 | 215, 208 | 277 |
| $[SeF_3]^+$ | melt | 781 | 381 | 743 | 275 | 277 |
| $[SeCl_3]^+$ | sol'n | 430 | 206 | 415 | 172 | 278 |
| $[SeBr_3]^+$ | solid | 265 | 127 | 247, 227 | 107 | 279 |
| $[TeCl_3]^+$ | sol'n | 399 | 170 | 385 | 150 | 280 |
| $[TeBr_3]^+$ | solid | 242 | 111 | 222 | 95, 85 | 280 |

Raman study of matrix-isolated $PrF_3$ showed that the infrared band at 542 $cm^{-1}$ was not a PrF stretching fundamental. Thus the structure of $PrF_3$ was concluded to be planar in agreement with the result of an electron diffraction study in the gaseous phase.[324] The $CH_3$ radical produced by the reaction of Li atoms with $CH_3Br$ or $CH_3I$ is planar.[321] For dimeric species such as $Al_2F_6$ and $Fe_2Cl_6$, see Sec. II-10. Normal

TABLE II-3*c*. VIBRATIONAL FREQUENCIES OF PYRAMIDAL $XO_3$ MOLECULES ($CM^{-1}$)

| Molecule | State | $\nu_1(A_1)$ | $\nu_2(A_1)$ | $\nu_3(E)$ | $\nu_4(E)$ | Refs. |
|---|---|---|---|---|---|---|
| $[SO_3]^{2-}$ | sol'n | 967 | 620 | 933 | 469 | 281 |
| $[SeO_3]^{2-}$ | sol'n | 807 | 432 | 737 | 374 | 282 |
| $[TeO_3]^{2-}$ | sol'n | 758 | 364 | 703 | 326 | 282 |
| $[ClO_3]^-$ | sol'n | 933 | 608 | 977 | 477 | 283 |
| | solid | 939 | 614 | 971 | 489 | 284 |
| $[BrO_3]^-$ | sol'n | 805 | 418 | 805 | 358 | 283 |
| | solid | 810 | 428 | 790 | 361 | 284 |
| $[IO_3]^-$ | sol'n | 805 | 358 | 775 | 320 | 283 |
| | solid | 796 | 348 | 745 | 306 | 284 |
| $XeO_3$ | sol'n | 780 | 344 | 833 | 317 | 285 |

coordinate analyses of planar $XY_3$ molecules have been carried out by many investigators.[325–330,318]

Table II-4*b* lists the vibrational frequencies of planar $XO_3$ type compounds. The infrared spectra of rare-earth orthoborates have been studied by Laperches and Tarte.[340] For metaborates, see Sec. II-12. As discussed in Sec. I-22, the spectra of calcite and aragonite are markedly different because of the difference in crystal structure. Recently, the $CO_3$ radical was produced by photolysis of $CO_2$, and its structure was concluded to be $\mathbf{C}_{2v}$. Of the following two structures:

O O
C=O C=O
O O

the one on the right was favored by normal coordinate analysis.[341]

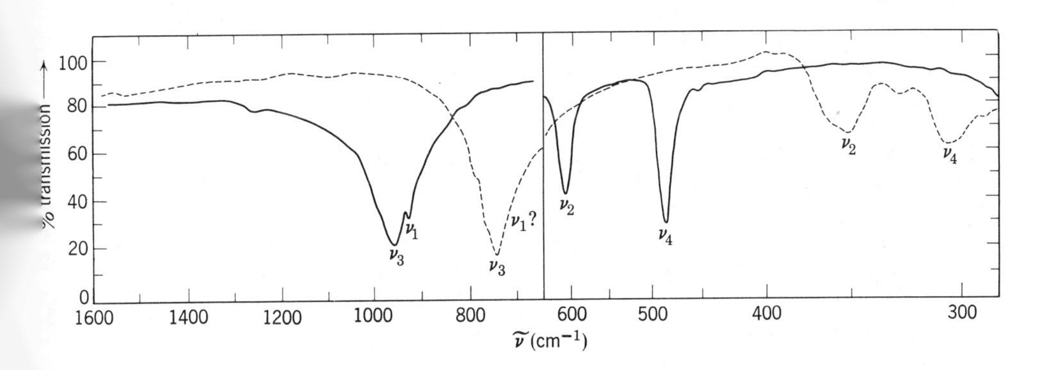


**Fig. II-5.** Infrared spectra of $KClO_3$ (solid line) and $KIO_3$ (broken line).

Table II-3*d*. Relationship Between $\mathbf{C}_{3v}$ and $\mathbf{C}_s$

| | | | | | | |
|---|---|---|---|---|---|---|
| $\mathbf{C}_{3v}$ XY$_3$ | $\nu_1(A_1)$ $\nu_s$(XY) | $\nu_2(A_1)$ $\delta_s$(YXY) | $\nu_3(E)$ $\nu_d$(XY) | | $\nu_4(E)$ $\delta_d$(YXY) | |
| | ↓ | ↓ | ↙ | ↘ | ↙ | ↘ |
| $\mathbf{C}_s$ ZXY$_2$ | $\nu_1(A')$ $\nu_s$(XZ) | $\nu_3(A')$ $\delta_s$(YXZ) | $\nu_2(A')$ $\nu_s$(XY) | $\nu_5(A'')$ $\nu_a$(XY) | $\nu_4(A')$ $\delta_s$(YXY) | $\nu_6(A'')$ $\delta_a$(YXZ) |

Table II-3*e*. Vibrational Frequencies of Pyramidal ZXY$_2$ Type Metal Halides (cm$^{-1}$)

| Z—X(Y)(Y) | $\nu_1(A')$ $\nu$(XZ) | $\nu_2(A')$ $\nu_s$(XY) | $\nu_3(A')$ $\delta_s$(YXZ) | $\nu_4(A')$ $\delta$(YXY) | $\nu_5(A'')$ $\nu_a$(XY) | $\nu_6(A'')$ $\delta_a$(YXZ) | Refs. |
|---|---|---|---|---|---|---|---|
| $HNF_2$ | 3193 | 972 | 500 | 1307 | 888 | 1424 | 288 |
| $HNCl_2$ | 3279 | 687 | — | 1002 | 695 | 1295 | 289 |
| $ClNH_2$ | 686 | — | 1553 | 1032 | 3380 | — | 289 |
| $ClNF_2$ | 692 | 918 | 552 | 366 | 842 | 382 | 290 |
| $ClPF_2$ | 545 | 864 | 411 | (308) | 852 | 259 | 291, 292 |
| $BrPF_2$ | 459 | 858 | 244 | 391 | 849 | 215 | 291, 292 |
| $FPCl_2$ | 838 | 525 | 328 | 203 | 525 | 267 | 291, 292 |
| $FPBr_2$ | 824 | 398 | 258 | 123 | 423 | 221 | 291, 292 |
| $IPF_2$ | 375 | 851 | 198 | 413 | 846 | 204 | 293 |
| $OSF_2$ | 1333 | 808 | 530 | (410) | 748 | 390 | 294 |
| $OSCl_2$ | 1251 | 492 | 194 | 344 | 455 | 284 | 295 |
| $OSBr_2$ | 1121 | 405 | 120 | 267 | 379 | 223 | 296 |
| $[FSO_2]^-$ | 598 | 1100 | 378 | 496 | 1170 | 280 | 297 |
| $[ClSO_2]^-$ | 214 | 1120 | 172 | 526 | 1312 | (103) | 297 |
| $[BrSO_2]^-$ | 203 | 1117 | 115 | 530 | 1308 | — | 297 |
| $[ISO_2]^-$ | 184 | 1112 | (55) | 530 | 1300 | — | 297 |
| $[FSeO_2]^-$ | 440 | 882 | 415 | 282 | 909 | 348 | 298 |
| $OSeF_2$ | 1049 | 667 | 362 | 282 | 637 | 253 | 299 |
| $OSeCl_2$ | 995 | 388 | 161 | 279 | 347 | 255 | 300 |
| $OSe(OH)_2$[a] | 831 | 702 | 430 | 336 | 690 | 364 | 301, 302 |
| $FClO_2$ | 630 | 1106 | 547 | 402 | 1271 | 367 | 303, 304 |
| $[OClF_2]^+$ | 1333 | 731 | 513 | 384 | 695 | 404 | 305, 306 |

[a] The OH group was assumed to be a single atom.

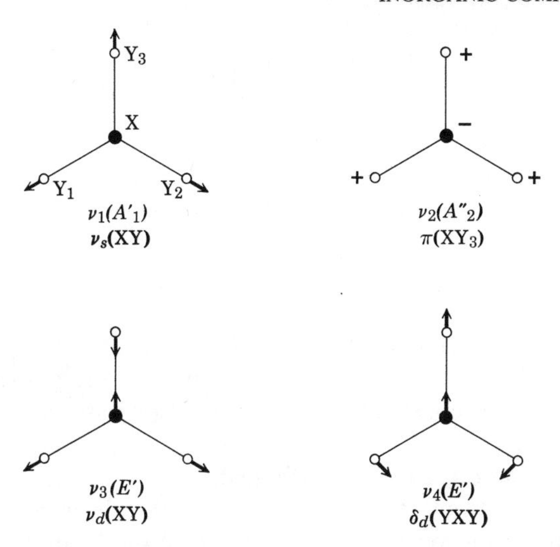


**Fig. II-6.** Normal modes of vibration of planar $XY_3$ molecules.

Crystals of $LiNO_3$, $NaNO_3$, and $KNO_3$ take the calcite structure (Sec. I-22). Nakagawa and Walter[337] carried out normal coordinate analyses on the whole Bravais lattices of these crystals. The spectra of anhydrous metal nitrates such as $Zn(NO_3)_2$[342] and $UO_2(NO_3)_2$[343] can be interpreted in terms of $\mathbf{C}_{2v}$ symmetry since the $NO_3$ group is covalently bonded to the metal (see Sec. III-8). Raman spectra of metal nitrates in the molten state[344,345] indicate that the degeneracy of the $\nu_3$ vibration is lost and the Raman inactive $\nu_2$ vibration appears. Apparently, the $\mathbf{D}_{3h}$ selection rule is violated because of cation–anion interaction. The infrared spectrum of monomeric lithium nitrate ($LiNO_3$) in an inert gas matrix shows a large splitting of the $\nu_3$ vibration (240 $cm^{-1}$) due to distortion of the $NO_3$ group.[346] Like $CO_3^{2-}$ and $NO_3^-$ ions, $CS_3^{2-}$ and $CSe_3^{2-}$ ions act as chelating ligands (see Sec. III-19). Normal coordinate analyses of planar $XY_3$ molecules have been made by many investigators. For some relatively recent work, see Refs. 347 and 348.

Some $XY_3$ type halides take the unusual T-shaped structure of $\mathbf{C}_{2v}$ symmetry shown below. This geometry is derived from a trigonal-bipyramidal structure in which two equatorial positions are occupied by two lone-pair electrons. Typical examples are $ClF_3$ and $BrF_3$. With the equatorial Y atom represented as Y′, the following assignments have been made for these molecules:[349] $\nu(XY')$, $A_1$, 754 and 672: $\nu(XY)$, $B_1$, 683.2 and 597; $\nu(XY)$, $A_1$, 523 and 547; $\delta$, $A_1$, 328 and 235; $\delta$, $B_1$, 431 and 347; $\pi$, $B_2$, 332 and 251.5 $cm^{-1}$ (for each mode, the former value is for $ClF_3$ and the latter is for $BrF_3$). The Raman spectrum of $XeF_4$ in $SbF_5$

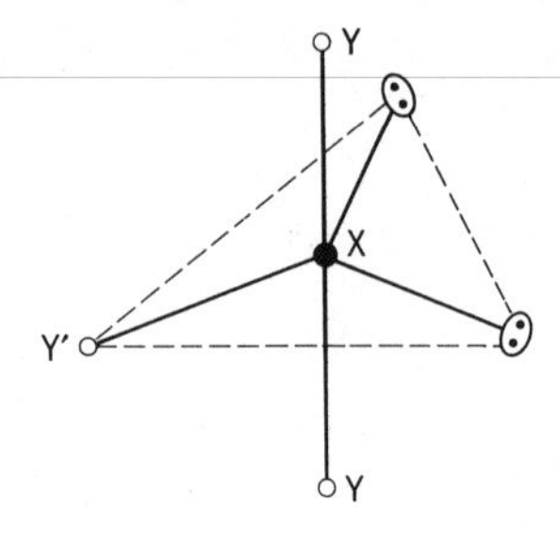


exhibits two strong polarized bands at 643 and 584 $cm^{-1}$, which were assigned to the T-shaped $XeF_3^+$ ion.[350] In an inert gas matrix $UO_3$ gives an infrared spectrum consistent with the T-shaped structure.[351]

### (2) $ZXY_2$ ($C_{2v}$) and ZXYW($C_s$) Molecules

If one of the Y atoms of a planar $XY_3$ molecule is replaced by a Z atom, the symmetry is lowered to $\mathbf{C}_{2v}$. If two of the Y atoms are replaced

TABLE II-4*a*. VIBRATIONAL FREQUENCIES OF PLANAR $XY_3$ MOLECULES ($CM^{-1}$)

| Molecule | State[a] | $\nu_1(A_1')$ | $\nu_2(A_2'')$ | $\nu_3(E')$ | $\nu_4(E')$ | Refs. |
|---|---|---|---|---|---|---|
| $^{10}BH_3$ | mat | (2623) | 1132 | 2820 | 1610 | 310 |
| $^{10}BF_3$ | gas | 888 | 718 | 1505 | 482 | 311, 312 |
| $^{11}BCl_3$ | liquid | 472.7 | — | 950.7 | 253.7 | 313, 314 |
| $^{11}BBr_3$ | liquid | 278.8 | — | 816 | 150.9 | 313, 315 |
| $^{11}BI_3$ | liquid | 192 | — | 691.8 | 101.0 | 313, 315 |
| $AlF_3$ | mat | (660) | 284 | 960 | 252 | 316 |
| $AlCl_3$ | gas | 371 | 160–170 | 615 | 146 | 317, 318 |
| $AlBr_3$ | gas | 228 | 107 | 450–500 | 93 | 317, 318 |
| $AlI_3$ | gas | 156 | 77 | 370–410 | 64 | 317, 318 |
| $GaCl_3$ | gas | 382 | 145 | 450 | 128 | 319, 318 |
| $GaBr_3$ | gas | 219, 237[b] | 95 | — | 84 | 319, 318 |
| $GaI_3$ | gas | 147 | 63 | 275 | 50 | 319, 318 |
| $InCl_3$ | gas | 350 | 110 | — | 94 | 317, 318 |
| $InBr_3$ | gas | 212 | 74 | 280 | 62 | 317, 318 |
| $InI_3$ | gas | 151 | 56 | 200–230 | 44 | 317, 318 |
| $TlBr_3$ | sol'n | 190 | 125 | 220 | 51 | 320 |
| $CH_3$ | mat | — | 730.3 | — | 1383 | 321 |
| $[CdCl_3]^-$ | sol'n | 265 | — | 287 | 90 | 322 |
| $[CdBr_3]^-$ | sol'n | 168 | — | 184 | 58 | 322 |
| $[CdI_3]^-$ | sol'n | 124 | — | 161 | 51 | 322 |
| $FeCl_3$ | mat | — | 116 | 464.8 | 102 | 323 |
| $PrF_3$ | mat | 526, 542 | 86 | 458 | 99 | 324 |

[a] mat: inert gas matrix.
[b] Fermi resonance.

TABLE II-4*b*. VIBRATIONAL FREQUENCIES OF PLANAR $XO_3$ AND RELATED COMPOUNDS IN THE CRYSTALLINE STATE ($cm^{-1}$)

| Compound | | $\nu_1(A_1')$ | $\nu_2(A_2'')$ | $\nu_3(E')$ | $\nu_4(E')$ | Refs. |
|---|---|---|---|---|---|---|
| $La[^{10}BO_3]$ | IR | 939 | 740·5 | 1330·0 | 606·2 | 331 |
| $H_3[BO_3]$ | IR | 1060 | 668, 648 | 1490–1428 | 545 | 332 |
| $Ca[CO_3]$ (calcite) | IR | — | 879 | 1429–1492 | 706 | 333 |
| | R | 1087 | — | 1432 | 714 | 333 |
| $Ca[CO_3]$ (aragonite) | IR | 1080 | 866 | 1504, 1492 | 711, 706 | 333 |
| | R | 1084 | 852 | 1460 | 704 | 333 |
| $Ba[CS_3]$ | R | 510 | 516[b] | 920 | 314 | 334, 335 |
| $Ba[CSe_3]$ | IR | 290 | 420 | 802 | 185 | 336, 335 |
| $Na[NO_3]$ | IR | — | 831 | 1405 | 692 | 337 |
| | R | 1068 | — | 1385 | 724 | 337 |
| $K[NO_3]$ | IR | — | 828 | 1370 | 695 | 337 |
| | R | 1049 | — | 1390 | 716 | 337 |
| $SO_3$[a] | IR | — | 484 | 1391 | 536 | 338 |
| | R | 1065 | 497.5 | 1390 | 530·2 | 339 |

[a] Gaseous state.
[b] Infrared.

by two different atoms, W and Z, the symmetry is lowered to $\mathbf{C}_s$. As a result, the selection rules are changed, as already shown in Table I-11. In both cases, all six vibrations become active in infrared and Raman spectra. Table II-4*c* lists the vibrational frequencies of planar $ZXY_2$ and ZXYW molecules. Although not listed in this table, the infrared spectra of binary mixed halides of boron have been measured in the gaseous phase.[362] The frequencies listed for the formate and acetate ions were obtained in aqueous solution. These frequencies are important when we discuss the vibrational spectra of metal salts of these anions (Sec. III-6).

## II-5. OTHER FOUR-ATOM MOLECULES

### (1) $X_2Y_2$ Molecules

Molecules like $O_2H_2$ take the nonplanar $\mathbf{C}_2$ structure (twisted about the O–O bond by ca. 90°), whereas $N_2F_2$ and $[N_2O_2]^{2-}$ exist in two forms: *trans*-planar ($\mathbf{C}_{2h}$) and *cis*-planar ($\mathbf{C}_{2v}$). Figure II-7 shows the six normal modes of vibration for the $\mathbf{C}_{2v}$ and $\mathbf{C}_2$ structures. The selection rules for these two structures are different only in the $\nu_6$ vibration, which is infrared inactive and Raman depolarized in the planar model but infrared active and Raman polarized in the nonplanar model.

TABLE II-4c. VIBRATIONAL FREQUENCIES OF PLANAR $ZXY_2$ AND ZXYW MOLECULES ($cm^{-1}$)

| $XY_3$ ($\mathbf{D}_{3h}$) | $\nu_1(A_1')$ $\nu_s(XY)$ | $\nu_2(A_2'')$ $\pi(XY_3)$ | $\nu_3(E')$ $\nu_d(XY)$ | | $\nu_4(E')$ $\delta_d(YXY)$ | | |
|---|---|---|---|---|---|---|---|
| $ZXY_2$ ($\mathbf{C}_{2v}$) | $\nu_1(A_1)$ $\nu(XZ)$ | $\nu_6(B_1)$ $\pi(ZXY_2)$ | $\nu_2(A_1)$ $\nu_s(XY)$ | $\nu_4(B_2)$ $\nu_a(XY)$ | $\nu_3(A_1)$ $\delta_s(ZXY)$ | $\nu_5(B_2)$ $\delta_a(ZXY)$ | |
| ZXYW ($\mathbf{C}_s$) | $\nu_1(A')$ $\nu(XZ)$ | $\nu_6(A'')$ $\pi$ | $\nu_2(A')$ $\nu(XY)$ | $\nu_4(A')$ $\nu(XW)$ | $\nu_3(A')$ $\delta(ZXY)$ | $\nu_5(A')$ $\delta(ZXW)$ | Refs. |
| $(HO)—NO_2$ | 886 | 765 | 1320 | 1710 | — | 583 | 352 |
| $F—NO_2$ | 822.4 | 742.0 | 1309.6 | 1791.5 | 559.6 | 567.8 | 353 |
| $Cl—NO_2$ | 792.6 | 652.3 | 1267.1 | 1684.6 | 408.1 | 369.6 | 353 |
| $[(HO)—CO_2]^-$ | 960 | 835 | 1338 | 1697 | 712 | 579 | 354 |
| $[H—CO_2]^-$ | 2803 | 1069 | 1351 | 1585 | 760 | 1383 | 355 |
| $[(CH_3)—CO_2]^-$ | 926 | 621 | 1413 | 1556 | 650 | 471 | 355 |
| $O{=}CF_2$ | 1928 | 774 | 965 | 1249 | 584 | 626 | 356 |
| $O{=}CCl_2$ | 1827 | 580 | 569 | 849 | 285 | 440 | 356 |
| $O{=}CBr_2$ | 1828 | 512 | 425 | 757 | 181 | 350 | 356 |
| O=CClF | 1868 | 667 | 776 | 1095 | 501 | 415 | 356 |
| O=CBrCl | 1828 | 547 | 517 | 806 | 240 | 372 | 356 |
| O=CBrF | 1874 | 620 | 721 | 1068 | 398 | 335 | 356 |
| O=CHF | 1837 | — | 2981 | 1065 | 1343 | 663 | 357 |
| $S{=}CF_2$ | 1368 | 622 | 787 | 1189 | 526 | 417 | 358 |
| $S{=}CCl_2$ | 1137 | 437 | 505 | 816 | — | — | 358 |
| $[Se—CS_2]^{2-}$ | 442 | 485 | 433 | 925 | 284 | 265 | 359 |
| $[O{=}NF_2]^-$ | 1862 | 715 | 897 | 1163 | 569 | 647 | 360, 361 |

Table II-5*a* lists the vibrational frequencies of $X_2Y_2$ type compounds. In a $N_2$ matrix $^{14}N_2O_2$ exists in three forms all containing the N–N linkage; their NO stretching frequencies ($cm^{-1}$) are as follows: *trans*, 1764; *cis* I, 1870, 1776; and *cis* II, 1870, 1785.[373] Ketelaar et al.[373a] observed the infrared frequencies of $S_2Cl_2$ and $S_2Br_2$ from *simultaneous transitions* in mixtures of each of these compounds with $CS_2$. These vibrations are similar to the combination bands in one molecule, but here the vibrations of two molecules are combined because of transient molecular collisions. The Raman spectra of $Se_2Cl_2$ and $Se_2Br_2$ in $CS_2$ and $CCl_4$ solutions indicate that the former takes the $\mathbf{C}_2$ structure whereas the latter takes the $\mathbf{C}_{2v}$ structure.[374]

Normal coordinate analyses of $O_2H_2$,[375] *trans*-$N_2O_2^{2-}$,[365] and $S_2X_2$ (X: a halogen)[372,376] have been made. Other $(XY)_2$ type compounds are known for dimeric metal halides such as $(LiF)_2$, which takes a planar-ring structure (Sec. II-1).

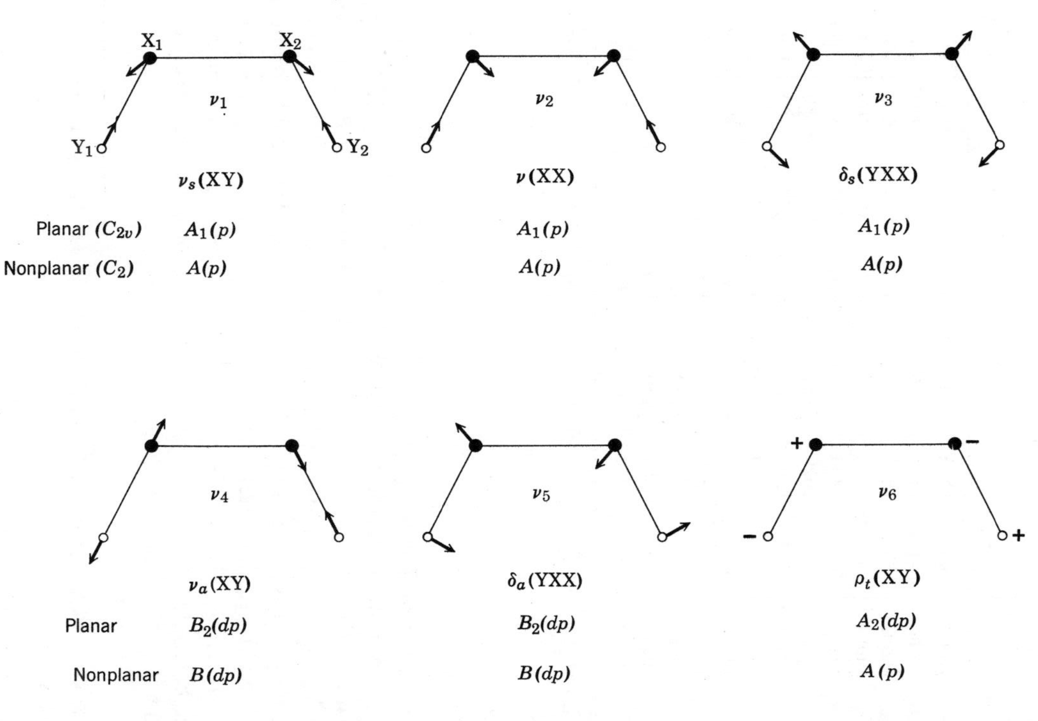


**Fig. II-7.** Normal modes of vibration of non-linear $X_2Y_2$ molecules (*p*:polarized; *dp*:depolarized).[358]

TABLE II-5*a*. VIBRATIONAL FREQUENCIES OF $X_2Y_2$ MOLECULES[a] ($CM^{-1}$)

| | | $\nu_1$ | $\nu_2$ | $\nu_3$ | $\nu_4$ | $\nu_5$ | $\nu_6$ | |
|---|---|---|---|---|---|---|---|---|
| *trans*-$X_2Y_2$ | $C_{2h}$ | $A_g$ | $A_g$ | $A_g$ | $B_u$ | $B_u$ | $A_u$ | |
| *cis*-$X_2Y_2$ | $C_{2v}$ | $A_1$ | $A_1$ | $A_1$ | $B_2$ | $B_2$ | $A_2$ | |
| twisted $X_2Y_2$ | $C_2$ | $A$ | $A$ | $A$ | $B$ | $B$ | $A$ | |
| Assignment | | $\nu_s$(XY) | $\nu$(XX) | $\delta_s$(YXX) | $\nu_a$(XY) | $\delta_a$(YXX) | $\pi$ | Refs. |
| *cis*-$[N_2O_2]^{2-}$ | solid | 830 | 1314 | 584 | 1047 | 330 | — | 363 |
| *trans*-$[N_2O_2]^{2-}$ | solid | (1419) | (1121) | (696) | 1031 | 371 | 492 | 364, 365 |
| *cis*-$N_2F_2$ | liquid | 896 | 1525 | 341 | 952 | 737 | (550) | 366 |
| *trans*-$N_2F_2$ | liquid | 1010 | 1522 | 600 | 990 | 423 | 364 | 367 |
| $O_2H_2$ | gas | 3607 | 1394 | 864 | 3608 | 1266 | 317 | 368 |
| $O_2D_2$ | gas | 2669 | 1029 | 867 | 2661 | 947 | 230 | 368 |
| $O_2F_2$ | matrix | 611 | 1290 | 366 | 624 | 459 | (202) | 369 |
| $S_2H_2$ | liquid | 2509 | 509 | (868)[b] | 2557[c] | 882 | — | 370 |
| $S_2F_2$ | gas | 717 | 615 | 320 | 681 | 301 | 183 | 371 |
| $S_2Cl_2$ | liquid | 449 | 540 | 206 | 436 | 240 | 102 | 372 |
| $S_2Br_2$ | liquid | 365 | 529 | 172 | 351 | 200 | 66 | 372 |
| $Se_2Cl_2$ | liquid | 367 | 288 | 130 | 367 | 146 | 87 | 372 |
| $Se_2Br_2$ | liquid | 265 | 292 | 107 | 265 | 118 | 50 | 372 |

[a] Except for $N_2F_2$ and $[N_2O_2]^{2-}$, all the molecules listed take the $C_2$ or $C_{2v}$ structure.
[b] Solid.
[c] Gas.

### (2) Other Planar Molecules ($C_s$)

Planar four-atom molecules of the WXYZ, XYZY, and $XY_3$ types have six normal modes of vibration, as shown in Fig. II-8. All these vibrations are both infrared and Raman active. In WXYZ compounds, the XYZ skeleton may be linear (HNCO, HOCN, HSCN) or nonlinear (HONO, HNSO). In the latter case, the whole molecule may take the *cis* or *trans* structure. Table II-5*b* lists the vibrational frequencies of molecules belonging to these types. Normal coordinate analyses have been made of $HN_3$[388] and HONO.[384]

## II-6. TETRAHEDRAL AND SQUARE-PLANAR FIVE-ATOM MOLECULES

### (1) Tetrahedral $XY_4$ Molecules ($T_d$)

Figure II-9 illustrates the four normal modes of vibration of a tetrahedral $XY_4$ molecule. All four vibrations are Raman active, whereas only $\nu_3$ and $\nu_4$ are infrared active. Fundamental frequencies of $XH_4$ type molecules are listed in Table II-6*a*. The trends $\nu_3 > \nu_1$ and $\nu_2 > \nu_4$ hold for

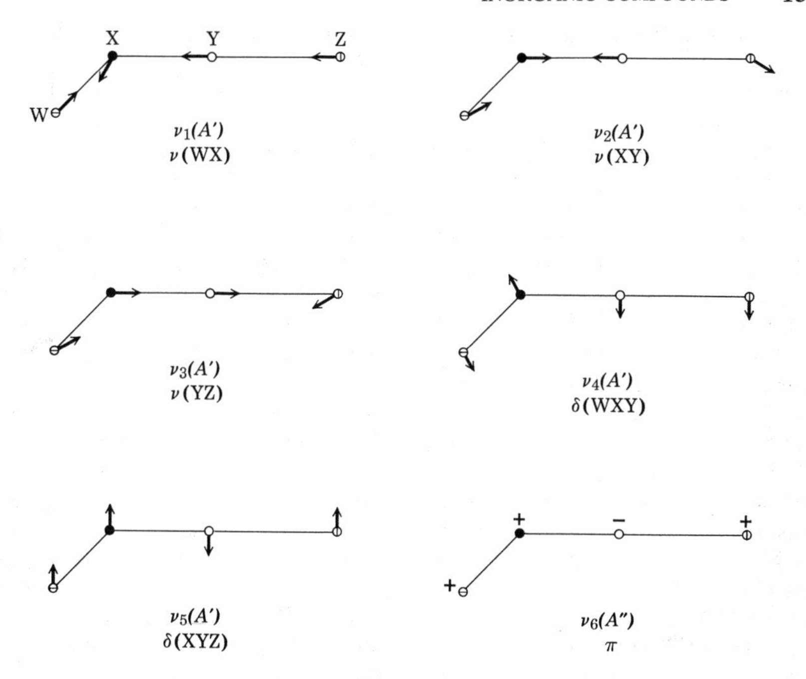


**Fig. II-8.** Normal modes of vibration of non-linear WXYZ molecules.

the majority of the compounds. The XH stretching frequencies may be lowered whenever the $XH_4$ ions form hydrogen bonds with counterions. In the same family of the periodic table, the XH stretching frequency decreases as the mass of the X atom increases. Shirk and Shriver[391] noted, however, that the $\nu_1$ frequency and the corresponding force constant show an unusual trend in Group IIIA:

| | $[BH_4]^-$ | $[GaH_4]^-$ | $[AlH_4]^-$ |
|---|---|---|---|
| $\tilde{\nu}_1(\mathrm{cm}^{-1})$ | 2270 | 1807 | 1757 |
| $F_{11}$(mdyn/Å) | 3.07 | 1.94 | 1.84 |

Figure II-10 shows the infrared spectrum of $NH_4Cl$, measured by Hornig et al.[403] They noted that the combination band between $\nu_4$ ($F_2$) and $\nu_6$ (rotatory lattice mode) is observed for $NH_4F$, $NH_4Cl$, and $NH_4Br$ because the $NH_4^+$ ion does not rotate freely in these crystals. In $NH_4I$ (phase I), however, this band is not observed because the $NH_4^+$ ion rotates freely.

Table II-6*b* lists the vibrational frequencies of a number of tetrahalogeno compounds. Except for $[TlBr_4]^-$ and $UF_4$, the trends $\nu_3 > \nu_1$

Table II-5b. Vibrational Frequencies of Planar Four-Atom Molecules ($cm^{-1}$)

| Molecule WXYZ | State[a] | $\nu_1$ $\nu$(WX) | $\nu_2$ $\nu$(XY) | $\nu_3$ $\nu$(YZ) | $\nu_4$ $\delta$(WXY) | $\nu_5$ $\delta$(XYZ) | $\nu_6$ $\pi$ | Refs. |
|---|---|---|---|---|---|---|---|---|
| HCNO | gas | 3336 | 1256 | 2198 | 331 | 538 | — | 377 |
| DCNO | gas | 2580 | 1254 | 2066 | — | — | — | 377 |
| HCNS | gas | 3539 | 1989 | 857 | 615 | 469 | 539 | 378 |
| DCNS | gas | 2645 | 1944 | 851 | 549 | 366 | 481 | 378 |
| HCNO | gas | 3531 | 2274 | 1527 | 777 | 660 | 578 | 379 |
| DCNO | gas | 2635 | 2235 | 1310 | 460 | 767 | 603 | 379 |
| HNCS | mat | 3505 | 1979 | 988 | 577 | 461 | — | 380 |
| DNCS | mat | 2623 | 1938 | — | 548 | 366 | — | 380 |
| HNNN | mat | 3324 | 2150 | 1273 | 1168 | 527 | 588 | 381 |
| DNNN | mat | 2466 | — | 1198 | 964 | 493 | — | 381 |
| HNSO | gas | 3345 | 1090 | 1261 | 911 | 453 | 759 | 382 |
| DNSO | gas | 2480 | 1055 | 1257 | 757 | 410 | 594 | 382 |
| HOCN | mat | 3506 | 1098 | 2294 | 1241 | 460 | 438 | 383 |
| DOCN | mat | 2590 | 1093 | 2292 | 957 | 437 | — | 383 |
| *cis*-HONO | mat | 3412 | 1633 | 1265 | 850 | 610 | 637 | 384 |
| *trans*-HONO | mat | 3558 | 1684 | 1298 | 815 | 625 | 583 | 385 |
| FNSO | liquid | 825 | 995 | 1230 | 228 | 600 | 395 | 386 |
| ClNSO | liquid | 526 | 989 | 1221 | 187 | 672 | 359 | 386 |
| BrNSO | liquid | 451 | 1000 | 1214 | 161 | 624 | 342 | 386 |
| INSO | liquid | 372 | 1028 | 1247 | 154 | 602 | 330 | 386 |
| ClSCN | sol'n | 520 | 678 | 2162 | — | 353 | — | 387 |
| BrSCN | sol'n | 451 | 676 | 2157 | — | 369 | — | 387 |
| ISCN | sol'n | 372 | 700 | 2130 | — | 362 | — | 387 |

[a] mat: inert gas matrix.

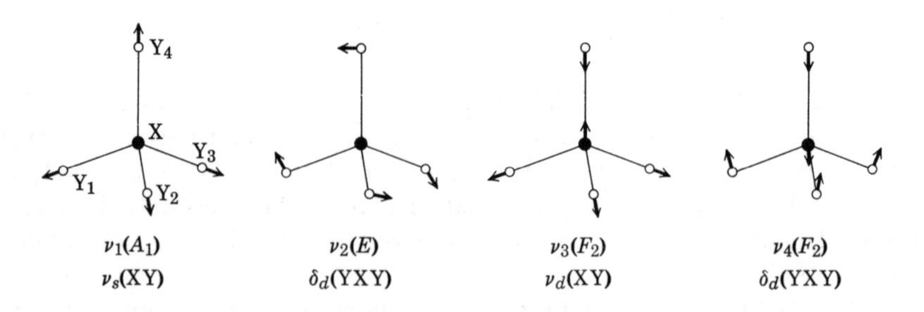


**Fig. II-9.** Normal modes of vibration of tetrahedral $XY_4$ molecules.

Table II-6*a*. Vibrational Frequencies of Tetrahedral $XH_4$ Molecules ($cm^{-1}$)

| Molecule | $\nu_1$ | $\nu_2$ | $\nu_3$ | $\nu_4$ | Refs. |
|---|---|---|---|---|---|
| $[^{10}BH_4]^-$ | 2270 | 1208 | 2250 | 1093 | 389,390 |
| $[^{10}BD_4]^-$ | 1604 | 856 | 1707 | 827 | 389,390 |
| $[AlH_4]^-$ | 1757 | 772 | 1678 | 760 or 766 | 391 |
| $[AlD_4]^-$ | 1256 | 549 | 1220 | 560 or 556 | 391 |
| $[GaH_4]^-$ | 1807 | — | — | — | 391 |
| $CH_4$ | 2917 | 1534 | 3019 | 1306 | 392 |
| $CD_4$ | 2085 | 1092 | 2259 | 996 | 393,394 |
| $SiH_4$ | 2180 | 970 | 2183 | 910 | 392,395 |
| $SiH_4$ | — | — | 2192 | 913 | 396 |
| $SiD_4$ | (1545) | (689) | 1597 | 681 | 397,395 |
| $GeH_4$ | 2106 | 931 | 2114 | 819 | 398,392 |
| $GeD_4$ | 1504 | 665 | 1522 | 596 | 398 |
| $SnH_4$ | — | 758 | 1901 | 677 | 399 |
| $SnD_4$ | — | 539 | 1368 | 487 | 399 |
| $[^{14}NH_4]^+$ | 3040 | 1680 | 3145 | 1400 | 392 |
| $[^{15}NH_4]^+$ | — | (1646) | 3137 | 1399 | 400 |
| $[ND_4]^+$ | 2214 | 1215 | 2346 | 1065 | 392 |
| $[NT_4]^+$ | — | 976 | 2022 | 913 | 400 |
| $[PH_4]^+$ | 2295 | 1086, 1026 | 2366, 2272 | 974, 919 | 401 |
| $[PD_4]^+$ | 1654 | 772, 725 | 1732 | 677 | 401 |
| $[AsH_4]^+$ | 2080 | 949 | 2142 | 818,813 | 402 |

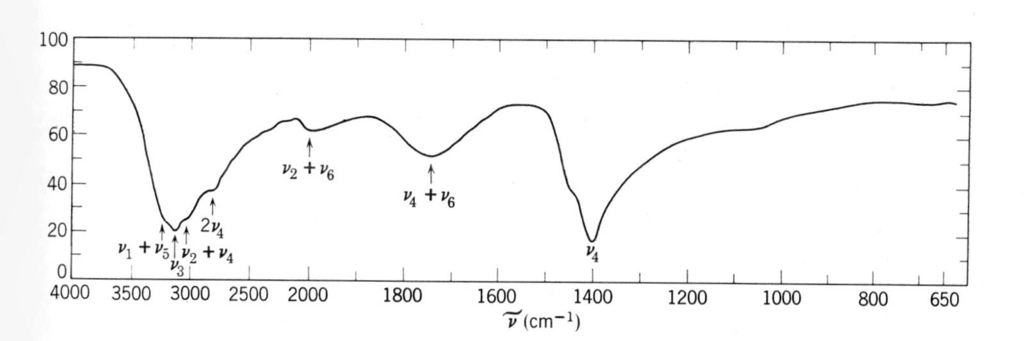


**Fig. II-10.** Infrared spectrum of $NH_4Cl$ ($\nu_5$, $\nu_6$; lattice modes).

Table II-6b. Vibrational Frequencies of Tetrahedral Halogeno Compounds ($cm^{-1}$)

| Molecule | $\nu_1$ | $\nu_2$ | $\nu_3$ | $\nu_4$ | Refs. |
|---|---|---|---|---|---|
| $[BeF_4]^{2-}$ | 547 | 255 | 800 | 385 | 404 |
| $[MgCl_4]^{2-}$ | 252 | 100 | 330 | 142 | 405 |
| $[MgBr_4]^{2-}$ | 150 | 61 | 290 | 90 | 405 |
| $[MgI_4]^{2-}$ | 107 | 42 | 259 | 60 | 405 |
| $[BF_4]^-$ | 777 | 360 | 1070 | 533 | 406 |
| $[BCl_4]^-$ | 396 | 196 | 723 | 275 | 407 |
| $[BBr_4]^-$ | 243 | 117 | 605 | 166 | 408 |
| $[AlF_4]^-$ | 622 | 210 | 760 | 322 | 409 |
| $[AlCl_4]^-$ | 348 | 119 | 498 | 182 | 410 |
| $[AlBr_4]^-$ | 212 | 98 | 394 | 114 | 411 |
| $[AlI_4]^-$ | 146 | 51 | 336 | 82 | 412 |
| $[GaCl_4]^-$ | 343 | 120 | 370 | 153 | 413 |
| $[GaBr_4]^-$ | 210 | 71 | 278 | 102 | 414 |
| $[GaI_4]^-$ | 145 | 52 | 222 | 73 | 415 |
| $[InCl_4]^-$ | 321 | 89 | 337 | 112 | 416 |
| $[InBr_4]^-$ | 197 | 55 | 239 | 79 | 417 |
| $[InI_4]^-$ | 139 | 42 | 185 | 58 | 415 |
| $[TlCl_4]^-$ | 312 | — | 293 | 93, 110 | 418 |
| $[TlBr_4]^-$ | 192 | — | 173, 185 | 78 | 418 |
| $[TlI_4]^-$ | 130 | — | 146 | 60 | 418 |
| $CF_4$ | 908.4 | 434.5 | 1283.0 | 631.2 | 419 |
| $CCl_4$ | 460.0 | 214.2 | 792, 765 | 313.5 | 419 |
| $CBr_4$ | 267 | 123 | 672 | 183 | 420 |
| $CI_4$ | 178 | 90 | 555 | 123 | 421 |
| $SiF_4$ | 800.8 | 264.2 | 1029.6 | 388.7 | 419 |
| $SiCl_4$ | 423.1 | 145.2 | 616.5 | 220.3 | 419 |
| $SiBr_4$ | 246.7 | 84.8 | 494.0 | 133.6 | 419 |
| $SiI_4$ | 166.3 | 57.6 | 405.0 | 90 | 419 |
| $GeF_4$ | 738 | 205 | 800 | 260 | 422 |
| $GeCl_4$ | 396.9 | 125.0 | 459.1 | 171.0 | 419 |
| $GeBr_4$ | 235.7 | 74.7 | 332.0 | 111.1 | 419 |
| $GeI_4$ | 156.0 | 51.6 | 273.0 | 77.3 | 419 |
| $SnCl_4$ | 369.1 | 95.2 | 408.2 | 126.1 | 419 |
| $SnBr_4$ | 222.1 | 59.4 | 284.0 | 85.9 | 419 |
| $SnI_4$ | 147.7 | 42.4 | 210 | 63.0 | 419 |
| $PbCl_4$ | 331 | 90 | 352 | 103 | 423 |
| $[NF_4]^+$ | 813 | 488 | 1159 | 611 | 424 |
| $[PCl_4]^+$ | 458 | 178 | 662 | 255 | 425 |

TABLE II-6*b* (*Continued*)

| Molecule | $\nu_1$ | $\nu_2$ | $\nu_3$ | $\nu_4$ | Refs. |
|---|---|---|---|---|---|
| $[PBr_4]^+$ | 254 | 116 | 503, 496 | 148 | 426, 427 |
| $[AsCl_4]^+$ | 422 | 156 | 500 | 187 | 428, 429 |
| $[CuCl_4]^{2-}$ | — | 77 | 267, 248 | 136, 118 | 430, 431 |
| $[CuBr_4]^{2-}$ | — | — | 216, 174 | 85 | 430 |
| $[ZnCl_4]^{2-}$ | 276 | 80 | 277 | 126 | 432 |
| $[ZnBr_4]^{2-}$ | 171 | — | 204 | 91 | 432 |
| $[ZnI_4]^{2-}$ | 118 | — | 164 | — | 432 |
| $[CdCl_4]^{2-}$ | 260 | — | 281 | 92 | 433 |
| $[CdBr_4]^{2-}$ | 159 | — | 188 | 63 | 433 |
| $[CdI_4]^{2-}$ | 115 | 40 | 141 | 49 | 433 |
| $[HgCl_4]^{2-}$ | 267 | 180 | 276 | 192 | 434 |
| $[HgI_4]^{2-}$ | 126 | 35 | 140 | 41 | 435 |
| $TiF_4$ | 712 | 185 | 793 | 209 | 436 |
| $TiCl_4$ | 389 | 114 | 498 | 136 | 437 |
| $TiBr_4$ | 231.5 | 68.5 | 393 | 88 | 437 |
| $TiI_4$ | 162 | 51 | 323 | 67 | 437 |
| $ZrF_4$ | (725 ~ 600) | (200 ~ 150) | 668 | 190 | 438 |
| $ZrCl_4$ | 377 | 98 | 418 | 113 | 437 |
| $ZrBr_4$ | 225.5 | 60 | 315 | 72 | 437 |
| $ZrI_4$ | 158 | 43 | 254 | 55 | 437 |
| $HfCl_4$ | 382 | 101.5 | 390 | 112 | 437 |
| $HfBr_4$ | 235.5 | 63 | 273 | 71 | 437 |
| $HfI_4$ | 158 | 55 | 224 | 63 | 437 |
| $VCl_4$ | 383 | 128 | 475 | 128, 150 | 439 |
| $[MnCl_4]^{2-}$ | 256 | — | 278, 301 | 120 | 440, 432 |
| $[MnBr_4]^{2-}$ | 195 | 65 | 209, 221 | 89 | 440, 432 |
| $[MnI_4]^{2-}$ | 108 | 46 | 188, 193 | 56 | 440, 432 |
| $[FeCl_4]^-$ | 330 | 114 | ~378 | ~136 | 432 |
| $[FeBr_4]^-$ | 200 | — | 290 | 95 | 432 |
| $[FeCl_4]^{2-}$ | 266 | 82 | 286 | 119 | 432 |
| $[FeBr_4]^{2-}$ | 162 | — | 219 | 84 | 432 |
| $[FeI_4]^{2-}$ | — | — | 186 | — | 430 |
| $[NiCl_4]^{2-}$ | 264 | — | 294, 280 | 119 | 441, 430 |
| $[NiBr_4]^{2-}$ | — | — | 228 | 81 | 441, 430 |
| $[NiI_4]^{2-}$ | 105 | — | 191 | — | 430 |
| $[CoCl_4]^{2-}$ | 269 | — | 311, 291 | 135 | 441 |
| $[CoBr_4]^{2-}$ | 166 | — | 231, 222 | 96 | 441 |
| $[CoI_4]^{2-}$ | 118 | — | 202, 194 | 56 | 430 |
| $UF_4$ | 614 | 340 | 420 | 180 | 442 |

and $\nu_4 > \nu_2$ hold for all the compounds. The latter trend is opposite to that found for $MH_4$ compounds. In the solid state, $\nu_3$ and $\nu_4$ may split into two or three bands because of the site effect. In some cases, the $MX_4$ ions are distorted to a flattened tetrahedron ($\mathbf{D}_{2d}$) or a structure of lower symmetry ($\mathbf{C}_s$).[430,443] According to X-ray analysis[444] the unit cell of $[(CH_3)_2CHNH_3]_2$ $[CuCl_4]$ contains two square-planar and four distorted tetrahedral $[CuCl_4]^{2-}$ ions. Willett et al.[445] demonstrated by using infrared spectroscopy that distorted tetrahedral ions can be pressed to square-planar ions under high pressure. In nitromethane, $(Et_4N)_2[CuCl_4]$ exhibits two bands at 278 and 237 cm$^{-1}$, indicating the distortion in solution.[446] A solution of $(Et_3NH)[GaCl_4]$ in 1,2-dichloroethylene exhibits three bands at 390, 383, and 359 cm$^{-1}$ due to lowering of symmetry caused by the NH····Cl hydrogen bonding.[447] Work and Good[448] studied the infrared spectra of long chain tertiary and quaternary ammonium salts of $[GaCl_4]^-$ and $[GaBr_4]^-$ ions in benzene solution, and found that the degree of distortion of these ions depends on the nature of the cation and the concentration. Distortion of $[MCl_4]^{2-}$ ions (M = Fe, Co, Ni and Zn) is also reported for their cesium and rubidium salts.[448a]

Table II-6*b* includes a number of data obtained by Clark et al. from their gas-phase Raman studies.[419,437] Some halides such as $TiI_4$,[449] $SnI_4$,[450] and $VCl_4$[451] have strong electronic absorption in the visible region, and their resonance Raman spectra have been measured in solution. As discussed in Sec. I-20, these compounds exhibit a series of $\nu_1$ overtones. Two important trends in frequency are noted in Table II-6*b*. First, the MX stretching frequency decreases as the halogen is changed in the order F > Cl > Br > I. The average values of $\nu(MBr)/\nu(MCl)$ and $\nu(MI)/\nu(MCl)$ calculated from all the compounds listed in Table II-6*b* are 0.76 and 0.62, respectively, for $\nu_3$, and 0.61 and 0.42, respectively, for $\nu_1$. These values are very useful when we assign the MX stretching bands of halogeno complexes (see Sec. III-16). Second, the effect of changing the oxidation state on the MX stretching frequency is seen in a pair such as $[FeX_4]^-$ and $[FeX_4]^{2-}$ (X = Cl and Br);[432] the MX stretching frequency increases as the oxidation state of the metal becomes higher. The ratio $\nu_3(FeX_4^-)/\nu_3(FeX_4^{2-})$ is 1.32 in this case.

A tetrahedral $MCl_4$ molecule in which M is isotopically pure and Cl is in natural abundance consists of five isotopic species because of the mixing of the $^{35}Cl$ (75.4%) and $^{37}Cl$ (24.6%) isotopes. Table II-6*c* lists their symmetries, percentages of natural abundance, and symmetry species of infrared active modes corresponding to the $\nu_3$ vibration of the $\mathbf{T}_d$ molecule. It has been established[452] that these nine bands overlap partially to give a "five-peak chlorine isotope pattern" whose relative

TABLE II-6c INFRARED ACTIVE VIBRATIONS OF $M^{35}Cl_n{}^{37}Cl_{4-n}$ TYPE MOLECULES

| Species | Symmetry | Abundance(%) | IR Active Modes |
|---|---|---|---|
| $M^{35}Cl_4$ | $\mathbf{T}_d$ | 32.5 | $F_2$ |
| $M^{35}Cl_3^{37}Cl$ | $\mathbf{C}_{3v}$ | 42.2 | $A_1, E$ |
| $M^{35}Cl_2^{37}Cl_2$ | $\mathbf{C}_{2v}$ | 20.5 | $A_1, B_1, B_2$ |
| $M^{35}Cl^{37}Cl_3$ | $\mathbf{C}_{3v}$ | 4.4 | $A_1, E$ |
| $M^{37}Cl_4$ | $\mathbf{T}_d$ | 0.4 | $F_2$ |

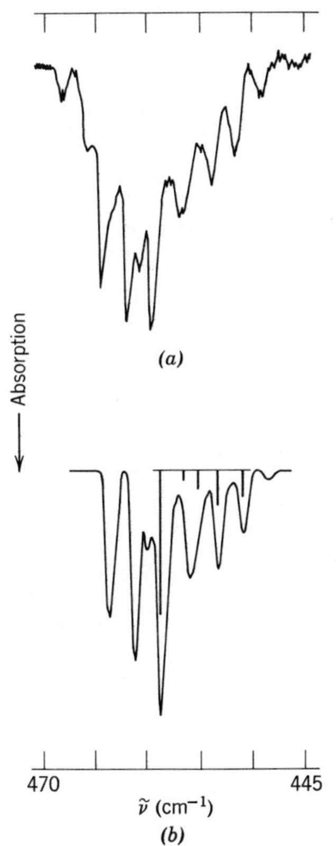


**Fig. II-11.** Matrix-isolation infrared (*a*) and computer-simulation spectra (*b*) of $GeCl_4$. Vertical lines in (*b*) show the five-peak chlorine isotope pattern of $^{74}GeCl_4$.

intensity is indicated by the vertical lines shown in Fig. II-11*b*. If M is isotopically mixed, the spectrum is too complicated to assign by the conventional method. For example, tin is a mixture of 10 isotopes, none of which is predominant. Thus 50 bands are expected to appear in the $\nu_3$ region of $SnCl_4$. It is almost impossible to resolve all these peaks, even in an inert gas matrix at 10° K. Königer and Müller,[453] therefore, prepared $^{116}SnCl_4$ and $^{116}Sn^{35}Cl_4$ on a milligram scale and measured their infrared spectra in Ar matrices. As expected, the former gave a "five-peak chlorine isotope pattern," whereas the latter showed a single peak at 409.8 $cm^{-1}$. This work was extended to $GeCl_4$, which consists of two Cl and five Ge isotopes. In this case, 25 peaks are expected to appear in the $\nu_3$ region. However, the observed spectrum (Fig. II-11*a*) shows about 10 bands. Königer et al.,[454] therefore, prepared $^{74}GeCl_4$ and $Ge^{35}Cl_4$ and measured their spectra in Ar matrices. As expected, both compounds showed a "five-peak" spectrum. The *ism* (isotope shift per unit mass difference) values for Cl and Ge were found to be 3.8 and 1.2 $cm^{-1}$, respectively. Using these values, it is now possible to calculate the frequencies of all other isotopic molecules. Furthermore, the relative intensity of individual peaks is known from the relative concentration of each isotopic molecule. On the basis of this information, Tevault et al.[455] obtained a computer-simulation infrared spectrum of $GeCl_4$ in natural abundance (Fig. II-11*b*).

Normal coordinate analyses of tetrahedral $XY_4$ molecules have been carried out by a number of investigators.[456,457] Thus far, Basile et al.[458] have made the most complete study; they calculated the force constants of 146 compounds by using GVF, UBF, and OVF fields (Sec. I-13), and discussed several factors that influence the values of the XY stretching force constants.

It has long been known that molecules such as $SF_4$, $SeF_4$, and $TeF_4$ assume a distorted tetrahedral structure ($\mathbf{C}_{2v}$) derived from a trigonal–bipyramidal geometry with a lone pair of electrons occupying an equatorial position:

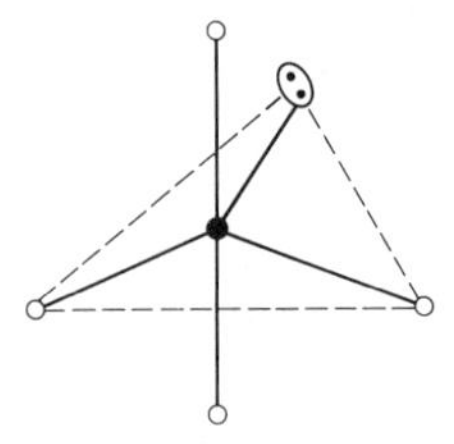

Table II-6*d* lists the vibrational frequencies of nine normal modes of such molecules. It should be noted that these compounds exhibit four stretch-

TABLE II-6*d*. VIBRATIONAL FREQUENCIES OF DISTORTED TETRAHEDRAL $XY_4$ MOLECULES[a] ($cm^{-1}$)

| $C_{2v}$ | $\nu_1$ $A_1$ | $\nu_2$ $A_1$ | $\nu_3$ $A_1$ | $\nu_4$ $A_1$ | $\nu_5$ $A_2$ | $\nu_6$ $B_1$ | $\nu_7$ $B_1$ | $\nu_8$ $B_2$ | $\nu_9$ $B_2$ | Refs. |
|---|---|---|---|---|---|---|---|---|---|---|
| $[SbF_4]^-$ | 596 | 449 | 285 | 163 | 220 | 431 | 257 | 566 | 180 | 459 |
| $[SbCl_4]^-$ | 339 | 296 | 147 | — | — | 321 | 199 | 246 | — | 459, 460 |
| $[SbBr_4]^-$ | 228 | 190 | — | — | — | 201 | 140 | 169 | — | 459, 460 |
| $[SbI_4]^-$ | 169 | — | 114 | — | — | 162 | 85 | 148 | — | 459, 460 |
| $SF_4$ | 892 | 558 | 465 | 226 | 414 | 730 | 532 | 867 | 353 | 461, 462 |
| $SeF_4$ | 747 | 571 | 409 | 156 | — | 622 | 361 | 733 | 250 | 463 |
| $TeF_4$ | 695 | 572 | 333 | (152) | — | 587 | 273 | 682 | (185) | 463 |
| $[ClF_4]^+$ | 796 | 568 | 511 | 237 | 473 | — | 536 | 827 | 395 | 464 |
| $[BrF_4]^+$ | 723 | 606 | 385 | 219 | — | 704 | 419 | 736 | 369 | 464 |
| $[IF_4]^+$ | 728 | 614 | 345 | 263 | — | — | 388 | 719 | 311 | 464 |

[a] Whereas $\nu_1$ and $\nu_8$ are stretching modes of equatorial bonds, $\nu_2$ and $\nu_6$ are stretching modes of axial bonds. For the normal modes of bending vibrations, see Ref. 461.

ing modes, two of which are polarized in the Raman. Adams and Downs[463] carried out normal coordinate analyses on $SeF_4$ and $TeF_4$ and found that the axial bonds are weaker than the equatorial bonds. The vibrational spectra of tetraalkylammonium salts of $[AsX_4]^-$ (X = Cl and Br), $[BiX_4]^-$ (X = Cl, Br, and I), and $[SbX_4]^-$ (X = Cl, Br, and I) have been studied in the solid state and in solution.[460] Except for solid $(Et_4N)[AsCl_4]$ and $[(n\text{-}Bu)_4N][SbI_4]$, all these ions assume the distorted tetrahedral structure shown above.

The $[ClF_4]^+$, $[BrF_4]^+$, and $[IF_4]^+$ ions were found in the following adducts:[464]

$$ClF_5\cdot(AsF_5) = [ClF_4][AsF_6]$$

$$BrF_5\cdot(SbF_5)_2 = [BrF_4][Sb_2F_{11}]$$

$$IF_5\cdot(SbF_5) = [IF_4][SbF_6]$$

It should be noted that $SeCl_4$, $SeBr_4$, $TeCl_4$, and $TeBr_4$ consist of the pyramidal $XY_3^+$ cation and the $Y^-$ anion in the solid state.[278–280]

Table II-6*e* lists the vibrational frequencies of tetrahedral $MO_4$, $MS_4$, and $MSe_4$ type compounds. The rules $\nu_3 > \nu_1$ and $\nu_4 > \nu_2$ hold for the

Table II-6e. Vibrational Frequencies of Tetrahedral $MO_4$, $MS_4$, and $MSe_4$ Type Compounds ($cm^{-1}$)

| Compound | $\nu_1$ | $\nu_2$ | $\nu_3$ | $\nu_4$ | Refs. |
|---|---|---|---|---|---|
| $[SiO_4]^{4-}$ | 819 | 340 | 956 | 527 | 465, 466 |
| $[PO_4]^{3-}$ | 938 | 420 | 1017 | 567 | 467, 468 |
| $[PS_4]^{3-}$ | 421 | 205 | 547 | 266 | 469 |
| $[AsO_4]^{3-}$ | 837 | 349 | 878 | 463 | 470 |
| $[AsS_4]^{3-}$ | 386 | 171 | 419 | 216 | 470 |
| $[SbS_4]^{3-}$ | 366 | 156 | 380 | 178 | 470 |
| $[SO_4]^{2-}$ | 983 | 450 | 1105 | 611 | 465 |
| $[SeO_4]^{2-}$ | 833 | 335 | 875 | 432 | 465 |
| $[ClO_4]^{-}$ | 928 | 459 | 1119 | 625 | 470 |
| $[BrO_4]^{-}$ | 801 | 331 | 878 | 410 | 471, 472 |
| $[IO_4]^{-}$ | 791 | 256 | 853 | 325 | 473 |
| $XeO_4$ | 775.7 | 267 | 879.2 | 305.9 | 474 |
| $[TiO_4]^{4-}$ | 761 | 306 | 770 | 371 | 475 |
| $[ZrO_4]^{4-}$ | 792 | 332 | 846 | 387 | 475 |
| $[HfO_4]^{4-}$ | 796 | 325 | 800 | 379 | 475 |
| $[VO_4]^{3-}$ | 826 | 336 | 804 | (336) | 476 |
| $[VO_4]^{4-}$ | 818 | 319 | 780 | 368 | 475 |
| $[VS_4]^{3-}$ | 404.5 | 193.5 | 470 | (193.5) | 477 |
| $[VSe_4]^{3-}$ | (232) | 121 | 365 | (121) | 477 |
| $[NbS_4]^{3-}$ | 408 | 163 | 421 | (163) | 477 |
| $[NbSe_4]^{3-}$ | 239 | 100 | 316 | (100) | 477 |
| $[TaS_4]^{3-}$ | 424 | 170 | 399 | (170) | 477 |
| $[TaSe_4]^{3-}$ | 249 | 103 | 277 | (103) | 477 |
| $[CrO_4]^{2-}$ | 846 | 349 | 890 | 378 | 476 |
| $[CrO_4]^{3-}$ | 834 | 260 | 860 | 324 | 475 |
| $[CrO_4]^{4-}$ | 806 | 353 | 855 | 404 | 475 |
| $[MoO_4]^{2-}$ | 897 | 317 | 837 | (317) | 476 |
| $[MoO_4]^{4-}$ | 792 | 328 | 808 | 373 | 475 |
| $[MoS_4]^{2-}$ | 458 | 184 | 472 | (184) | 478 |
| $[MoSe_4]^{2-}$ | 255 | 120 | 340 | 120 | 479 |
| $[WO_4]^{2-}$ | 931 | 325 | 838 | (325) | 476 |
| $[WO_4]^{4-}$ | 821 | 323 | 840 | 367 | 475 |
| $[WS_4]^{2-}$ | 479 | 182 | 455 | (182) | 478 |
| $[WSe_4]^{2-}$ | 281 | 107 | 309 | (107) | 477 |
| $[MnO_4]^{-}$ | 839 | 360 | 914 | 430 | 476 |
| $[MnO_4]^{2-}$ | 812 | 325 | 820 | 332 | 475 |
| $[MnO_4]^{3-}$ | 810 | 324 | 839 | 349 | 475 |
| $[TcO_4]^{-}$ | 912 | 325 | 912 | 336 | 476 |
| $[ReO_4]^{-}$ | 971 | 331 | 920 | (331) | 476 |
| $[ReO_4]^{3-}$ | 808 | 264 | 853 | 319 | 475 |

TABLE II-6*e* (*Continued*)

| Compound | $\nu_1$ | $\nu_2$ | $\nu_3$ | $\nu_4$ | Refs. |
|---|---|---|---|---|---|
| $[ReS_4]^-$ | 501 | 200 | 486 | (200) | 480 |
| $[FeO_4]^{2-}$ | 832 | 340 | 790 | 322 | 475 |
| $[FeO_4]^{3-}$ | 776 | 265 | 805 | 335 | 475 |
| $[FeO_4]^{4-}$ | 762 | 257 | 857 | 314 | 475 |
| $RuO_4$ | 885.3 | ~319 | 921 | 336 | 481 |
| $[RuO_4]^-$ | 830 | 339 | 845 | 312 | 475 |
| $[RuO_4]^{2-}$ | 840 | 331 | 804 | 336 | 475 |
| $OsO_4$ | 965.2 | 333.1 | 960.1 | 322.7 | 482 |
| $[CoO_4]^{4-}$ | 790 | 300 | 855 | 340 | 475 |
| $[B(OH)_4]^{-\,a}$ | 754 | 379 | 945 | 533 | 483 |
| $[Al(OH)_4]^{-\,a}$ | 615 | 310 | (720) | (310) | 484 |
| $[Zn(OH)_4]^{2-\,a}$ | 470 | 300 | (570) | (300) | 484 |

[a] Only $MO_4$ skeletal vibrations are listed for this ion.

majority of the compounds. It should be noted that $\nu_2$ and $\nu_4$ are often too close to be observed as separate bands in Raman spectra. Weinstock et al.[476] showed that, in Raman spectra, $\nu_2$ should be stronger than $\nu_4$, and that $\nu_4$ is hidden by $\nu_2$ in $[MoO_4]^{2-}$ and $[ReO_4]^-$ ions.

Table II-6*e* includes several series of oxoanions in which the oxidation state of the metal is changed. These series show that both stretching frequencies ($\nu_1$ and $\nu_3$) decrease as the oxidation state is lowered, for example:

| | $[CrO_4]^{2-}$ | | $[CrO_4]^{3-}$ | | $[CrO_4]^{4-}$ |
|---|---|---|---|---|---|
| $\tilde{\nu}_1$ | 846 | > | 834 | > | 806 |
| $\tilde{\nu}_3$ | 890 | > | 860 | > | 855 |

However, exceptions are seen for $[MnO_4]^- > [MnO_4]^{2-} < [MnO_4]^{3-}$ ($\nu_1$, $\nu_3$) and $[FeO_4]^{2-} < [FeO_4]^{3-} < [FeO_4]^{4-}$ ($\nu_3$). A gradual decrease in frequency is also seen in isoelectronic series such as $[OsO_4] > [ReO_4]^- > [WO_4]^{2-}$ and $[CrO_4]^{2-} > [VO_4]^{3-} > [TiO_4]^{4-}$. Although these trends are obvious in terms of the frequency, the same results are expected in terms of the force constant[475] since the mass effect is nonexistent or very small in these series.

Resonance Raman spectra of highly colored ions such as $CrO_4^{2-}$,[485] $MoS_4^{2-}$,[486] $VS_4^{3-}$,[486a] and $MnO_4^-$,[485] have been measured. The $\nu_3$ (infrared) bands of gaseous $RuO_4$[481] and $XeO_4$[487] exhibit complicated band contours consisting of individual isotope peaks of the central metal.

TABLE II-6g. VIBRATIONAL SPECTRA OF $ZXY_3$ MOLECULES ($cm^{-1}$)

| $C_{3v}$ $ZXY_3$ | $\nu_1(A_1)$ $\nu(XY_3)$ | $\nu_2(A_1)$ $\nu(XZ)$ | $\nu_3(A_1)$ $\delta(XY_3)$ | $\nu_4(E)$ $\nu(XY_3)$ | $\nu_5(E)$ $\delta(XY_3)$ | $\nu_6(E)$ $\rho_r(XY_3)$ | Refs. |
|---|---|---|---|---|---|---|---|
| $ICCl_3$ | 390 | 684 | 224 | 755 | 284 | 188 | 489 |
| $H^{28}SiF_3$ | 855.8 | 2315.6 | 425.3 | 997.8 | 843.6 | 306.2 | 490 |
| $FSiCl_3$ | 465 | 948 | 239 | 640 | 282 | 167 | 491 |
| $FSiBr_3$ | 318 | 912 | 163 | 520 | 226 | 110 | 491 |
| $FSiI_3$ | 242 | 894 | 115 | 424 | 194 | 71 | 491 |
| $BrSiF_3$ | 858 | 505 | 288 | 940 | 338 | 200 | 491 |
| $HGeCl_3$ | 418.4 | 2155.7 | 181.8 | 708.6 | 454 | 145.0 | 492 |
| $ONF_3$ | 743 | 1691 | 528 | 883 | 558 | 400 | 493 |
| $[ClPBr_3]^+$ | 285 | 587 | 149 | 500 | 172 | 120 | 494 |
| $[BrPCl_3]^+$ | 399 | 582 | 217 | 657 | 235 | 159 | 494 |
| $OPF_3$ | 873 | 1415 | 473 | 990 | 485 | 345 | 495 |
| $OPCl_3$ | 486 | 1290 | 267 | 581 | 337 | 193 | 496 |
| $OPBr_3$ | 340 | 1261 | 173 | 488 | 267 | 118 | 497, 496 |
| $SPF_3$ | 980 | 694 | 445 | 944 | 405 | 276 | 498 |
| $SPCl_3$ | 435 | 753 | 250 | 542 | 250 | 167 | 499 |
| $SPBr_3$ | 299 | 718 | 165 | 438 | 179 | 115 | 500 |
| $[OSF_3]^+$ | 909 | 1540, 1532 | 535 | 1063 | 508 | 387 | 501 |
| $NSF_3$ | 768 | 1512 | 520 | 811 | 430 | 340 | 502 |
| $[FSO_3]^-$ | 1142 | 862 | 571 | 1302 | 619 | 424 | 503 |
| $[ClSO_3]^-$ | 1042 | 381 | 601 | 1300 | 553 | 312 | 504 |
| $[SSO_3]^-$ | 995 | 446 | 669 | 1123 | 541 | 335 | 505 |
| $[FSeO_3]^-$ | 885 | 685 | 495 | 922, 965, 985 | 555 | 365 | 503 |
| $FClO_3$ | 1062.8, 1060.9 | 716.8, 706.6 | 548.8 | 1314 | 573 | 414 | 506 |
| $FBrO_3$ | 875.2 | 605.0 | (354) | 974 | (376) | (296) | 506 |
| $[NClO_3]^{2-}$ | 815 | 1256 | 594 | 870 | 623 | 457 | 507 |
| $OVF_3$ | 722 | 1058 | 258 | 806 | 308 | 204 | 508 |
| $OVCl_3$ | 408 | 1035 | 165 | 504 | 249 | 129 | 509 |
| $OVBr_3$ | 271 | 1025 | 120 | 400 | 83 | 212 | 510 |
| $ONbCl_3$ | 395 | 997 | 106 | 448 | 225 | 110 | 509 |
| $[FCrO_3]^-$ | 911 | 635 | 338 | 955 | 370 | 261 | 511 |
| $[ClCrO_3]^-$ | 907 | 438 | 295 | 954 | 365 | 209 | 512 |
| $[BrCrO_3]^-$ | 906 | 395 | 242 | 948 | 364 | 200 | 513 |
| $[OMoS_3]^{2-}$ | 461 | 862 | 183 | 470 | 183 | 263 | 514 |
| $[OMoSe_3]^{2-}$ | 293 | 858 | 120 | 355 | 120 | 188 | 514 |
| $[SMoO_3]^{2-}$ | 882 | 475 | 331 | 833 | 314 | 239 | 488 |
| $[SMoSe_3]^{2-}$ | — | 471 | 121 | 342 | 121 | — | 488 |
| $[SeMoS_3]^{2-}$ | 349 | 458 | — | 473 | 150 | 183 | 514a |
| $[OWS_3]^{2-}$ | 474 | 878 | 182 | 451 | 182 | 264 | 514 |
| $[OWSe_3]^{2-}$ | 292 | 878 | (120) | 312 | (120) | 194 | 514 |
| $[SWSe_3]^{2-}$ | 468 | 281 | 108 | 311 | 150 | 108 | 514a |
| $FMnO_3$ | 905 | 721 | 338 | 953 | 374 | 264 | 515 |
| $ClMnO_3$ | 890 | 456 | — | 950 | — | — | 515 |
| $FTcO_3$ | 962 | 696 | 317 | 951 | 347 | 231 | 516 |
| $ClTcO_3$ | 948 | 451 | 300 | 932 | 342 | 197 | 517 |
| $FReO_3$ | 1009 | 666 | 321 | 980 | (403) | (174) | 518 |
| $ClReO_3$ | 1001 | 435 | 293 | 961 | 344 | 196 | 513 |
| $BrReO_3$ | 997 | 350 | 195 | 963 | 332 | 168 | 513 |
| $[NReO_3]^{2-}$ | 878 | 1022 | 315 | 830 | 273 | 380 | 519 |
| $[SReO_3]^-$ | 948 | 528 | 322 | 906 | 322 | (240) | 513 |
| $[NOsO_3]^-$ | 898 | 1029 | 309 | 872 | 302 | 374 | 520 |

### TABLE II-6*h*. VIBRATIONAL FREQUENCIES OF $Z_2XY_2$ MOLECULES ($cm^{-1}$)

| $Z_2XY_2$ | $\nu_1(A_1)$ $\nu(XY)$ | $\nu_2(A_1)$ $\nu(XZ)$ | $\nu_3(A_1)$ $\delta(XY_2)$ | $\nu_4(A_1)$ $\delta(XZ_2)$ | $\nu_5(A_2)$ $\rho_t(XY_2)$ | $\nu_6(B_1)$ $\nu(XY)$ | $\nu_7(B_1)$ $\rho_w(XY_2)$ | $\nu_8(B_2)$ $\nu(XZ)$ | $\nu_9(B_2)$ $\rho_r(XY_2)$ | Refs. |
|---|---|---|---|---|---|---|---|---|---|---|
| $F_2CI_2$ | 605 | 1064 | 272 | 112 | 251 | 740 | 200 | 1110 | 278 | 527 |
| $H_2SiCl_2$ | 942 | 2221 | 514 | 188 | 710 | 868 | 566 | 2221 | 602 | 528 |
| $F_2SiBr_2$ | 414 | 891 | 270 | 115 | 187 | 540 | 241 | 974 | 257 | 529 |
| $H_2GeF_2$ | 860 | 2155 | 720 | (270) | (664) | 814 | 720 | 2174 | 596 | 530 |
| $H_2GeCl_2$ | 840 | 2132 | 404 | 163 | 648 | 772 | 420 | 2155 | 533 | 528 |
| $H_2GeBr_2$ | 848 | 2122 | 298 | 104 | 105 | 757 | 324 | 2138 | 492 | 530 |
| $H_2GeI_2$ | 821 | 2090 | 220 | 96 | 628 | 706 | 294 | 2110 | 451 | 530 |
| $[Cl_2PBr_2]^+$ | 326 | 584 | 191 | 132 | 150 | 518 | 173 | 616 | 201 | 513 |
| $[O_2PF_2]^-$ | 910 | 1179 | 269 | 567 | — | 962 | 492 | 1269 | 528 | 531 |
| $O_2SF_2$ | 849 | 1270 | 385 | 553 | — | 886 | 544 | 1503 | 540 | 532 |
| $O_2SCl_2$ | 405 | 1182 | 218 | 560 | 388 | 362 | 380 | 1414 | 282 | 533 |
| $[O_2VF_2]^-$ | 664 | 970 | — | 330 | — | 631 | 295 | 962 | 295 | 534 |
| $[O_2VCl_2]^-$ | 453 | 972 | — | 316 | — | 435 | 232 | 961 | 232 | 535 |
| $[O_2NbS_2]^{3-}$ | 464 | 897 | 246 | 356 | — | 514 | (297) | 872 | (271) | 536 |
| $O_2CrF_2$ | 727 | 1006 | 208 | 364 | (259) | 789 | 304 | 1016 | 274 | 537, 538 |
| $O_2CrCl_2$ | 475 | 995 | 140 | 356 | (224) | 500 | 257 | 1002 | 215 | 539 |
| $[O_2^{92}MoS_2]^{2-}$ | 473 | 819 | 200 | 307 | 267 | 506 | 267 | 801 | 246 | 540 |
| $[O_2MoSe_2]^{2-}$ | 283 | 864 | 114 | 339 | 251 | 353 | 251 | 834 | — | 536 |
| $[O_2WS_2]^{2-}$ | 454 | 886 | 196 | 310 | 280 | 442 | 280 | 848 | 235 | 536 |
| $[O_2WSe_2]^{2-}$ | 282 | 888 | 116 | 319 | 235 | 329 | 235 | 845 | 156 | 536 |

### TABLE II-6*i*. VIBRATIONAL FREQUENCIES OF $ZWXY_2$ MOLECULES ($cm^{-1}$)

| $ZWXY_2$ [a] | $\nu_1(A')$ $\nu_s(XY_2)$ | $\nu_2(A')$ $\nu(XW)$ | $\nu_3(A')$ $\nu(XZ)$ | $\nu_4(A')$ $\delta_s(XY_2)$ | $\nu_5(A')$ $\delta_s(WXY)$ | $\nu_6(A')$ $\delta_s(ZXY_2)$ | $\nu_7(A'')$ $\nu_a(XY_2)$ | $\nu_8(A'')$ $\delta(ZXW)$ | $\nu_9(A'')$ $\delta_a(ZXY_2)$ | Refs. |
|---|---|---|---|---|---|---|---|---|---|---|
| $OClPF_2$ | 900 | 623 | 1384 | (419) | (274) | 412 | 960 | 274 | 419 | 545, 292 |
| $SClPF_2$ | 949 | 549 | 735 | 394 | 363 | 209 | 925 | 252 | 316 | 545, 292 |
| $OBrPF_2$ | 884 | 561 | 1380 | (413) | (240) | 316 | 947 | 240 | 413 | 545, 292 |
| $SBrPF_2$ | 938 | 477 | 719 | 389 | 288 | 175 | 911 | 231 | 297 | 545, 292 |
| $OFPCl_2$ | 546 | 907 | 1358 | 205 | 330 | 382 | 626 | 253 | 374 | 545, 292 |
| $SFPCl_2$ | 479 | 915 | 750 | 193 | 328 | 268 | 574 | 193 | 317 | 545, 292 |
| $OFPBr_2$ | 472 | 888 | 1337 | 133 | 273 | 304 | 536 | 220 | 290 | 545, 292 |
| $SFPBr_2$ | 377 | 887 | 713 | 129 | 274 | 218 | 470 | 162 | 254 | 545, 292 |
| $OBrPCl_2$ | 545 | 432 | 1285 | 242 | 172 | 285 | 580 | 161 | 327 | 546 |
| $SBrPCl_2$ | 493 | 372 | 743 | (230) | 150 | 206 | 536 | 150 | 230 | 546 |
| $OClPBr_2$ | 391 | 552 | 1275 | 130 | 209 | 291 | 492 | 157 | 271 | 546 |
| $SClPBr_2$ | 333 | 500 | 729 | 121 | 196 | 190 | 436 | 136 | 205 | 546 |
| $[OSeMoS_2]^{2-}$ | 478[b] | 355 | 869 | 190 | — | 273 | 467[b] | — | 273 | 547, 548 |
| $[OSeWS_2]^{2-}$ | 473 | 320 | 879 | 190 | — | 265[b] | 458 | — | 255[b] | 549, 548 |
| $[OSMoSe_2]^{2-}$ | 360[b] | 461 | 865 | — | — | — | 320[b] | — | — | 550 |
| $[OSWSe_2]^{2-}$ | 317[b] | 459 | 882 | — | — | — | 312[b] | — | — | 550 |

[a] X denotes the central atom.
[b] These assignments may be interchanged.

TABLE II-6*f*. CORRELATION TABLE FOR $\mathbf{T}_d$, $\mathbf{C}_{3v}$, $\mathbf{C}_{2v}$, AND $\mathbf{C}_1$

| Point group | $\nu_1$ | $\nu_2$ | $\nu_3$ | $\nu_4$ |
|---|---|---|---|---|
| $\mathbf{T}_d$ | $A_1$(R) | $E$(R) | $F_2$(IR, R) | $F_2$(IR, R) |
| $\mathbf{C}_{3v}$ | $A_1$(IR, R) | $E$(IR, R) | $A_1$(IR, R) + $E$(IR, R) | $A_1$(IR, R) + $E$(IR, R) |
| $\mathbf{C}_{2v}$ | $A_1$(IR, R) | $A_1$(IR, R) + $A_2$(R) | $A_1$(IR, R) + $B_1$(IR, R) + $B_2$(IR, R) | $A_1$(IR, R) + $B_1$(IR, R) + $B_2$(IR, R) |
| $\mathbf{C}_1$ | $A$(IR, R) | 2 $A$(IR, R) | 3 $A$(IR, R) | 3 $A$(IR, R) |

Schmidt and Müller[488] reviewed the vibrational spectra of transition metal chalcogen compounds. Basile et al.[458] carried out normal coordinate analyses of more than 60 compounds of these types.

### (2) Tetrahedral $ZXY_3$, $Z_2XY_2$, and $ZWXY_2$ Molecules

If one of the Y atoms of an $XY_4$ molecule is replaced by a Z atom, the symmetry of the molecule is lowered to $\mathbf{C}_{3v}$. If two Y atoms are replaced, the symmetry becomes $\mathbf{C}_{2v}$. In $ZWXY_2$ and ZWXYU types (X: central atom), the symmetry is further lowered to $\mathbf{C}_1$. As a result, the selection rules are changed as shown in Table II-6*f*. The number of infrared active vibrations is six for $ZXY_3$ and eight for $Z_2XY_2$. Table II-6*g* lists the vibrational frequencies of $ZXY_3$ type molecules. The SO stretching frequency of the $[OSF_3]^+$ ion in $[OSF_3]SbF_6$ (1536 $cm^{-1}$) is the highest that has been observed, and corresponds to a force constant of 14.7 mdyn/Å.[501]

Vibrational spectra have been reported for a number of mixed halogeno complexes. Some recent references are as follows: $[AlCl_nBr_{4-n}]^-$ (521), $SiF_nCl_{4-n}$ (522), $SiCl_nBr_{4-n}$ (523), and $[FeCl_nBr_{4-n}]^-$ (524). It is interesting to note that the SiFClBrI molecule exhibits the SiF, SiCl, SiBr, and SiI stretching bands at 910, 587, 486, and 333 $cm^{-1}$, respectively.[525] The vibrational spectrum of $OClF_3$ suggests a trigonal-bipyramidal structure in which the oxygen, the fluorine, and an electron pair occupy three equatorial positions.[526]

Table II-6*h* lists the vibrational frequencies of tetrahedral $Z_2XY_2$ molecules. The vibrational spectrum of $O_2XeF_2$ can be interpreted on the basis of a trigonal-bipyramidal structure in which two F atoms are axial and two O atoms and a pair of electrons are equatorial.[541] The structure of $[O_2ClF_2]^-$ [542] is similar to that of $O_2XeF_2$, but that of $[O_2ClF_2]^+$,[543] is pseudotetrahedral. The gas-phase Raman spectrum of $Cl_2TeBr_2$ at 310°C

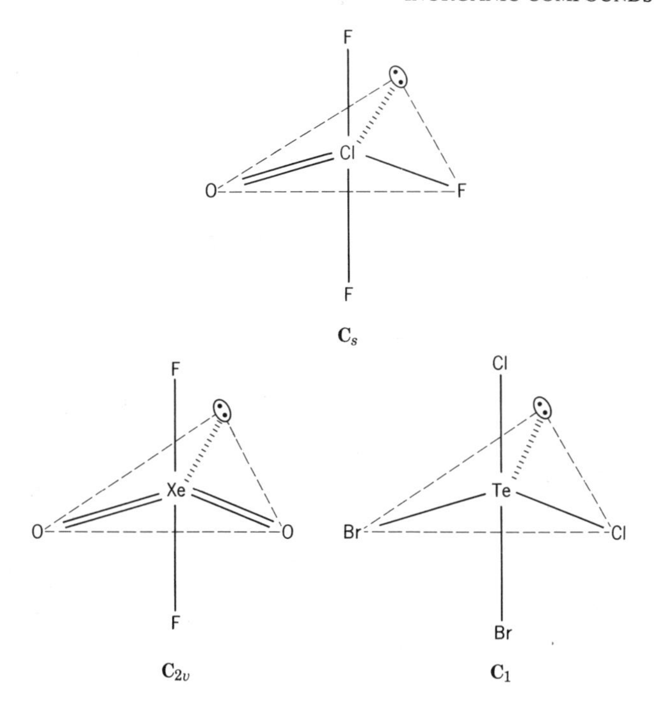


is indicative of the $\mathbf{C}_1$ symmetry shown in the above diagram.[544] Table II-6*i* lists the vibrational frequencies of tetrahedral $ZWXY_2$ molecules. Other references are as follows: $SFPCl_2$ (551), $FOPCl_2$ (552), $FOPBr_2$ (553), $ClOPBr_2$ and $BrOPCl_2$ (496), $FBrSO_2$ (554), and $FClCrO_2$ (555).

**(3) Square-Planar $XY_4$ Molecules ($D_{4h}$)**

Figure II-12 shows the seven normal modes of vibration of square-planar $XY_4$ molecules. Vibrations $\nu_3$, $\nu_6$, and $\nu_7$ are infrared active, whereas $\nu_1$, $\nu_2$, and $\nu_4$ are Raman active. Table II-6*j* lists the vibrational frequencies of some ions belonging to this group; $XeF_4$ (Sec. I-10) is an unusual example of a neutral molecule which takes a square-planar structure. Bosworth and Clark[561] measured the relative intensities of Raman active fundamentals of some of these ions and calculated their bond polarizability derivatives and bond anisotropies. They[562] also measured the resonance Raman spectra of several $[AuBr_4]^-$ salts in the solid state, and observed progressions such as $n\nu_1$ ($n = 1$–$9$) and $\nu_2 + n\nu_1$ ($n = 1$–$5$). For normal coordinate analyses of square-planar $XY_4$ molecules, see Refs. 557 and 563.

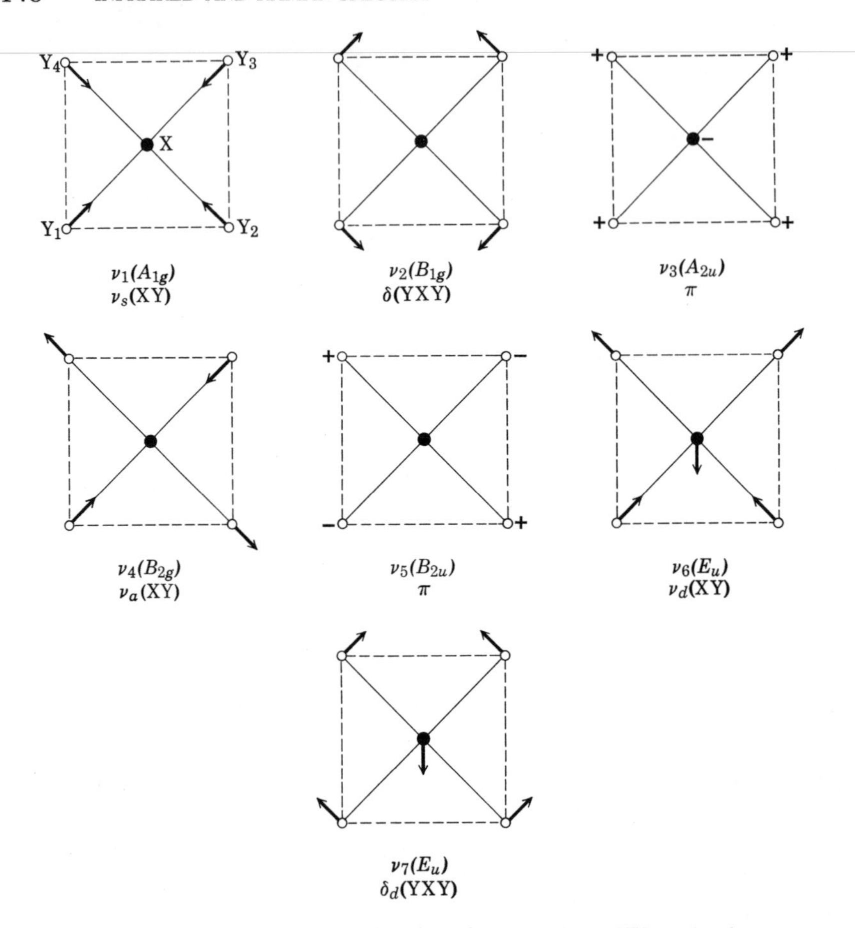


**Fig. II-12.** Normal modes of vibration of square-planar $XY_4$ molecules.

## II-7. TRIGONAL-BIPYRAMIDAL AND TETRAGONAL-PYRAMIDAL $XY_5$ AND RELATED MOLECULES

An $XY_5$ molecule may be a trigonal bipyramid ($\mathbf{D}_{3h}$) or a tetragonal pyramid ($\mathbf{C}_{4v}$). If it is trigonal-bipyramidal, only two stretching vibrations ($A_2''$ and $E'$) are infrared active. If it is tetragonal-pyramidal, three stretching vibrations (two $A_1$ and $E$) are infrared active. As discussed in Sec. I-10, however, it is not always possible to make clear-cut distinctions of these structures based on selection rules since practical difficulties arise in counting the number of fundamental vibrations in infrared and Raman spectra.

Table II-6*j*. Vibrational Frequencies of Square-Planar $XY_4$ Molecules ($cm^{-1}$)[a]

| $XY_4$ | $\nu_1(A_{1g})$ $\nu_s(XY)$ | $\nu_2(B_{1g})$ $\delta(XY_2)$ | $\nu_3(A_{2u})$ $\pi$ | $\nu_4(B_{2g})$ $\nu_a(XY)$ | $\nu_6(E_u)$ $\nu_d(XY)$ | $\nu_7(E_u)$ $\delta_d(XY_2)$ | Refs. |
|---|---|---|---|---|---|---|---|
| $[ClF_4]^-$ | 505 | 288 | 425 | 417 | 680–500 | — | 556 |
| $[BrF_4]^-$ | 523 | 246 | 317 | 449 | 580–410 | (194) | 556 |
| $[ICl_4]^-$ | 288 | 128 | — | 261 | 266 | — | 557 |
| $XeF_4$ | 554.3 | 218 | 291 | 524 | 586 | (161) | 558 |
| $[AuCl_4]^-$ | 347 | 171 | — | 324 | 350 | 179 | 559 |
| $[AuBr_4]^-$ | 212 | 102 | — | 196 | 252[b] | ~110[b] | 559 |
| $[AuI_4]^-$ | 148 | 75 | — | 110 | 192 | 113 | 559 |
| $[PdCl_4]^{2-}$ | 303 | 164 | 150 | 275 | 321 | 161 | 560 |
| $[PdBr_4]^{2-}$ | 188 | 102 | 114 | 172 | 243 | 104 | 560, 559 |
| $[PtCl_4]^{2-}$ | 330 | 171 | 147 | 312 | 313 | 165 | 560 |
| $[PtBr_4]^{2-}$ | 208 | 106 | 105 | 194 | 227 | 112 | 560 |
| $[PtI_4]^{2-}$ | 155 | 85 | 105 | 142 | 180 | 127 | 560, 559 |

[a] For these molecules $\nu_5$ is inactive. The designations $B_{1g}$ and $B_{2g}$ may be interchanged, depending on the definition of symmetry axes involved.
[b] From Ref. 562.

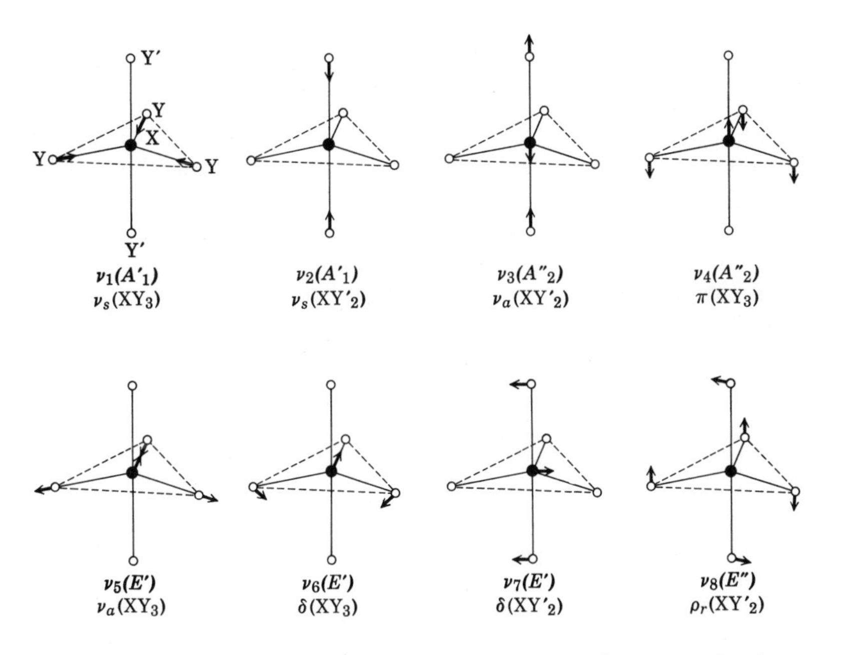


**Fig. II-13.** Normal modes of vibration of trigonal-bipyramidal $XY_5$ molecules.

### (1) Trigonal-Bipyramidal $XY_5$ Molecules ($\mathbf{D}_{3h}$)

Figure II-13 shows the eight normal vibrations of a trigonal-bipyramidal $XY_5$ molecule. Six of these eight ($A_1'$, $E'$, and $E''$) are Raman active and five ($A_2''$ and $E'$) are infrared active. Three stretching vibrations ($\nu_1$, $\nu_2$, and $\nu_5$) are allowed in the Raman, whereas two ($\nu_3$ and $\nu_5$) are allowed in the infrared. Table II-7*a* lists the observed frequencies and band assignments of trigonal-bipyramidal $XY_5$ molecules.

The majority of the compounds show the frequency order $\nu_5 > \nu_3 > \nu_1 > \nu_2$. All the data for $[XY_5]^{n-}$ were obtained either in solution or in the solid state where the ions are monomeric. Most neutral $XY_5$ molecules are dimerized or polymerized in the condensed phases. The molecules $MoCl_5$,[577] $NbCl_5$,[575,578], $TaCl_5$,[579] and $WCl_5$[580] are dimeric in the liquid and solid states ($\mathbf{D}_{2h}$), whereas $NbF_5$ and $TaF_5$ are known to be tetrameric in the crystalline state.[581] Some of these molecules are

TABLE II-7*a*. VIBRATIONAL FREQUENCIES OF TRIGONAL-BIPYRAMIDAL $XY_5$ MOLECULES ($cm^{-1}$)

| Molecule | Phase | $\nu_1$ | $\nu_2$ | $\nu_3$ | $\nu_4$ | $\nu_5$ | $\nu_6$ | $\nu_7$ | $\nu_8$ | Refs. |
|---|---|---|---|---|---|---|---|---|---|---|
| $[SiF_5]^-$ | sol'n[a] | 708 | 519 | 785 | 481 | 874 | 449 | — | — | 564 |
| $[SiCl_5]^-$ | sol'n[a] | 372 | — | 395 | 271 | 550 | 250 | — | — | 565 |
| $[GeCl_5]^-$ | solid | 348 | 236 | 310 | 200 | 395 | 200 | — | — | 566 |
| $[SnCl_5]^-$ | solid | 340 | — | 314 | 160 | 350 | 150 | 66 | 169 | 567 |
| $[SnBr_5]^-$ | solid | — | — | 208 | 106 | 256 | 111 | — | — | 568 |
| $PF_5$ | gas | 817 | 640 | 944 | 575 | 1026 | 532 | 300 | 514 | 569 |
| $PCl_5$ | sol'n[a] | 394 | 385 | 444 | 299 | 580 | 278 | 98 | 261 | 570 |
| $AsF_5$ | gas | 733 | 642 | 784 | 400 | 809 | 366 | 123 | 388 | 571 |
| $SbCl_5$ | gas | 355 | 309 | — | — | 400 | 173 | 58 | 120 | 572 |
| $[CuCl_5]^{3-}$ | solid | 260 | — | 268 | — | 170[b] | 95[c] | — | — | 573 |
| $[CdCl_5]^{3-}$ | solid | 251 | — | 236 | — | 157[b] | 98[c] | — | — | 573 |
| $[TiCl_5]^-$ | solid | — | — | 346 | 170 | 385 | 212 | (83) | — | 568 |
| $VF_5$ | gas | 719 | 608 | 784 | 331 | 810 | 282 | (200) | 350 | 574 |
| $NbCl_5$ | matrix | (349) | (293) | 396 | 126 | 444 | 159 | 99 | (139) | 575 |
| $NbBr_5$ | gas | 234 | 178 | — | — | — | 119 | 67 | 101 | 572 |
| $TaCl_5$ | gas | 406 | 324 | — | — | — | 181 | 54 | 127 | 572 |
| $TaBr_5$ | gas | 240 | 182 | — | — | — | 110 | 70 | 93 | 572 |
| $MoF_5$ | matrix | — | — | 683 | — | 713 | 261 | 112 | — | 576 |
| $MoCl_5$ | gas | 390 | 313 | — | — | 418 | 200 | 100 | 175 | 572 |

[a] Nonaqueous solution.
[b] May be assigned to $\nu_7$.
[c] May be assigned to $\nu_8$.

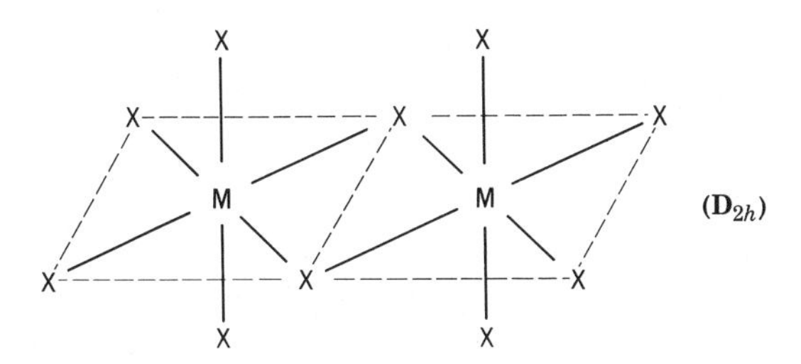


polymerized even in the gaseous phase. For example, $SbF_5$ is monomeric ($\mathbf{D}_{3h}$) at 350°C but polymeric at 140°C in the gaseous phase,[582] and $NbF_5$ and $TaF_5$ are polymeric in the gaseous phase if the temperature is below 350°C.[583] Although $PCl_5$ exists as a $\mathbf{D}_{3h}$ molecule in the gaseous and liquid states, it has an ionic structure consisting of $[PCl_4]^+[PCl_6]^-$ units in the crystalline state, as proved by Raman spectroscopy.[584] The importance of the $\nu_7$ vibration in the intramolecular conversion of pentacoordinate molecules has been discussed by Holmes.[585]

Normal coordinate analyses on trigonal-bipyramidal $XY_5$ molecules have been carried out by several investigators. Some recent studies are reported in Refs. 586–588. These calculations show that equatorial bonds are stronger than axial bonds. Vibrational spectra of mixed halogeno compounds such as $PF_nCl_{5-n}$[589] and $PF_3X_2$ (X = Cl and Br)[590] have been assigned.

### (2) Tetragonal-Pyramidal $XY_5$ and $ZXY_4$ Molecules ($C_{4v}$)

Figure II-14 shows the nine normal modes of vibration of a tetragonal-pyramidal $ZXY_4$ molecule. Only $A_1$ and $E$ vibrations are infrared active, whereas all nine vibrations belonging to the $A_1$, $B_1$, $B_2$, and $E$ species are Raman active. Table II-7*b* lists the vibrational frequencies of tetragonal-pyramidal $XY_5$ and $ZXY_4$ molecules. In the majority of $XY_5$ molecules, the axial stretching frequency ($\nu_1$) is higher than the equatorial stretching frequencies ($\nu_2$, $\nu_4$, and $\nu_7$). This is opposite to the trend found for trigonal-bipyramidal $XY_5$ molecules, discussed in the preceding section. For normal coordinate analyses on these compounds, see Refs. 594, 595, 602, and 603.

## II-8. OCTAHEDRAL MOLECULES

### (1) Octahedral $XY_6$ Molecules ($O_h$)

Figure II-15 illustrates the six normal modes of vibration of an octahedral $XY_6$ molecule. Vibrations $\nu_1$, $\nu_2$, and $\nu_5$ are Raman active,

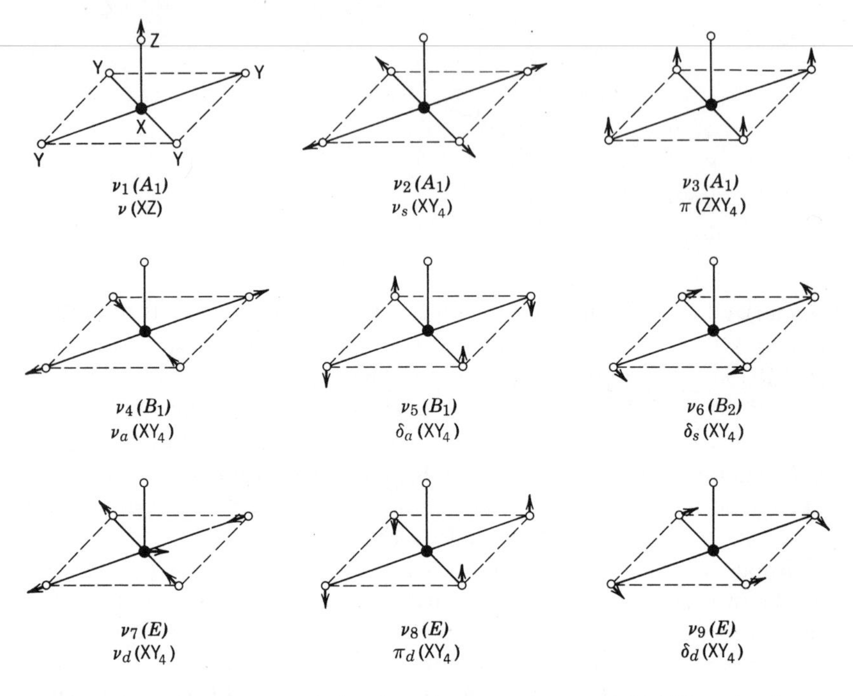


**Fig. II-14.** Normal modes of vibration of tetragonal-pyramidal $ZXY_4$ molecules.

whereas only $\nu_3$ and $\nu_4$ are infrared active. Since $\nu_6$ is inactive in both, its frequency is estimated from an analysis of combination and overtone bands.

Table II-8*a* lists the vibrational frequencies of a number of hexa-halogeno compounds. In general, the order of the stretching frequencies is $\nu_1 > \nu_3 \gg \nu_2$ or $\nu_1 < \nu_3 \gg \nu_2$, depending on the compound. The order of the bending frequencies is $\nu_4 > \nu_5 > \nu_6$ in most cases. In the same family of the periodic table, the stretching frequencies decrease as the mass of the central atom increases, for example:

| | $[AlF_6]^{3-}$ | | $[GaF_6]^{3-}$ | | $[InF_6]^{3-}$ | | $[TlF_6]^{3-}$ |
|---|---|---|---|---|---|---|---|
| $\tilde{\nu}_1$ (cm$^{-1}$) | 541 | > | 535 | > | 497 | > | 478 |
| $\tilde{\nu}_3$ (cm$^{-1}$) | 568 | > | 481 | > | 447 | > | 412 |

The trend in $\nu_1$ directly reflects the trend in the stretching force constant (and bond strength) since the central atom is not moving in this mode. In $\nu_3$, however, both X and Y atoms are moving, and the mass effect of the

TABLE II-7b. VIBRATIONAL FREQUENCIES OF TETRAGONAL-PYRAMIDAL $XY_5$ AND $ZXY_4$ MOLECULES ($cm^{-1}$)

| $XY_5$ or $ZXY_4$ | $\nu_1$ | $\nu_2$ | $\nu_3$ | $\nu_4$ | $\nu_5$ | $\nu_6$ | $\nu_7$ | $\nu_8$ | $\nu_9$ | Refs. |
|---|---|---|---|---|---|---|---|---|---|---|
| $[InCl_5]^{2-}$ | 294 | 283 | 140 | 287 | 193 | 165 | 274 | 108 | 143 | 591 |
| $[SbF_5]^{2-}$ | 557 | 427 | 278 | 388 | — | 220 | 375, 347 | 307 | 142 | 592 |
| $[SbCl_5]^-$ | 445 | 285 | 180 | 420 | — | 117 | 300 | 255 | 90 | 593 |
| $[SF_5]^-$ | 796 | 522 | 469 | (435) | 269 | 342 | 590 | (435) | 241 | 594 |
| $[SeF_5]^-$ | 666 | 515 | 332 | 460 | 236 | 282 | 480 | 399 | 202 | 594 |
| $[TeF_5]^-$ | 624 | 517 | 291 | 579 | — | 243 | 488 | 345 | — | 592 |
| $ClF_5$ | 709 | 538 | 480 | 480 | (346) | 375 | 732 | — | 296 | 595, 596 |
| $BrF_5$ | 682 | 570 | 365 | 535 | (281) | 312 | 644 | 414 | 237 | 595, 597 |
| $IF_5$ | 698 | 593 | 315 | 575 | (257) | 273 | 640 | 374 | 189 | 595, 592 |
| $NbF_5$ | 740 | 686 | 513 | — | — | — | 729 | 261 | 103 | 598 |
| $[OTeF_4]^{2-}$ | 837 | 461 | — | 390 | — | 190 | 335 | — | 129 | 599 |
| $[OClF_4]^-$ | 1216 | 462 | 339 | 350 | — | 283 | 600, 550 | 415, 394 | 213 | 600 |
| $OXeF_4$ | 920 | 567 | 285 | 527 | (230) | 233 | 608 | 365 | 161 | 595 |
| $OMoF_4$ | 1048 | 714 | 264 | — | — | — | 720 | 294 | 236 | 601, 601a |
| $[OMoCl_4]^-$ | 1008 | 354 | 184 | 327 | 158 | 167 | 364 | 240 | 114 | 602 |
| $OWF_4$ | 1055 | 733 | 248 | 631 | 328 | 291 | 698 | 298 | 236 | 601, 601a |
| $[NRuCl_4]^-$ | 1092 | 346 | 197 | 304 | 154 | 172 | 378 | 267 | 163 | 602 |
| $[NRuBr_4]^-$ | 1088 | 224 | 156 | 187 | 103 | 128 | 304 | 211 | 98 | 602 |
| $[NOsCl_4]^-$ | 1123 | 358 | 184 | 352 | 149 | 174 | 365 | 271 | 132 | 602 |
| $[NOsBr_4]^-$ | 1119 | 162 | 122 | 156 | 110 | 120 | 220 | 273 | 98 | 602 |

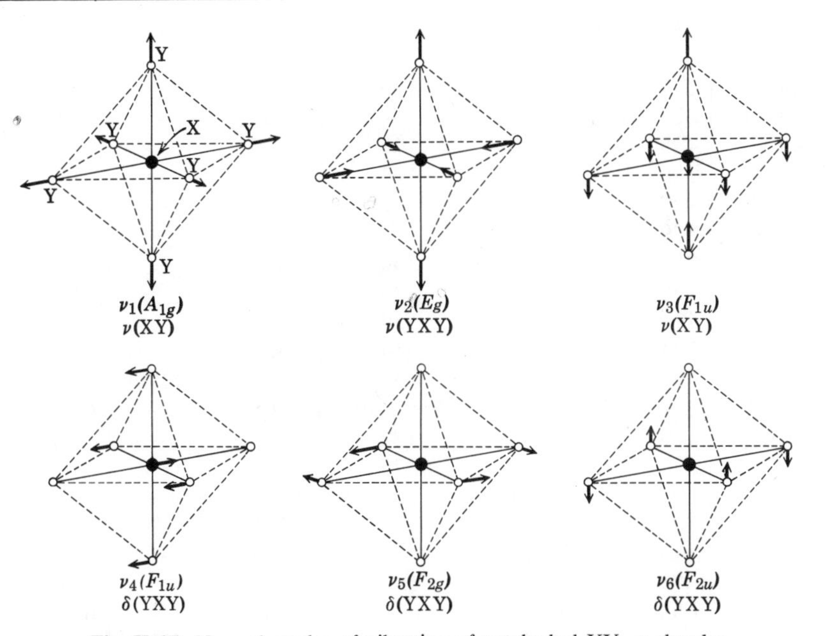


**Fig. II-15.** Normal modes of vibration of octahedral $XY_6$ molecules.

Table II-8*a*. Vibrational Frequencies of Octahedral $XY_6$ Molecules ($cm^{-1}$)

| Molecule | $\nu_1$ | $\nu_2$ | $\nu_3$ | $\nu_4$ | $\nu_5$ | $\nu_6$[a] | Refs. |
|---|---|---|---|---|---|---|---|
| $[AlF_6]^{3-}$ | 541 | — | 568 | 387 | 322 | — | 604 |
| $[GaF_6]^{3-}$ | 535 | — | 481 | 298 | 281 | — | 604 |
| $[InF_6]^{3-}$ | 497 | — | 447 | 226 | 229 | — | 604 |
| $[InCl_6]^{3-}$ | 277 | 193 | 250 | 157 | (149) | — | 605 |
| $[TlF_6]^{3-}$ | 478 | — | 412 | 202 | 209 | — | 604 |
| $[TlCl_6]^{3-}$ | 280 | 262 | 294 | 222, 246 | 155 | (136) | 606 |
| $[TlBr_6]^{3-}$ | 161 | 153 | 190, 195 | 134, 156 | 95 | (80) | 606 |
| $[SiF_6]^{2-}$ | 663 | 477 | 741 | 483 | 408 | — | 607 |
| $[GeF_6]^{2-}$ | 624 | 471 | 603 | 339, 359 | 335 | — | 607 |
| $[GeCl_6]^{2-}$ | 318 | 213 | 310 | 213 | 191 | — | 608 |
| $[SnF_6]^{2-}$ | 592 | 477 | 559 | 300 | 252 | — | 607 |
| $[SnCl_6]^{2-}$ | 311 | 229 | 303 | 166 | 158 | — | 609 |
| $[SnBr_6]^{2-}$ | 190 | 144 | 224 | 118 | 109 | — | 610 |
| $[SnI_6]^{2-}$ | 122 | 93 | 161 | 84 | 78 | — | 611 |
| $[PbCl_6]^{2-}$ | 281 | 209 | 262 | 142 | 139 | — | 609 |
| $[PF_6]^-$ | 745 | 572 | 840 | 555 | 465 | (402) | 612 |
| $[PCl_6]^-$ | 360 | 283 | 444 | 285 | 238 | — | 608 |
| $[AsF_6]^-$ | 685 | 576 | 699 | 392 | 372 | — | 607 |
| $[AsCl_6]^-$ | 337 | 289 | 333 | 220 | 202 | — | 608 |
| $[SbF_6]^-$ | 668 | 558 | 669 | 350 | 294 | — | 607 |
| $[SbCl_6]^-$ | 330 | 282 | 353 | 180 | 175 | — | 613 |
| $[SbCl_6]^{3-}$ | 327 | 274 | — | — | 137 | — | 662 |
| $[SbBr_6]^-$ | 195 | 164 | 241 | 120 | 120 | — | 614 |
| $[SbBr_6]^{3-}$ | 180 | 153 | 180 | 107 | 73 | — | 615, 616 |
| $[SbI_6]^{3-}$ | 107 | 96 | 108 | 82 | 54 | — | 615, 616 |
| $[BiF_6]^-$ | 579 | 526 | — | — | 231 or 241 | — | 617 |
| $[BiCl_6]^{3-}$ | 259 | 215 | 172 | 130 | 115 | — | 615, 616 |
| $[BiBr_6]^{3-}$ | 156 | 130 | 128 | 75 | 62 | — | 615, 616 |
| $[BiI_6]^{3-}$ | 114 | 103 | 96 | (59) | 54 | — | 615, 616 |
| $SF_6$ | 774 | 642 | 939 | 614 | 523 | (347) | 618, 619 |
| $SeF_6$ | 708 | 658 | 780 | 437 | 403 | (264) | 618, 619 |
| $[SeCl_6]^{2-}$ | 299 | 255 | 280 | 160–140 | 165 | — | 605 |
| $[SeBr_6]^{2-}$ | 179 | 157 | 225 | 122 | 105 | — | 620 |
| $TeF_6$ | 698 | 672 | 752 | 325 | 312 | (197) | 618, 619 |
| $[TeCl_6]^{2-}$ | 301 | 253 | 243 | 139 | 150 | — | 621 |
| $[TeCl_6]^{3-}$ | 264 | 192 | 230 | 146 | (135) | — | 605 |
| $[TeBr_6]^{2-}$ | 174 | 153 | 198 | — | 75 | — | 621 |
| $[ClF_6]^+$ | 679 | 630 | 890 | 582 | 513 | — | 622 |
| $[BrF_6]^+$ | 658 | 660 | — | — | 405 | — | 623 |
| $[BrF_6]^-$ | 568 | 454 | 400 | 204, 184 | 250 | (138) | 624 |

TABLE II-8*a* (*Continued*)

| Molecule | $\nu_1$ | $\nu_2$ | $\nu_3$ | $\nu_4$ | $\nu_5$ | $\nu_6$[a] | Refs. |
|---|---|---|---|---|---|---|---|
| $[AuF_6]^-$ | 595 | 520 | — | — | 224 | — | 625 |
| $[ScF_6]^{3-}$ | 504 | 370 | — | — | 240 | — | 626 |
| $[CeCl_6]^{2-}$ | 295 | 265 | 268 | 117 | 120 | (86) | 627 |
| $[TiF_6]^{2-}$ | — | — | 585 | 278, 312 | — | — | 628 |
| $[TiCl_6]^{2-}$ | 320 | 271 | 316 | 183 | 173 | — | 629 |
| $[TiCl_6]^{3-}$ | 322 | 278 | 304, 290 | — | 175 | — | 630 |
| $[TiBr_6]^{2-}$ | 192 | — | 244 | 119 | 115 | — | 629 |
| $[ZrF_6]^{2-}$ | 581 | — | — | — | 228 | — | 631 |
| $[ZrCl_6]^{2-}$ | 327 | 237 | 290 | 150 | 153 | — | 632 |
| $[ZrBr_6]^{2-}$ | 194 | 144 | 223 | 106 | 99 | — | 633 |
| $[HfF_6]^{2-}$ | 613 | — | — | — | 275 | — | 628 |
| $[HfCl_6]^{2-}$ | 326 | 257 | 275 | 145 | 156 | (80) | 609 |
| $[HfBr_6]^{2-}$ | 197 | 142 | 189 | 102 | 101 | — | 633 |
| $[VF_6]^-$ | 676 | 538 | 646 | 300 | 330 | — | 634 |
| $[VF_6]^{2-}$ | 584 | — | 578 | 273 | — | — | 634 |
| $[VF_6]^{3-}$ | 533 | — | 511 | 292 | — | — | 634 |
| $[VCl_6]^{2-}$ | — | — | 355, 305 | — | — | — | 635 |
| $[NbF_6]^-$ | 683 | 562 | 602 | 244 | 280 | — | 636 |
| $[NbCl_6]^-$ | 368 | 288 | 333 | 162 | 183 | — | 609 |
| $[NbCl_6]^{2-}$ | — | — | 314 | 165 | — | — | 637 |
| $[NbBr_6]^-$ | 219 | 179 | 239 | 112 | — | — | 638 |
| $[NbBr_6]^{2-}$ | — | — | 236 | 112 | — | — | 637 |
| $[NbI_6]^-$ | — | — | 180 | 70, 66 | — | — | 639 |
| $[TaF_6]^-$ | 692 | 581 | 560 | 240 | 272 | (192) | 640 |
| $[TaCl_6]^-$ | 378 | 298 | 330 | 158 | 180 | — | 609 |
| $[TaCl_6]^{2-}$ | — | — | 297 | 160 | — | — | 609 |
| $[TaBr_6]^-$ | 230 | 179 | 213 | 106 | 114 | — | 638 |
| $[TaBr_6]^{2-}$ | — | — | 217 | 109 | — | — | 637 |
| $[TaI_6]^-$ | — | — | 160 | 80 | — | — | 639 |
| $CrF_6$ | (720) | (650) | 790 | (266) | (309) | (110) | 641 |
| $[CrCl_6]^{3-}$ | 286 | 237 | 315 | 199 | 162 | 182 | 642 |
| $MoF_6$ | 742 | 651 | 741 | 264 | 318 | (116) | 618, 619 |
| $[MoCl_6]^{2-}$ | — | — | 300 | — | — | — | 635 |
| $WF_6$ | 770 | 676 | 711 | 258 | 321 | (127) | 618, 619 |
| $WCl_6$ | 437 | 331 | 373 | 160 | 182 | — | 609 |
| $[WCl_6]^-$ | 378 | 318 | 330 | 158 | — | — | 609 |
| $[WCl_6]^{2-}$ | — | — | 303 | 167 | — | — | 609 |
| $[MnF_6]^{2-}$ | 592 | 508 | 620 | 335 | 308 | — | 643 |
| $TeF_6$ | 713 | (639) | 748 | 265 | (297) | (145) | 619 |
| $ReF_6$ | 754 | (671) | 715 | 257 | (295) | (147) | 619 |

TABLE II-8*a* (*Continued*)

| Molecule | $\nu_1$ | $\nu_2$ | $\nu_3$ | $\nu_4$ | $\nu_5$ | $\nu_6$[a] | Refs. |
|---|---|---|---|---|---|---|---|
| $[ReCl_6]^-$ | — | — | 318 | 161 | — | — | 644 |
| $[ReCl_6]^{2-}$ | 346 | (275) | 313 | 172 | 159 | — | 609, 645 |
| $[ReBr_6]^{2-}$ | 213 | (174) | 217 | 118 | 104 | — | 646 |
| $[NiF_6]^{2-}$ | 562 | 520 | 658 | 345 | 310 | (220) | 647 |
| $[PdCl_6]^{2-}$ | 318 | 289 | 346 | 200 | 178 | — | 648 |
| $[PdBr_6]^{2-}$ | 198 | 176 | 253 | 130 | 100 | — | 649 |
| $PtF_6$ | 656 | (601) | 705 | 273 | (242) | (211) | 619 |
| $[PtF_6]^{2-}$ | 611 | 576 | 571 | 281 | 210 | (143) | 650 |
| $[PtCl_6]^{2-}$ | 348 | 318 | 342 | 183 | 171 | (88) | 610 |
| $[PtBr_6]^{2-}$ | 213 | 190 | 243 | 146 | 137 | — | 648 |
| $[PtI_6]^{2-}$ | — | — | 186 | 46 | — | — | 651 |
| $[FeF_6]^{3-}$ | 538 | 374 | — | — | 253 | — | 626 |
| $RuF_6$ | (675) | (624) | 735 | 275 | (283) | (186) | 619 |
| $[RuCl_6]^{2-}$ | — | — | 346 | 188 | — | — | 652 |
| $OsF_6$ | 731 | (668) | 720 | 268 | (276) | (205) | 619 |
| $[OsCl_6]^{2-}$ | 352 | — | 304 | 174 | 177 | — | 645, 653 |
| $[OsBr_6]^{2-}$ | 218 | 162 | 211 | — | — | — | 645, 653 |
| $RhF_6$ | (634) | (595) | 724 | 283 | (269) | (192) | 619 |
| $[RhCl_6]^{2-}$ | — | — | 329 | 187 | — | — | 654 |
| $IrF_6$ | 702 | 645 | 719 | 276 | 267 | (206) | 619 |
| $[IrCl_6]^{2-}$ | 352 | (225) | 333 | 184 | 190 | — | 609 |
| $[IrCl_6]^{3-}$ | — | — | 296 | 200 | — | — | 651 |
| $[IrBr_6]^{2-}$ | — | — | 235 | 82 | — | — | 654 |
| $[ThCl_6]^{2-}$ | 294 | 255 | 259 | — | 114 | — | 655 |
| $UF_6$ | 667 | 533 | 624 | 186 | 202 | (142) | 618, 619, 619a |
| $[UF_6]^-$ | — | — | 525 | 173 | — | — | 656 |
| $[UCl_6]^-$ | 343 | 273 | 310 | 122 | 136 | — | 656, 657 |
| $[UCl_6]^{2-}$ | 299 | 237 | 262 | 114 | 121 | (80) | 655 |
| $[UBr_6]^-$ | — | — | 214 | 87 | — | — | 656 |
| $NpF_6$ | 654 | 535 | 624 | 199 | 208 | (164) | 619 |
| $[NpCl_6]^{2-}$ | 310 | — | 265 | 117 | 128 | — | 658 |
| $PuF_6$ | (628) | (523) | 616 | 206 | (211) | (173) | 619 |

[a] The value of $\nu_6$ can also be estimated by the empirical relation $\nu_6 = \nu_5/\sqrt{2}$ (Ref. 659).

X atom cannot be ignored completely. Across the periodic table, the stretching frequencies increase as the oxidation state of the central atom becomes higher. Thus we have:

| | $[AlF_6]^{3-}$ | | $[SiF_6]^{2-}$ | | $[PF_6]^-$ | | $SF_6$ |
|---|---|---|---|---|---|---|---|
| $\tilde{\nu}_1$ (cm$^{-1}$) | 541 | < | 663 | < | 745 | < | 774 |
| $\tilde{\nu}_3$ (cm$^{-1}$) | 568 | < | 741 | < | 840 | < | 939 |

The effect of lowering the oxidation state is clearly seen in a series such as $[VF_6]^{n-}$ ($n = 1$, 2, and 3) and $[WCl_6]^{n-}$ ($n = 0$, 1, and 2):

| | $[VF_6]^-$ | | $[VF_6]^{2-}$ | | $[VF_6]^{3-}$ |
|---|---|---|---|---|---|
| $\tilde{\nu}_1$ (cm$^{-1}$) | 676 | > | 584 | > | 533 |
| $\tilde{\nu}_3$ (cm$^{-1}$) | 646 | > | 578 | > | 511 |

As in many other cases, the higher the oxidation state, the higher is the frequency. In some cases, however, the effect of changing the oxidation state is not obvious (e.g., $TaBr_6^-$ and $TaBr_6^{2-}$). The bending frequencies do not exhibit clear-cut trends. The effect of changing the halogen is seen in a number of series, for example:

| | $[SnF_6]^{2-}$ | | $[SnCl_6]^{2-}$ | | $[SnBr_6]^{2-}$ | | $[SnI_6]^{2-}$ |
|---|---|---|---|---|---|---|---|
| $\tilde{\nu}_1$ (cm$^{-1}$) | 592 | > | 311 | > | 190 | > | 127 |
| $\tilde{\nu}_3$ (cm$^{-1}$) | 559 | > | 303 | > | 224 | > | 161 |

The stretching force constants also follow the same order. The ratios $\nu$(MBr)/$\nu$(MCl) and $\nu$(MI)/$\nu$(MCl) are about 0.61 and 0.42, respectively, for $\nu_1$, and about 0.76 and 0.62, respectively, for $\nu_3$.

In the $[MCl_6]^{3-}$ series, $\nu_3$ and $\nu_4$ change as follows:

| | $Cr^{3+}(d^3)$ | $Mn^{3+}(d^4)$ | $Fe^{3+}(d^5)$ | $In^{3+}$ |
|---|---|---|---|---|
| $\tilde{\nu}_3$ (cm$^{-1}$) | 315 | 342 | 248 | 248 |
| $\tilde{\nu}_4$ (cm$^{-1}$) | 200 | 183 | 184 | 161 |

All these metals are in the high spin states. For $Fe^{3+}(t_{2g}^3 e_g^2)$, occupation of the antibonding orbitals lowers $\nu_3$ drastically in relation to the $Cr^{3+}$ complex; its $\nu_3$ is comparable to that of the $In^{3+}$ complex, whose $\nu_3$ is lowered because of the increased mass of the metal. On the other hand, the $\nu_3$ of $[MnCl_6]^{3-}$ is higher than that of $[CrCl_6]^{3-}$ because the static Jahn–Teller effect of the $Mn^{3+}$ ion causes a tetragonal distortion.[627]

The Raman intensity of an $XY_6$ molecule normally follows the order $I(\nu_1) > I(\nu_2) > I(\nu_5)$. Adams and Downs[660] noted that $I(\nu_2)/I(\nu_1)$ is 0.5–1 for $[TeCl_6]^{2-}$ and $[TeBr_6]^{2-}$, although it normally ranges from 0.05 to 0.1. Furthermore, they perceived $\nu_3$, which is not allowed in Raman spectra. From these and other items of evidence, they proposed that the $\mathbf{O}_h$ selection rule breaks down in $[TeX_6]^{2-}$ because less symmetrical electronic excited states perturb the $\mathbf{O}_h$ ground state. They also noted the distortion of $[SbX_6]^{3-}$ ions to $\mathbf{C}_{3v}$ symmetry from their Raman spectra in solution. Woodward and Creighton[661] noted that $I(\nu_2) > I(\nu_1)$ holds in the aqueous Raman spectra of $Na_2PtX_6$ (X = Cl and Br) and $Na_2PdCl_6$, and attributed this unusual trend to the presence of six nonbonding *d*-electrons in the valence shell. Bosworth and Clark[662] carried out a Raman study on 17 $[XY_6]^{n-}$ type metal halides, and discussed the results in terms of the preresonance Raman effect. Infrared spectra show the distortion of $\mathbf{O}_h$ symmetry of $[SbCl_6]^{3-}$,[663] $[TiX_6]^{3-}$ (X = Cl and Br),[664] and $[MF_6]^{2-}$ (M = Tc and Re)[665] in the solid state.

Weinstock et al.[666] noted that the combination bands $(\nu_1+\nu_3)$ and $(\nu_2+\nu_3)$ appear with similar frequencies, intensities, and shapes in the infrared spectra of $MoF_6(d^\circ)$ and $RhF_6(d^3)$. As is shown in Fig. II-16, however, $(\nu_2+\nu_3)$ was very broad and weak in $TcF_6(d^1)$, $ReF_6(d^1)$, $RuF_6(d^2)$, and $OsF_6(d^2)$. This anomaly was attributed to a dynamic Jahn–Teller effect. The static Jahn–Teller effect does not seem to operate in these compounds since no splittings of the triply degenerate fundamentals were observed. Perhaps the most fascinating $XY_6$ type molecule is $XeF_6$. In their earlier work, Claassen et al.[667] suggested the distortion of $XeF_6$ from $\mathbf{O}_h$ symmetry since they observed two stretching bands in infrared and three stretching bands in Raman spectra. It was not possible, however, to determine the precise structure of $XeF_6$ until they[668] carried out a detailed infrared, Raman, and electronic spectral study of $XeF_6$ vapor as a function of temperature. They were then able to show that $XeF_6$ consists of the three electronic isomers shown in Fig. II-17, and to explain subtle differences in spectra at different temperatures as a shift of equilibrium among these three isomers.

The $^{35}Cl$ NQR spectra provide information about the $\sigma$ and $\pi$ contributions to the covalent M–Cl bonding in $[MCl_6]^{n-}$ type ions, and these can be correlated with the force constants obtained from infrared and Raman studies.[609] Both infrared and NQR spectra suggest low site symmetry of the $[MCl_6]^{3-}$ ion in $K_3[MCl_6]\cdot H_2O$ (M = Ir and Rh) crystals.[669]

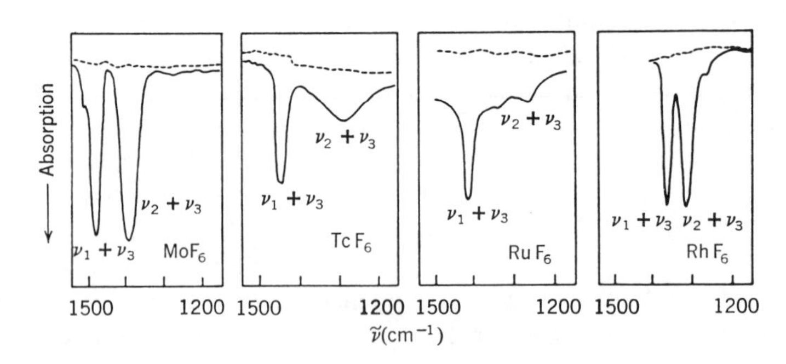


**Fig. II-16.** Band profiles for $(\nu_1+\nu_3)$ and $(\nu_2+\nu_3)$ for the 4*d* transition series hexafluorides.[666]

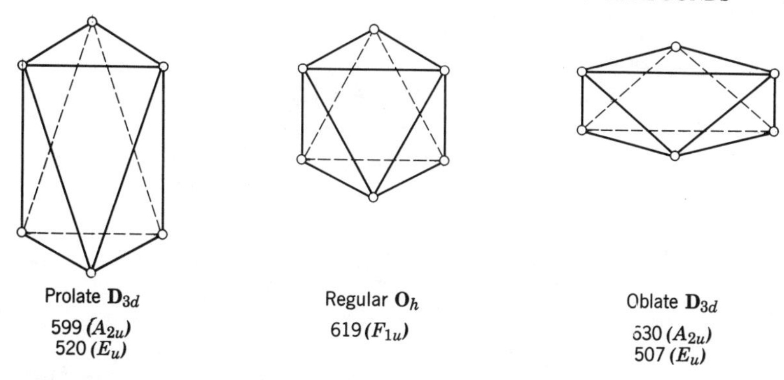


**Fig. II-17.** Structures of three isomers of $XeF_6$ and their IR active XeF stretching frequencies ($cm^{-1}$). The Xe atom at the center is not shown.

Normal coordinate analyses on octahedral hexahalogeno compounds have been made by a number of investigators. Kim et al.[670] calculated the force constants of 15 metal fluorides using the UBF and OVF fields (see Sec. I-13), and found that the latter is better than the former. Labonville et al.[671] calculated the force constants of 62 metal halides by using the UBF, OVF, GVF, modified UBF, and modified OVF fields, and found that the modified OVF field gives the best overall agreement with the observed frequencies. They also discussed the dependence of force constants on the mass of the halogen, the oxidation state of the metal, the number of nonbonding electrons in the valence shell, and the crystal field stabilization energy.

Table II-8*b* lists the vibrational frequencies of $MO_6$ type ions. Hauck and Fadini calculated the force constants of these ions.[672,673]

TABLE II-8*b*. VIBRATIONAL FREQUENCIES OF OCTAHEDRAL $MO_6$ MOLECULES ($CM^{-1}$)

| Compound | $\nu_1$ | $\nu_2$ | $\nu_3$ | $\nu_4$ | $\nu_5$ | Refs. |
|---|---|---|---|---|---|---|
| $Li_6[TeO_6]$ | 700 | 540 | 640 | 470 | 355 | 672 |
| $Li_6[WO_6]$ | 740 | 450 | 620 | 425 | 360 | 672 |
| $\alpha$-$Li_6[ReO_6]$ | 680 | 505 | 620 | 425 | 360 | 672 |
| $Ca_4[PtO_6]$ | — | 530 | 575 | 425 | 345 | 673 |
| $\alpha$-$Li_6[TeO_6]$ | 700 | 540 | 640 | 470 | 355 | 673 |
| $Ca_5[IO_6]_2$ | 771 | 490, 538 | 765, 695 | 451, 435 | — | 673 |

TABLE II-8c. VIBRATIONAL FREQUENCIES OF OCTAHEDRAL $XY_5Z$ AND $XY_4WZ$ MOLECULES ($cm^{-1}$)

| $XY_5Z$ or $XY_4WZ$ | $\nu_1(A_1)$ $\nu(XZ)$ | $\nu_2(A_1)$ $\nu(XW)^a$ | $\nu_3(A_1)$ $\nu(XY_4)$ | $\nu_4(A_1)$ $\pi(XY_4)$ | $\nu_5(B_1)$ $\nu(XY_4)$ | $\nu_6(B_1)$ $\pi(XY_4)$ | $\nu_7(B_2)$ $\delta(XY_4)$ | $\nu_8(E)$ $\nu(XY_4)$ | $\nu_9(E)$ $\rho_w(XW)^a$ | $\nu_{10}(E)$ $\rho_w(XZ)$ | $\nu_{11}(E)$ $\delta(XY_4)$ | Refs. |
|---|---|---|---|---|---|---|---|---|---|---|---|---|
| $[SbCl_5Br]^-$ | 219 | 308 | 334 | 151 | 287 | — | — | 344 | — | — | — | 674, 674a |
| $[SbBr_5Cl]^-$ | 305 | 206 | 192 | — | 186 | — | — | 239 | — | — | — | 674 |
| $SF_5Cl$ | 402 | 855 | 707 | 602 | 625 | 271 | 505 | 909 | 597 | 399 | 441 | 675 |
| $[SF_5O]^-$ | 1153 | 722 | 697 | 506 | 541 | 472 | 452 | 780 | 607 | 530 | 325 | 676 |
| $SeF_5Cl$ | 729 | 654 | 440 | 384 | 636 | — | 380 | 745 | 421 | 334 | 213 | 677 |
| $[SeF_5O]^-$ | 919 | 559 | 649 | — | 556 | — | — | 639 | — | — | — | 678 |
| $IF_5O$ | 927 | 680 | 640 | 363 | 647 | 307 | 330 | 710 | 372 | 343 | 205 | 679 |
| $[TiF_5O]^{3-}$ | 920 | 379 | 520 | 290 | — | — | — | 520 | 138 | 335 | 235 | 680 |
| $[VF_5O]^{3-}$ | 943 | 383 | 525 | 317 | — | — | — | 525 | 139 | 342 | 237 | 680 |
| $[NbCl_5Br]^-$ | 210 | 310 | 365 | 181 | 285 | 120 | 134 | 352 | 161 | 153 | 75 | 681 |
| $[TaCl_5Br]^-$ | 204 | 318 | 368 | 183 | 300 | (120) | 168 | 325 | 151 | 143 | 73 | 681 |
| $[TaBr_5Cl]^-$ | 323 | 231 | 187 | 110 | 180 | (73) | 96 | 214 | 123 | 144 | 76 | 681 |
| $[MoCl_5O]^{2-}$ | 998 | 318 | 331 | 168 | 336 | 159 | 164 | 321 | 233 | 137 | 147 | 682 |
| $WF_5Cl$ | 407 | 744 | 703 | 257 | 644 | 182 | 377 | 661 | 290 | 227 | 307 | 683 |
| $ReF_5O$ | 990 | 739 | 643 | 309 | 652 | 234 | 334 | 713 | 260 | 365 | 125 | 679 |
| $[RuCl_5N]^-$ | 1048 | 284 | 318 | 192 | 307 | 168 | 184 | 337 | 233 | 154 | 174 | 682 |
| $[RuBr_5N]^-$ | 1046 | 201 | 207 | 156 | 181 | 136 | 147 | 204 | 257 | 110 | 144 | 682 |
| $OsF_5O$ | 963 | 716 | 644 | 281 | 644 | 210 | 332 | 701 | 263 | 367 | 164 | 679 |
| $[OsCl_5N]^{2-}$ | 1084 | 324 | 348 | 189 | 334 | 169 | 181 | 336 | 264 | 146 | 172 | 682 |
| $[OsBr_5N]^{2-}$ | 1085 | 192 | 198 | 156 | 172 | 136 | 149 | 234 | 217 | 115 | 144 | 682 |
| $[MoCl_4OBr]^{2-}$ | 235 | 964 | 301 | 149 | 288 | — | — | 320 | 229 | 92 | 162 | 684 |
| $[WCl_4OBr]^{2-}$ | 233 | 960 | 326 | 149 | — | — | 187 | 298 | 230 | 92 | 162 | 684 |

[a] For $XY_5Z$, W is regarded as Y *trans* to Z.

**(2) Octahedral $XY_nZ_{6-n}$ Molecules**

The $XY_5Z$ molecule belongs to the $\mathbf{C}_{4v}$ point group, and its 11 normal vibrations are classified into 4 $A_1$, 2 $B_1$, $B_2$, and 4 $E$ modes, of which only $A_1$ and $E$ are infrared active; all are Raman active. Table II-8*c* lists the observed frequencies of $XY_5Z$ type molecules. Less complete assignments are reported for $[MX_5O]^{2-}$, where M is Nb, Mo, or W, and X is Cl or Br.[685] Table II-8*c* also includes $XY_4WZ$ type compounds, in which W and Z are in the *trans* position.

The $XY_4Z_2$ molecule may be *cis* ($\mathbf{C}_{2v}$) or *trans* ($\mathbf{D}_{4h}$). The *cis*-isomer is expected to give four XY stretching $(2A+B_1+B_2)$ and two XZ stretching $(A_1+B_1)$ modes, all of which are infrared as well as Raman active. The *trans*-isomer is expected to give three XY stretching $(A_{1g}+B_{1g}+E_u)$ and two XZ stretching $(A_{1g}+A_{2u})$ modes, of which $E_u$ and $A_{2u}$ are infrared active and $A_{1g}$ and $B_{1g}$ are Raman active. The selection rules for other $XY_nZ_{6-n}$ molecules are tabulated in Appendix III. These selection rules can be used to distinguish the structures of stereo isomers on the basis of their vibrational spectra. For example, Clark et al.[629] measured the infrared and Raman spectra of mixed halogeno ions of the $[MX_4Y_2]^{2-}$ type (M=Ti and Sn; X=Cl, Br, and I) and concluded that most and probably all of these ions take the *cis* structure in the solid state. Some recent reports on vibrational spectra of $XY_nZ_{6-n}$ type compounds are as follows: $[OsBr_nI_{6-n}]^{2-}$ (686) and $[MO_nF_{6-n}]^{3-}$ (M= Ti, V, Nb, and Mo) (687). Ref 688 describes normal coordinate analyses on these types of compounds.

## II-9. $XY_7$ AND $XY_8$ MOLECULES

The $XY_7$ type molecules are very rare. Both $IF_7$ and $ReF_7$ are known to be pentagonal-bipyramidal ($\mathbf{D}_{5h}$), and their vibrational spectra have been assigned completely, as shown in Table II-9. The $A_1'$, $E_1''$, and $E_2'$ vibrations are Raman active, and the $A_2''$ and $E_1'$ vibrations are infrared active. According to Eysel and Seppelt,[690] $IF_7$ undergoes minor dynamic distortions from $\mathbf{D}_{5h}$ symmetry which cause violation of the $\mathbf{D}_{5h}$ selection rules for combination bands but not for the fundamentals. Normal coordinate analysis[690] shows that the axial bonds are definitely stronger and shorter than the equatorial ones. Table II-9 also lists the frequencies and assignments of the $[UO_2F_5]^{2-}$ ion of $\mathbf{D}_{5h}$ symmetry.[691]

The $XY_8$ type molecule may take the form of (I) a cube ($\mathbf{O}_h$), (II) an archimedean antiprism ($\mathbf{D}_{4d}$), (III) a dodecahedron ($\mathbf{D}_{2d}$), or (IV) a face-centered trigonal prism ($\mathbf{C}_{2v}$). Although $XY_8$ molecules are rare, X-ray analysis indicates that $[TaF_8]^{3-}$ and $[CrO_8]^{3-}$ ions take structures II and III, respectively.[692,693] The infrared and Raman spectra of crystalline

TABLE II-9. VIBRATIONAL FREQUENCIES[a] OF $XY_7$ AND $XY_5Z_2$ MOLECULES ($cm^{-1}$)

| $D_{5h}$ | $ReF_7$[689] | $IF_7$[689] | $IF_7$[690] | $UF_5O_2^{3-}$[691] | Assignment[b,690] |
|---|---|---|---|---|---|
| $\nu_1(A_1')$ | 736 | 676 | 675 | (668) | $\nu_s(MF_a)$ |
| $\nu_2(A_1')$ | 645 | 635 | 629 | 816 | $\nu_s(MF_e)$ |
| $\nu_3(A_2'')$ | 703 | 670 | 672 | 873 | $\nu_a(MF_a)$ |
| $\nu_4(A_2'')$ | 299 | 365 | 257 | 380 | $\delta(F_eMF_a)$ |
| $\nu_5(E_1')$ | 703 | 746 | 746 | 740 | $\nu_d(MF_e)$ |
| $\nu_6(E_1')$ | 353 | 425 | 425 | 425 | $\delta_d(F_eMF_e)$ |
| $\nu_7(E_1')$ | 217 | 257 | 363 | 240 | $\delta_d(F_aMF_a)$ |
| $\nu_8(E_1'')$ | 597 | 510 | 308 | — | $\delta_d(F_eMF_a)$ |
| $\nu_9(E_2')$ | 489 | 352 | 509 | — | $\nu_d(MF_e)$ |
| $\nu_{10}(E_2')$ | 352 | 310 | 342 | — | $\delta_d(F_eMF_e)$ |

[a] In these molecules $\nu_{11}(E_2'')$ is inactive.
[b] $F_a$ and $F_e$ denote the axial and equatorial F atoms, respectively.

$Na_3[TaF_8]$ are in accord with structure II, proposed by X-ray analysis.[694] For normal coordinate analyses of a cubic and an archimedean antiprism $XY_8$ molecule, see Refs. 695 and 696, respectively.

## II-10. $X_2Y_4$ AND $X_2Y_6$ MOLECULES

### (1) $X_2Y_4$ Molecules

Depending on the common angle of the two $XY_2$ planes, the $XY_2$—$XY_2$ molecule may be completely planar ($\mathbf{D}_{2h}$), staggered ($\mathbf{D}_{2d}$), or intermediate between the two ($\mathbf{D}_2$). The $\mathbf{D}_{2h}$ structure may be confirmed if the infrared and Raman mutual exclusion rule holds. The $\mathbf{D}_{2d}$ and $\mathbf{D}_2$ structures can be distinguished by comparing the number of fundamentals with that predicted for each structure; 8 for $\mathbf{D}_2$ and 5 for $\mathbf{D}_{2d}$ in the infrared, and 12 for $\mathbf{D}_2$ and 9 for $\mathbf{D}_{2d}$ in the Raman.

$B_2F_4$[697] and $B_2Cl_4$[698] are staggered in the gaseous and liquid phases and planar in the solid state, whereas $B_2Br_4$[699] is staggered in all phases. Apparently, steric hindrance plays a main role in determining the conformation. Both $B_2F_4$ and $B_2Cl_4$ are also staggered in Ar matrices.[700] The vibrational spectra of $N_2O_4$ have been assigned on the basis of $\mathbf{D}_{2h}$ symmetry.[701,702] In a $N_2$ matrix, $N_2O_4$ is a mixture of the $\mathbf{D}_{2h}$, $\mathbf{D}_{2d}$, and ONO—$NO_2$ isomers.[78] The vibrational spectra of the oxalato ion ($C_2O_4^{2-}$) have been assigned based on $\mathbf{D}_{2d}$,[703] $\mathbf{D}_{2h}$,[704] and $\mathbf{D}_2$[705] symmetry.

Molecules like $N_2H_4$ and $N_2F_4$ take the *trans* ($\mathbf{C}_{2h}$), *gauche* ($\mathbf{C}_2$), or *cis* ($\mathbf{C}_{2v}$) structure (Fig. II-18), depending on the angle of internal rotation.

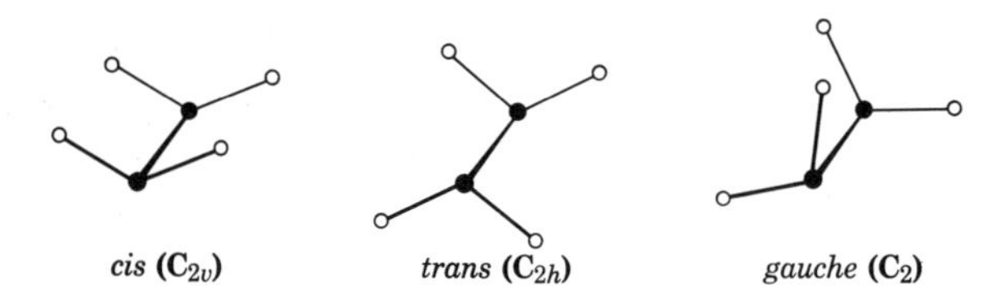


**Fig. II-18.** Various conformations of hydrazine.

Most of these compounds exist as the *trans*- or *gauche*-isomer or as a mixture of both. The *trans*-isomer shows 6, whereas the *gauche*-isomer shows 12, fundamentals in the infrared.

In the gaseous, liquid, and solid states, $N_2F_4$ is a mixture of the *trans*- and *gauche*-isomers, and complete vibrational assignments have been made on each isomer.[706] $N_2H_4$ is pure *gauche* in all physical states.[707] The infrared spectrum of $P_2H_4$ in the gaseous state has been assigned on the basis of the *gauche* structure.[708] However, it is *trans* in the solid state.[708a] The *trans* structure has been deduced from the vibrational spectra of $P_2F_4$ (all states),[708b] $P_2Cl_4$ (all states),[709] and $P_2I_4$ (solid and solution).[710]

The infrared spectra of dimeric metal halides $(MX_2)_2$ isolated in inert gas matrices have been assigned on the basis of a planar-cyclic ring structure (Sec. II-2).

### (2) Bridged $X_2Y_6$ Molecules ($D_{2h}$)

Figure II-19 illustrates the 18 normal modes of vibration[711] and band assignments for nonplanar bridging $X_2Y_6$ type molecules. The $A_g$, $B_{1g}$, $B_{2g}$, and $B_{3g}$ vibrations are Raman active, whereas the $B_{1u}$, $B_{2u}$, and $B_{3u}$ vibrations are infrared active. Table II-10*a* lists the vibrational frequencies of molecules belonging to this type. In most compounds, the $\nu_1$, $\nu_8$, $\nu_{11}$, and $\nu_{16}$ vibrations are largely due to the terminal $XY_2$ stretching motions, and their frequencies are higher than those of $\nu_2$, $\nu_6$, $\nu_{13}$, and $\nu_{17}$, which are mainly due to the vibrations of the bridging $X_2Y'_2$ group. Normal coordinate analyses of these molecules have been made by several investigators.[715–719] Adams and Churchill[716] showed that, except for $Ga_2I_6$ and $Al_2I_6$, the values of the bridging stretching force constants are only 40–45% of those of the terminal force constants. It should be noted that $In_2Cl_6$ and $In_2Br_6$ are polymeric and $In_2I_6$ is dimeric in the crystalline state, as shown by their spectra.[720]

Table II-10*b* lists eight stretching frequencies of planar $X_2Y_6$ ions and molecules. In the $[M_2X_6]^{2-}$ (M = Pd and Pt; X = Cl, Br, and I) series,

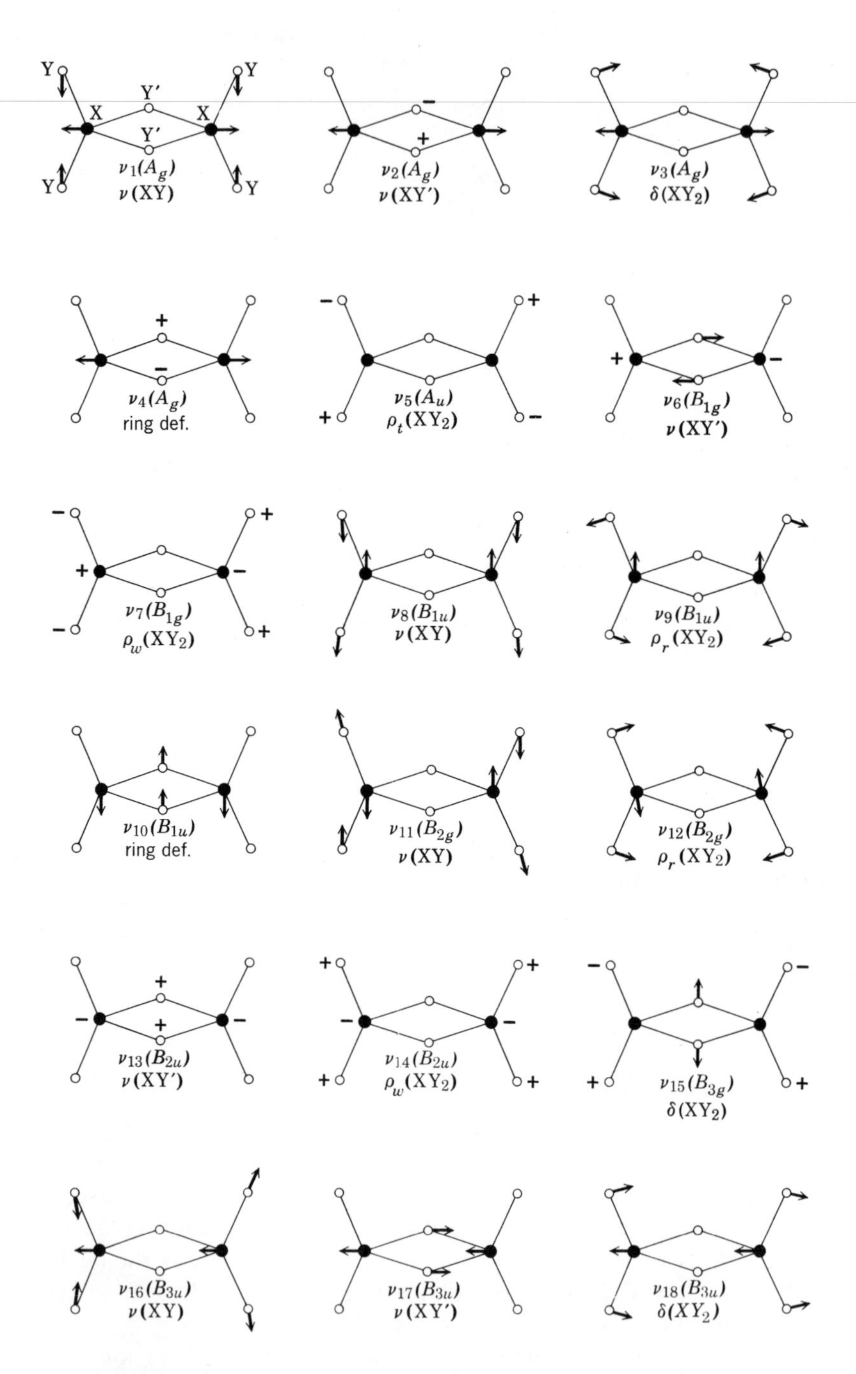


**Fig. II-19.** Normal modes of vibration of bridged $X_2Y_6$ molecules.[711]

TABLE II-10*a*. VIBRATIONAL FREQUENCIES OF NONPLANAR BRIDGING $X_2Y_6$ MOLECULES ($cm^{-1}$)

| | | $^{11}B_2H_6$ | $^{11}B_2D_6$ | $Al_2F_6$ | $Al_2Cl_6$ | $Al_2Br_6$ | $Al_2I_6$ | $Ga_2Cl_6$ | $Ga_2Br_6$ | $Ga_2I_6$ | $In_2I_6$ | $Fe_2Cl_6$ |
|---|---|---|---|---|---|---|---|---|---|---|---|---|
| $A_g$ | $\nu_1$ | 2524 | 1850 | — | 501 | 410 | 348 | 413 | 291 | 229 | 187 | (396) |
| | $\nu_2$ | 2104 | 1500 | — | 336 | 212 | 148 | 318 | 204 | 143 | 134 | (275) |
| | $\nu_3$ | 1180 | 915 | — | 217 | 142 | 95 | 167 | 119 | 85 | 69 | (148) |
| | $\nu_4$ | 794 | 700 | — | 115 | 70 | 56 | 100 | 64 | 50 | 40 | (73) |
| $A_u$ | $\nu_5$ | (829) | (592) | — | — | — | — | — | — | — | — | (50) |
| $B_{1g}$ | $\nu_6$ | 1768 | 1270 | — | 438 | 354 | — | 243 | 241 | 195 | 114 | 315 |
| | $\nu_7$ | (1035) | (860) | — | 168 | 82 | 82 | 125 | 85 | 64 | 55 | (130) |
| $B_{1u}$ | $\nu_8$ | 2612 | 1985 | 995 | 625 | 500 | 415 | 464 | 347 | 273 | 228 | 468 |
| | $\nu_9$ | (950) | 730 | 340 | — | — | — | — | 102 | — | — | 119 |
| | $\nu_{10}$ | 368 | 250 | — | — | — | — | — | — | — | — | 24 |
| $B_{2g}$ | $\nu_{11}$ | 2591 | 1975 | — | 608 | 491 | 405 | 462 | 339 | 265 | 232 | (466) |
| | $\nu_{12}$ | (920) | (725) | — | — | 116 | 63 | 117 | 74 | 55 | 49 | (66) |
| $B_{2u}$ | $\nu_{13}$ | 1915 | 1460 | 600 | 420 | 341 | 291 | 318 | 232 | 189 | 158 | 328 |
| | $\nu_{14}$ | 973 | 720 | — | — | 90 | 64 | 114 | 82 | 61 | 49 | 99 |
| $B_{3g}$ | $\nu_{15}$ | 1012 | 730 | — | — | — | 54 | 215 | 158 | 68 | 44 | (96) |
| $B_{3u}$ | $\nu_{16}$ | 2525 | 1840 | 805 | 484 | 373 | 320 | 390 | 269 | 213 | 178 | 406 |
| | $\nu_{17}$ | 1602 | 1199 | 575 | — | 198 | 140 | 282 | 188 | 134 | 125 | 280 |
| | $\nu_{18}$ | 1177 | 873 | 300 | — | 110 | 81 | 156 | 90 | 77 | 59 | 116 |
| Refs. | | 712 | 713 | 714 | 715 | 716 | 716 | 716,717 | 716 | 716 | 716 | 718 |

TABLE II-10*b*. VIBRATIONAL FREQUENCIES[a] OF PLANAR BRIDGING $X_2Y_6$ MOLECULES ($cm^{-1}$)

| | | $[Pd_2Cl_6]^{2-}$ | $[Pd_2Br_6]^{2-}$ | $[Pd_2I_6]^{2-}$ | $[Pt_2Cl_6]^{2-}$ | $[Pt_2Br_6]^{2-}$ | $[Pt_2I_6]^{2-}$ | $I_2Cl_6$ | $Au_2Cl_6$ |
|---|---|---|---|---|---|---|---|---|---|
| $A_g$ | $\nu_1$, $\nu(XY_t)$ | 346 | 262 | 219 | 349 | 241 | 196 | 344 | 379 |
| | $\nu_2$, $\nu(XY_b)$ | 302 | 194 | 143 | 316 | 211 | 160 | 198 | 328 |
| $B_{1g}$ | $\nu_6$, $\nu(XY_t)$ | 328 | 253 | — | 333 | 238 | 196 | 314 | 366 |
| | $\nu_7$, $\nu(XY_b)$ | 265 | 173 | 130 | 294 | 193 | 145 | 142 | 289 |
| $B_{2u}$ | $\nu_{12}$, $\nu(XY_t)$ | 335 | 257 | 218 | 330 | 236 | 196 | 327 | 364 |
| | $\nu_{13}$, $\nu(XY_b)$ | 262 | 178 | — | 300 | 192 | 147 | 170 | 309 |
| $B_{3u}$ | $\nu_{16}$, $\nu(XY_t)$ | 343 | 264 | 218 | 341 | 239 | 196 | 340 | 374 |
| | $\nu_{17}$, $\nu(XY_b)$ | 297 | 192 | 140 | 312 | 210 | 157 | 205 | 309 |
| Refs. | | 721 | 721 | 721 | 721 | 721 | 721 | 722 | 722 |

[a] $XY_t$ and $XY_b$ denote terminal and bridging XY stretching modes, respectively. When these two modes couple strongly, distinction between them is not clear (see Ref. 721).

Goggin[721] showed that the distinction between terminal and bridging vibrations is meaningless except for X = Cl, since these vibrations couple so strongly with each other. According to Forneris et al.,[722] the terminal and bridging stretching force constants of $Au_2Cl_6$ are 2.22 and 1.15 mdyn/Å, respectively, and those of $I_2Cl_6$ are 1.70 and 0.40 mdyn/Å, respectively (modified UBF). On the other hand, Adams and Churchill[716] report values of 2.419 and 1.482 mdyn/Å, respectively, for the terminal and bridging force constants of $Au_2Cl_6$ (GVF).

**(3) Ethane-Type $X_2Y_6$ Molecules ($D_{3d}$)**

The ethane type $X_2Y_6$ molecule may be staggered ($\mathbf{D}_{3d}$), eclipsed ($\mathbf{D}_{3h}$), or *gauche* ($\mathbf{D}_3$). Figure II-20 shows the 12 normal modes of vibration of the staggered $X_2Y_6$ molecule. The $A_{1g}$ and $E_g$ vibrations are Raman active, and the $A_{2u}$ and $E_u$ vibrations are infrared active. Table II-10*c* lists the observed frequencies and band assignments based on $\mathbf{D}_{3d}$ symmetry. It should be noted that neutral $Ga_2X_6$ (X: a halogen)

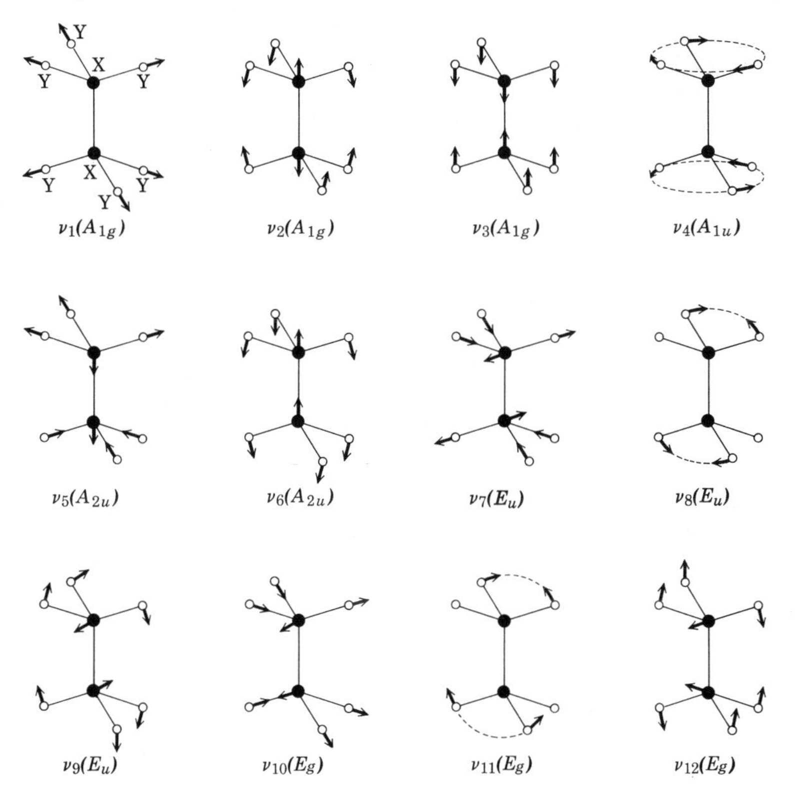


**Fig. II-20.** Normal modes of vibration of ethane-type $X_2Y_6$ molecules.

TABLE II-10c. VIBRATIONAL FREQUENCIES OF ETHANE TYPE $X_2Y_6$ MOLECULES ($cm^{-1}$)

| | | Molecule ($\mathbf{D}_{3d}$) | | | | | | | | | | | | |
|---|---|---|---|---|---|---|---|---|---|---|---|---|---|---|
| | | $C_2H_6$ | $Si_2H_6$ | $Ge_2H_6$ | $N_2H_6^{2+}$ | $P_2O_6^{4-}$ | $S_2O_6^{2-}$ | $Ga_2Cl_6^{2-}$ | $Ga_2Br_6^{2-}$ | $Ga_2I_6^{2-}$ | $Si_2F_6$ | $Si_2Cl_6$ | $Si_2Br_6$ | $Si_2I_6$ |
| $A_{1g}$ | $\nu_1$, $\nu(XY_3)$ | 2899 | 2152 | (2070) | 2650 | 1062 | 1102 | 375 | 316 | 285 | 910 | 351 | 223 | 154 |
| | $\nu_2$, $\delta(XY_3)$ | (1375) | 909 | 765 | 1524 | 670 | 710 | 106 | 70 | 42,48 | 220 | 127 | 80 | (51) |
| | $\nu_3$, $\nu(XX)$ | 993 | 434 | 229 | 1027 | 275 | 293 | 233 | 164 | 118 | 541 | 624 | 562 | 510 |
| $A_{1u}$ | $\nu_4$, $\rho_t(XY_3)$ | 275 | — | 144 | 455 | — | — | — | — | — | — | — | — | — |
| $A_{2u}$ | $\nu_5$, $\nu(XY_3)$ | 2954 | 2154 | 2078 | 2600 | 942 | 1000 | 302 | 201 | 155 | 819 | 460 | 329 | 255 |
| | $\nu_6$, $\delta(XY_3)$ | 1379 | 844 | 755 | 1485 | 562 | 577 | 151 | 110 | 88 | 403 | 241 | 168 | 116 |
| $E_u$ | $\nu_7$, $\nu(XY_3)$ | 2994 | 2179 | 2114 | 2739 | 1085 | 1240 | 327 | 237,228 | 200 | 971 | 603 | 479 | 388 |
| | $\nu_8$, $\delta(XY_3)$ | 1486 | 940 | 898 | 1613 | 494 | 516 | 141 | 92 | 74 | 340 | 178 | 114 | 81 |
| | $\nu_9$, $\rho_r(XY_3)$ | 821 | 379 | 407 | 1096 | 200 | 204 | 89–66 | 64 | (50) | 203 | 74 | 50 | (31) |
| $E_g$ | $\nu_{10}$, $\nu(XY_3)$ | 2963 | 2155 | 2150 | 2745 | 1168 | 1216 | 314 | 228 | 184 | 985 | 590 | 473 | 398 |
| | $\nu_{11}$, $\delta(XY_3)$ | 1460 | 929 | 875 | 1599 | 508 | 556 | 116 | 84 | 75 | 306 | 211 | 139 | 94 |
| | $\nu_{12}$, $\rho_r(XY_3)$ | (1155) | 625 | 417 | 1105 | 323 | 320 | 146 | 102 | 84 | 135 | 132 | 89 | (53) |
| Refs. | | 723 | 724 | 725 | 726 | 727 | 728 | 729 | 729 | 729 | 729a | 730 | 730 | 730 |

molecules take the bridging $\mathbf{D}_2h$ structure (Table II-10a), whereas $[Ga_2X_6]^{2-}$ ions take the ethane-like $\mathbf{D}_{3d}$ structure.

The structure of $Si_2Cl_6$ has been controversial; Griffiths[731] prefers the $\mathbf{D}_{3h}$ or $\mathbf{D}'_{3h}$ structure* for liquid $Si_2Cl_6$, whereas Ozin[732] favors the $\mathbf{D}_{3d}$ model for all phases. The $\mathbf{D}_{3h}$ and $\mathbf{D}_{3d}$ selection rules are similar except that the $E_u$ modes of $\mathbf{D}_{3d}$ which are infrared active become both infrared and Raman active ($E'$) in $\mathbf{D}_{3h}$. The SiSi stretching mode was assigned at 354 $cm^{-1}$ by Griffiths and at 627 $cm^{-1}$ by Ozin. According to Höfler et al.,[730] the MM stretching force constant increases in the order $Si_2H_6 < Si_2I_6 < Si_2Br_6 < Si_2Cl_6$. Normal coordinate analyses have also been made on $Si_2Cl_6$[732] and $[Ga_2Cl_6]^{2-}$.[729]

## II-11. $X_2Y_7$, $X_2Y_8$, $X_2Y_9$, AND $X_2Y_{10}$ MOLECULES

### (1) $X_2Y_7$ Molecules

The $X_2Y_7$ ($XY_3$—Y—$XY_3$) type molecule belongs to the $\mathbf{C}_s$, $\mathbf{C}_{2v}$, or $\mathbf{C}_1$ point group, depending on the relative orientation of the two $XY_3$ groups:

```
Y       Y       Y
  \   /   \   /
Y—X         X—Y
  /           \
Y               Y
```

Seventeen vibrations are infrared active in $\mathbf{C}_{2v}$, while all 21 vibrations are infrared active in $\mathbf{C}_s$ and $\mathbf{C}_1$ symmetry. The 21 normal vibrations of the $X_2Y_7$ molecule may be classified into in-phase and out-of-phase coupling motions of terminal $XY_3$ group vibrations and the skeletal vibrations of the XYX bridge. Table II-11 lists the observed frequencies of these bridging vibrations.

On the basis of normal coordinate analyses, Brown and Ross[734] have made complete assignments of the vibrational spectra of the $S_2O_7^{2-}$, $Se_2O_7^{2-}$, $V_2O_7^{4-}$, and $Cr_2O_7^{2-}$ ions ($\mathbf{C}_{2v}$ symmetry). The molecule $Re_2O_7$ is monomeric in the gaseous and liquid states and polymeric in the solid state. Beattie and Ozin[735] assigned the spectra of gaseous $Re_2O_7$ on the basis of $\mathbf{C}_{2v}$ symmetry. Vibrational analysis of $Cl_2O_7$ has been made by assuming $\mathbf{C}_2$[738] or $\mathbf{C}_{2v}$[736] symmetry. On the assumption of a linear OPO bridge, the vibrational assignments of divalent metal pyrophosphates ($M_2P_2O_7$) have been made in terms of $\mathbf{D}_{3h}$ or $\mathbf{D}_{3d}$ symmetry.[739,740] According to Wing and Callahan,[741] the MOM bridge angle is always larger than 115°, and $\nu_a$(MO) is at least 215 $cm^{-1}$ higher than $\nu_s$(MO).

* If free rotation about the SiSi bond occurs, the symmetry plane ($\sigma_h$) is lost from $\mathbf{D}_{3h}$, and the molecule belongs to $\mathbf{D}'_{3h}$. Its selection rules are essentially the same as those of $\mathbf{D}_{3h}$.

TABLE II-11. YXY BRIDGING FREQUENCIES OF $X_2Y_7$ MOLECULES ($cm^{-1}$)

| Compound | $\nu_a(YX_2)$ | $\nu_s(YX_2)$ | $\delta(YX_2)$ | Refs. |
|---|---|---|---|---|
| $Na_4[P_2O_7]$ | 915 | 730 | — | 733 |
| $Na_4[As_2O_7]$ | 735 | 550 | 245 | 733 |
| $Na_2[S_2O_7]$ | 825 | 725 | 182 or 116 | 734 |
| $Na_2[Se_2O_7]$ | 707 | 556 | — | 734 |
| $Na_4[V_2O_7]$ | 710 | 533 | 200 | 734 |
| $Na_2[Cr_2O_7]$ | 770 | 565, 554 | 220 | 734 |
| $Re_2O_7$ | 804 | 456 | 50 | 735 |
| $Cl_2O_7$ | 785 | 700 | (165) | 736 |
| $[Ga_2Cl_7]^-$ | 286 | 276 | — | 737 |
| $[Ga_2Br_7]^-$ | 222 | 195 | — | 737 |

This separation increases as the MOM angle increases. In the $M_2O_2$ ring bridge, the MOM angle is 80–90°, and the separation is much smaller.

**(2) $X_2Y_8$ Molecules**

The symmetry of the $X_2Y_8$ ($XY_3$—Y—Y—$XY_3$) molecule may be low enough to activate all 24 normal vibrations in both infrared and Raman spectra. Thus far, the XYYX bridging frequencies have been assigned at 988 [$\nu_a$(XYYX)], 784 [$\nu_s$(XYYX)], 890 [$\nu$(YY)], and 397 and 328 [$\delta$(XYYX)] for the $[P_2O_8]^{4-}$ ion, and at 1062, 834, 854, and 328 and 236 $cm^{-1}$, respectively, for the $[S_2O_8]^{4-}$ ion.[742]

**(3) $X_2Y_9$ Molecules**

The $X_2Y_9$ molecule shown below belongs to the point group $\mathbf{D}_{3h}$, and its 27 normal vibrations are classified into 4 $A_1'$(R), $A''$(i.a.), $A_2'$(i.a.), 3 $A_2''$(IR), 5 $E'$(IR,R), and 4 $E''$(R). Two $XY_t$ stretching ($A_2''$ and $E'$) and two $XY_b$ stretching ($A_2''$ and $E'$) vibrations are infrared active. Ziegler and Risen[743] carried out normal coordinate analyses of $[Cr_2Cl_9]^{3-}$ and $[W_2Cl_9]^{3-}$ ions. Both X-ray and vibrational analyses show that direct M–M interaction is present in the latter but not in the former ion. Beattie

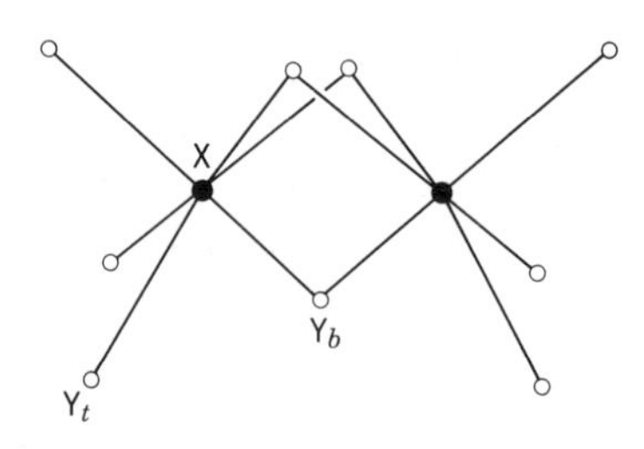

et al.[638] made complete assignments of the infrared and Raman spectra of the $[Tl_2Cl_9]^{3-}$ ion based on $\mathbf{D}_{3h}$ symmetry. The stretching frequencies are reported for other ions such as $[Ti_2Cl_9]^{3-}$, $[V_2Cl_9]^{3-}$, and $[Cr_2Cl_9]^{3-}$,[744] and $[Rh_2Cl_9]^{3-}$.[745]

### (4) $X_2Y_{10}$ Molecules

The $X_2Y_{10}$ molecule may take either one of the following structures:

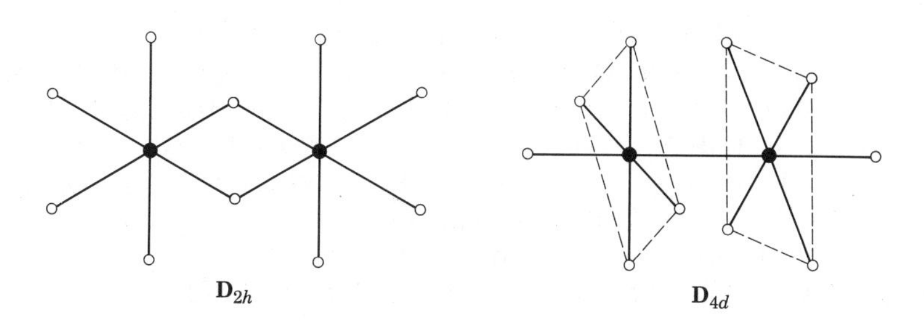


The 30 normal vibrations of the $\mathbf{D}_{2h}$ molecule are classified into 6 $A_g$, 2 $A_u$, 4 $B_{1g}$, 4 $B_{1u}$, 3$B_{2g}$, 4 $B_{2u}$, 2 $B_{3g}$, and 5 $B_{3u}$, of which 13 vibrations ($B_{1u}$, $B_{2u}$, and $B_{3u}$) are infrared active. These include four terminal and two bridging stretching modes. Beattie et al.[638] carried out normal coordinate analysis to assign the vibrational spectra of $Nb_2X_{10}$ and $Ta_2X_{10}$ (X = Cl and Br) based on $\mathbf{D}_{2h}$ symmetry. The infrared frequencies of $Re_2Cl_{10}$ and $Nb_2Cl_{10}$ have been reported.[746]

The symmetry of the single-bridge $XY_5$—$XY_5$ type molecule may be $\mathbf{D}_{4d}$ (staggered) or $\mathbf{D}_{4h}$ (eclipsed), and these two structures cannot be distinguished by the selection rules. Under $\mathbf{D}_{4d}$ symmetry, the 30 normal vibrations are classified into 4 $A_1$, $B_1$, 3 $B_2$, 4 $E_1$, 3 $E_2$, and 4 $E_3$ species, of which $B_2$ and $E_1$ are infrared active and $A_1$, $E_2$, and $E_3$ are Raman active. Wilmhurst and Bernstein[747] carried out normal coordinate analyses on $S_2F_{10}$, and Dodd et al.[748] assigned the spectra of $S_2F_{10}$ and $Te_2F_{10}$ based on $\boldsymbol{D}_{4d}$ symmetry.

## II-12. COMPOUNDS OF OTHER TYPES

Many compounds do not belong to any of the types discussed in the preceding sections. For these, the reader should consult *Spectroscopic Properties of Inorganic and Organometallic Compounds*, Vols. 1–9 (1968–1975), published by the Chemical Society, London, and other reference books cited at the end of Part I. Here, references of some typical compounds are classified according to their central elements, and review articles are cited whenever available.

**(1) Compounds of Group IIIA Elements**

Several review articles are available on the vibrational spectra of boron compounds. Lehmann and Shapiro[749] reviewed $\nu$(BH), $\delta(BH_2)$, $\nu$(BC), $\nu$(BB), and so forth of alkylboranes, and Bellamy et al.[750] discussed $\nu$(BN), $\nu$(BH), $\nu$(BX), $\nu$(BC), and so forth. Meller[751] lists $\nu$(BN), $\nu$(BH), $\nu$(BX), and $\nu$(BC) of a number of organoboron–nitrogen compounds. For vibrational spectra of boron compounds containing the B–P bond, see a review by Verkade.[752] The group frequency charts shown in Appendix VI include $\nu$(BH), $\nu$(BO), and $\nu$(BX).

For individual compounds, only references are given: $B_2O_2$ (753), $B_2O_3$ (753), $BO_2^-$ (754), HBO (755), $H_3B_3O_3$ (756), $B_3O_3(OH)_3$ (757), $H_2B_2O_3$ (758), $H_3BO_3$ (759), $B_2O_5^{4-}$ (760), $Al(BH_4)_3$ (760a), $B_5H_9$ (761), $B_{10}H_{14}$ (762), $B_3H_8^-$ (763), $B_9Cl_9$ (764), $BX_3$—$NEt_3$ (765), $B_3N_3H_6$ (766, 767), $B_3N_3H_3X_3$(768), $AlH_3\cdot$(amine)(769), $MX_3\cdot$(amine) (M = Al and Ga) (770), $InX_3L_2$ (L = $OEt_2$, $SEt_2$)( 771), $GaX_3L$ (L = $PH_3$, $PMe_3$, etc.) (772), and $M[N(SiMe_3)_2]_3$ (M = Al, Ga, and In) (773).

**(2) Compounds of Group IVA Elements**

The vibrational spectra of silicon compounds have been reviewed by Smith.[774] Aylett[775] reviewed the vibrational spectra of silicon hydrides. Campbell-Ferguson and Ebsworth[776] reviewed the spectra of halogenosilane-amine adducts, and Schumann[777] reviewed $\nu$(GeP) and $\nu$(SnP) of a large number of compounds. Appendix VI gives group frequency charts for $\nu$(SiH), $\nu$(GeH), $\nu$(SiO), $\nu$(SiX), and $\nu$(GeX).

For individual compounds, only references are cited: $C_3O_2$ (778, 779), $C_3S_2$ (780), quartz (781, 782), $Si_3O_9^{6-}$ (783), silicates (784, 785), germanates (785), silicon oxides (786), germanium oxides (787), tin oxides (788), lead oxides (789), $(SiH_3)_2O$ (790), $(SiH_3)_2S$ (791), $(SiH_3)_3N$ (792), $(GeH_3)_3As$ (793), $(GeH_3)_3Sb$ (793), $SiH_3SiF_3$ (794), $Ge_2Br_6$ (795), and $Sn_3F_8$ (796).

**(3) Compounds of Group VA Elements**

The vibrational spectra of phosphorus compounds have been reviewed by Corbridge,[797,798] Thomas,[799] and others.[800,801] Figure II-21 shows a group frequency chart based on these reviews. Appendix VI also gives group frequency charts for $\nu$(PH), $\nu$(PO), and $\nu$(PX).

For individual compounds, only references are given: $N_2O_3$ (802), solid $N_2O_5$ (803), $N_2O_5$ vapor (804), $H_2N_2O_2$ (805), $HNO_3$ (806), $FNO_3$ and $ClNO_3$ (807), $NH_2OH$ (808, 809), $[NH_3OH]X$ (810), $NI_3NH_3$ (811), $P_4$ (812), $As_4$ (813), $P_4O_6$ (814), $As_4O_6$ (813, 814), $P_4O_{10}$ (814), $HPO_4^{2-}$ and $H_2PO_4^-$ (815), $HPO_3^{2-}$and $H_2PO_3^-$ (816), $P_3O_9^{3-}$ (817), $As_3O_9^{3-}$ (818), $P_4O_{12}^{4-}$ (819), $H_2P_2O_4^{2-}$ (820), $P_4S_3$ (821), $P_4S_9$ (822), $As_4S_4$ (823),

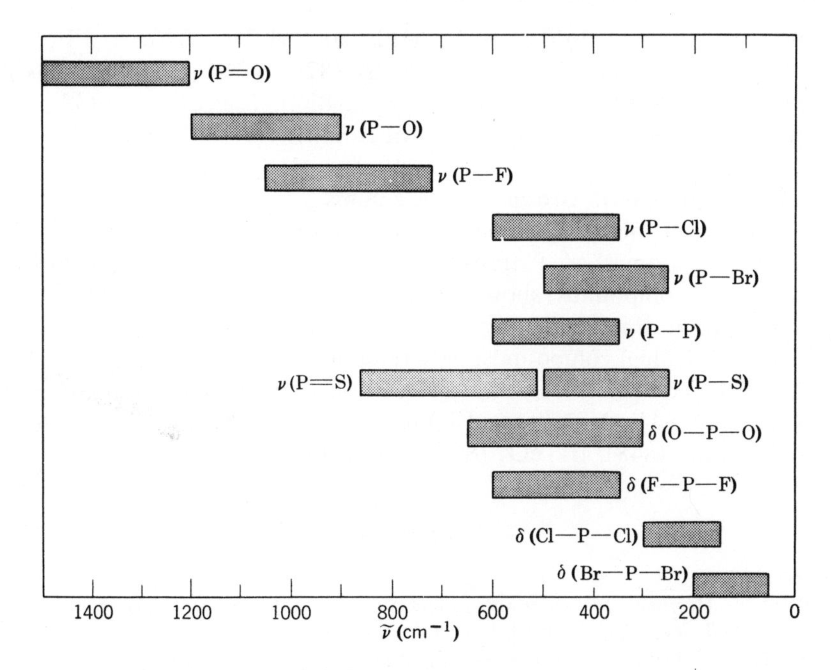


**Fig. II-21.** Characteristic frequencies of phosphorus compounds.

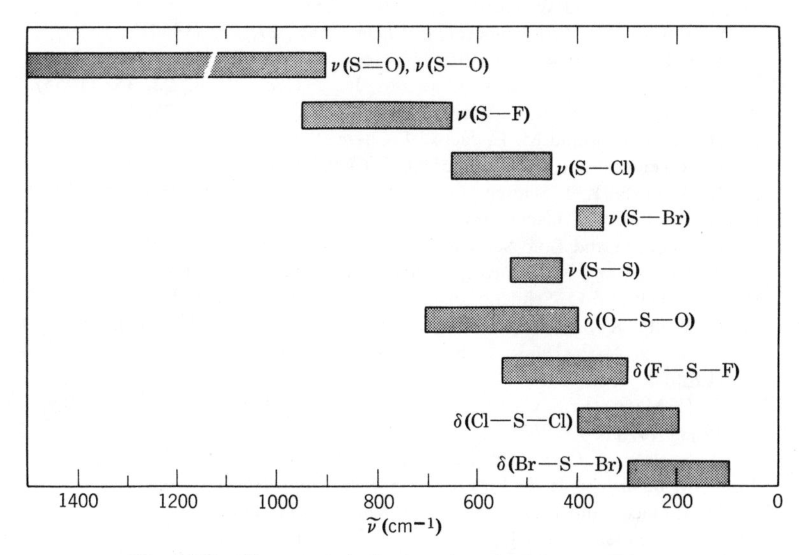


**Fig. II-22.** Characteristic frequencies of sulfur compounds.

$P_nAs_{4-n}$ (824), $S_nPO_{4-n}^{3-}$ (825), $HPF_4$ and $H_2PF_3$ (826), $H_3PF_2$ (827), $PF_3X_2$ (X = Cl and Br) (828), $SPF_2Br$ (829), $(PNCl_2)_3$ (830), $(PNBr_2)_3$ (831), $(PO_2NH)_n^{n-}$ ($n = 3$ and 4) (817, 832), $N[P(CF_3)_2]_3$ (833), $BiX_4^-$, $BiX_5^{2-}$, and $Bi_2X_9^{3-}$ (834), and $\alpha$-$BiF_5$ (835).

### (4) Compounds of Group VIA Elements

Müller et al.[488,836] reviewed the vibrational spectra of thio and seleno compounds of $d^0$ transition metal ions. The group frequency chart of sulfur compounds shown in Fig. II-22 is based on the literature appearing in this book.

For individual compounds, only references are cited: $S_6$ (837), $S_7$ (838), $S_8$ (839), $H_2S_3$ (840), $S_2O_5^{2-}$ (841), $H_2SO_4$ (842), adducts of $S_2N_2$ (843) and $S_4N_4$ (844), $NS_4^-$ (845), $ClOSO_2F$ (846), $XNSF_2$ (X = F, Cl, and Br) (847), $Se_4^{2+}$ (848), $\beta$-$TeO_3$ (849), and $TeO_4^{2-}$ (850).

## References


1. G. Herzberg, *Molecular Spectra and Molecular Structure*, Vol. I: *Spectra of Diatomic Molecules*, Van Nostrand, Princeton, N.J. (1950).
2. A. E. Douglas, *Can. J. Phys.*, **35,** 71 (1957).
3. D. H. Rank, D. P. Eastman, B. S. Rao, and T. A. Wiggins, *J. Opt. Soc. Am.*, **52,** 1 (1962).
4. W. Klemperer, W. G. Norris, A. Büchler, and A. G. Emslie, *J. Chem. Phys.*, **33,** 1534 (1960).
5. S. E. Veazey and W. Gordy, *Phys. Rev.*, **138,** *A*, 1303 (1965).
6. S. A. Rice and W. Klemperer, *J. Chem. Phys.*, **27,** 573 (1957).
7. V. I. Baikov and K. P. Vasilevskii, *Opt. Spectrosc.*, **22,** 364 (1967).
8. W. V. F. Brooks and B. L. Crawford, Jr., *J. Chem. Phys.*, **23,** 363 (1955).
9. J. A. Horsley and R. F. Barrow, *Trans. Faraday Soc.*, **63,** 32 (1967).
10. D. E. Milligan and M. E. Jacox, *J. Chem. Phys.*, **52,** 2594 (1970).
11. K. Rosengren and G. C. Pimentel, *J. Chem. Phys.*, **43,** 507 (1965).
12. N. Acquista, L. J. Schoen, and D. R. Lide, Jr., *J. Chem. Phys.*, **48,** 1534 (1968).
13. W. R. Busing, *J. Chem. Phys.*, **23,** 933 (1955).
14. N. Acquista and L. J. Schoen, *J. Chem. Phys.*, **53,** 1290 (1970).
15. A. Snelson and K. S. Pitzer, *J. Phys. Chem.*, **67,** 882 (1963).
16. S. Schlick and O. Schnepp, *J. Chem. Phys.*, **41,** 463 (1964).
17. D. White, K. S. Seshadri, D. F. Dever, D. E. Mann, and M. J. Linevsky, *J. Chem. Phys.*, **39,** 2463 (1963); K. S. Seshadri, D. White, and D. E. Mann, *ibid.*, **45,** 4697 (1966).
18. D. E. Mann, G. V. Calder, K. S. Seshadri, D. White, and M. J. Linevsky, *J. Chem. Phys.*, **46,** 1138 (1967).
19. A. Snelson, *J. Phys. Chem.*, **71,** 3202 (1967).
20. J. M. Brom and H. F. Franzen, *J. Chem. Phys.*, **54,** 2874 (1971).
21. M. E. Jacox and D. E. Milligan, *J. Chem. Phys.*, **50,** 3252 (1969).
22. R. A. Penneman and L. H. Jones, *J. Chem. Phys.*, **24,** 293 (1956).
23. D. E. Milligan and M. E. Jacox, *J. Chem. Phys.*, **47,** 278 (1967).

24. H. Dubost and A. Abouaf-Marquin, *Chem. Phys. Lett.*, **17,** 269 (1972).
25. See Table II–1a.
26. R. Steudel, *Z. Anorg. Allg. Chem.*, **361,** 180 (1968).
27. J. S. Anderson and J. S. Ogden, *J. Chem. Phys.*, **51,** 4189 (1969).
28. J. S. Ogden and M. J. Ricks, *J. Chem. Phys.*, **52,** 352 (1970).
29. C. P. Marino, J. D. Guerin, and E. R. Nixon, *J. Mol. Spectrosc.*, **51,** 160 (1974).
30. J. S. Ogden and M. J. Ricks, *J. Chem. Phys.*, **53,** 896 (1970).
31. J. S. Ogden and M. J. Ricks, *J. Chem. Phys.*, **56,** 1658 (1972).
32. M. E. Jacox and D. E. Milligan, *J. Chem. Phys.*, **46,** 184 (1967); and **40,** 2461 (1964).
33. D. E. Tevault and L. Andrews, *J. Phys. Chem.*, **77,** 1640 (1973).
34. D. E. Milligan and M. E. Jacox, *J. Chem. Phys.*, **55,** 3404 (1971).
35. W. A. Guillory and C. E. Hunter, *J. Chem. Phys.*, **50,** 3516 (1969).
36. V. P. Babaeva and V. Ya. Rosolovskii, *Russ. J. Inorg. Chem.*, **16,** 471 (1971).
37. G. A. Ozin, *Chem. Commun.*, 1325 (1969).
38. L. Andrews and R. Smardzewski, *J. Chem. Phys.*, **58,** 2258 (1973).
39. See Table II–1a.
40. D. E. McKee and N. Bartlett, *Inorg. Chem.*, **12,** 2738 (1973).
41. L. Andrews and J. I. Raymond, *J. Chem. Phys.*, **55,** 3078 (1971).
42. W. Holzer, W. F. Murphy, and H. J. Bernstein, *J. Mol. Spectrosc.*, **32,** 13 (1969).
43. R. E. Barletta, H. H. Claassen, and R. L. McBeth, *J. Chem. Phys.*, **55,** 5409 (1971).
44. W. F. Howard and L. Andrews, *J. Am. Chem. Soc.*, **95,** 3045 (1973).
45. G. A. Olah and M. B. Comisarow, *J. Am. Chem. Soc.*, **91,** 2172 (1969).
46. W. F. Howard and L. Andrews, *J. Am. Chem. Soc.*, **95,** 2056 (1973).
47. M. R. Clarke and G. Mamantov, *Inorg. Nucl. Chem. Lett.*, **7,** 993 (1971).
48. M. M. Rochkind and G. C. Pimentel, *J. Chem. Phys.*, **46,** 4481 (1967); W. A. Alcock and G. C. Pimentel, *ibid.*, **48,** 2373 (1968).
49. W. Holzer, W. F. Murphy, and H. J. Bernstein, *J. Chem. Phys.*, **52,** 399 (1970).
50. R. J. Gillespie and M. J. Morton, *Chem. Commun.*, 1565 (1968).
51. R. J. Gillespie and M. J. Morton, *J. Mol. Spectrosc.*, **30,** 178 (1969).
52. W. Weltner and D. McLeod, *J. Phys. Chem.*, **69,** 3488 (1965).
53. W. Weltner and D. McLeod, *J. Chem. Phys.*, **42,** 882 (1965).
54. W. Weltner and D. McLeod, *J. Mol. Spectrosc.*, **17,** 276 (1965).
55. D. W. Green, W. Korfmacher, and D. M. Gruen, *J. Chem. Phys.*, **58,** 404 (1973).
56. S. D. Gabelnick, G. T. Reedy, and M. G. Chasanov, *J. Chem. Phys.*, **58,** 4468 (1973).
57. A. J. Edwards, W. E. Falconer, J. E. Griffiths, W. A. Sunder, and M. J. Vasile, *J. Chem. Soc.*, Dalton, 1129 (1974).
58. M. J. Linevsky, *J. Chem. Phys.*, **34,** 587 (1961).
59. A. Snelson, *J. Chem. Phys.*, **46,** 3652 (1966).
60. S. Schlick and O. Schnepp, *J. Chem. Phys.*, **41,** 463 (1964).
61. M. Freiberg, A. Ron, and O. Schnepp, *J. Phys. Chem.*, **72,** 3526 (1968).
62. J. M. Brom, Jr., and H. F. Franzen, *J. Chem. Phys.*, **54,** 2874 (1971).
63. J. R. Durig, K. K. Lau, G. Nagarajan, M. Walker, and J. Bragin, *J. Chem. Phys.*, **50,** 2130 (1969).
64. K. S. Seshadri, D. White, and D. E. Mann, *J. Chem. Phys.*, **45,** 4697 (1966).
65. V. E. Bondybey and J. W. Nibler, *J. Chem. Phys.*, **58,** 2125 (1973).
66. D. Sellmann, A. Brandl, and R. Endell, *Angew. Chem.*, Int. Ed., **12,** 1019 (1973).
67. J. L. Hollenberg, *J. Chem. Phys.*, **46,** 3099 (1967); D. F. Smith, *ibid.*, **48,** 1429 (1968).
68. P. A. Giguère and N. Zengin, *Can. J. Chem.*, **36,** 1013 (1958).

69. D. F. Hornig and W. E. Osberg, *J. Chem. Phys.*, **23,** 662 (1955); G. L. Hiebert and D. F. Hornig, *ibid.*, **28,** 316 (1958).
70. A. J. Barnes, H. E. Hallam, and G. F. Scrimshaw, *Trans. Faraday Soc.*, **65,** 3150, 3159, 3172 (1969).
71. V. Lorenzelli, *Compt. Rend.*, **258,** 5386 (1964).
72. W. B. Person, R. E. Humphrey, W. A. Deskin, and A. I. Popov, *J. Am. Chem. Soc.*, **80,** 2049 (1958); W. B. Person, R. E. Erickson, and R. E. Buckes, *ibid.*, **82,** 29 (1960); A. I. Popov, R. E. Humphrey, and W. B. Person, *ibid.*, **82,** 1850 (1960).
73. P. Klaboe, *J. Am. Chem. Soc.*, **89,** 3667 (1967).
74. L. H. Jones, *J. Chem. Phys.*, **22,** 217 (1954).
75. W. D. Stalleup and D. Williams, *J. Chem. Phys.*, **10,** 199 (1942).
76. R. A. Penneman and L. H. Jones, *J. Chem. Phys.*, **28,** 169 (1958).
77. L. H. Jones, *J. Mol. Spectrosc.*, **45,** 55 (1973); **49,** 82 (1974).
78. W. G. Fateley, H. A. Bent, and B. Crawford, Jr., *J. Chem. Phys.*, **31,** 204 (1959).
79. J. W. Hastie, R. H. Hauge, and J. L. Margrave, *Chem. Commun.*, 1452 (1969).
80. I. Eliezer and A. Reger, *Coord. Chem. Rev.*, **9,** 189 (1972–73).
81. J. W. Hastie, R. H. Hauge, and J. L. Margrave, *High Temp. Sci.*, **1,** 76 (1969).
82. A. Snelson, *J. Phys. Chem.*, **72,** 250 (1968).
83. D. E. Mann, G. V. Calder, K. S. Seshadri, D. White, and M. J. Linevsky, *J. Chem. Phys.*, **46,** 1138 (1967).
84. D. White, G. V. Calder, S. Hemple, and D. E. Mann, *J. Chem. Phys.*, **59,** 6645 (1973).
85. G. V. Calder, D. E. Mann, K. S. Seshadri, M. Allavena, and D. White, *J. Chem. Phys.*, **51,** 2093 (1969).
86. D. E. Milligan and M. E. Jacox, *J. Chem. Phys.*, **48,** 2265 (1968).
87. L. Andrews, *J. Chem. Phys.*, **48,** 979 (1968).
88. L. Andrews and T. G. Carver, *J. Chem. Phys.*, **49,** 896 (1968).
89. J. W. Hastie, R. H. Hauge, and J. L. Margrave, *J. Am. Chem. Soc.*, **91,** 2536 (1969).
90. D. E. Milligan and M. E. Jacox, *J. Chem. Phys.*, **49,** 1938 (1968).
91. G. Maass, R. H. Hauge, and J. L. Margrave, *Z. Anorg. Allg. Chem.*, **392,** 295 (1972).
92. J. W. Hastie, R. H. Hauge, and J. L. Margrave, *J. Phys. Chem.*, **72,** 4492 (1968).
93. L. Andrews and D. L. Frederick, *J. Am. Chem. Soc.*, **92,** 775 (1970).
94. R. H. Hauge, J. W. Hastie, and J. L. Margrave, *J. Mol. Spectrosc.*, **45,** 420 (1973).
95. G. A. Ozin and A. Vander Voet, *J. Chem. Phys.*, **56,** 4768 (1972).
96. I. R. Beattie and R. O. Peny, *J. Chem. Soc.*, *A*, 2429 (1970).
97. M. D. Harmony and R. J. Myers, *J. Chem. Phys.*, **37,** 636 (1962).
98. J. K. Burdett, L. Hodges, V. Dunning, and J. H. Current, *J. Phys. Chem.*, **74,** 4053 (1970).
99. L. Andrews and D. L. Frederick, *J. Phys. Chem.*, **73,** 2774 (1969).
100. M. M. Rochkind and G. C. Pimentel, *J. Chem. Phys.*, **42,** 1361 (1965).
101. W. H. Kirchhoff and D. R. Johnson, *J. Mol. Spectrosc.*, **48,** 157 (1973).
102. H. Stammreich, R. Forneris, and K. Sone, *J. Chem. Phys.*, **23,** 972 (1955).
103. G. A. Ozin and A. Vander Voet, *Chem. Commun.*, 896 (1970).
104. H. H. Claassen, G. L. Goodman, J. C. Malm, and F. Schreiner, *J. Chem. Phys.*, **42,** 1229 (1965).
105. P. A. Agron, G. M. Begun, H. A. Levy, A. A. Mason, C. G. Jones, and D. F. Smith, *Science*, **139,** 24 (1962).
106. L. Y. Nelson and G. C. Pimentel, *Inorg. Chem.*, **6,** 1758 (1967).
107. D. N. Waters and B. Basak, *J. Chem. Soc.*, *A*, 2733 (1971).
108. P. Braunstein and R. J. H. Clark, *J. Chem. Soc.*, Dalton, 1845 (1973).

109. A. Loewenschuss, A. Ron, and O. Schnepp, *J. Chem. Phys.*, **49,** 272 (1968); **50,** 2502 (1969).
110. J. W. Hastie, R. Hauge, and J. L. Margrave, *J. Chem. Phys.*, **51,** 2648 (1969).
111. M. E. Jacox and D. E. Milligan, *J. Chem. Phys.*, **51,** 4143 (1969).
112. K. R. Thompson and K. D. Carlson, *J. Chem. Phys.*, **49,** 4379 (1968).
113. D. L. Cocke, C. A. Chang, and K. A. Gingerich, *Appl. Spectrosc.*, **27,** 260 (1973).
114. H. Huber, E. P. Kündig, G. A. Ozin, and A. Vander Voet, *Can. J. Chem.*, **52,** 95 (1974).
115. L. Andrews, *J. Phys. Chem.*, **73,** 3922 (1969).
116. A. Sommer, D. White, M. J. Linevsky, and D. E. Mann, *J. Chem. Phys.*, **38,** 87 (1963).
117. M. J. Linevsky, D. White, and D. E. Mann, *J. Chem. Phys.*, **41,** 542 (1964).
118. A. Snelson, *J. Phys. Chem.*, **74,** 2574 (1970).
119. A. J. Hinchcliffe and J. S. Ogden, *J. Phys. Chem.*, **77,** 2537 (1973); *Chem. Commun.*, 1053 (1969).
120. J. H. Taylor, W. S. Benedict, and J. Strong, *J. Chem. Phys.*, **20,** 1884 (1952).
121. A. H. Nielsen and R. J. Lagemann, *J. Chem. Phys.*, **22,** 36 (1954).
122. T. Wentink, *J. Chem. Phys.*, **29,** 188 (1958).
123. G. W. King and K. Srikameswaran, *J. Mol. Spectrosc.*, **29,** 491 (1969).
124. J. S. Anderson, A. Bos, and J. S. Ogden, *Chem. Commun.*, 1381 (1971).
125. D. E. Milligan and M. E. Jacox, *J. Chem. Phys.*, **55,** 3404 (1971).
126. R. V. St. Louis and B. L. Crawford, Jr., *J. Chem. Phys.*, **42,** 857 (1965).
127. J. W. Nebgen, A. D. McElroy, and H. F. Klodowski, *Inorg. Chem.*, **4,** 1796 (1965).
128. D. F. Smith, Jr., J. Overend, R. C. Spiker, and L. Andrews, *Spectrochim. Acta*, **28A** 87 (1972).
129. L. Andrews and R. C. Spiker, *J. Chem. Phys.*, **59,** 1863 (1973); D. M. Thomas and L. Andrews, *J. Mol. Spectrosc.*, **50,** 220 (1974).
130. A. Barbe, C. Secroun, and P. Jouve, *J. Mol. Spectrosc.*, **49,** 171 (1974).
131. A. G. Hopkins, S. Tang, and C. W. Brown, *J. Am. Chem. Soc.*, **95,** 3486 (1973).
132. T. Chivers and I. Drummond, *Inorg. Chem.*, **11,** 2525 (1972).
133. J. W. Hastie, R. H. Hauge, and J. L. Margrave, *J. Inorg. Nucl. Chem.*, **31,** 281 (1969).
134. A. Barbe and P. Jouve, *J. Mol. Spectrosc.*, **38,** 273 (1971).
135. D. E. Milligan and M. E. Jacox, *J. Chem. Phys.*, **55,** 1003 (1971).
136. S. N. Cesaro, M. Spoliti, A. J. Hinchcliffe, and J. S. Ogden, *J. Chem. Phys.*, **55,** 5834 (1971).
137. M. Spoliti, S. N. Cesaro, and E. Coffari, *J. Chem. Thermodyn.*, **4,** 507 (1972).
138. R. D. Spratley, J. J. Turner, and G. C. Pimentel, *J. Chem. Phys.*, **44,** 2063 (1966).
139. J. S. Ogden and J. J. Turner, *J. Chem. Soc.*, *A*, 1483 (1967).
140. M. M. Rochkind and G. C. Pimentel, *J. Chem. Phys.*, **42,** 1361 (1965).
141. A. Arkell and I. Schwager, *J. Am. Chem. Soc.*, **89,** 5999 (1967).
142. B. Tanguy, B. Frit, G. Turrell, and P. Hagenmuller, *Compt. Rend.*, **264C,** 301 (1967).
142a. A. H. Nielsen and P. J. H. Woltz, *J. Chem. Phys.*, **20,** 1878 (1952).
143. K. O. Christe, C. J. Schack, D. Pilipovich, and W. Sawodny, *Inorg. Chem.*, **8,** 2489 (1969).
144. C. Campbell, J. P. M. Jones, and J. J. Turner, *Chem. Commun.*, 888 (1968).
145. S. D. Gabelnick, G. T. Reedy, and M. G. Chasanov, *J. Chem. Phys.*, **60,** 1167 (1974).
146. N. S. McIntyre, K. R. Thompson, and W. Weltner, *J. Phys. Chem.*, **75,** 3243 (1971).
147. W. Weltner, and D. McLeod, *J. Chem. Phys.*, **42,** 882 (1965).
148. W. Weltner and D. McLeod, *J. Mol. Spectroc.*, **17,** 276 (1965).

149. S. D. Gabelnick, G. T. Reedy, and M. G. Chasanov, *J. Chem. Phys.*, **58,** 4468 (1973).
150. L. H. Jones and R. A. Penneman, *J. Chem. Phys.*, **21,** 542 (1953); L. H. Jones, *ibid.*, **23,** 2105 (1955).
151. W. P. Griffith and J. D. Wickins, *J. Chem. Soc.*, *A*, 400 (1968).
152. G. Pausewang and K. Dehnicke, *Z. Anorg. Allg. Chem.*, **369,** 265 (1969).
153. F. W. Moore and R. E. Rice, *Inorg. Chem.*, **7,** 2510 (1968).
154. B. Šoptrajanov, A. Nickolovski, and I. Petrov, *Spectrochim. Acta*, **24A** 1617 (1968).
155. L. H. Jones and R. A. Penneman, *J. Chem. Phys.*, **21,** 542 (1953).
156. L. H. Jones, *Spectrochim. Acta*, **11,** 409 (1959).
157. S. P. McGlynn, J. K. Smith, and W. C. Neely, *J. Chem. Phys.*, **35,** 105 (1961).
158. H. R. Hoestra, *Inorg. Chem.*, **2,** 492 (1963).
159. J. I. Bullock, *J. Chem. Soc.*, *A*, 781 (1969).
160. K. Ohwada, *Spectrochim. Acta*, **24A** 595 (1968).
161. R. J. Gillespie and M. J. Morton, *Inorg. Chem.*, **9,** 616 (1970).
162. G. Mamantov, E. J. Vasini, M. C. Moulton, D. G. Vickroy, and T. Maekawa, *J. Chem. Phys.*, **54,** 3419 (1971).
163. K. O. Christe, W. Sawodny, and J. P. Guertin, *Inorg. Chem.*, **6,** 1159 (1967).
164. R. J. Gillespie and M. J. Morton, *Inorg. Chem.*, **9,** 811 (1970).
165. H. A. Carter and F. Aubke, *Can. J. Chem.*, **48,** 3456 (1970).
166. K. O. Christe and C. J. Schack, *Inorg. Chem.*, **9,** 2296 (1970).
167. T. Surles, L. A. Quarterman and H. H. Hyman, *J. Inorg. Nucl. Chem.*, **35,** 668 (1973).
168. K. O. Christe and W. Sawodny, *Inorg. Chem.*, **8,** 212 (1969).
169. J. C. Evans and G. Y-S. Lo, *J. Chem. Phys.*, **44,** 3638 (1966).
170. W. Gabes and H. Gerding, *J. Mol. Struct.*, **14,** 267 (1972).
171. R. Forneris and Y. Tavares-Forneris, *J. Mol. Struct.*, **23,** 241 (1974).
171a. J. Shamir and R. Rafaeloff, *Spectrochim. Acta*, **29A,** 873 (1973).
172. W. W. Wilson and F. Aubke, *Inorg. Chem.*, **13,** 326 (1974).
173. D. H. Boal and G. A. Ozin, *J. Chem. Phys.*, **55,** 3598 (1971).
174. A. G. Maki and R. Forneris, *Spectrochim. Acta*, **23A,** 867 (1967).
175. M. Couzi, J. C. Cornut, and P. V. Huong, *J. Chem. Phys.*, **56,** 426 (1972).
176. J. J. Rush, L. W. Schroeder, and A. J. Melveger, *J. Chem. Phys.*, **56,** 2793 (1972).
177. J. C. Evans and G. Y-S. Lo, *J. Phys. Chem.*, **70,** 543 (1966).
178. G. C. Stirling, C. J. Ludman, and T. C. Waddington, *J. Chem. Phys.*, **52,** 2730 (1970).
179. J. W. Nibler and G. C. Pimentel, *J. Chem. Phys.*, **47,** 710 (1967).
180. J. C. Evans and G. Y-S. Lo, *J. Phys. Chem.*, **71,** 3942 (1967).
181. D. E. Milligan and M. E. Jacox, *J. Chem. Phys.*, **55,** 2550 (1971).
182. P. N. Noble, *J. Chem. Phys.*, **56,** 2088 (1972).
183. K. Kaya, N. Mikami, Y. Udagawa, and M. Ito, *Chem. Phys. Lett.*, **16,** 151 (1972).
184. W. Kiefer and H. J. Bernstein, *Chem. Phys. Lett.*, **16,** 5 (1972).
185. M. E. Heyde, L. Rimai, R. G. Kilponen, and D. Gill, *J. Am. Chem. Soc.*, **94,** 5222 (1972).
186. G. R. Smith and W. A. Guillory, *J. Chem. Phys.*, **56,** 1423 (1972).
187. D. E. Milligan and M. E. Jacox, *J. Chem. Phys.*, **43,** 4487 (1965).
188. T. Nakata and S. Matsushita, *J. Phys. Chem.*, **72,** 458 (1968).
189. A. Müller, R. Kebabcioglu, B. Krebs, P. Bouclier, J. Portier, and P. Hagenmuller, *Z. Anorg. Allg. Chem.*, **368,** 31 (1969).
190. W. S. Benedict, N. Gailar, and E. K. Plyler, *J. Chem. Phys.*, **24,** 1139 (1956).
191. G. E. Walrafen, *J. Chem. Phys.*, **40,** 3749 (1964).

192. C. Haas and D. F. Hornig, *J. Chem. Phys.*, **32,** 1763 (1960); D. F. Hornig, H. F. White, and F. P. Reding, *Spectrochim. Acta*, **12,** 338 (1958).
193. S. Pinchas and M. Halmann, *J. Chem. Phys.*, **31,** 1692 (1959).
194. P. A. Staats, H. W. Morgan, and J. H. Goldstein, *J. Chem. Phys.*, **24,** 916 (1956).
195. H. C. Allen and E. K. Plyler, *J. Chem. Phys.*, **25,** 1132 (1956).
196. A. J. Tursi and E. R. Nixon, *J. Chem. Phys.*, **53,** 518 (1970).
197. F. P. Reding and D. F. Hornig, *J. Chem. Phys.*, **27,** 1024 (1957).
198. A. H. Nielsen and H. H. Nielsen, *J. Chem. Phys.*, **5,** 277 (1937).
199. D. M. Cameron, W. C. Sears, and H. H. Nielsen, *J. Chem. Phys.*, **7,** 994 (1939).
200. E. Greinacher, W. Lüttke, and R. Mecke, *Z. Elektrochem.*, **59,** 23 (1955).
201. M. L. Josien and P. Saumagne, *Bull. Soc. Chim. Fr.*, 937 (1956).
202. W. A. Senior and W. K. Thompson, *Nature*, **205,** 170 (1965).
203. J. E. Bertie, H. J. Labbé, and E. Whalley, *J. Chem. Phys.*, **49,** 775, 2141 (1968); J. E. Bertie and E. Whalley, *ibid.*, **40,** 1637 (1964).
204. H. C. Allen, E. D. Tidwell, and E. K. Plyler, *J. Chem. Phys.*, **25,** 302 (1956).
205. J. Pacansky and G. V. Calder, *J. Phys. Chem.*, **76,** 454 (1972).
206. P. A. Staats, H. W. Morgan, and J. H. Goldstein, *J. Chem. Phys.*, **25,** 582 (1956).
207. R. E. Dodd and R. Little, *Spectrochim. Acta*, **16,** 1083 (1960).
208. W. O. Freitag and E. R. Nixon, *J. Chem. Phys.*, **24,** 109 (1956).
209. T. B. Freedman and E. R. Nixon, *J. Chem. Phys.*, **56,** 698 (1972).
210. S. Hemple and E. R. Nixon, *J. Chem. Phys.*, **47,** 4273 (1967).
211. Z. K. Ismail, R. H. Hauge, and J. L. Margrave, *J. Mol. Spectrosc.*, **45,** 304 (1973).
212. D. E. Milligan and M. E. Jacox, *J. Chem. Phys.*, **47,** 278 (1967).
213. Z. K. Ismail, R. H. Hauge, and J. L. Margrave, *J. Chem. Phys.*, **57,** 5137 (1972).
214. E. R. Lory and R. F. Porter, *J. Am. Chem. Soc.*, **93,** 6301 (1971).
215. M. E. Jacox, D. E. Milligan, N. G. Mall, and W. E. Thompson, *J. Chem. Phys.*, **43,** 3734 (1965).
216. A. Maki, E. K. Plyler, and E. D. Tidwell, *J. Res. Nat. Bur. Stand.*, **66A** 163 (1962).
217. T. Wentink Jr., *J. Chem. Phys.*, **29,** 188 (1958); **30,** 105 (1959).
218. W. Beck, *Chem. Ber.*, **95,** 341 (1962); **98,** 298, (1965).
219. K. O. Christe, R. D. Wilson, and W. Sawodny, *J. Mol. Struct.*, **8,** 245 (1971).
220. D. E. Milligan, M. E. Jacox, and A. M. Bass, *J. Chem. Phys.*, **43,** 3149 (1965).
221. G. M. Begun and W. H. Fletcher, *J. Chem. Phys.*, **28,** 414 (1958).
222. D. E. Milligan and M. E. Jacox, *J. Chem. Phys.*, **44,** 2850 (1966).
223. D. E. Milligan and M. E. Jacox, *J. Chem. Phys.*, **47,** 5157 (1967).
224. O. H. Ellestad, P. Klaeboe, E. E. Tucker, and J. Songstad, *Acta Chem. Scand.*, **26,** 172, 1724 (1972).
225. P. Gray and T. C. Waddington, *Trans. Faraday Soc.*, **53,** 901 (1957).
226. P. O. Kinell and B. Strandberg, *Acta Chem. Scand.*, **13,** 1607 (1959).
227. H. W. Morgan, *J. Inorg. Nucl. Chem.*, **16,** 368 (1960).
228. D. E. Milligan and M. E. Jacox, *J. Chem. Phys.*, **39,** 712 (1963); **51,** 277 (1969).
229. J. F. Ogilvie, *Spectrochim. Acta*, **23A** 737 (1967).
230. D. E. Milligan and M. E. Jacox, *J. Chem. Phys.*, **38,** 2627 (1963); *J. Mol. Spectrosc.*, **42,** 495 (1972).
231. J. A. Goleb, H. H. Claassen, M. H. Studier, and E. H. Appelman, *Spectrochim. Acta*, **28A** 65 (1972).
232. I. Schwager and A. Arkell, *J. Am. Chem. Soc.*, **89,** 6006 (1967).
233. R. J. Isabel and W. A. Guillory, *J. Chem. Phys.*, **57,** 1116 (1972).
234. R. R. Smardzewski and W. B. Fox, *J. Am. Chem. Soc.*, **96,** 304 (1974); *J. Chem. Phys.*, **60,** 2104 (1974).

235. A. Müller, G. Nagarajan, O. Glemser, and J. Wegener, *Spectrochim. Acta*, **23A,** 2683 (1967); A. Müller, N. Mohan, S. J. Cyvin, N. Weinstock, and O. Glemser, *J. Mol. Spectrosc.*, **59,** 161 (1976).
236. R. R. Ryan and L. H. Jones, *J. Chem. Phys.*, **50,** 1492 (1969); L. H. Jones, L. B. Asprey, and R. R. Ryan, *ibid.*, **47,** 3371 (1967).
237. L. H. Jones, R. R. Ryan, and L. B. Asprey, *J. Chem. Phys.*, **49,** 581 (1968).
238. J. Laane, L. H. Jones, R. R. Ryan, and L. B. Asprey, *J. Mol. Spectrosc.*, **30,** 485 (1969).
239. D. E. Milligan, M. E. Jacox, A. M. Bass, J. J. Comeford, and D. E. Mann, *J. Chem. Phys.*, **42,** 3187 (1965).
240. M. E. Jacox and D. E. Milligan, *J. Chem. Phys.*, **43,** 866 (1965).
241. R. D. Spratley, J. J. Turner, and G. C. Pimentel, *J. Chem. Phys.*, **44,** 2063 (1966).
242. L. Andrews, F. K. Chi, and A. Arkell, *J. Am. Chem. Soc.*, **96,** 1997 (1974).
243. H. Richert and O. Glemser, *Z. Anorg. Allg. Chem.*, **307,** 328 (1961).
244. C. E. Smith, D. E. Milligan, and M. E. Jacox, *J. Chem. Phys.*, **54,** 2780 (1971).
245. S. C. Peake and A. J. Downs, *J. Chem. Soc.*, Dalton, 859 (1974).
246. G. A. Ozin and A. Vander Voet, *J. Chem. Phys.*, **56,** 4768 (1972).
247. G. Herzberg, *Infrared and Raman Spectra of Polyatomic Molecules*, Van Nostrand, Princeton, N.J., 1945. p. 295.
248. F. P. Reding and D. F. Hornig, *J. Chem. Phys.*, **19,** 594 (1951); **22,** 1926 (1954).
249. H. W. Morgan, P. A. Staats, and J. H. Goldstein, *J. Chem. Phys.*, **27,** 1213 (1957).
250. S. Sundaram and F. F. Cleveland, *J. Mol. Spectrosc.*, **5,** 61 (1960).
251. E. Lee and C. K. Wu, *Trans. Faraday Soc.*, **35,** 1366 (1939).
252. W. H. Haynie and H. H. Nielsen, *J. Chem. Phys.*, **21,** 1839 (1953).
253. P. V. Huong and B. Desbat, *J. Raman Spectrosc.*, **2,** 373 (1974).
254. R. C. Taylor and G. L. Vidale, *J. Am. Chem. Soc.*, **78,** 5999 (1956).
255. R. Savoie and P. A. Giguère, *J. Chem. Phys.*, **41,** 2698 (1964).
256. T. Birchall and I. Drummond, *J. Chem. Soc.*, *A*, 3162 (1971).
257. R. D. Gillard and G. Wilkinson, *J. Chem. Soc.*, 1640 (1964).
258. S. Sundaram, F. Suszek, and F. F. Cleveland, *J. Chem. Phys.*, **32,** 251 (1960).
259. G. DeAlti, G. Costa, and V. Galasso, *Spectrochim. Acta*, **20,** 965 (1964).
260. M. Pariseau, E. Wu, and J. Overend, *J. Chem. Phys.*, **39,** 217 (1963).
261. J. G. Contreras, J. S. Poland, and D. G. Tuck, *J. Chem. Soc.*, Dalton, 922 (1973).
262. D. E. Milligan, M. E. Jacox, and J. J. Comeford, *J. Chem. Phys.*, **44,** 4058 (1966).
263. D. E. Milligan, M. E. Jacox, and W. A. Guillory, *J. Chem. Phys.*, **49,** 5330 (1968).
264. M. E. Jacox and D. E. Milligan, *J. Chem. Phys.*, **49,** 3130 (1968).
265. W. A. Guillory and C. E. Smith, *J. Chem. Phys.*, **53,** 1661 (1970).
266. P. S. Poskozim and A. L. Stone, *J. Inorg. Nucl. Chem.*, **32,** 1391 (1970).
267. I. Wharf and D. F. Shriver, *Inorg. Chem.*, **8,** 914 (1969).
268. J. Shamir and H. H. Hyman, *Spectrochim. Acta*, **23A** 1899 (1967).
269. P. J. Hendra and J. R. Mackenzie, *Chem. Commun.*, 760 (1968).
270. R. J. H. Clark and D. M. Rippon, *J. Mol. Spectrosc.*, **52,** 58 (1974).
271. H. Stammreich, R. Forneris, and Y. Tavares, *J. Chem. Phys.*, **25,** 580 (1956).
272. C. J. Adams and A. J. Downs, *J. Chem. Soc.*, *A*, 1534 (1971).
273. E. Denchik, S. C. Nyburg, G. A. Ozin, and J. T. Szymanski, *J. Chem. Soc.*, *A*, 3157 (1971).
274. V. A. Maroni and P. T. Cunningham, *Appl. Spectrosc.*, **27,** 428 (1973).
275. T. R. Manley and D. A. Williams, *Spectrochim. Acta*, **21,** 1773 (1965).
276. H. E. Doorenbos, J. C. Evans, and R. O. Kagel, *J. Phys. Chem.*, **74,** 3385 (1970).
277. J. A. Evans and D. A. Long, *J. Chem. Soc.*, *A*, 1688 (1968).

278. E. A. Robinson and J. A. Ciruna, *Can. J. Chem.*, **46,** 3197 (1968).
279. J. W. George, N. Katsaros, and K. J. Wynne, *Inorg. Chem.*, **6,** 903 (1967).
280. D. M. Adams and P. J. Lock, *J. Chem. Soc.*, *A*, 145 (1967).
281. J. C. Evans and H. J. Bernstein, *Can. J. Chem.*, **33,** 1270 (1955).
282. H. Siebert, *Z. Anorg. Allg. Chem.*, **275,** 225 (1955).
283. D. J. Gardiner, R. B. Girling, and R. E. Hester, *J. Mol. Struct.*, **13,** 105 (1972).
284. W. Sterzel and W. D. Schnee, *Z. Anorg. Allg. Chem.*, **383,** 231 (1971).
285. H. H. Claassen and G. Knapp, *J. Am. Chem. Soc.*, **86,** 2341 (1964).
286. C. Rocchiciolli, *Compt. Rend.*, **242,** 2922 (1956); **244,** 2704 (1957); **247,** 1108 (1958); **249,** 236 (1959).
287. W. E. Dasent and T. C. Waddington, *J. Chem. Soc.*, 2429, 3350 (1960).
288. L. J. Schoen and D. R. Lide, *J. Chem. Phys.*, **38,** 461 (1963).
289. G. E. Moore and R. M. Badger, *J. Am. Chem. Soc.*, **74,** 6076 (1952).
290. J. J. Comeford, *J. Chem. Phys.*, **45,** 3463 (1966); J. J. Comeford, D. E. Mann, L. J. Schoen, and D. R. Lide, *J. Chem. Phys.*, **38,** 461 (1963).
291. A. Müller, E. Niecke, B. Krebs, and O. Glemser, *Z. Naturforsch.*, **23b,** 588 (1968).
292. A. Müller, K. Königer, S. J. Cyvin, and A. Fadini, *Spectrochim. Acta*, **29A,** 219 (1973).
293. C. R. S. Dean, A. Finch, and P. N. Crates, *J. Chem. Soc.*, Dalton, 1384 (1972).
294. J. K. O'Loane and M. K. Wilson, *J. Chem. Phys.*, **23,** 1313 (1955).
295. D. E. Martz and R. T. Lagemann, *J. Chem. Phys.*, **22,** 1193 (1954).
296. H. Stammreich, R. Forneris, and Y. Tavares, *J. Chem. Phys.*, **25,** 1277 (1956).
297. D. F. Burow, *Inorg. Chem.*, **11,** 573 (1972).
298. R. Paetzold and K. Aurich, *Z. Anorg. Allg. Chem.*, **335,** 281 (1965).
299. L. E. Alexander and I. R. Beattie, *J. Chem. Soc.*, Dalton, 1745 (1972).
300. J. A. Rolfe and L. A. Woodward, *Trans. Faraday Soc.*, **51,** 779 (1955).
301. A. Simon and R. Paetzold, *Z. Anorg. Allg. Chem.*, **301,** 246 (1959); *Naturwissenschaften*, **44,** 108 (1957).
302. M. Falk and P. A. Giguère, *Can. J. Chem.*, **34,** 1680 (1958).
303. D. F. Smith, G. M. Begun, and W. H. Fletcher, *Spectrochim. Acta*, **20,** 1763 (1964).
304. A. J. Arvia and P. J. Aymonino, *Spectrochim. Acta*, **19,** 1449 (1963).
305. K. O. Christe, E. C. Curtis, and C. J. Schack, *Inorg. Chem.*, **11,** 2212 (1972).
306. R. Bougon, J. Isabey, and P. Plurien, *Compt. Rend.*, **273C,** 415 (1971).
307. A. Simon and R. Paetzold, *Z. Anorg. Allg. Chem.*, **303,** 39, 46, 53, 72, 79 (1960); *Z. Elektrochem.*, **64,** 209 (1960).
308. R. Steudel and D. Lautenbach, *Z. Naturforsch.*, **24b,** 350 (1969).
309. M. Goldstein and G. C. Tok, *J. Chem. Soc.*, *A*, 2303 (1971).
310. A. Kaldor and R. F. Porter, *J. Am. Chem. Soc.*, **93,** 2140 (1971).
311. J. Vanderryn, *J. Chem. Phys.*, **30,** 331 (1959).
312. D. A. Dows, *J. Chem. Phys.*, **31,** 1637 (1959).
313. R. J. H. Clark and P. D. Mitchell, *J. Chem. Phys.*, **56,** 2225 (1972).
314. D. A. Dows and G. Bottger, *J. Chem. Phys.*, **34,** 689 (1961).
315. T. Wentink Jr. and V. H. Tiensuu, *J. Chem. Phys.*, **28,** 826 (1958).
316. A. Snelson, *J. Phys. Chem.*, **71,** 3202 (1967).
317. I. R. Beattie and J. R. Horder, *J. Chem. Soc.*, *A*, 2655 (1969).
318. D. F. Wolfe and G. L. Humphrey, *J. Mol. Struct.*, **3,** 293 (1969).
319. G. K. Selivanov and A. A. Mal'tsev, *Zh. Strukt. Khim.*, **14,** 943 (1973).
320. J. E. D. Davies and D. A. Long, *J. Chem. Soc.*, *A*, 2050 (1968).
321. L. Andrews and G. C. Pimentel, *J. Chem. Phys.*, **47,** 3637 (1967).
322. J. E. D. Davies and D. A. Long, *J. Chem. Soc.*, *A*, 2054 (1968).

323. R. A. Frey, R. D. Werder, and H. H. Günthard, *J. Mol. Spectrosc.*, **35,** 260 (1970).
324. R. D. Wesley and C. W. DeKock, *J. Chem. Phys.*, **55,** 3866 (1971); M. Lesiecki, J. W. Nibler, and C. W. DeKock, *ibid.*, **57,** 1352 (1972).
325. K. Shimizu and H. Shingu, *Spectrochim. Acta*, **22,** 1999 (1966).
326. S. Konaka, Y. Murata, K. Kuchitsu, and Y. Morino, *Bull. Chem. Soc. Jap.*, **39,** 1134 (1966).
327. L. Beckmann, L. Gutjahr, and R. Mecke, *Spectrochim. Acta*, **21,** 141 (1965).
328. I. W. Levin and S. Abramowitz, *J. Chem. Phys.*, **43,** 4213 (1965).
329. J. L. Duncan, *J. Mol. Spectrosc.*, **13,** 338 (1964).
330. C. D. Bass, L. Lynds, T. Wolfram, and R. E. DeWames, *J. Chem. Phys.*, **40,** 3611 (1964).
331. W. C. Steele and J. C. Decius, *J. Chem. Phys.*, **25,** 1184 (1956).
332. P. E. Bethell and N. Sheppard, *Trans. Faraday Soc.*, **51,** 9 (1959).
333. S. Bhagavantum and T. Venkatarayudu, *Proc. Indian Acad. Sci.*, **9A,** 224 (1939).
334. A. Müller and M. Stockburger, *Z. Naturforsch.*, **20A,** 1242 (1965).
335. A. Müller, N. Mohan, P. Cristophliemk, I. Tossidis, and M. Dräger, *Spectrochim. Acta*, **29A,** 1345 (1973).
336. A. Müller, G. Gattow, and H. Seidel, *Z. Anorg. Allg. Chem.*, **347,** 24 (1966).
337. I. Nakagawa and J. L. Walter, *J. Chem. Phys.*, **51,** 1389 (1969).
338. K. Stopperka, *Z. Anorg. Allg. Chem.*, **345,** 277 (1966).
339. A. Kalder, A. G. Maki, A. J. Dorney, and I. M. Mills, *J. Mol. Spectrosc.*, **45,** 247 (1973).
340. J. P. Laperches and P. Tarte, *Spectrochim. Acta*, **22,** 1201 (1966).
341. M. E. Jacox and D. E. Milligan, *J. Chem. Phys.*, **54,** 919 (1971).
342. C. C. Addison and B. M. Gatehouse, *J. Chem. Soc.*, 613 (1960).
343. J. R. Ferraro and A. Walker, *J. Chem. Phys.*, **45,** 550 (1966).
344. G. E. Walrafen and D. E. Irish, *J. Chem. Phys.*, **40,** 911 (1964).
345. G. J. Janz and T. R. Kozlowski, *J. Chem. Phys.*, **40,** 1699 (1964).
346. D. Smith, D. W. James, and J. P. Devlin, *J. Chem. Phys.*, **54,** 4437 (1971).
347. C. J. Peacock, A. Müller, and R. Kobabcioglu, *J. Mol. Struct.*, **2,** 163 (1968).
348. P. Thirugnanasambandam and G. J. Srinivasan, *J. Chem. Phys.*, **50,** 2467 (1969).
349. R. A. Frey, R. L. Redington, and A. L. K. Aljbury, *J. Chem. Phys.*, **54,** 344 (1971).
350. R. J. Gillespie, B. Landa, and G. J. Schrobilgen, *Chem. Commun.*, 1543 (1971).
351. S. D. Gabelnick, G. T. Reedy, and M. G. Chasanov, *J. Chem. Phys.*, **59,** 6397 (1973).
352. H. Cohn, C. K. Ingold, and H. G. Poole, *J. Chem. Soc.*, 4272 (1952).
353. D. L. Bernitt, R. H. Miller, and I. C. Hisatsune, *Spectrochim. Acta*, **23A,** 237 (1967).
354. D. L. Bernitt, K. O. Hartman, and I. C. Hisatsune, *J. Chem. Phys.*, **42,** 3553 (1965).
355. K. Itoh and H. J. Bernstein, *Can. J. Chem.*, **34,** 170 (1956).
356. J. Overend and J. C. Evans, *Trans. Faraday Soc.*, **55,** 1817 (1959).
357. R. F. Stratton and A. H. Nielsen, *J. Mol. Spectrosc.*, **4,** 373 (1960).
358. A. J. Downs, *Spectrochim. Acta*, **19,** 1165 (1963).
359. A. Müller, N. Mohan, P. Cristophliemk, I. Tossidis, and M. Dräger, *Spectrochim. Acta*, **29A,** 1345 (1973).
360. C. A. Wamser, W. B. Fox, B. Sukornick, J. R. Holmes, B. B. Stewart, R. Juurik, N. Vanderkooi, and D. Gould, *Inorg. Chem.*, **8,** 1249 (1969).
361. A. Allan, J. L. Duncan, J. H. Holloway, and D. C. McKean, *J. Mol. Spectrosc.*, **31,** 368 (1969).
362. D. F. Wolfe and G. L. Humphrey, *J. Mol. Struct.*, **3,** 293 (1969).
363. J. Goubeau and K. Laitenberger, *Z. Anorg. Allg. Chem.*, **320,** 78 (1963).

364. J. E. Rauch and J. C. Decius, *Spectrochim. Acta*, **22,** 1963 (1966).
365. G. E. McGraw, D. L. Bernitt, and I. C. Hisatsune, *Spectrochim. Acta*, **23A,** 25 (1967).
366. S. T. King and J. Overend, *Spectrochim. Acta*, **23A,** 61 (1967).
367. S. T. King and J. Overend, *Spectrochim. Acta*, **22,** 689 (1966).
368. P. A. Giguère and T. K. K. Srinivasan, *J. Raman Spectrosc.*, **2,** 125 (1974).
369. D. J. Gardiner, N. J. Lawrence, and J. J. Turner, *J. Chem. Soc.*, *A*, 400 (1971).
370. N. Zengin and P. A. Giguère, *Can. J. Chem.*, **37,** 632 (1959).
371. R. D. Brown and G. P. Pez, *Spectrochim. Acta*, **26A,** 1375 (1970).
372. R. Forneris and C. E. Hennies, *J. Mol. Struct.*, **5,** 449 (1970).
373. W. A. Guillory and C. E. Hunter, *J. Chem. Phys.*, **50,** 3516 (1969).
373a. J. A. A. Ketelaar, F. N. Hooge, and G. Blasse, *Rec. Trav. Chim.*, **75,** 220 (1956); F. N. Hooge and J. A. A. Ketelaar, *ibid.*, **77,** 902 (1958).
374. W. Kiefer, *Spectrochim. Acta*, **27A,** 1285 (1971).
375. P. A. Giguère and O. Bain, *J. Phys. Chem.*, **56,** 340 (1952).
376. C. A. Frenzel and K. E. Blick, *J. Chem. Phys.*, **55,** 2715 (1971).
377. W. Beck and K. Feldl, *Agnew. Chem.*, **78,** 746 (1966); W. Beck, P. Swoboda, K. Feldl, and R. S. Tobias, *Chem. Ber.*, **104,** 533 (1971).
378. G. R. Draper and R. L. Werner, *J. Mol. Spectrosc.*, **50,** 369 (1974).
379. D. A. Dows and G. C. Pimental, *J. Chem. Phys.*, **23,** 1258 (1955).
380. J. R. Durig and D. W. Wertz, *J. Chem. Phys.*, **46,** 3069 (1967).
381. C. B. Moore and K. Rosengren, *J. Chem. Phys.*, **44,** 4108 (1966).
382. H. Richert, *Z. Anorg. Allg. Chem.*, **309,** 171 (1961).
383. M. E. Jacox and D. E. Milligan, *J. Chem. Phys.*, **40,** 2457 (1964).
384. W. A. Guillory and C. E. Hunter, *J. Chem. Phys.*, **54,** 598 (1971).
385. R. T. Hall and G. C. Pimentel, *J. Chem. Phys.*, **38,** 1889 (1963).
386. H. H. Eysel, *J. Mol. Struct.*, **5,** 275 (1970).
387. M. J. Nielsen and A. D. E. Pullin, *J. Chem. Soc.*, 604 (1960).
388. W. T. Thompson and W. H. Fletcher, *Spectrochim. Acta*, **22,** 1907 (1966).
389. A. R. Emery and R. C. Taylor, *J. Chem. Phys.*, **28,** 1029 (1958).
390. C. J. H. Schutte, *Spectrochim. Acta*, **16,** 1054 (1960); J. A. A. Ketelaar and C. J. H. Schutte, *ibid.*, **17,** 1240 (1961).
391. A. E. Shirk and D. F. Shriver, *J. Am. Chem. Soc.*, **95,** 5904 (1973).
392. Landolt-Börnstein, *Physikalisch-chemische Tabellen*, Vol. 2, 1951.
393. G. E. MacWood and H. C. Urey, *J. Chem. Phys.*, **4,** 402 (1936).
394. H. M. Kaylor and A. H. Nielsen, *J. Chem. Phys.*, **23,** 2139 (1955).
395. I. F. Kovalev, *Opt. Spektrosk.*, **2,** 310 (1957).
396. R. E. Wilde, T. K. K. Srinivasan, R. W. Harral, and S. G. Sankar, *J. Chem. Phys.*, **55,** 5681 (1971).
397. J. H. Meal and M. K. Wilson, *J. Chem. Phys.*, **24,** 385 (1956).
398. L. P. Lindemann and M. K. Wilson, *J. Chem. Phys.*, **22,** 1723 (1954).
399. I. W. Levin and H. Ziffer, *J. Chem. Phys.*, **43,** 4023 (1965).
400. H. W. Morgan, P. A. Staats, and J. H. Goldstein, *J. Chem. Phys.*, **27,** 1212 (1957).
401. J. R. Durig, D. J. Antion, and F. G. Baglin, *J. Chem. Phys.*, **49,** 666 (1968).
402. J. R. Durig, C. B. Pate, and Y. S. Li, *J. Chem. Phys.*, **54,** 1033 (1971).
403. E. L. Wagner and D. F. Hornig, *J. Chem. Phys.*, **18,** 296, 305 (1950); R. C. Plumb and D. F. Hornig, *ibid.*, **21,** 366 (1953); **23,** 947 (1955); W. Vedder and D. F. Hornig, *ibid.*, **35,** 1560 (1961).
404. A. S. Quist, J. B. Bates, and G. E. Boyd, *J. Phys. Chem.*, **76,** 78 (1972).
405. V. A. Maroni, *J. Chem. Phys.*, **55,** 4789 (1971).

406. A. S. Quist, J. B. Bates, and G. E. Boyd, *J. Chem. Phys.*, **54,** 4896 (1971).
407. J. I. Bullock, N. J. Taylor, and F. W. Parrett, *J. Chem. Soc.*, Dalton, 1843 (1972).
408. J. A. Creighton, *J. Chem. Soc.*, 6589 (1965).
409. B. Gilbert, G. Mamantov, and G. M. Begun, *Inorg. Nucl. Chem. Lett.*, **10,** 1123 (1974).
410. E. Rytter and H. A. Øye, *J. Inorg. Nucl. Chem.*, **35,** 4311 (1973).
411. D. H. Brown and D. T. Stewart, *Spectrochim. Acta,* **26A,** 1344 (1970).
412. G. M. Begun, C. R. Boston, G. Torsi, and G. Mamantov, *Inorg. Chem.*, **10,** 886 (1971).
413. H. A. Øye and W. Bues, *Inorg. Nucl. Chem. Lett.*, **8,** 31 (1972).
414. L. A. Woodward and A. A. Nord, *J. Chem. Soc.*, 2655 (1955).
415. L. A. Woodward and G. H. Singer, *J. Chem. Soc.*, 716 (1958).
416. L. A. Woodward and M. J. Taylor, *J. Chem. Soc.*, 4473 (1960).
417. L. A. Woodward and P. T. Bill, *J. Chem. Soc.*, 1699 (1955).
418. D. M. Adams and D. M. Morris, *J. Chem. Soc.*, *A*, 694 (1968).
419. R. J. H. Clark and D. M. Rippon, *Chem. Commun.*, 1295 (1971); R. J. H. Clark and P. D. Mitchell, *J. Chem. Soc.*, Faraday II, **71,** 515 (1975).
420. R. R. Haun and W. D. Harkins, *J. Am. Chem. Soc.*, **54,** 3917 (1932).
421. H. Stammreich, Y. Tavares, and D. Bassi, *Spectrochim. Acta*, **17,** 661 (1961).
422. A. D. Caunt, L. N. Short, and L. A. Woodward, *Trans. Faraday Soc.*, **48,** 873 (1952); Nature, **168,** 557 (1951).
423. R. J. H. Clark and B. K. Hunter, *J. Mol. Struct.*, **9,** 354 (1971).
424. K. O. Christe, J. P. Guertin, A. E. Pavlath, and W. Sawodny, *Inorg. Chem.*, **6,** 533 (1967).
425. P. Van Huong and B. Desbat, *Bull. Soc. Chim. Fr.*, 2631 (1972).
426. M. Delahaye, P. Dhamelincourt, and J. C. Merlin, *Compt. Rend.*, **272B,** 370 (1971).
427. W. Gabes and H. Gerding, *Rec. Trav. Chim.*, **90,** 157 (1971); W. Gabes, K. Olie, and H. Gerding, *ibid.*, **91,** 1367 (1972).
428. A. Müller and A. Fadini, *Z. Anorg. Allg. Chem.*, **349,** 164 (1967).
429. J. Weidlein and K. Dehnicke, *Z. Anorg. Allg. Chem.*, **337,** 113 (1965).
430. A. Sabatini and L. Sacconi, *J. Am. Chem. Soc.*, **86,** 17 (1964).
431. I. R. Beattie, T. R. Gilson, and G. A. Ozin, *J. Chem. Soc.*, *A*, 534 (1969).
432. J. S. Avery, C. D. Burbridge, and D. M. L. Goodgame, *Spectrochim. Acta*, **24A,** 1721 (1968).
433. J. E. D. Davies and D. A. Long, *J. Chem. Soc.*, *A*, 2054 (1968).
434. G. J. Janz and D. W. James, *J. Chem. Phys.*, **38,** 905 (1963).
435. D. A. Long and J. Y. H. Chau, *Trans Faraday Soc.*, **58,** 2325 (1962).
436. L. E. Alexander and I. R. Beattie, *J. Chem. Soc.*, Dalton, 1745 (1972).
437. R. J. H. Clark, B. K. Hunter, and D. M. Rippon, *Inorg. Chem.*, **11,** 56 (1972); R. J. H. Clark and D. M. Rippon, *J. Mol. Spectrosc.*, **44,** 479 (1972).
438. A. Büchler, J. B. Berkowitz-Mattuck, and D. H. Dugre, *J. Chem. Phys.*, **34,** 2202 (1961).
439. M. F. A. Dove, J. A. Creighton, and L. A. Woodward, *Spectrochim Acta*, **18,** 267 (1962).
440. H. G. M. Edwards, M. J. Ware, and L. A. Woodward, *Chem. Commun.*, 540 (1968).
441. H. G. M. Edwards, L. A. Woodward, M. J. Gall, and M. J. Ware, *Spectrochim. Acta*, **26A,** 287 (1970).
442. W. Krasser and H. W. Nürnberg, *Spectrochim. Acta*, **26A,** 1059 (1970).
443. D. M. Adams and P. J. Lock, *J. Chem. Soc.*, *A*, 620 (1967).
444. D. N. Anderson and R. D. Willett, *Inorg. Chem. Acta*, **8,** 167 (1974).

445. R. D. Willett, J. R. Ferraro, and M. Choca, *Inorg. Chem.*, **13,** 2919 (1974).
446. D. Forster, *Chem. Commun.*, 113 (1967).
447. P. L. Goggin and T. G. Buick, *Chem. Commun.*, 290 (1967).
448. R. A. Work and M. L. Good, *Spectrochim. Acta*, **28A,** 1537 (1972).
448a. J. T. R. Dunsmuir and A. P. Lane, *J. Chem. Soc.*, *A*, 404, 2781 (1971).
449. R. J. H. Clark and P. D. Mitchell, *J. Am. Chem. Soc.*, **95,** 8300 (1973); *J. Raman Spectrosc.*, **2,** 399 (1974).
450. R. J. H. Clark and P. D. Mitchell, *Chem. Commun.*, 762 (1973).
451. T. Kamisuki and S. Maeda, *Chem. Phys. Lett.*, **21,** 330 (1973).
452. S. T. King, *J. Chem. Phys.*, **49,** 1321 (1968).
453. F. Königer and A. Müller, *J. Mol. Spectrosc.*, **56,** 200 (1975).
454. F. Königer, A. Müller, and K. Nakamoto, *Z. Naturforsch.*, **30b,** 456 (1975).
455. D. Tevault, J. D. Brown, and K. Nakamoto, *Appl. Spectrosc.*, **30,** 461 (1976).
456. W. A. Yeranos and J. D. Graham, *Spectrochim. Acta*, **23A,** 732 (1967).
457. A. Müller and B. Krebs, *J. Mol. Spectrosc.*, **24,** 180 (1967); A. Müller and A. Fadini, *Z. Anorg. Allg. Chem.*, **349,** 164 (1967).
458. L. J. Basile, J. R. Ferraro, P. LaBonville, and M. C. Wall, *Coord. Chem. Rev.*, **11,** 21 (1973).
459. C. J. Adams and A. J. Downs, *J. Chem. Soc.*, *A*, 1534 (1971).
460. G. Y. Ahlijah and M. Goldstein, *J. Chem. Soc.*, *A*, 326 (1970); *Chem. Commun.*, 1356 (1968).
461. L. E. Alexander and I. R. Beattie, *J. Chem. Soc.*, Dalton, 1745 (1972).
462. I. W. Levin, *J. Chem. Phys.*, **55,** 5393 (1971).
463. C. J. Adams and A. J. Downs, *Spectrochim. Acta*, **28A,** 1841 (1972).
464. K. O. Christe and W. Sawodny, *Inorg. Chem.*, **12,** 2879 (1973).
465. Landolt-Börnstein, *Physikalisch-Chemische Tabellen*, Vol. 2, 1951.
466. D. Fortnum and J. O. Edwards, *J. Inorg. Nucl. Chem.*, **2,** 264 (1956).
467. E. Steger and K. Herzog, *Z. Anorg. Allg. Chem.*, **331,** 169 (1964).
468. E. Steger and W. Schmidt, *Ber. Bunsenges. Phys. Chem.*, **68,** 102 (1964).
469. B. Krebs and A. Müller, *Spectrochim. Acta*, **22,** 1532 (1966); A. Müller, N. Mohan, P. Christophliemk, I. Tossidis, and M. Dräger, *ibid.*, **29A** 1345 (1973).
470. H. Siebert, *Z. Anorg. Allg. Chem.*, **275,** 225 (1954).
471. L. C. Brown, G. M. Begun, and G. E. Boyd, *J. Am. Chem. Soc.*, **91,** 2250 (1969).
472. E. H. Appelman, *Inorg. Chem.*, **8,** 223 (1969).
473. H. Siebert, *Z. Anorg. Allg. Chem.*, **273,** 21 (1953).
474. R. S. McDowell and L. B. Asprey, *J. Chem. Phys.*, **57,** 3062 (1972).
475. F. Gonzalez-Vilchez and W. P. Griffith, *J. Chem. Soc.*, Dalton, 1416 (1972).
476. N. Weinstock, H. Schulze, and A. Müller, *J. Chem. Phys.*, **59,** 5063 (1973).
477. A. Müller, K. H. Schmidt, K. H. Tytko, J. Bouwma, and F. Jellinek, *Spectrochim. Acta*, **28A,** 381 (1972).
478. A. Müller, N. Weinstock, and H. Schulze, *Spectrochim. Acta*, **28A,** 1075 (1972); K. H. Schmidt and A. Müller, *ibid.*, **28A,** 1829 (1972).
479. A. Müller, B. Krebs, R. Kebabcioglu, M. Stockburger, and O. Glemser, *Spectrochim. Acta*, **24A,** 1831 (1968).
480. A. Müller, E. Diemann, and U. V. K. Rao, *Chem. Ber.*, **103,** 2961 (1970).
481. R. S. McDowell, L. B. Asprey, and L. C. Hoskins, *J. Chem. Phys.*, **56,** 5712 (1972).
482. R. S. McDowell, *Inorg. Chem.*, **6,** 1759 (1967); R. S. McDowell and M. Goldblatt, *ibid.*, **10,** 625 (1971).
483. J. O. Edwards, G. C. Morrison, V. F. Ross, and J. W. Schultz, *J. Am. Chem. Soc.*, **77,** 266 (1955).

484. E. R. Lippincott, J. A. Psellos, and M. C. Tobin, *J. Chem. Phys.*, **20,** 536 (1952).
485. W. Kiefer and H. J. Bernstein, *Mol. Phys.*, **23,** 835 (1972).
486. A. Ranade and M. Stockburger, *Chem. Phys. Lett.*, **22,** 257 (1973).
486a. A. Ranade, W. Krasser, A. Müller, and E. Ahlborn, *Spectrochim. Acta*, **30A,** 139 (1974).
487. R. S. McDowell and L. B. Asprey, *J. Chem. Phys.*, **57,** 3062 (1972).
488. K. H. Schmidt and A. Müller, *Coord. Chem. Rev.*, **14,** 115 (1974).
489. R. H. Mann and P. M. Harris, *J. Mol. Spectrosc.*, **45,** 65 (1973).
490. H. Bürger, S. Biedermann, and A. Ruoff, *Spectrochim. Acta*, **30A,** 1655 (1974).
491. J. Goubeau, F. Haenschke, and A. Ruoff, *Z. Anorg. Allg. Chem.*, **366,** 113 (1969).
492. A. Ruoff, H. Bürger, S. Biedermann, and J. Cichon, *Spectrochim. Acta*, **30A,** 1647 (1974).
493. E. C. Curtis, D. Philipovich, and W. H. Maberly, *J. Chem. Phys.*, **46,** 2904 (1967).
494. A. Finch, P. N. Gates, F. J. Ryan, and F. F. Bentley, *J. Chem. Soc.*, Dalton, 1863 (1973).
495. H. S. Gutowsky and A. D. Liehr, *J. Chem. Phys.*, **26,** 329 (1957).
496. M. L. Delwaulle and F. François, *Compt. Rend.*, **220,** 817 (1945).
497. H. Gerding and M. van Driel, *Rec. Trav. Chim.*, **61,** 419 (1942).
498. A. Müller, B. Krebs, E. Niecke, and A. Ruoff, *Ber. Bunsenges. Phys. Chem.*, **71,** 571 (1967).
499. H. Gerding and R. Westrik, *Rec. Trav. Chim.*, **61,** 842 (1942).
500. M. L. Delwaulle and F. François, *Compt. Rend.*, **226,** 896 (1948).
501. M. Brownstein, P. A. W. Dean, and R. J. Gillespie, *Chem. Commun.*, **9,** (1970).
502. A. Müller, A. Ruoff, B. Krebs, O. Glemser, and W. Koch, *Spectrochim. Acta*, **25A,** 199 (1969).
503. C. S. Alleyne, K. O. Mailer, and R. C. Thompson, *Can. J. Chem.*, **52,** 336 (1974).
504. D. J. Stufkens and H. Gerding, *Rec. Trav. Chim.*, **89,** 417 (1970).
505. E. Steger, I. C. Ciurea, and A. Fadini, *Z. Anorg. Allg. Chem.*, **350,** 225 (1967).
506. H. H. Claassen and E. H. Appelman, *Inorg. Chem.* **9,** 622 (1970).
507. J. Goubeau, E. Kilcioglu, and E. Jacob, *Z. Anorg. Allg. Chem.*, **357,** 190 (1968).
508. H. Selig and H. H. Claassen, *J. Chem. Phys.*, **44,** 1404 (1966).
509. G. A. Ozin and D. J. Reynolds, *Chem. Commun.*, 884 (1969).
510. F. A. Miller and W. K. Baer, *Spectrochim. Acta*, **17,** 114 (1961).
511. H. Stammreich, O. Sala, and D. Bassi, *Spectrochim. Acta*, **19,** 593 (1963).
512. H. Stammreich, O. Sala, and K. Kawai, *Spectrochim. Acta*, **17,** 226 (1961).
513. A. Müller, K. H. Schmidt, E. Ahlborn, and C. J. L. Lock, *Spectrochim. Acta*, **29A,** 1773 (1973).
514. K. H. Schmidt and A. Müller, *Spectrochim. Acta*, **28A,** 1829 (1972).
514a. A. Müller, K. H. Schmidt, and U. Zint, *Spectrochim. Acta*, **32A,** 901 (1976).
515. P. J. Aymonino, H. Schulze, and A. Müller, *Z. Naturforsch.*, **24b,** 1508 (1969); M. J. Reisfeld, L. B. Asprey, and N. A. Matwiyoff, *Spectrochim. Acta*, **27A,** 765 (1971).
516. J. Binenboym, U. El-Gad, and H. Selig, *Inorg. Chem.*, **13,** 319 (1974).
517. A. Guest, H. E. Howard-Lock, and C. J. L. Lock, *J. Mol. Spectrosc.*, **43,** 273 (1972).
518. H. Selig and U. El-Gad, *J. Inorg. Nucl. Chem.*, **35,** 3517 (1973).
519. A. Müller, B. Krebs, and W. Höltje, *Spectrochim. Acta*, **23A,** 2753 (1967).
520. R. J. Collin, W. P. Griffith, and D. Pawson, *J. Mol. Struct.*, **19,** 531 (1973).
521. R. H. Bradley, P. N. Brier, and D. E. H. Jones, *J. Chem. Soc.*, A, 1397 (1971).
522. K. Hamada, G. A. Ozin, and E. A. Robinson, *Bull. Chem. Soc. Jap.*, **44,** 2555 (1971).

523. F. Höfler, *Z. Naturforsch.*, **26a,** 547 (1971).
524. C. A. Clausen and M. L. Good, *Inorg. Chem.*, **9,** 220 (1970).
525. F. Höfler and W. Veigl, *Agnew. Chem. Int. Ed.*, **10,** 919 (1971).
526. K. O. Christe and E. C. Curtis, *Inorg. Chem.*, **11,** 2196 (1972).
527. I. McAlpine and H. Sutcliffe *Spectrochim. Acta*, **25A,** 1723 (1969).
528. J. E. Drake, C. Riddle, and D. E. Rogers, *J. Chem. Soc. A*, 910 (1969).
529. M.-L. Dubois, M.-B. Delhaye, and F. Wallart, *Compt. Rend.*, **269B,** 260 (1969).
530. J. E. Drake and C. Riddle, *J. Chem. Soc.*, A, 2114 (1969).
531. J. Weidlein, *Z. Anorg. Allg. Chem.*, **358,** 13 (1968).
532. S. Sportouch, R. J. H. Clark, and R. Gaufres, *J. Raman Spectrosc.*, **2,** 153 (1974).
533. D. E. Martz and R. T. Langeman, *J. Chem. Phys.*, **22,** 1193 (1954).
534. E. Ahlborn, E. Diemann, and A. Müller, *Chem. Commun.*, 378 (1972).
535. E. Ahlborn, E. Diemann, and A. Müller, *Z. Anorg. Allg. Chem.*, **394,** 1 (1972).
536. M. Muller, M. J. F. Leroy, and R. Rohmer, *Compt. Rend.*, **270C,** 1458 (1970).
537. S. D. Brown, G. L. Gard, and T. M. Loehr, *J. Chem. Phys.*, **64,** 1219 (1976).
538. W. E. Hobbs, *J. Chem. Phys.*, **28,** 1220 (1958).
539. M. Spoliti, J. H. Thirtle, and T. M. Dunn, *J. Mol. Spectrosc.*, **52,** 146 (1974).
540. A. Müller, N. Weinstock, K. H. Schmidt, K. Nakamoto, and C. W. Schläpfer, *Spectrochim. Acta*, **28A,** 2289 (1972).
541. H. H. Claassen, E. L. Gasner, and H. Kim, *J. Chem. Phys.*, **49,** 253 (1968).
542. K. O. Christe and E. C. Curtis, *Inorg. Chem.*, **11,** 35 (1972).
543. K. O. Christe, R. D. Wilson, and E. C. Curtis, *Inorg. Chem.*, **12,** 1358 (1973).
544. G. A. Ozin and A. Vander Voet, *Chem. Commun.*, 1489 (1970).
545. A. Müller, B. Krebs, E. Niecke, and A. Ruoff, *Ber. Busenges. Phys. Chem.*, **71,** 571 (1967).
546. M. L. Delwaulle and F. François, *J. Chim. Phys.*, **46,** 87 (1949); *Compt. Rend.*, **226,** 894 (1948).
547. A. Müller and E. Diemann, *Z. Naturforsch.*, **B24,** 353 (1969).
548. A. Müller and E. Diemann, *Chem. Ber.*, **102,** 2603 (1969).
549. E. Diemann and A. Müller, *Inorg. Nucl. Chem. Lett.*, **5,** 339 (1969).
550. A. Müller and E. Diemann, *Z. Anorg. Allg. Chem.*, **373,** 57 (1970).
551. J. R. Durig and J. W. Clark, *J. Chem. Phys.*, **46,** 3057 (1967).
552. M. L. Delwaulle and F. François, *Compt. Rend.*, **222,** 1193 (1946).
553. A. Müller, E. Niecke, and O. Glemser, *Z. Anorg. Allg. Chem.*, **350,** 246 (1967).
554. T. T. Crow and R. T. Lagemann, *Spectrochim. Acta*, **12,** 143 (1958).
555. G. D. Flesch and H. J. Svec, *J. Am. Chem. Soc.*, **80,** 3189 (1958).
556. K. O. Christe and C. J. Schack, *Inorg. Chem.*, **9,** 1852 (1970).
557. H. Stammreich and R. Forneris, *Spectrochim. Acta*, **16,** 363 (1960).
558. P. Tsao, C. C. Cobb, and H. A. Claassen, *J. Chem. Phys.*, **54,** 5247 (1971).
559. P. J. Hendra, *J. Chem. Soc.*, *A*, 1298 (1967); *Spectrochim. Acta*, **23A,** 2871 (1967).
560. P. L. Goggin and J. Mink, *J. Chem. Soc.*, Dalton, 1479 (1974).
561 Y. M. Bosworth and R. J. H. Clark, *Inorg. Chem.*, **14,** 170 (1975).
562. Y. M. Bosworth and R. J. H. Clark, *J. Chem. Soc.*, Dalton, 381 (1975).
563. J. Hiraishi and T. Shimanouchi, *Spectrochim. Acta*, **22,** 1483 (1966).
564. H. C. Clark, K. R. Dixon, and J. G. Nicolson, *Inorg. Chem.*, **8,** 450 (1969).
565. I. R. Beattie and K. M. Livingston, *J. Chem. Soc.*, *A*, 859 (1969).
566. I. R. Beattie, T. Gilson, K. Livingston, V. Fawcett, and G. A. Ozin, *J. Chem. Soc.*, *A*, 712 (1967).
567. J. I. Bullock, N. J. Taylor, and F. W. Parrett, *J. Chem. Soc.*, *Dalton*, 1843 (1972).

568. J. A. Creighton and J. H. S. Green, *J. Chem. Soc., A*, 808 (1968).
569. I. R. Beattie, K. M. S. Livingston, and D. J. Reynolds, *J. Chem. Phys.*, **51,** 4269 (1969).
570. P. van Huong and B. Desbat, *Bull. Soc. Chim. Fr.*, 2631 (1972).
571. L. C. Hoskins and R. C. Lord, *J. Chem. Phys.*, **46,** 2402 (1967).
572. I. R. Beattie and G. A. Ozin, *J. Chem. Soc., A*, 1691 (1969).
573. T. V. Long, A. W. Herlinger, E. F. Epstein, and I. Bernal, *Inorg. Chem.*, **9,** 459 (1970).
574. H. H. Claassen and H. Selig, *J. Chem. Phys.*, **44,** 4039 (1965).
575. R. D. Werder, R. A. Frey, and H. Günthard, *J. Chem. Phys.*, **47,** 4159 (1967).
576. N. Acquista and S. Abramowitz, *J. Chem. Phys.*, **58,** 5484 (1973).
577. D. E. Sands and A. Zalkin, *Acta Crystallogr.*, **12,** 723 (1959).
578. A. Zalkin and D. E. Sands, *Acta Crystallogr.*, **11,** 615 (1958).
579. R. A. Walton and B. J. Brisdon, *Spectrochim. Acta*, **23A,** 2489 (1967).
580. P. M. Boorman, N. N. Greenwood, M. A. Hildon, and H. J. Whitfield, *J. Chem. Soc., A*, 2017 (1967).
581. A. J. Edwards, *J. Chem. Soc.*, 3714 (1964).
582. L. E. Alexander and I. R. Beattie, *J. Chem. Phys.*, **56,** 5829 (1972).
583. L. E. Alexander, I. R. Beattie, and P. J. Jones, *J. Chem. Soc.*, Dalton, 210 (1972).
584. H. Gerding and H. Houtgraaf, *Rec. Trav. Chim.*, **74,** 5 (1955).
585. R. R. Holmes, *Acc. Chem. Res.*, **5,** 296 (1972).
586. R. R. Holmes, R. M. Deiters, and J. A. Golen, *Inorg. Chem.*, **8,** 2612 (1969).
587. I. W. Levin, *J. Mol. Spectrosc.*, **33,** 61 (1970).
588. H. Selig, J. H. Holloway, J. Tyson, and H. H. Claassen, *J. Chem. Phys.*, **53,** 2559 (1970).
589. R. R. Holmes, *J. Chem. Phys.*, **46,** 3718, 3724, 3730 (1967).
590. J. A. Salthouse and T. C. Waddington, *Spectrochim. Acta*, **23A,** 1069 (1967).
591. D. M. Adams and R. R. Smardzewski, *J. Chem. Soc.*, **A,** 714 (1971).
592. L. E. Alexander and I. R. Beattie, *J. Chem. Soc.*, **A,** 3091 (1971).
593. H. A. Szymanski, R. Yelin, and L. Marabella, *J. Chem. Phys.*, **47,** 1877 (1967).
594. K. O. Christe, E. C. Curtis, C. J. Schack, and D. Pilipovich, *Inorg. Chem.*, **11,** 1679 (1972).
595. G. M. Begun, W. H. Fletcher, and D. F. Smith, *J. Chem. Phys.*, **42,** 2236 (1965).
596. K. O. Christe, *Spectrochim. Acta*, **27A,** 631 (1971).
597. R. A. Frey, R. L. Redington, and A. L. Khidir Aljbury, *J. Chem. Phys.* **54,** 344 (1971).
598. N. Acquista and S. Abramowitz, *J. Chem. Phys.*, **56,** 5221 (1972).
599. J. B. Milne and D. Moffett, *Inorg. Chem.*, **12,** 2240 (1973).
600. K. O. Christe and E. C. Curtis, *Inorg. Chem.*, **11,** 2209 (1972).
601. L. E. Alexander, I. R. Beattie, A. Bukovszky, P. J. Jones, C. J. Marsden, and G. J. Van Schalkwyk, *J. Chem. Soc.*, Dalton, 81 (1974).
601a. R. T. Paine and R. S. McDowell, *Inorg. Chem.*, **13,** 2366 (1974).
602. R. J. Collin, W. P. Griffith, and D. Pawson, *J. Mol. Struct.*, **19,** 531 (1973).
603. M. G. Krishna Pillai and P. Parameswaran Pillai, *Can. J. Chem.*, **46,** 2393 (1968).
604. M. J. Reisfeld, *Spectrochim. Acta*, **29A,** 1923 (1973).
605. T. Barrowcliffe, I. R. Beattie, P. Day, and K. Livingston, *J. Chem. Soc.*, **A,** 1810 (1967).
606. T. G. Spiro, *Inorg. Chem.*, **6,** 569 (1967).
607. G. M. Begun and A. C. Rutenberg, *Inorg. Chem.*, **6,** 2212 (1967).

608. I. R. Beattie, T. Gilson, K. Livingston, V. Fawcett, and G. A. Ozin, *J. Chem. Soc.*, **A,** 712 (1967).
609. T. L. Brown, W. G. McDugle, Jr., and L. G. Kent, *J. Am. Chem. Soc.*, **92,** 3645 (1970).
610. M. Debeau and M. Krauzman, *Compt. Rend.*, **264B,** 1724 (1967).
611. I. Wharf and D. F. Shriver, *Inorg. Chem.*, **8,** 914 (1969).
612. H. G. Mayfield and W. E. Bull, *J. Am. Chem. Soc.*, **A,** 2280 (1971).
613. M. Burgard and J. MacCordick, *Inorg. Nucl. Chem. Lett.*, **6,** 599 (1970).
614. C. J. Adams and A. J. Downs, *J. Inorg. Nucl. Chem.*, **34,** 1829 (1972).
615. M. A. Hooper and D. W. James, *Aust. J. Chem.*, **26,** 1401 (1973).
616. M. A. Hooper and D. W. James, *J. Inorg. Nucl. Chem.*, **35,** 2335 (1973).
617. T. Surles, L. A. Quarterman, and H. H. Hyman, *J. Inorg. Nucl. Chem.*, **35,** 670 (1973).
618. Y. M. Bosworth, R. J. H. Clark, and D. M. Rippon, *J. Mol. Spectrosc.*, **46,** 240 (1973).
619. H. H. Claassen, G. L. Goodman, J. H. Holloway, and H. Selig, *J. Chem. Phys.*, **53,** 341 (1970).
619a. R. T. Paine, R. S. McDowell, L. B. Asprey, and L. H. Jones, *J. Chem. Phys.*, **64,** 3081 (1976).
620. P. J. Hendra and Z. Jović, *J. Chem. Soc.*, **A,** 600 (1968).
621. D. M. Adams and D. M. Morris, *J. Chem. Soc.*, **A,** 2067 (1967).
622. K. O. Christe, *Inorg. Chem.*, **12,** 1580 (1973).
623. R. J. Gillespie and G. J. Schrobilgen, *Inorg. Chem.*, **13,** 1230 (1974).
624. R. Bougon, P. Charpin, and J. Soriano, *Compt. Rend.*, **272C,** 565 (1971).
625. K. Leary and N. Bartlett, *Chem. Commun.*, 903 (1972).
626. K. Wieghardt and H. H. Eysel, *Z. Naturforsch.*, **25b,** 105 (1970).
627. D. M. Adams and D. M. Morris, *J. Chem. Soc.*, **A,** 694 (1968).
628. D. H. Brown, K. R. Dixon, C. M. Livingston, R. H. Nuttall, and D. W. A. Sharp, *J. Chem. Soc.*, **A,** 100 (1967).
629. R. J. H. Clark, L. Maresca, and R. J. Puddephatt, *Inorg. Chem.*, **7,** 1603 (1968).
630. P. C. Crouch, G. W. A. Fowles, and R. A. Walton, *J. Chem. Soc.*, **A,** 972 (1969).
631. P. A. W. Dean and D. F. Evans, *J. Chem. Soc.*, **A,** 698 (1967).
632. D. M. Adams and D. C. Newton, *J. Chem. Soc.*, **A,** 2262 (1968).
633. W. von Bronswyk, R. J. H. Clark, and L. Maresca, *Inorg. Chem.*, **8,** 1395 (1969).
634. R. Becker and W. Sawodney, *Z. Naturforsch.*, **28b,** 360 (1973).
635. R. A. Walton and B. J. Brisdon, *Spectrochim. Acta*, **23A,** 2222 (1967).
636. O. L. Keller, *Inorg. Chem.*, **2,** 783 (1963).
637. S. M. Horner, R. J. H. Clark, B. Crociani, D. B. Copley, W. W. Horner, F. N. Collier, and S. Y. Tyree, *Inorg. Chem.*, **7,** 1859 (1968).
638. I. R. Beattie, T. R. Gilson, and G. A. Ozin, *J. Chem. Soc.*, 2765 (1968).
639. G. A. Ozin, G. W. A. Fowles, D. J. Tidmarsh, and R. A. Walton, *J. Chem. Soc.*, **A,** 642 (1969).
640. O. L. Keller and A. Chetham-Strode, *Inorg. Chem.*, **5,** 367 (1966).
641. B. Weinstock and G. L. Goodman, *Adv. Chem. Phys.*, **11,** 169 (1965).
642. H. H. Eysel, *Z. Anorg. Allg. Chem.*, **390,** 210 (1972).
643. C. D. Flint, *J. Mol. Spectrosc.*, **37,** 414 (1971).
644. P. W. Frais, C. J. L. Lock, and A. Guest, *Chem. Commun.*, 1612 (1970).
645. G. L. Bottger and C. V. Damsgard, *Spectrochim. Acta*, **28A,** 1631 (1972).
646. L. A. Woodward and M. J. Ware, *Spectrochim. Acta*, **20,** 711 (1964).
647. M. J. Reisfeld, *J. Mol. Spectrosc.*, **29,** 120 (1969).

648. M. Debeau and H. Poulet, *Spectrochim. Acta*, **25A,** 1553 (1969).
649. P. J. Hendra and P. J. D. Park, *Spectrochim. Acta*, **23A,** 1635 (1967).
650. L. A. Woodward and M. J. Ware, *Spectrochim. Acta*, **19,** 775 (1963).
651. D. M. Adams and H. A. Gebbie, *Spectrochim. Acta*, **19,** 925 (1963).
652. J. M. Fletcher, W. E. Gardner, A. C. Fox, and G. Topping, *J. Chem. Soc.*, **A,** 1038. (1967).
653. D. A. Kelley and M. L. Good, *Spectrochim. Acta*, **28A,** 1529 (1972).
654. M. Debeau, *Spectrochim. Acta*, **25A,** 1311 (1969).
655. L. A. Woodward and M. J. Ware, *Spectrochim. Acta*, **24A,** 921 (1968).
656. J. L. Ryan, *J. Inorg. Nucl. Chem.*, **33,** 153 (1971).
657. E. Stumpp and G. Piltz, *Z. Anorg. Allg. Chem.*, **409,** 53 (1974).
658. B. W. Berringer, J. B. Gruber, T. M. Loehr, and G. P. O'Leary, *J. Chem. Phys.*, **55,** 4608 (1971).
659. D. M. Yost, C. S. Steffens, and S. T. Gross, *J. Chem. Phys.*, **2,** 311 (1934).
660. C. J. Adams and A. J. Downs, *Chem. Commun.*, 1699 (1970).
661. L. A. Woodward and J. A. Creighton, *Spectrochim. Acta*, **17,** 594 (1961).
662. Y. M. Bosworth and R. J. H. Clark, *J. Chem. Soc.*, Dalton, 1749 (1974).
663. E. Martineau and J. B. Milne, *J. Chem. Soc.*, **A,** 2971 (1970).
664. H. P. Fritz, W. P. Griffith, G. Stefaniak, and R. S. Tobias, *Z. Naturforsch.*, **25b,** 1087 (1970).
665. W. Krasser and K. Schwochau, *Z. Naturforsch.*, **25a,** 206 (1970).
666. B. Weinstock, H. H. Claassen, and C. L. Chernick, *J. Chem. Phys.*, **38,** 1470 (1963).
667. H. Kim, H. H. Claassen, and E. Pearson, *Inorg. Chem.*, **7,** 616 (1968).
668. H. H. Claassen, G. L. Goodman, and H. Kim, *J. Chem. Phys.*, **56,** 5042 (1972).
669. P. J. Cresswell, J. E. Fergusson, B. R. Penfold, and D. E. Scaife, *J. Chem. Soc.*, Dalton, 254 (1972).
670. H. Kim, P. A. Souder, and H. H. Claassen, *J. Mol. Spectrosc.*, **26,** 46 (1968).
671. P. LaBonville, J. R. Ferraro, M. C. Wall, and L. J. Basile, *Coord. Chem. Rev.*, **7,** 257 (1972).
672. J. Hauck and A. Fadini, *Z. Naturforsch.*, **25b,** 422 (1970).
673. J. Hauck, *Z. Naturforsch.*, **25b,** 224, 468, 647 (1970).
674. C. J. Adams and A. J. Downs, *J. Inorg. Nucl. Chem.*, **34,** 1829 (1972).
674a. G. Goetz, M. Deneux, and M. J. F. Leroy, *Bull. Soc. Chim. Fr.*, 29 (1971).
675. J. E. Griffith, *Spectrochim. Acta*, **23A,** 2145 (1967).
676. K. O. Christe, C. J. Schack, D. Pilipovich, E. C. Curtis, and W. Sawodny, *Inorg. Chem.*, **12,** 620 (1973).
677. K. O. Christe, C. J. Schack, and E. C. Curtis, *Inorg. Chem.*, **11,** 583 (1972).
678. K. Seppelt, *Z. Anorg. Allg. Chem.*, **399,** 87 (1973).
679. J. H. Holloway, H. Selig, and H. H. Claassen, *J. Chem. Phys.*, **54,** 4305 (1971).
680. K. Dehnicke, G. Pausewang, and W. Rüdorff, *Z. Anorg. Allg. Chem.*, **366,** 64 (1969).
681. G. A. Ozin, G. W. A. Fowles, D. J. Tidmarsh, and R. A. Walton, *J. Chem. Soc.*, **A,** 642 (1969).
682. R. J. Collin, W. P. Griffith, and D. Pawson, *J. Mol. Struct.*, **19,** 531 (1973).
683. D. M. Adams, G. W. Fraser, D. M. Morris, and R. D. Peacock, *J. Chem. Soc.*, **A,** 1131 (1968).
684. J. P. Brunette and M. J. F. Leroy, *J. Inorg. Nucl. Chem.*, **36,** 289 (1974).
685. A. Sabatini and I. Bertini, *Inorg. Chem.*, **5,** 204 (1966).
686. W. Preetz and H. J. Walter, *J. Inorg. Nucl. Chem.*, **33,** 3179 (1971).
687. G. Pausewang, W. Rüdorff, and K. Dehnicke, *Z. Anorg. Allg. Chem.*, **364,** 83 (1969).
688. J. N. Murrell, *J. Chem. Soc.*, **A,** 297 (1969).

689. H. H. Claassen, E. L. Gasner, and H. Selig, *J. Chem. Phys.*, **49,** 1803 (1968).
690. H. H. Eysel and K. Seppelt, *J. Chem. Phys.*, **56,** 5081 (1972).
691. Nguyen-Quy-Dao, *Bull. Soc. Chim. Fr.*, 3976 (1968).
692. J. L. Hoard, W. G. Martin, M. E. Smith, and J. E. Whitney. *J. Am. Chem. Soc.*, **76,** 3820 (1954).
693. R. Stomberg and C. Brosset, *Acta Chem. Scand.*, **14,** 441 (1960).
694. K. O. Hartman and F. A. Miller, *Spectrochim. Acta*, **24A,** 669 (1968).
695. C. W. F. T. Pistorius, *Bull. Soc. Chim. Belg.*, **68,** 630 (1959).
696. H. L. Schlaefer and H. F. Wasgestian, *Theor. Chim. Acta.* **1,** 369 (1963).
697. J. R. Durig, J. W. Thompson, J. D. Witt, and J. D. Odom, *J. Chem. Phys.*, **58,** 5339 (1973).
698. D. E. Mann and L. Fano, *J. Chem. Phys.*, **26,** 1665 (1957).
699. J. D. Odom, J. E. Saunders, and J. R. Durig, *J. Chem. Phys.*, **56,** 1643 (1972).
700. L. A. Minon, K. S. Seshadri, R. C. Taylor, and D. White, *J. Chem. Phys.*, **53,** 2416 (1970).
701. G. M. Begun and W. H. Fletcher, *J. Mol. Spectrosc.*, **4,** 388 (1960).
702. I. C. Hisatsune, J. P. Devlin, and Y. Wada, *J. Chem. Phys.*, **33,** 714 (1960).
703. G. M. Begun and W. H. Fletcher, *Spectrochim. Acta*, **19,** 1343 (1963).
704. H. Murata and K. Kawai, *J. Chem. Phys.*, **25,** 589, 796 (1956).
705. R. E. Hester and R. A. Plane, *Inorg. Chem.*, **3,** 513 (1964).
706. J. R. Durig and J. W. Clark, *J. Chem. Phys.*, **48,** 3216 (1968).
707. J. R. Durig, S. F. Bush, and E. E. Mercer, *J. Chem. Phys.*, **44,** 4238 (1966).
708. E. R. Nixon, *J. Phys. Chem.*, **60,** 1054 (1956).
708a. S. G. Frankiss, *Inorg. Chem.*, **7,** 1931 (1968).
708b. K. H. Rhee, A. M. Snider, and F. A. Miller, *Spectrochim. Acta*, **29A,** 1029 (1973).
709. S. G. Frankiss and F. A. Miller, *Spectrochim. Acta*, **21,** 1235 (1965).
710. S. G. Frankiss, F. A. Miller, H. Stammreich, and Th. T. Sans, *Spectrochim. Acta*, **23A,** 543 (1967).
711. R. P. Bell and H. C. Longuet-Higgins, *Proc. Roy. Soc.*, **A183,** 357 (1945).
712. R. C. Lord and E. Nielsen, *J. Chem. Phys.*, **19,** 1 (1951).
713. W. J. Lehmann, J. F. Ditter, and I. Shapiro, *J. Chem. Phys.*, **29,** 1248 (1958).
714. A. Snelson, *J. Phys. Chem.*, **71,** 3202 (1967).
715. V. A. Maroni, D. M. Gruen, R. L. McBeth, and E. J. Cairns, *Spectrochim. Acta*, **26A,** 418 (1970).
716. D. M. Adams and R. G. Churchill, *J. Chem. Soc.*, **A,** 697 (1970).
717. I. R. Beattie, T. Gilson, and P. Cocking, *J. Chem. Soc.*, **A,** 702 (1967).
718. R. A. Frey, R. D. Werder, and H. H. Günthard, *J. Mol. Spectrosc.*, **35,** 260 (1970).
719. T. Onishi and T. Shimanouchi, *Spectrochim. Acta*, **20,** 721 (1964).
720. N. N. Greenwood, D. J. Prince, and B. P. Straughan, *J. Chem. Soc.*, **A,** 1694 (1968).
721. P. L. Goggin, *J. Chem. Soc.*, Dalton, 1483 (1974).
722. R. Forneris, J. Hiraishi, F. A. Miller, and M. Uehara, *Spectrochim. Acta.* **26A,** 581 (1970).
723. B. L. Crawford, W. H. Avery, and J. W. Linnett, *J. Chem. Phys.*, **6,** 682 (1938).
724. G. W. Bethke and M. K. Wilson, *J. Chem. Phys.*, **26,** 1107 (1957).
725. D. A. Dows and R. M. Hexter, *J. Chem. Phys.*, **24,** 1029 (1956).
726. R. G. Snyder and J. C. Decius, *Spectrochim. Acta*, **13,** 280 (1959).
727. W. G. Palmer, *J. Chem. Soc.*, 1552 (1961).
728. K. Buijs, *J. Chem. Phys.*, **36,** 861 (1962).
729. C. A. Evans, K. H. Tan, S. P. Tapper, and M. J. Taylor, *J. Chem. Soc.*, Dalton, 988 (1973).

729a. F. Höfler, S. Waldhör, and E. Hengge, *Spectrochim. Acta*, **28A,** 29 (1972).
730. F. Höfler, W. Sawodny, and E. Hengge, *Spectrochim. Acta*, **26A,** 819 (1970).
731. J. E. Griffiths, *Spectrochim. Acta*, **25A,** 965 (1969).
732. G. A. Ozin, *J. Chem. Soc.*, **A,** 2952 (1969).
733. W. Bues, K. Buchler, and P. Kuhnle, *Z. Anorg. Allg. Chem.*, **325,** 8 (1963).
734. R. G. Brown and S. D. Ross, *Spectrochim. Acta*, **28,** 1263 (1972).
735. I. R. Beattie and G. A. Ozin, *J. Chem. Soc.*, **A,** 2615 (1969).
736. J. Roziere, J.-L. Pascal, and A. Potier, *Spectrochim. Acta*, **29A,** 169 (1973).
737. A. Grodzicki and A. Potier, *J. Inorg. Nucl. Chem.*, **35,** 61 (1973).
738. J. D. Witt and R. M. Hammaker, *J. Chem. Phys.*, **58,** 303 (1973).
739. A. Hezel and S. D. Ross, *Spectrochim. Acta*, **23A,** 1583 (1967).
740. R. W. Mooney and R. L. Goldsmith, *J. Inorg. Nucl. Chem.*, **31,** 933 (1969).
741. R. M. Wing and K. P. Callahan, *Inorg. Chem.*, **8,** 871 (1969).
742. A. Simon and H. Richter, *Z. Anorg. Allg. Chem.*, **304,** 1 (1960); **315,** 196 (1962).
743. R. J. Ziegler and W. M. Risen, Jr., *Inorg. Chem.*, **11,** 2796 (1972).
744. P. C. Crouch, G. W. A. Fowles, and R. A. Walton, *J. Chem. Soc.*, **A,** 972 (1969).
745. R. A. Work and M. L. Good, *Inorg, Chem.*, **9,** 956 (1970).
746. D. A. Edwards and R. T. Ward, *J. Chem. Soc.*,**A,** 1617 (1970).
747. J. K. Wilmshurst and H. J. Bernstein, *Can. J. Chem.*, **35,** 191 (1957).
748. R. E. Dodd, L. A. Woodward, and H. L. Roberts, *Trans. Faraday Soc.*, **53,** 1545 (1957).
749. W. J. Lehmann and I. Shapiro, *Spectrochim. Acta*, **17,** 396 (1961).
750. L. J. Bellamy, W. Gerrard, M. F. Lappert, and R. L. Williams, *J. Chem. Soc.*, 2412 (1958).
751. A. Meller, *Organometal. Chem. Rev.*, **2,** 1 (1967).
752. J. G. Verkade, *Coord. Chem. Rev.*, **9,** 1 (1972).
753. W. Weltner, Jr., and J. R. Warn, *J. Chem. Phys.*, **37,** 292 (1962).
754. I. Hisatsune, *Inorg. Chem.*, **3,** 168 (1964).
755. E. R. Lory and R. F. Porter, *J. Am. Chem. Soc.*, **93,** 6301 (1971).
756. A. Kaldor and R. F. Porter, *Inorg. Chem.*, **10,** 775 (1971).
757. J. L. Parsons, *J. Chem. Phys.*, **33,** 1860 (1960).
758. F. A. Grimm and R. F. Porter, *Inorg. Chem.*, **8,** 731 (1969).
759. J. R. Durig, W. H. Green, and A. L. Marston, *J. Mol. Struct.*, **2,** 19 (1968).
760. W. Bues, G. Foerster, and R. Schmitt, *Z. Anorg. Allg. Chem.*, **344,** 148 (1966).
760a. D. A. Coe and J. W. Nibler, *Spectrochim. Acta*, **29A,** 1789 (1973).
761. H. J. Hrostowski, and G. C. Pimentel, *J. Am. Chem. Soc.*, **76,** 998 (1954).
762. W. E. Keller and H. L. Johnston, *J. Chem. Phys.*, **20,** 1749 (1952).
763. F. Klanberg, E. L. Muetterties, and L. J. Guggenberger, *Inorg. Chem.*, **7,** 2272 (1968).
764. G. F. Lanthier and A. G. Massey, *J. Inorg. Nucl. Chem.*, **32,** 1807 (1970).
765. P. G. Davies, M. Goldstein, and H. A. Willis, *Inorg. Nucl. Chem. Lett.*, **3,** 249 (1967).
766. T. Totani and H. Watanabe, *Spectrochim. Acta*, **25A,** 585 (1969).
767. K. E. Blick, J. W. Dawson, and K. Niedenzu, *Inorg. Chem.*, **9,** 1416 (1970).
768. K. E. Blick, K. Niedenzu, W. Sawodny, T. Takasuka, T. Totani, and H. Watanabe, *Inorg. Chem.*, **10,** 1133 (1971).
769. N. N. Greenwood and B. S. Thomas, *J. Chem. Soc.*, **A,** 814 (1971).
770. I. R. Beattie and G. A. Ozin, *J. Chem. Soc.*, **A,** 2372 (1968).
771. C. A. Evans and M. J. Taylor, *J. Chem. Soc.*, **A,** 1343 (1969).
772. A. Balls, N. N. Greenwood, and B. P. Straughan, *J. Chem. Soc.*, **A,** 753 (1968).

773. H. Burger, J. Cichon, V. Goetze, U. Wannagat, and H. J. Wismar, *J. Organometal. Chem.*, **33,** 1 (1971).
774. A. L. Smith, *Spectrochim. Acta*, **16,** 87 (1960).
775. B. J. Aylett, *Adv. Inorg. Chem. Radiochem.*, **11,** 262 (1968).
776. H. J. Campbell-Ferguson and E. A. V. Ebsworth, *J. Chem. Soc.*, **A,** 705 (1967).
777. H. Schumann, *Angew. Chem. Int. Ed.*, **8,** 937 (1969).
778. L. A. Carreira, R. O. Carter, J. R. Durig, R. C. Lord, and C. C. Milionis, *J. Chem. Phys.*, **59,** 1028 (1973).
779. W. H. Smith and J. B. Bates, *J. Raman Spectrosc.*, **1,** 83 (1973).
780. J. B. Bates and W. H. Smith, *Chem. Phys. Lett.*, **14,** 362 (1972).
781. J. B. Bates and A. S. Quist, *J. Chem. Phys.* **56,** 1528 (1972).
782. S. M. Shapiro, D. C. O'Shea, and H. Z. Cummins, *Phys. Rev. Lett.* **19,** 361 (1967).
783. W. P. Griffith, *J. Chem. Soc.*, **A,** 905 (1967).
784. J. Etchepare, *Spectrochim. Acta*, **26A,** 2147 (1970).
785. P. Tarte, M. J. Pottier, and A. M. Procès, *Spectrochim. Acta*, **29A,** 1017 (1973).
786. J. S. Anderson and J. S. Ogden, *J. Chem. Phys.*, **51,** 4189 (1969).
787. J. S. Ogden and M. J. Ricks, *J. Chem. Phys.*, **52,** 352 (1970).
788. J. S. Ogden and M. J. Ricks, *J. Chem. Phys.*, **53,** 896 (1970).
789. J. S. Ogden and M. J. Ricks, *J. Chem. Phys.*, **56,** 1658 (1972).
790. D. C. McKean, *Spectrochim. Acta*, **26A,** 1833 (1970).
791. H. R. Linton and E. R. Nixon, *J. Chem. Phys.*, **29,** 921 (1958).
792. F. A. Miller, J. Perkins, G. A. Gibbon, and B. A. Swisshelm, *J. Raman Spectrosc.*, **2,** 93 (1974).
793. E. A. V. Ebsworth, D. W. H. Rankin, and G. M. Sheldrick, *J. Chem. Soc.*, **A,** 2828 (1968).
794. D. Solan and A. B. Burg, *Inorg. Chem.*, **11,** 1253 (1972).
795. M. D. Curtis and P. Wolber, *Inorg. Chem.*, **11,** 431 (1972).
796. M. F. Dove, R. King, and T. J. King, *Chem. Commun*, 944 (1973).
797. D. E. C. Corbridge, in *Topics in Phosphorus Chemistry*, M. Grayson and E. J. Griffith, eds., Vol. 6, Interscience, New York, 1969, p. 235.
798. D. E. C. Corbridge, *The Structural Chemistry of Phosphorus*, Elsevier, Amsterdam, 1974.
799. L. C. Thomas, *Interpretation of the Infrared Spectra of Organophosphorus Compounds*, Heyden, London, 1974.
800. E. Steger, *Z. Chem.*, **12,** 52 (1972).
801. L. C. Thomas and R. A. Chittenden, *Spectrochim. Acta*, **26A,** 781 (1970).
802. C. H. Bibart and G. E. Ewing, *J. Chem. Phys.*, **61,** 1293 (1974).
803. R. Teranishi and J. C. Decius, *J. Chem. Phys.*, **22,** 896 (1954); **21,** 116 (1953).
804. I. C. Hisatsune, J. P. Devlin, and Y. Wada, *Spectrochim. Acta*, **18,** 1641 (1962).
805. G. E. McGraw, D. L. Bernitt, and I. C. Hisatsune, *Spectrochim. Acta*, **23A,** 25 (1967).
806. G. E. McGraw, D. L. Bernitt, and I. C. Hisatsune, *J. Chem. Phys.*, **42,** 237 (1965).
807. K. O. Christe, C. J. Schack, and R. D. Wilson, *Inorg. Chem.*, **13,** 2811 (1974).
808. R. E. Nightingale and E. L. Wagner, *J. Chem. Phys.*, **22,** 203 (1954).
809. K. Tamagake, Y. Hamada, J. Yamaguchi, A. Y. Hirakawa, and M. Tsuboi, *J. Mol. Spectrosc.*, **49,** 232 (1974).
810. D. L. Frasco and E. L. Wagner, *J. Chem. Phys.*, **30,** 1124 (1959).
811. K. Knuth, J. Jander, and U. Engelhardt, *Z. Anorg. Allg. Chem.*, **392,** 279 (1972).
812. Y. M. Bosworth, R. J. H. Clark, and D. M. Rippon, *J. Mol. Spectrosc.*, **46,** 240 (1973).

813. S. B. Brumback and G. M. Rosenblatt, *J. Chem. Phys.*, **56,** 3110 (1972).
814. I. R. Beattie, K. M. S. Livingston, G. A. Ozin, and P. J. Reynolds, *J. Chem. Soc.*, **A,** 449 (1970).
815. A. C. Chapman and L. E. Thirlwell, *Spectrochim. Acta*, **20,** 937 (1964).
816. M. Tsuboi, *J. Am. Chem. Soc.*, **79,** 1351 (1957).
817. W. P. Griffith and K. J. Rutt, *J. Chem. Soc.*, **A,** 2331 (1968).
818. W. P. Griffith, *J. Chem. Soc.*, **A,** 905 (1967).
819. E. Steger and A. Simon, *Z. Anorg. Allg. Chem.*, **291,** 76 (1957); **294,** 1, 147 (1958).
820. M. Baudler and M. Mengel, *Z. Anorg. Allg. Chem.*, **373,** 285 (1970).
821. M. Gardner, *J. Chem. Soc.*, Dalton, 691 (1973).
822. M. Meisel and H. Grunze, *Z. Anorg. allg. hem.*, **373,** 265 (1970).
823. T. J. Bastow and H. J. Whitfield, *J. Chem. Soc.*, Dalton, 1739 (1973).
824. G. A. Ozin, *J. Chem. Soc.*, **A,** 2307 (1970).
825. D. B. Powell and J. G. V. Scott, *Spectrochim. Acta*, **28A,** 1067 (1972).
826. R. R. Holmes and C. J. Hora, Jr., *Inorg. Chem.*, **11,** 2506 (1972).
827. F. Seel and K. Velleman, *Z. Anorg. Allg. Chem.*, **385,** 123 (1971).
828. J. A. Salthouse and T. C. Waddington, *Spectrochim. Acta*, **23A,** 1069 (1967).
829. A. Müller, F. Königer, S. J. Cyvin, and A. Fadini, *Spectrochim. Acta*, **29A,** 219 (1973).
830. I. C. Hisatsune, *Spectrochim. Acta*, **21,** 1899 (1965).
831. T. R. Manley and D. A. Williams, *Spectrochim. Acta*, **23A,** 149 (1967).
832. E. Steger and K. Lunkwitz, *J. Mol. Struct.*, **3,** 67 (1969).
833. P. J. Hendra, R. A. Johns, C. T. S. Miles, and C. J. Vear, *Spectrochim. Acta.* **26A,** 2169 (1970).
834. R. A. Work and M. L. Good, *Spectrochim. Acta*, **29A,** 1547 (1973).
835. I. R. Beattie, K. M. S. Livingston, G. A. Ozin, and D. J. Reynolds, *J. Chem. Soc.*, **A,** 958 (1969).
836. E. Diemann and A. Müller, *Coord. Chem. Rev.*, **10,** 79 (1973).
837. L. A. Nimon, V. D. Neff, R. E. Cantley, and R. O. Buttlar, *J. Mol. Spectrosc.*, **22,** 105 (1967).
838. M. Gardner and A. Rogstad, *J. Chem. Soc.*, Dalton, 599 (1973).
839. G. A. Ozin, *J. Chem. Soc.*, **A,** 116 (1969).
840. H. Wieser, P. J. Krueger, E. Muller, and J. B. Hyne, *Can. J. Chem.*, **47,** 1633 (1969).
841. A. W. Herlinger and T. V. Long, *Inorg. Chem.*, **8,** 2661 (1969).
842. K. Stopperka and F. Kilz, *Z. Anorg. Allg. Chem.*, **370,** 49 (1969).
843. R. L. Patton and W. L. Jolly, *Inorg. Chem.*, **8,** 1392, 1389 (1969).
844. P. J. Ashley and E. G. Torrible, *Can. J. Chem.*, **47,** 2587 (1969).
845. T. Chivers and I. Drummond, *Inorg. Chem.*, **13,** 1222 (1974).
846. K. O. Christe, C. J. Schack, and E. C. Curtis, *Spectrochim. Acta*, **26A,** 2367 (1970).
847. R. Kebabcioglu, R. Mews, and O. Glemser, *Spectrochim. Acta*, **28A,** 1593 (1972).
848. R. J. Gillespie and G. P. Pez, *Inorg. Chem.*, **8,** 1229 (1969).
849. J. Loub, *Z. Anorg. Allg. Chem.*, **362,** 98 (1968).
850. N. E. Erickson and A. G. Maddock, *J. Chem. Soc.*, **A,** 1665 (1970).

# Coordination Compounds

## *Part III*

## III-1. AMMINE, AMIDO, AND RELATED COMPLEXES

### (1) Ammine ($NH_3$) Complexes

Vibrational spectra of metal ammine complexes have been studied extensively, and these are reviewed by Schmidt and Müller.[1] Figure III-1 shows the infrared spectra of typical hexammine complexes in the high frequency region. To assign these $NH_3$ group vibrations, it is convenient to use the six normal modes of vibration of a simple 1:1 (metal/ligand) complex model such as that shown in Fig. III-2. Table III-1 lists the infrared frequencies and band assignments of hexammine complexes. It is seen that the antisymmetric and symmetric $NH_3$ stretching, $NH_3$ degenerate deformation, $NH_3$ symmetric deformation, and $NH_3$ rocking vibrations appear in the regions of 3400–3000, 1650–1550, 1370–1000, and 950–590 $cm^{-1}$, respectively.

The $NH_3$ stretching frequencies of the complexes are lower than those of the free $NH_3$ molecule for two reasons.[18] One is the effect of coordination. Upon coordination, the N–H bond is weakened and the $NH_3$ stretching frequencies are lowered. The stronger the M–N bond, the weaker is the N–H bond and the lower are the $NH_3$ stretching frequencies if other conditions are equal. Thus the $NH_3$ stretching frequencies may be used as a rough measure of the M–N bond strength. The other reason is the effect of the counter-ion. The $NH_3$ stretching frequencies of the chloride are much lower than those of the perchlorate, for example. This is attributed to the weakening of the N–H bond due to the formation of the N–H . . . Cl type hydrogen bond in the former.

The effects of coordination and hydrogen bonding mentioned above shift the $NH_3$ deformation and rocking modes to higher frequencies. Among them, the $NH_3$ rocking mode is most sensitive, and the degenerate deformation is least sensitive, to these effects. Thus the $NH_3$ rocking frequency is often used to compare the strength of the M–N bond in a series of complexes of the same type and anion.[18]

To assign the skeletal modes such as the MN stretching and NMN bending modes, it is necessary to consider the normal modes of the octahedral $MN_6$ skeleton ($\mathbf{O}_h$ symmetry). The MN stretching mode in the low frequency region is of particular interest since it provides direct information about the structure of the MN skeleton and the strength of the M–N bond. However, its band assignments have been controversial for some compounds; $[Co(NH_3)_6]Cl_3$ is a typical example. As shown in Fig. III-3, the infrared spectrum of this complex exhibits three very weak bands at 498, 477, and 449 $cm^{-1}$. Nakamoto et al.[19] originally assigned these bands to the components of the infrared active $F_{1u}$ CoN stretching mode, which was split because of lowering of symmetry in the crystalline

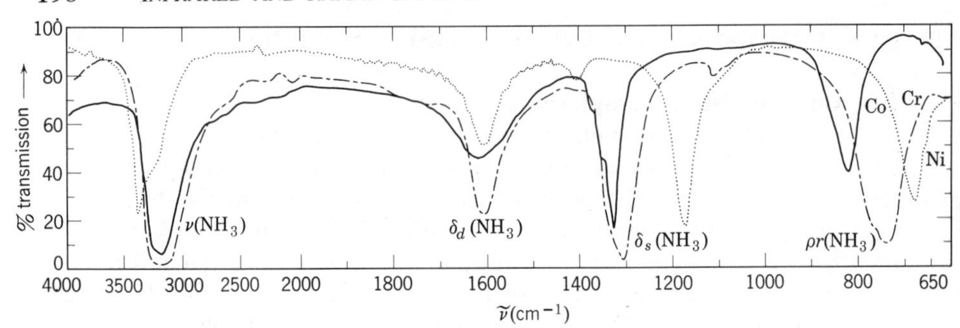


**Fig. III-1.** Infrared spectra of hexammine complexes: $[Co(NH_3)_6]Cl_3$, solid line; $[Cr(NH_3)_6]Cl_3$, dot-dash line; $[Ni(NH_3)_6]Cl_2$, dotted line.

state. On the other hand, Siebert and Eysel[12] proposed assigning them to the $A_{1g}$, $F_{1u}$, and $E_g$ stretching modes, respectively, and Swaddle et al.[20] assigned them to the $A_{1g}$, $A_{1g}$, and $E_g$ modes, all of which are only Raman active. These assignments are supported by the observation of two Raman bands at 500 and 445 $cm^{-1}$. Müller et al.[5,11] prefer the original assignments of Nakamoto et al. because, among other reasons,

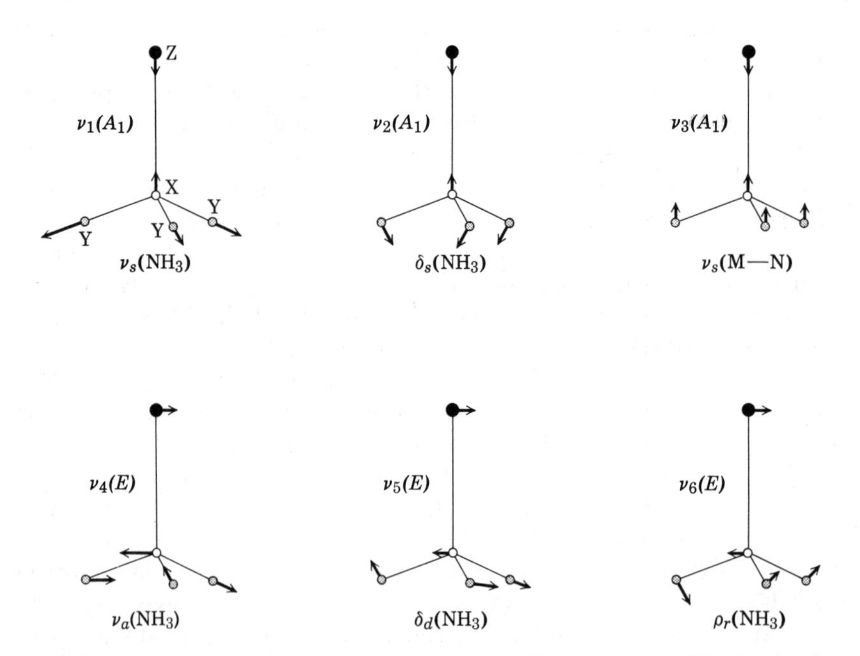


**Fig. III-2.** Normal modes of vibration of tetrahedral $ZXY_3$ molecules. (The band assignment is given for an $M—NH_3$ group.)

TABLE III-1. INFRARED FREQUENCIES OF OCTAHEDRAL HEXAMMINE COMPLEXES ($cm^{-1}$)[a]

| Complex | $\nu_a(NH_3)$ | $\nu_s(NH_3)$ | $\delta_a(HNH)$ | $\delta_s(HNH)$ | $\rho_r(NH_3)$ | $\nu(MN)$ IR | $\nu(MN)$ Raman | $\delta(NMN)$ | Refs. |
|---|---|---|---|---|---|---|---|---|---|
| $[Mg(NH_3)_6]Cl_2$ | 3353 | 3210 | 1603 | 1170 | 660 | 363 | 335 ($A_{1g}$) | 198 | 2 |
| | | | | | | | 243 ($E_g$) | | |
| $[Cr(NH_3)_6]Cl_3$ | 3257 | 3185 | 1630 | 1307 | 748 | 495<br>473 | 465 ($A_{1g}$) | — | 3, 4 |
| | | 3130 | | | | 456 | 412 ($E_g$) | | |
| $[^{50}Cr(NH_3)_6](NO_3)_3$ | 3310 | 3250 | 1627 | 1290 | 770 | 471 | — | 270 | 5 |
| | | 3190 | | | | | | | |
| $[Mn(NH_3)_6]Cl_2$ | 3340 | 3160 | 1608 | 1146 | 592 | 302 | 330 ($A_{1g}$) | 165 | 1, 6 |
| $[Fe(NH_3)_6]Cl_2$ | 3335 | 3175 | 1596 | 1156 | 633 | 315 | — | 170 | 1, 6 |
| $[Ru(NH_3)_6]Cl_2$ | 3300 | 3195 | 1610 | 1220 | 769 | — | — | — | 7 |
| $[Ru(NH_3)_6]Cl_3$ | 3077 | | 1618 | 1368 | 788 | 463 | 500 ($A_{1g}$) | 283 | 8 |
| | | | | 1342 | | | 475 ($E_g$) | 263 | |
| $[Os(NH_3)_6]OsBr_6$ | 3125 | | 1595 | 1339 | 818 | 452 | — | 256 | 8 |
| $[Co(NH_3)_6]Cl_2$ | 3330 | 3250 | 1602 | 1163 | 654 | 325 | 357 ($A_{1g}$) | 92 | 9, 6, 10 |
| | | | | | | | 255 ($E_g$) | | |
| $[Co(NH_3)_6]Cl_3$ | 3240 | 3160 | 1619 | 1329 | 831 | 498 | 500 ($A_{1g}$) | 331 | 11, 12, 13 |
| | | | | | | 477 | 445 ($E_g$) | | |
| | | | | | | 449 | | | |
| $[Co(ND_3)_6]Cl_3$ | 2440 | 2300 | 1165 | 1020 | 667 | 462 | — | 294 | 5 |
| | | | | | | 442 | | | |
| | | | | | | 415 | | | |
| $[Rh(NH_3)_6]Cl_3$ | 3200 | | 1618 | 1352 | 845 | 472 | 515 ($A_{1g}$) | 302 | 8, 14 |
| | | | | | | | 480 ($E_g$) | | |
| $[Ir(NH_3)_6]Cl_3$ | 3155 | | 1587 | 1350 | 857 | 475 | 527 ($A_{1g}$) | 279 | 8, 14 |
| | | | | 1323 | | | 500 ($E_g$) | 264 | |
| $[^{58}Ni(NH_3)_6]Cl_2$ | 3345 | 3190 | 1607 | 1176 | 685 | 335 | 370 ($A_{1g}$) | 217 | 11, 15 |
| | | | | | | | 265 ($E_g$) | | |
| $[Zn(NH_3)_6]Cl_2$ | 3350 | 3220 | 1596 | 1145 | 645 | 300 | — | — | 1, 10 |
| $[Cd(NH_3)_6]Cl_2$ | — | — | 1585 | 1091 | 613 | 298 | 342 ($A_{1g}$) | — | 10, 6 |
| $[Pt(NH_3)_6]Cl_4$ | 3150 | 3050 | 1565 | 1370 | 950 | 530 | 569 ($A_{1g}$) | 318 | 16, 17 |
| | | | | | | 516 | 545 ($E_g$) | | |

[a] All infrared frequencies are those of the $F_{1u}$ species.

these three bands are shifted to lower frequencies by the $^{14}N$–$^{15}N$ and H–D substitutions, as theoretically predicted for the $F_{1u}$ mode. According to Nakagawa and Shimanouchi,[21] the intensity of the MN stretching mode in the infrared increases as the M–N bond becomes more ionic and as the MN stretching frequency becomes lower. Relative to the Co(III)–N bond of the $[Co(NH_3)_6]^{3+}$ ion, the Co(II)–N bond of the $[Co(NH_3)_6]^{2+}$ ion is more ionic, and its stretching frequency is much lower (325 $cm^{-1}$). This may be responsible for the strong appearance of the Co(II)–N stretching band in the infrared.

As listed in Table III-1, two Raman active MN stretching modes ($A_{1g}$ and $E_g$) are observed for the octahedral hexammine salts. In general, $\nu(A_{1g})$ is higher than $\nu(E_g)$. However, the relative position of $\nu(F_{1u})$ with respect to these two vibrations changes from one compound to another.

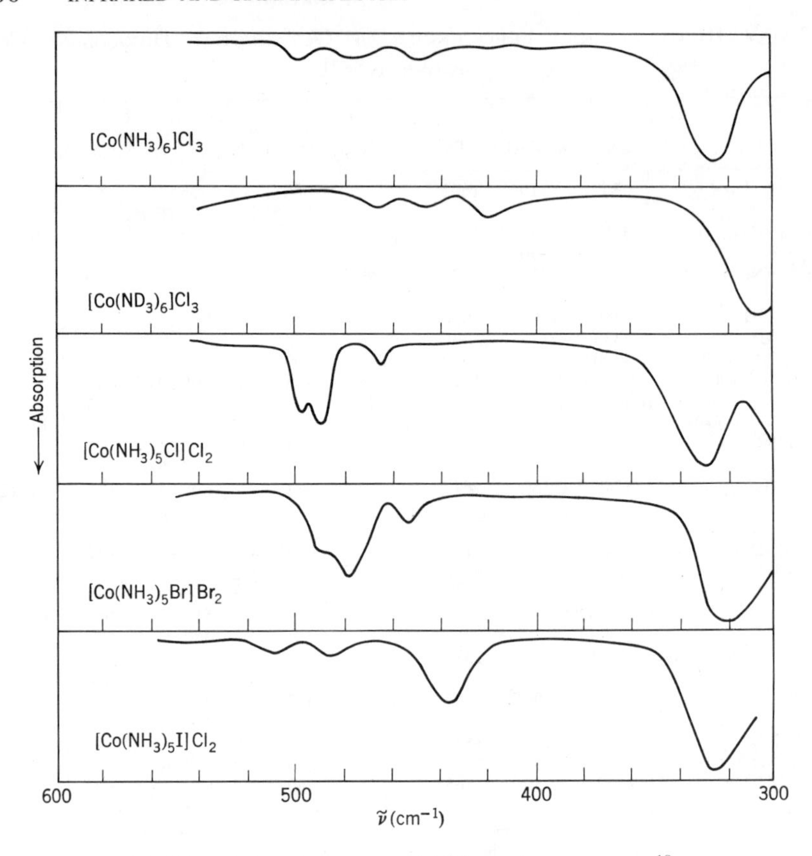


**Fig. III-3.** Infrared spectra of Co(III) ammine complexes.[19]

Another obvious trend in $\nu$(MN) is $\nu(M^{4+}—N) > \nu(M^{3+}—N) > \nu(M^{2+}—N)$. This holds for all symmetry species. Table III-1 shows that the $NH_3$ rocking frequency also follows the same trend as above.

Normal coordinate analyses on metal ammine complexes have been carried out by many investigators. Among them, Nakagawa, Shimanouchi, and co-workers[9,17,21] have made the most comprehensive study, using the UBF field. The MN stretching force constants of the hexammine complexes follow this order:

$$\begin{array}{ccccc} \text{Pt(IV)} \gg & \text{Co(III)} > & \text{Cr(III)} > & \text{Ni(II)} \approx & \text{Co(II)} \\ 2.13 & 1.05 & 0.94 & 0.34 & 0.33 \ \text{mdyn/Å} \end{array}$$

Terrasse et al.[14] report a value of 1.6 mdyn/Å for the Rh-N stretching force constant of the $[Rh(NH_3)_6]^{3+}$ ion in the UBF field. On the other hand, Müller et al.[5,6,11,15] calculated the GVF constants of a number of ammine complexes by using the point mass model (i.e., the $NH_3$ ligand is regarded as a single atom having the mass of $NH_3$), and refined their values with isotope shift data (H–D, $^{14}N$–$^{15}N$, and metal isotopes). For the hexammine series, they obtained the following order:

$$\begin{array}{cccccccccccc} Pt^{4+} & > & Ir^{3+} & > & Os^{3+} & > & Rh^{3+} & > & Ru^{3+} & > & Co^{3+} & > \\ 2.75 & & 2.28 & & 2.13 & & 2.10 & & 2.01 & & 1.86 & \end{array}$$

$$\begin{array}{ccccccccccccc} Cr^{3+} & > & Ni^{2+} & > & Co^{2+} & > & Fe^{2+} & \sim & Cd^{2+} & > & Zn^{2+} & > & Mn^{2+} \\ 1.66 & & 0.85 & & 0.80 & & & 0.73 & & & 0.69 & & 0.67\ \text{mdyn/Å} \end{array}$$

For a series of divalent metals, the above order is parallel to the Irving–Williams series ($Mn^{2+} < Fe^{2+} < Co^{2+} < Ni^{2+} > Cu^{2+} > Zn^{2+}$). Schmidt and Müller[1] discussed the relationship between the MN stretching force constant and the stability constant or the bond energy.

Table III-2 lists the observed infrared frequencies and band assignments of tetrahedral, square-planar, and linear metal ammine complexes. The Raman active MN stretching frequencies are also included in Table III-2. Normal coordinate analyses have been made by Nakagawa et al.[9,17,21] by using the UBF field; the following values were obtained for the MN stretching force constants:

$$\begin{array}{ccccccc} Hg^{2+} & > & Pt^{2+} & > & Pd^{2+} & > & Cu^{2+} \\ 2.05 & & 1.92 & & 1.71 & & 0.84\ \text{mdyn/Å} \end{array}$$

On the other hand, Schmidt and Müller[6] obtained the following values by using the GVF field and the point mass approximation:

$$\begin{array}{ccccccccccc} Pt^{2+} & > & Pd^{2+} & \gg & Co^{2+} & \sim & Zn^{2+} & \sim & Cu^{2+} & > & Cd^{2+} \\ 2.54 & & 2.15 & & 1.44 & & 1.43 & & 1.42 & & 1.24\ \text{mdyn/Å} \end{array}$$

Vibrational spectra of metal ammine complexes in the crystalline state exhibit lattice vibrations below 200 $cm^{-1}$. Assignments of lattice modes have been made for the hexammine salts of Mg(II),[2] Co(II),[10,31] Ni(II),[10,21,31,30] $[Co(NH_3)_6][Co(CN)_6]$,[32] and $[Pt(NH_3)_4]Cl_2$.[25] The Magnus green salt, $[Pt(NH_3)_4][PtCl_4]$, is of particular interest. Originally, Hiraishi et al.[17] assigned the infrared band at 200 $cm^{-1}$ to a lattice mode which corresponds to the stretching mode of the Pt—Pt—Pt chain. This high frequency was justified on the basis of the strong Pt–Pt interaction in this salt. Adams and Hall,[33] on the other hand, assigned this mode at 81 $cm^{-1}$, and the 201 $cm^{-1}$ band to a $NH_3$ torsion. The vibrational frequencies of

TABLE III-2. INFRARED FREQUENCIES OF OTHER AMMINE COMPLEXES ($cm^{-1}$)

| Complex | $\nu_a(NH_3)$ | $\nu_s(NH_3)$ | $\delta_a(HNH)$ | $\delta_s(HNH)$ | $\rho_r(NH_3)$ | $\nu(MN)$ IR | $\nu(MN)$ Raman | $\delta(NMN)$ | Refs. |
|---|---|---|---|---|---|---|---|---|---|
| Tetrahedral | | | | | | | | | |
| $[Co(NH_3)_4](ReO_4)_2$ | 3340 | 3260 | 1610 | 1240 | 693 | 430 | 405 ($A_1$) | 195 | 22 |
| $[^{64}Zn(NH_3)_4]I_2$ | 3275 | 3150 | 1596 | 1253 | 685 | — | 432 ($A_1$) | 156 | 23 |
| | 3233 | | | 1239 | | | 412 ($F_2$) | | |
| $[Cd(NH_3)_4](ReO_4)_2$ | 3354 | 3267 | 1617 | 1176 | 670 | 370 | — | 166 | 1 |
| | | | | | | | | 160 | |
| Square-planar | | | | | | | | | |
| $[^{104}Pd(NH_3)_4]Cl_2 \cdot H_2O$ | 3270 | 3170 | 1630 | 1279 | 849 | 495 | 510 ($A_{1g}$) | 291 | 5, 24 |
| | | | | | 802 | | 468 ($B_{1g}$) | 238 | |
| $[Pt(NH_3)_4]Cl_2$ | 3236 | 3156 | 1563 | 1325 | 842 | 510 | 524 ($A_{1g}$) | 297 | 17, 25 |
| | | | | | | | 508 ($B_{1g}$) | 235 | |
| $[Cu(NH_3)_4]SO_4 \cdot H_2O$ | 3327 | 3169 | 1669 | 1300 | 735 | 426 | 420 ($A_{1g}$) | 256 | 26, 5 |
| | 3253 | | 1639 | 1283 | | | 375 ($B_{1g}$) | 227 | |
| Linear | | | | | | | | | |
| $[Ag(NH_3)_2]_2SO_4$ | 3320 | 3150 | 1642 | 1236 | 740 | 476 | 372 ($A_1$) | 221 | 27, 28 |
| | 3230 | | 1626 | 1222 | 703 | 400 | | 211 | |
| $[Hg(NH_3)_2]Cl_2$ | 3265 | 3197 | 1605 | 1268 | 719 | 513 | 412 | — | 29, 28 |

lattice modes and low-frequency internal modes of hexammine complexes have also been studied by Janik et al.,[34,35] using the neutron inelastic scattering technique. Attempts have been made to determine the vibrational frequencies in the electronic excited states by the analysis of vibronic spectra; the absorption,[4,36] emission,[37] and luminescence[38] spectra of $[Cr(NH_3)_6]^{3+}$ salts ($^2E_g \leftarrow {}^4A_{2g}$ transition) have yielded the $NH_3$ rocking and NCrN bending frequencies of the $F_{2u}$ vibrations, which are forbidden in both infrared and Raman spectra of octahedral hexammine complexes.

**(2) Halogenoammine Complexes**

If the $NH_3$ groups of a hexammine complex are partly replaced by other groups, the degenerate vibrations are split because of lowering of symmetry, and new bands belonging to other groups appear. Here we discuss only halogenoammine complexes. The infrared spectra of $[Co(NH_3)_5X]^{2+}$ and *trans*-$[Co(NH_3)_4X_2]^+$ type complexes have been studied by Nakagawa and Shimanouchi.[9,21] Table III-3 lists the observed frequencies and band assignments obtained by these workers. The infrared spectra of some of these complexes in the CoN stretching region are shown in Fig. III-3. Normal coordinate analyses on these complexes[9] have yielded the following UBF stretching force constants (mdyn/Å): $K$(Co—N), 1.05; $K$(Co—F), 0.99; $K$(Co—Cl), 0.91; $K$(Co—Br), 1.03; and $K$(Co—I) 0.62. Raman spectra of some chloroammine Co(III) complexes have been assigned.[39] For halogenoammine complexes of Cr(III), see Refs. 40 and 3.

TABLE III-3. SKELETAL VIBRATIONS OF PENTAMMINE AND *trans*-TETRAMMINE Co(III) COMPLEXES ($cm^{-1}$)[9,21]

| Complex | | $\nu$(CoN) | $\nu$(CoX) | Skeletal Bending |
|---|---|---|---|---|
| Pentammine ($\mathbf{C}_{4v}$ symmetry) | | | | |
| $[Co(NH_3)_5F]^{2+}$ | $A_1$ | 480, 438 | 343 | 308 |
| | $E$ | 498 | — | 345, 290, 219 |
| $[Co(NH_3)_5Cl]^{2+}$ | $A_1$ | 476, 416 | 272 | 310 |
| | $E$ | 498 | — | 292, 287, 188 |
| $[Co(NH_3)_5Br]^{2+}$ | $A_1$ | 475, 410 | 215 | 287 |
| | $E$ | 497 | — | 290, 263, 146 |
| $[Co(NH_3)_5I]^{2+}$ | $A_1$ | 473, 406 | 168 | 271 |
| | $E$ | 498 | — | 290, 259, 132 |
| *trans*-Tetrammine ($\mathbf{D}_{4h}$ symmetry) | | | | |
| $[Co(NH_3)_4Cl_2]^+$ | $A_{2u}$ | — | 353 | 186 |
| | $E_u$ | 501 | — | 290, 167 |
| $[Co(NH_3)_4Br_2]^+$ | $A_{2u}$ | — | 317 | 227 |
| | $E_u$ | 497 | — | 280, 120 |

Detailed vibrational assignments have been made for halogenoammine complexes of Os(III)[41] and of Ru(III), Rh(III), Os(III), and Ir(III).[42]

In regard to $M(NH_3)_4X_2$ and $M(NH_3)_3X_3$ type complexes, the main interest has been the distinction of stereoisomers by vibrational spectroscopy. As shown in Appendix III, *trans*-$MN_4X_2$ ($\mathbf{D}_{4h}$) exhibits one MN stretching ($E_u$) and one MX stretching ($A_{2u}$), while *cis*-$MN_4X_2$ ($\mathbf{C}_{2v}$) shows four MN stretching (two $A_1$, $B_1$, and $B_2$) and two MX stretching ($A_1$ and $B_1$) vibrations in the infrared. For *mer*-$MN_3X_3$ ($\mathbf{C}_{2v}$), three MN stretching and three MX stretching vibrations are infrared active, whereas only two MN stretching and two MX stretching vibrations are infrared active for *fac*-$MN_3X_3$ ($\mathbf{C}_{3v}$). James and Nolan[16] have measured and assigned the Raman spectra of a series of $[Pt(NH_3)_nCl_{6-n}]^{(n-2)+}$ type complexes.

Vibrational spectra of the planar $M(NH_3)_2X_2$ type complexes [M = Pt(II) and Pd(II)] have been studied by many investigators. Table III-4 summarizes the observed frequencies and band assignments of their skeletal vibrations. Figure III-4 shows the infrared spectra of *cis*- and *trans*-$[Pd(NH_3)_2Cl_2]$ obtained by Layton et al.[44] As expected, both the PdN and PdCl stretching bands split into two in the *cis*- isomer. Durig et al.[45] found that the PdN stretching frequencies range from 528 to 436 $cm^{-1}$, depending on the nature of other ligands in the complex. In general, the PtN stretching band shifts to a lower frequency as a ligand of stronger *trans*-influence is introduced in the position *trans* to the Pt–N bond.[46]

TABLE III-4. SKELETAL FREQUENCIES OF SQUARE-PLANAR $M(NH_3)_2X_2$ TYPE COMPLEXES ($CM^{-1}$)[a]

| Complex | $\nu$(MN) | $\nu$(MX) | Bending | Refs. |
|---|---|---|---|---|
| *trans*-$[Pd(NH_3)_2Cl_2]$ | | | | |
| IR | 496 | 333 | 245, 222, 162, 137 | 43 |
| R | 494 | 295 | | |
| *cis*-$[Pd(NH_3)_2Cl_2]$ | | | | |
| IR | 495, 476 | 327, 306 | 245, 218, 160, 135 | 43 |
| *trans*-$[Pd(NH_3)_2Br_2]$ | | | | |
| IR | 490 | — | 220, 220, 122, 101 | 43 |
| *cis*-$[Pd(NH_3)_2Br_2]$ | | | | |
| IR | 480, 460 | 258 | 225, 225, 120, 100 | 43 |
| *trans*-$[Pd(NH_3)_2I_2]$ | | | | |
| IR | 480 | 191 | 263, 218, 109 | 43 |
| *trans*-$[Pt(NH_3)_2Cl_2]$ | | | | |
| IR | 572 | 365 | 220, 195 | 24 |
| R | 529 | 318 | — | 24 |
| *cis*-$[Pt(NH_3)_2Cl_2]$ | | | | |
| IR | 510 | 330, 323 | 250, 198, 155, 123 | 25 |
| *trans*-$[Pt(NH_3)_2Br_2]$ | | | | |
| IR | 504 | 206 | 230 | 24 |
| R | 535 | 206 | — | 24 |
| *trans*-$[Pt(NH_3)_2I_2]$ | | | | |
| R | 532 | 153 | — | 24 |

[a] For band assignments, see also Refs. 17 and 48.

Using infrared spectroscopy, Durig and Mitchell[47] studied the isomerization of *cis*-$[Pd(NH_3)_2X_2]$ to its *trans* isomer.

### (3) Amido ($NH_2$) Complexes

The vibrational spectra of amido complexes may be interpreted in terms of the normal vibrations of a pyramidal $ZXY_2$ type molecule. Mizushima et al.[49] and Niwa et al.[50] carried out normal coordinate analysis on the $[Hg(NH_2)_2]_\infty^+$ ion (infinite chain polymer); the results of the latter workers are given in Table III-5. Brodersen and Becher[51] studied the infrared spectra of a number of compounds containing Hg–N bonds and assigned the HgN stretching bands at 700–400 $cm^{-1}$. The infrared spectrum of the $NH_2^-$ ion in alkali metal salts has been measured.[52] Alkylamido complexes of the type $M(NR_2)_{4,5}$ (M = Ti, Zr, Hf, V, Nb, and Ta) exhibit their MN stretching bands in the 700–530 $cm^{-1}$ region.[53]

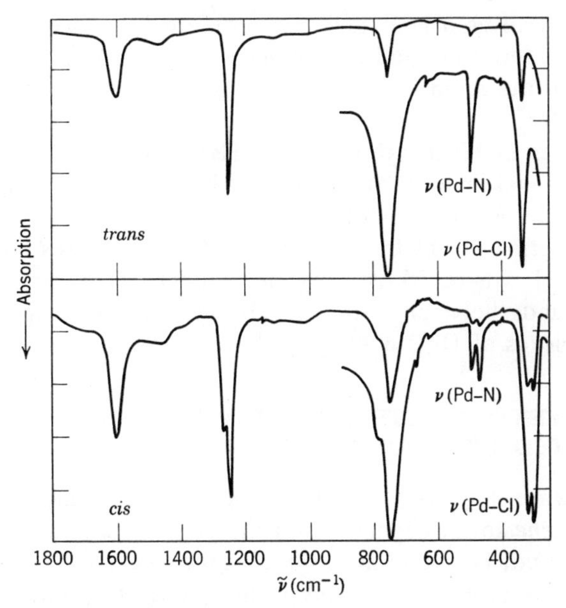


**Fig. III-4.** Infrared spectra of *trans*- and *cis*-$[Pd(NH_3)_2Cl_2]$.[44]

### (4) Alkylamine Complexes

Infrared spectra of methylamine complexes, $[Pt(CH_3NH_2)_2X_2]$ (X: a halogen), have been studied by Watt et al.[54] and Kharitonov et al.[55] Far-infrared spectra of $[M(R_2NH)_2X_2]$[M = Zn(II) or Cd(II); R = ethyl or *n*-propyl; X = Cl or Br] type complexes have also been reported.[56] Chatt and co-workers[57] studied the effect of hydrogen bonding on the NH stretching frequencies of *trans*-$[Pt(RNH_2)Cl_2L]$ type complexes (R = Me Et, etc.; L = $C_2H_4$, $PEt_3$, etc) in organic solvents such as chloroform and dioxane. Their study revealed that the complexes of primary amines have a strong tendency to associate through intermolecular hydrogen bonds of the NH · · · Cl type, whereas those of secondary amines have

Table III-5. Infrared Frequencies and Band Assignments of Amido Complexes ($cm^{-1}$)[50]

| Compound | $\nu(NH_2)$ | $\delta(NH_2)$ | $\rho_w(NH_2)$ | $\rho_r(NH_2)$ | $\nu(HgN)$ |
|---|---|---|---|---|---|
| $[Hg(NH_2)]^+_\infty(Cl)^-_\infty$ | 3200, 3175 | 1540 | 1025 | 673 | 573 |
| $[Hg(NH_2]^+_\infty(Br)^-_\infty$ | 3220, 3180 | 1525 | 1008 | 652 | 560 |

little tendency to associate. Later, this difference was explained on the basis of steric repulsion and intramolecular interaction between the NH hydrogen and the nonbonding *d*-electrons of the metal.[58]

### (5) Complexes of Hydroxylamine and Hydrazine

The vibrational spectra of hydroxylamine ($NH_2OH$) complexes have been studied by Kharitonov et al.[59] Sacconi and Sabatini[60] reported the infrared spectra of hydrazine ($NH_2NH_2$) complexes of the type $[M(N_2H_4)_2Cl_2]$, where M is a divalent metal, and assigned the MN stretching bands between 440 and 330 $cm^{-1}$. Infrared spectra of hydrazine complexes of Hg[61] and Ni[62,63] have been reported.

## III-2. COMPLEXES OF ETHYLENEDIAMINE AND RELATED COMPOUNDS

As is shown in Fig. III-5, 1,2-disubstituted ethane may exist in the *cis*, *trans*, or *gauche* form, depending on the angle of internal rotation. The *cis* form may not be stable in the free ligand because of steric repulsion between two X groups. The *trans* form belongs to point group $\mathbf{C}_{2h}$, in which only the *u* vibrations (antisymmetric with respect to the center of

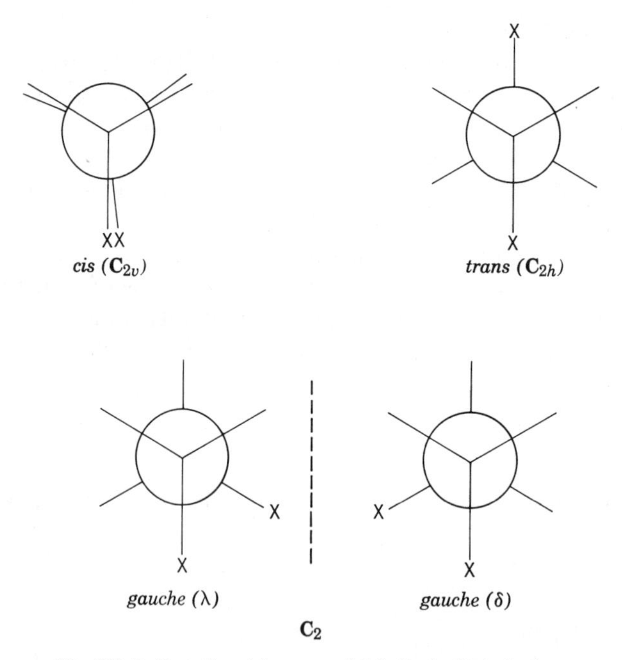


**Fig. III-5.** Rotational isomers of 1,2-disubstituted ethane.

symmetry) are infrared active. On the other hand, both *gauche* forms belong to point group $\mathbf{C}_2$, in which all the vibrations are infrared active. Thus the *gauche* form exhibits more bands than the *trans* form. Mizushima and co-workers[64] have shown that 1,2-dithiocyanatoethane ($NCS—CH_2—CH_2—SCN$) in the crystalline state definitely exists in the *trans* form, because no infrared frequencies coincide with Raman frequencies (mutual exclusion rule). By comparing the spectrum of the crystal with that of a $CHCl_3$ solution, they concluded that several extra bands observed in solution can be attributed to the *gauche* form. Table III-6 summarizes the infrared frequencies and band assignments obtained by Mizushima et al. It is seen that the $CH_2$ rocking vibration provides the most clear-cut diagnosis of conformation: one band ($A_u$) at 749 $cm^{-1}$ for the *trans* form, and two bands ($A$ and $B$) at 918 and 845 $cm^{-1}$ for the *gauche* form.

The compound 1,2-dithiocyanatoethane may take the *cis* or *gauche* form when it coordinates to a metal through the S atoms. The chelate ring formed will be completely planar in the *cis*, and puckered in the *gauche*, form. The *cis* and *gauche* forms can be distinguished by comparing the spectrum of a metal chelate with that of the ligand in $CHCl_3$ solution (*gauche* + *trans*). Table III-6 compares the infrared spectrum of 1,2-dithiocyanatoethanedichloroplatinum(II) with that of the free ligand in a $CHCl_3$ solution. Only the bands characteristic of the *gauche* form are observed in the Pt(II) complex. This result definitely indicates that the chelate ring in the Pt(II) complex is *gauche*. The method described above has also been applied to the metal complexes of 1,2-dimethylmercaptoethane ($CH_3S—CH_2—CH_2—SCH_3$).[65] In this case, the free ligand exhibits one $CH_2$ rocking at 735 $cm^{-1}$ in the crystalline state (*trans*), whereas the metal complex always exhibits two $CH_2$ rockings at 920–890 and 855–825 $cm^{-1}$ (*gauche*).

The conformation of ethylenediamine (en) in metal complexes is of particular interest in coordination chemistry. Unfortunately, the $CH_2$ rocking mode discussed above does not provide a clear-cut diagnosis in this case, since it couples strongly with the $NH_2$ rocking and CN stretching modes that appear in the same frequency region. However, X-ray analyses on *trans*-$[Co(en)_2Cl_2]Cl \cdot HCl \cdot 2H_2O$[66] and other complexes indicate that the chelating ethylenediamine takes the *gauche* conformation without exception. Powell and Sheppard[67] carried out an extensive infrared study on ethylenediamine and its metal complexes. Other references are as follows: $[M(en)_3]^{2+}$ [M = Zn(II), Cd(II), and Hg(II)] (68), $[M(en)_3]^{3+}$ [M = Cr(III), Co(III), and Rh(III)] (69), $[M(en)_2]^{2+}$ [M = Ni(II), Pd(II), and Pt(II)] (70), and $M(en)X_2$ [X = Cl, I (71) and SCN (72)]. Omura et al.[73] carried out normal coordinate analysis on the $[M(en)_2]^{2+}$

TABLE III-6. INFRARED SPECTRA OF 1,2-DITHIOCYANATO-ETHANE AND ITS Pt(II) COMPLEX ($cm^{-1}$)[64]

| Ligand | | | |
|---|---|---|---|
| Crystal (*trans*) | $CHCl_3$ Solution (*gauche* + *trans*) | Pt Complex (*gauche*) | Assignment |
| — | 2170 (*g*) | 2165 (*g*) | $\nu(C\equiv N)$ |
| 2155 (*t*) | 2170 (*t*) | — | |
| 1423 (*t*) | 1423 (*t*) | — | $\delta(CH_2)$ |
| — | 1419 (*g*) | 1410 (*g*) | |
| 1291[a] (*t*) | — | — | $\rho_w(CH_2)$ |
| — | 1285 (*g*) | 1280 (*g*) | |
| 1220 (*t*) | 1215 (*t*) | — | |
| 1145 (*t*) | 1140 (*t*) | — | $\rho_t(CH_2)$ |
| — | 1100 (*g*) | 1110 (*g*) | |
| — | —[b] (*g*) | 1052 (*g*) | $\nu(CC)$ |
| 1037[a] | — | — | |
| — | 918 (*g*) | 929 (*g*) | $\rho_r(CH_2)$ |
| — | 845 (*g*) | 847 (*g*) | |
| 749 (*t*) | —[b] | — | |
| 680 (*t*) | 677 (*t*) | — | $\nu(CS)$ |
| 660 (*t*) | 660 (*t*) | — | |

[a] Raman frequencies in the crystalline state.
[b] Hidden by $CHCl_3$ absorption.

ion ($\mathbf{C}_{2h}$ symmetry), and assigned the MN stretching bands at the following frequencies ($cm^{-1}$):

| M = Cu(II) | | Pd(II) | Pt(II) |
|---|---|---|---|
| $A_u$ | 472 | 518 | 544 |
| $B_u$ | 412 | 470 | 475 |

Lever and Mantovani[74] assigned the MN stretching bands of $M(N—N)_2X_2$ [M = Cu(II), Co(II), and Ni(II); N—N = en, dimethyl-en, etc.; X = Cl, Br, etc.] type complexes by using the metal isotope technique. For these compounds, the CoN and NiN stretching bands have been assigned to 400–230 $cm^{-1}$,[74] and the CuN stretching vibrations have been located in the 420–360 $cm^{-1}$ range.[75] A straight line relationship between the square of the CuN stretching frequency and the energy of the main electronic *d-d* band was found,[76] with some exceptions.[75]

The infrared spectra of *cis*-and *trans*-$[M(en)_2X_2]^+$ [M = Co(III), Cr(III), Ir(III), and Rh(III); X = Cl, Br, etc.] have been studied extensively.[77–82]

These isomers can be distinguished by comparing the spectra in the regions of 1700–1500 ($NH_2$ bending), 950–850 ($CH_2$ rocking), and 610–500 $cm^{-1}$ (MN stretching).

As stated in Sec. III-1, the band at 200 $cm^{-1}$ of Magnus green salt, $[Pt(NH_3)_4][PtCl_4]$, has been assigned to the Pt–Pt interaction (lattice) mode since the Pt—Pt distance is very short (3.25Å). Omura et al.[83] and Berg and Rasmussen[84] could not find an analogous band in the infrared spectra of Magnus type $[Pt(en)_2][PtCl_4]$ and its analogs. The latter workers, therefore, concluded that no correlation exists between the color of a Magnus type salt and the presence of a lattice vibration near 200 $cm^{-1}$.

The most interesting information obtained from infrared studies is that ethylenediamine takes the *trans* form when it functions as a bridging group between two metal atoms. Powell and Sheppard[85] were the first to suggest that ethylenediamine in $(C_2H_4)Cl_2Pt(en)PtCl_2(C_2H_4)$ is likely to be *trans*, since the infrared spectrum of this compound is simpler than that of other complexes in which ethylenediamine is *gauche*. Similar results have been obtained for $(AgCl)_2en$,[86] $(AgSCN)_2en$,[87] $(AgCN)_2en$,[87] $Hg(en)Cl_2$,[88] and $M(en)Cl_2$ (M = Zn or Cd).[86] The structure of these complexes may be depicted as follows:

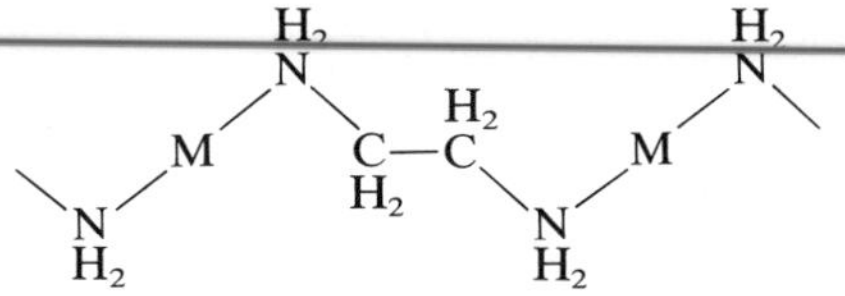

```
              H2                      H2
              N                       N
            /   \       H2          /   \
          M      C — C            M
 \      /        H2     \       /
   N                      N
   H2                     H2
```

A more complete study, including the infrared and Raman spectra, of $M(en)X_2$ type complexes [M = Zn(II), Cd(II) and Hg(II); X = Cl, Br, and SCN] has been made by Iwamoto and Shriver.[89] Mutual exclusion of infrared and Raman spectra, along with other evidence, supports the $\mathbf{C}_{2h}$ bridging structure of the en ligand in the Cd and Hg complexes (see Fig. III-6).

The infrared spectra of diethylenetriamine (dien) complexes have been reported for [Pd(dien)X]X (X = Cl, Br, and I)[90] and $[Co(dien)(en)Cl]^{2+}$.[91] The latter exists in the four isomeric forms shown in Fig. III-7. Their infrared spectra revealed that the $\omega$- and $\kappa$-isomers contain dien in the *mer*-configuration; the $\pi$- and $\varepsilon$-isomers, in the *fac*-configuration. The *mer*- and *fac*-isomers of $[M(dien)X_3]$ [M = Cr(III), Co(III), and Rh(III); X: a halogen] can also be distinguished by infrared spectra.[92]

The infrared spectra of $\beta$, $\beta'$, $\beta''$-triaminotriethylamine (tren) complexes with Co(III)[93] and lanthanides[94] have been reported. Buckingham and Jones[95] measured the infrared spectra of $[M(trien)X_2]^+$, where trien is triethylenetetramine, M is Co(III), Cr(III), or Rh(III), and X is a halogen

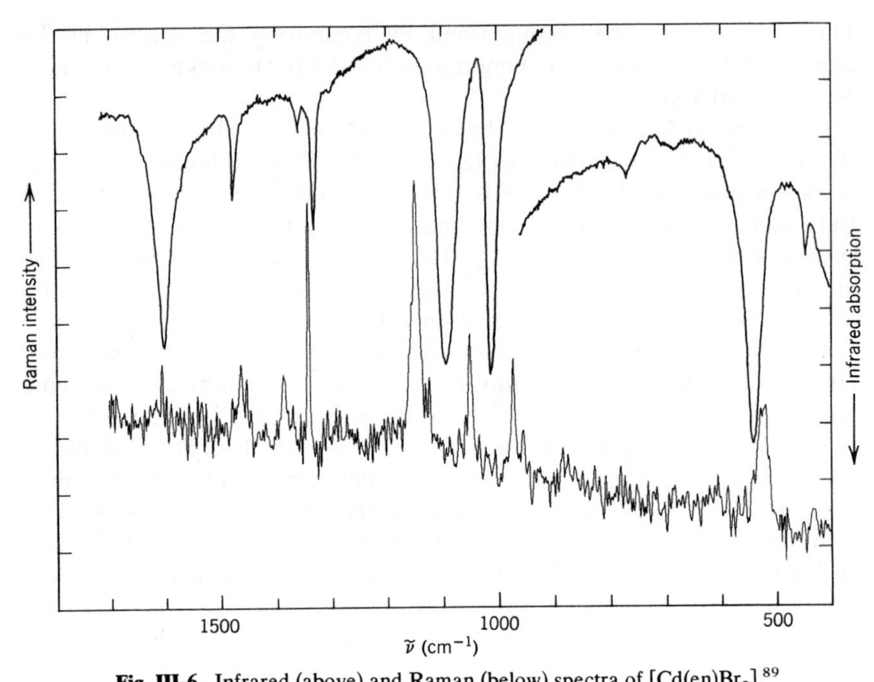


**Fig. III-6.** Infrared (above) and Raman (below) spectra of $[Cd(en)Br_2]$.[89]

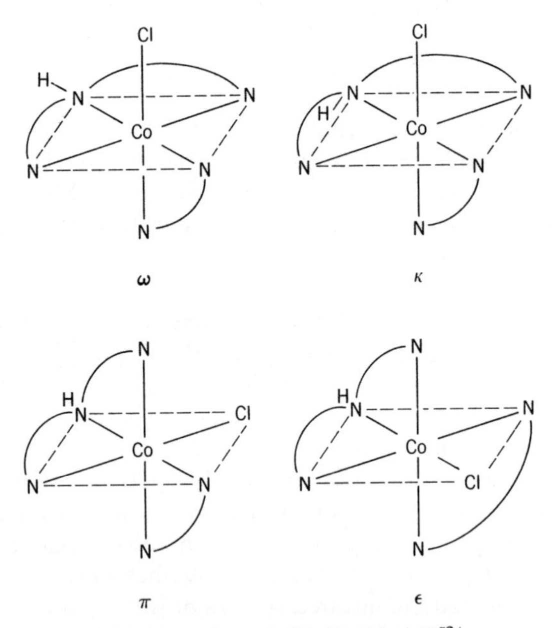


**Fig. III-7.** Structures of $[Co(dien)(en)C1]^{2+}$.

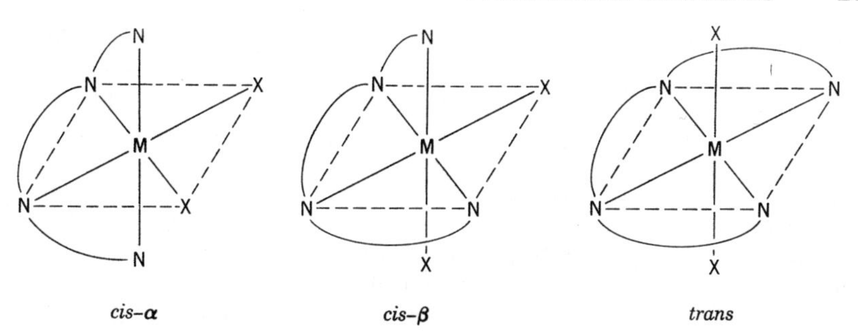


**Fig. III-8.** Structures of $[M(trien)X_2]^+$ type ions.

or an acido anion. These compounds give three isomers (Fig. III-8) which can be distinguished, for example, by the $CH_2$ rocking vibrations in the 920–860 $cm^{-1}$ region. For $[Co(trien)Cl_2]$ $ClO_4$, *cis*-α-isomer exhibits two strong bands at 905 and 871 $cm^{-1}$, *cis*-β-isomer shows four bands at 918, 898, 868, and 862 $cm^{-1}$ whereas *trans*-isomer gives only one band at 874 $cm^{-1}$ with a weak band at 912 $cm^{-1}$. Far-infrared spectra of some of these trien complexes have been reported.[96]

## III-3. COMPLEXES OF PYRIDINE, BIPYRIDINE, AND RELATED COMPOUNDS

### (1) Pyridine and Its Derivatives

Upon complex formation, the pyridine (py) vibrations in the high frequency region are not shifted appreciably, whereas those at 604 (in-plane ring deformation) and 405 $cm^{-1}$ (out-of-plane ring deformation) are shifted to higher frequencies. Clark and Williams[97] carried out an extensive far-infrared study on metal pyridine complexes. Table III-7 lists the observed frequencies of these metal-sensitive py vibrations and metal—py stretching vibrations. Clark and Williams showed that ν(M—py) and ν(MX) (X: a halogen) are very useful in elucidating the stereochemistry of these py complexes. For example, *fac*-$[Rh(py)_3Cl_3]$ exhibits two ν(Rh—py) ($\mathbf{C}_{3v}$ symmetry), whereas *mer*-$Rh(py)_3Cl_3$ shows three ν(Rh—py) ($\mathbf{C}_{2v}$ symmetry) near 250 $cm^{-1}$. The infrared spectra of these two compounds are given in Fig. III-43.

The metal isotope technique has been used to assign the ν(M—py) and ν(MX) vibrations of $Zn(py)_2X_2$[98] and $Ni(py)_4X_2$.[99] The former vibrations have been located in the 225–160 and 250–225 $cm^{-1}$ regions, respectively, for the Zn(II) and Ni(II) complexes. Figure III-9 shows the infrared and Raman spectra of $[^{64}Zn(py)_2Cl_2]$ and its $^{68}Zn$ analog. As expected from its $\mathbf{C}_{2v}$ symmetry, two ν(Zn—py) and two ν(ZnCl) are

TABLE III-7. VIBRATIONAL FREQUENCIES OF PYRIDINE COMPLEXES ($cm^{-1}$)[97]

| Complex | Structure | py | py | $\nu$(M—py) |
|---|---|---|---|---|
| $Co(py)_2Cl_2^a$ | monomeric, tetrahedral | 642 | 422 | 253[b] |
| $Ni(py)_2I_2$ | monomeric, tetrahedral | 643 | 428 | 240 |
| $Cr(py)_2Cl_2$ | polymeric, octahedral | 640 | 440 | 219 |
| $Cu(py)_2Cl_2$ | polymeric, octahedral | 644 | 441 | 268 |
| $Co(py)_2Cl_2^a$ | polymeric, octahedral | 631 | 429 | 243,235[b] |
| *mer*-$[Rh(py)_3Cl_3]$ | monomeric, octahedral | 650 | 468 | 265, 245, 230 |
| *fac*-$[Rh(py)_3Cl_3]$ | monomeric, octahedral | 643 | 464 | 266, 245 |
| *trans*-$[Ni(py)_4Cl_2]$ | monomeric, octahedral | 626 | 426 | 236 |
| *trans*-$[Ir(py)_4Cl_2]Cl$ | monomeric, octahedral | 650 | 469 | 260, (255) |
| *cis*-$[Ir(py)_4Cl_2]Cl$ | monomeric, octahedral | 656 | 468 | 287, 273 |
| *trans*-$[Pt(py)_2Br_2]$ | monomeric, square-planar | 656 | 476 | 297 |
| *cis*-$[Pt(py)_2Br_2]$ | monomeric, square-planar | 659 644 | 448 | 260, 234 |

[a] $Co(py)_2Cl_2$ exists in two forms: blue complex (monomeric, tetrahedral) and violet complex (polymeric, octahedral). The far-infrared spectra of these two forms are shown in Fig. III-44 (Sec. III-16).
[b] Assignments made by Postmus et al.[102].

metal isotope sensitive. Far-infrared spectra of metal pyridine nitrate complexes, $M(py)_x(NO_3)_y$, have been reported.[100,101]

The infrared spectra of metal complexes with alkyl pyridines have been studied extensively.[103–107] Using the metal isotope technique, Lever and Ramaswamy[108] assigned the M—pic stretching bands of $M(pic)_2X_2$[M = Ni(II) and Cu(II); pic = picoline; X = Cl, Br, and I] in the 300–230 $cm^{-1}$ region. The infrared spectra of metal complexes with halogenopyridines have been reported.[104,109,110] Infrared spectra have been used to determine whether coordination occurs through the nitrile or the pyridine nitrogen in cyanopyridine complexes. It was found that 3- and 4-cyanopyridines coordinate to the metal via the pyridine nitrogen,[111–113] whereas 2-cyanopyridine coordinates to the metal via the nitrile nitrogen.[111,114]

Infrared spectra of aromatic amine N-oxides and their metal complexes have been reviewed by Garvey et al.[115] The N=O stretching band of pyridine N-oxide (1265 $cm^{-1}$) is shifted by 70–30 $cm^{-1}$ to a lower frequency upon complexation. The following references are given for three complexes: Fe(II) (116), Hg(II) (117), and Fe(III) (118). Metal complexes with various imidazole derivatives are known, and the infrared spectra of the following compounds have been studied: complexes of imidazole,[119–121]

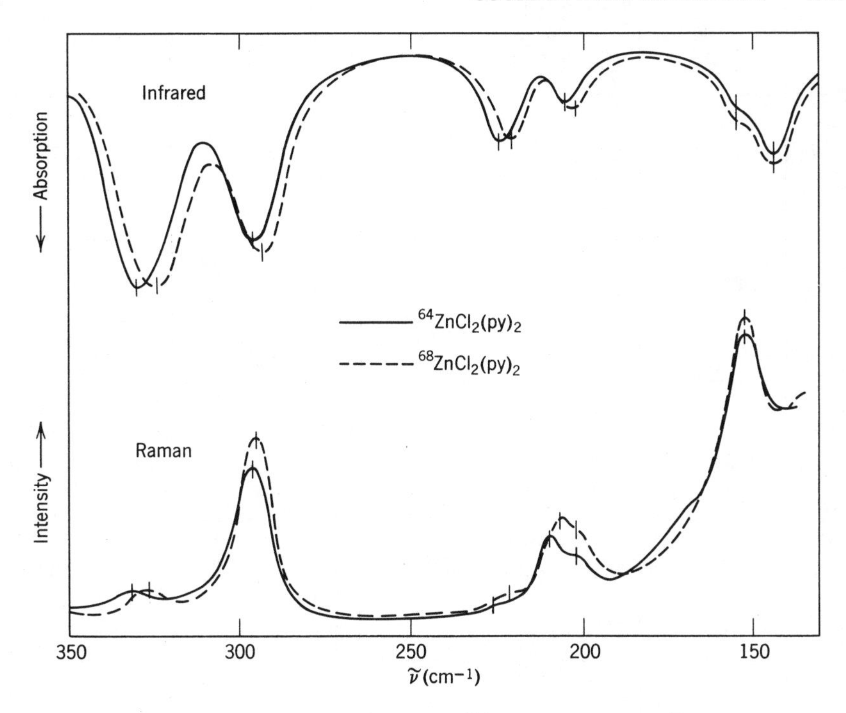


**Fig. III-9.** Infrared and Raman spectra of $^{64}Zn(py)_2Cl_2$ and its $^{68}Zn$ analog.

2-methylimidazole,[122,123] *N*-methylimidazole,[124] 4- and 5-bromoimidazole,[125] and benzimidazole.[126,127]

## (2) Complexes of 2,2′-Bipyridine and Related Ligands

Infrared spectra of metal complexes of 2,2′-bipyridine (bipy) and 1,10-phenanthroline (phen) have been studied extensively. In general, the bands in the high frequency region are not metal-sensitive since they originate in the heterocyclic or aromatic ring of the ligand. Thus the main interest has been focused on the low frequency region, where $\nu$(MN) and other metal-sensitive vibrations appear. It has been difficult, however, to assign $\nu$(MN) empirically since several ligand vibrations also appear in the same frequency region. This difficulty has been overcome by using the metal isotope technique. Hutchinson et al.[128] first applied this method to the tris-bipy and phen complexes of Fe(II), Ni(II), and Zn(II). Later, this work was extended to other metals in various oxidation states.[129] Table III-8 lists $\nu$(MN), magnetic moments, and the electronic configuration of these tris-bipy complexes. The results revealed several interesting relationships between $\nu$(MN) and the electronic structure.

TABLE III-8. MN STRETCHING FREQUENCIES AND ELECTRONIC STRUCTURES IN $[M(BIPY)_3]^{n+}$ TYPE COMPOUNDS ($cm^{-1}$)[a]

| | −I | 0 | I | II | III |
|---|---|---|---|---|---|
| $d^3$ | | | | 374<br>V 335<br>(3.67)<br>$(t_{2g})^3$ | 385<br>Cr 349<br>(3.78)<br>$(t_{2g})^3$ |
| $d^4$ | | 374<br>Ti 339<br>(0)<br>$(t_{2g})^4$-*ls* | | 351<br>Cr 343<br>(2.9)<br>$(t_{2g})^4$-*ls* | |
| $d^5$ | 365<br>Ti 322<br>(1.74)<br>$(t_{2g})^5$-*ls* | 371<br>V 343<br>(1.68)<br>$(t_{2g})^5$-*ls* | 371<br>Cr 343<br>(2.0)<br>$(t_{2g})^5$-*ls* | 224<br>Mn 191<br>(5.95)<br>$(t_{2g})^3(e_g)^2$-*hs* | 384<br>Fe 367<br>(?) |
| $d^6$ | | 382<br>Cr 308<br>(0)<br>$(t_{2g})^6$ | | 386<br>Fe 376<br>(0)<br>$(t_{2g})^6$ | 378[b]<br>Co 370<br>(0)<br>$(t_{2g})^6$ |
| $d^7$ | | 258<br>Mn 227<br>(4.10)<br>$(t_{2g})^5(e_g)^2$ | | 266<br>Co 228<br>(4.85)<br>$(t_{2g})^5(e_g)^2$ | |
| $d^8$ | 235<br>Mn 184<br>(3.71)<br>$(t_{2g})^6(e_g)^2$ | | 244<br>Co 194<br>(3.3)<br>$(t_{2g})^6(e_g)^2$ | 282<br>Ni 258<br>(3.10)<br>$(t_{2g})^6(e_g)^2$ | |
| $d^9$ | | 280<br>Co 257<br>(2.23)<br>$(t_{2g})^6(e_g)^3$ | | 291<br>Cu 268<br>(?)<br>$(t_{2g})^6(e_g)^3$ | |
| $d^{10}$ | | | | 230<br>Zn 184<br>(0)<br>$(t_{2g})^6(e_g)^4$ | |

[a] The numbers at the upper right of each group indicate the MN stretching frequencies ($cm^{-1}$). The number in parentheses gives the observed magnetic moment in Bohr magnetons. ls = low spin; hs = high spin.
[b] Values for $[Co(phen)_3](ClO_4)_3$.

1. In terms of simple MO theory, Cr(III), Cr(II), Cr(I), Cr(0), V(II), V(0), Ti(0), Ti(-I), Fe(III), Fe(II), and Co(III) have filled or partly filled $t_{2g}$ (bonding) and empty $e_g$ (antibonding) orbitals. The $\nu$(MN) of these metals (Group A) are in the 300–390 $cm^{-1}$ region.
2. On the other hand, Co(II), Co(I), Co(0), Mn(II), Mn(0), Mn(-I), Ni(II), Cu(II), and Zn(II) have filled or partly filled $e_g$ orbitals. The $\nu$(MN) of these metals (Group B) are in the 180–290 $cm^{-1}$ region.
3. Thus no marked changes in frequencies are seen in the Cr(III)–Cr(0) and Co(II)–Co(0) series, although a dramatic decrease in frequencies is observed in going from Co(III) to Co(II).
4. The fact that the $\nu$(MN) do not change appreciably in the former two series indicates that the M–N bond strength remains approximately the same.

These results also suggest that, as the oxidation state is lowered, increasing numbers of electrons of the metal reside in essentially ligand orbitals which do not affect the M–N bond strength.

Other work on bipy and phen complexes includes a far-infrared study of tris-bipy complexes with low oxidation state metals [Cr(0), V(-I), Ti(0), etc.],[130] the assignments of infrared spectra of $M(bipy)Cl_2$ and $M(phen)Cl_2$ (M = Cu, Ni, etc.) by the metal isotope technique,[131] normal coordinate analysis on $Pd(bipy)Cl_2$ and its bipy-$d_8$ analog,[132] Raman spectra of phen complexes with Zn(II) and Hg(II),[133] and a resonance Raman study of Fe(II) complexes with bipy and other $\alpha$-diimines.[134]

The metal isotope technique has been used to study the effect of magnetic crossover on the low-frequency spectrum of $Fe(phen)_2(NCS)_2$. This compound exists as a high-spin complex at 298°K and as a low-spin complex at 100°K. Figure III-10 shows the infrared spectra of $^{54}Fe(phen)_2(NCS)_2$ obtained by Takemoto and Hutchinson.[135] On the basis of observed isotopic shifts, along with other evidence, they made the following assignments:

| | $\nu$(Fe—NCS) | $\nu$[Fe—N(phen)] |
|---|---|---|
| High-spin | 252 (4.0) | 222 (4.5) |
| Low-spin | 532.6 (1.6) | 379 (5.0) |
| | 528.5 (1.7) | 371 (6.0) |

The numbers in parenthesis indicate the isotope shift, $\nu(^{54}Fe) - \nu(^{57}Fe)$. Both vibrations show large shifts to higher frequencies in going from the high to the low-spin complex. This result suggests the marked strengthening of these coordinate bonds in going from the high to the low-spin complex as confirmed by X-ray analysis.[136] The work of Takemoto and

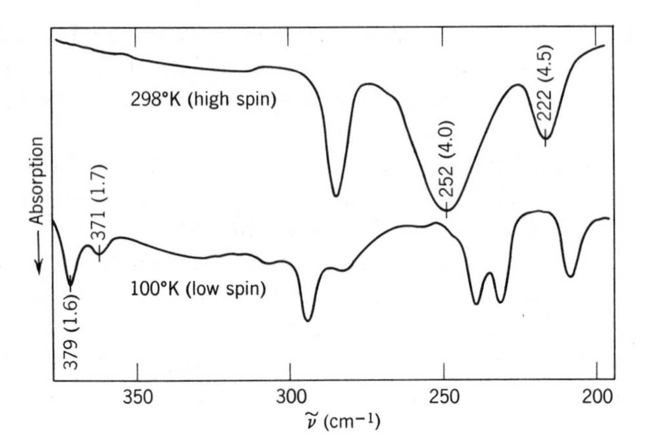


**Fig. III-10.** Infrared spectra of $^{54}Fe(phen)_2(NCS)_2$. The number in parentheses indicates the isotope shift, $\tilde{\nu}(^{54}Fe)$-$\tilde{\nu}(^{57}Fe)$.

Hutchinson has been extended to $Fe(bipy)_2(NCS)_2$ and $Fe(phen)_2(NCSe)_2$.[137] It has also been shown by infrared spectroscopy that a partial high → low spin conversion occurs under high pressure.[138] Barnard et al.[139] studied the vibrational spectra of bis[tri-(2-pyridyl)amine]Co(II) perchlorate in the high-spin (293°K) and low-spin (100°K) states. In the infrared, the CoN stretching band is at 263 $cm^{-1}$ for the high-spin complex, whereas it splits and shifts to 312 and 301 $cm^{-1}$ in the low-spin complex.

The metal isotope technique has been used to assign the MN vibrations of metal complexes with many other ligands. For example, Takemoto[140] assigned the $NiN_2$ and $NiN_1$ stretching vibrations of $[Ni(DAPD)_2]^{2-}$ at 416–341 and 276 $cm^{-1}$, respectively.

DAPDH: 2,6-diacetylpyridine dioxime

DMNAPY: 2,7-dimethyl-1,8-naphthyridine

In $Ni(DAPD)_2$, where the Ni atom is in the +IV state, the $NiN_2$ and $NiN_1$ stretching bands are located at 509.8–472.0 and 394.8 $cm^{-1}$, respectively. These high frequency shifts in going from Ni(II) ($d^8$) to Ni(IV) ($d^6$, diamagnetic) have also been observed for diarsine complexes (Sec. III-

18). Hutchinson and Sunderland[141] have noted that the MN stretching frequencies of the Ni(II) and Zn(II) complexes of 2,7-dimethyl-1,8-napthyridine(DMNAPY), shown above, are lower than those of the corresponding tris-bipy complexes by 16 to 24%. This was attributed to the weakening of the M–N bond due to the strain in the four-membered chelate rings of the DMNAPY complexes. Normal coordinate analysis has been carried out on the $M(DMG)_2$ series (DMG: dimethylglyoximate ion)[142] and the MN streching force constants (mdyn/Å) have been found to be as follows:

| Pt(II) | | Pd(II) | | Cu(II) | | Ni(II) | |
|---|---|---|---|---|---|---|---|
| 3.77 | > | 2.84 | > | 1.92 | > | 1.88 | (GVF) |

This work was extended to bis(glyoximato) complexes of Pt(II), Pd(II), and Ni(II).[143] The metal isotope technique has been used to assign the MN stretching vibrations of metal complexes with 8-hydroxyquinoline[144] and 1,8-naphthyridine.[145]

### (3) Metalloporphyrins and Related Compounds

Infrared studies have been made on more complicated metal chelate systems such as those of dipyrromethane and porphin derivatives. Again, the main interest has been focused on the low frequency region, where metal-sensitive bands appear. Murakami and co-workers made empirical assignments of infrared spectra of metal complexes with dipyrromethene[146] and tetradehydrocorrin derivatives.[147] Kobayashi et al.[148] reported the infrared spectra of metal phthalocyanines and assigned the metal–ligand vibrations in the 200–100 $cm^{-1}$ region. Boucher and Katz[149] noted that the bands near 970–920, 530–500, and 350 $cm^{-1}$ in metal chelates of protoporphyrin IX dimethyl ester and hematoporphyrin IX dimethyl ester are metal-sensitive. Ogoshi et al.[150] assigned the $\nu$(MN) of octaethylporphyrin (OEP) complexes as follow:

| Pd(II) | Ni(II) | Co(II) | Cu(II) | Zn(II) | Mg(II) | |
|---|---|---|---|---|---|---|
| 275 | 287 | 264 | 234 | 203 | 214 | $cm^{-1}$ |

Ogoshi et al.[151] also prepared porphin complexes with $^{64}Zn$, $^{68}Zn$, Cu, and Ni, and carried out normal coordinate analysis to assign 18 infrared active in-plane vibrations. Figures III-11 and 12 illustrate the infrared spectra of free porphin and its metal complexes in the high and low frequency regions, respectively. It has been found that the bands between

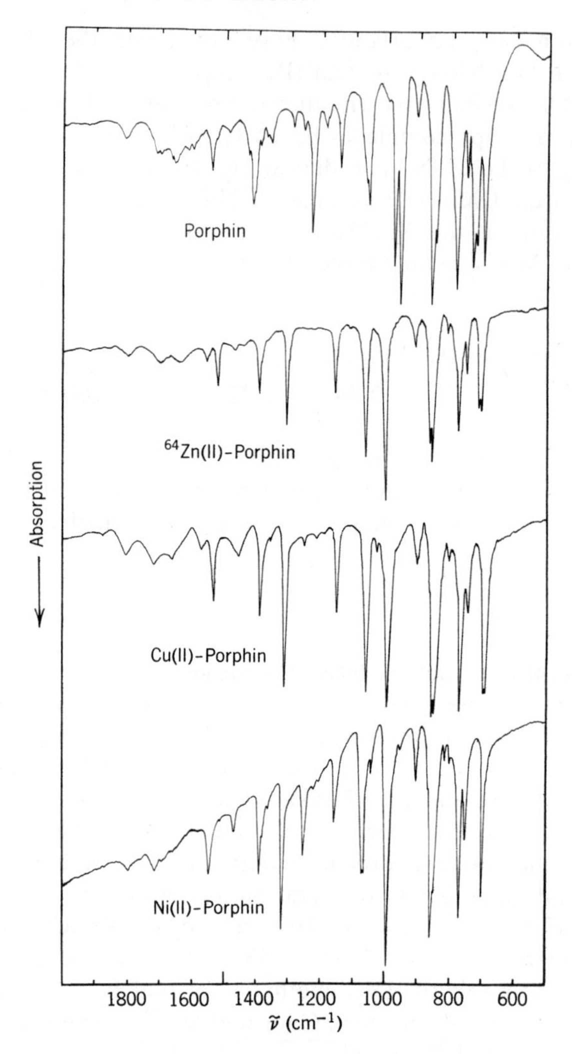


**Fig. III-11.** Infrared spectra of porphin and its $^{64}$Zn, Cu, and Ni complexes.

1700 and 950 cm$^{-1}$ are due to $\nu$(CC), $\nu$(CN), $\delta$(CH), $\delta$(CCN), or coupled vibrations between these modes, and that the largest contribution of $\nu$(MN) is in the vibrations at ca. 203, 246, and 295 cm$^{-1}$ of the Zn, Cu, and Ni complexes, respectively. The 203 cm$^{-1}$ band of the Zn complex gives a shift of 3.5 cm$^{-1}$ to a lower frequency by the $^{64}$Zn–$^{68}$Zn substitution.

The metal isotope technique has also been used to assign $\nu$(MN) of

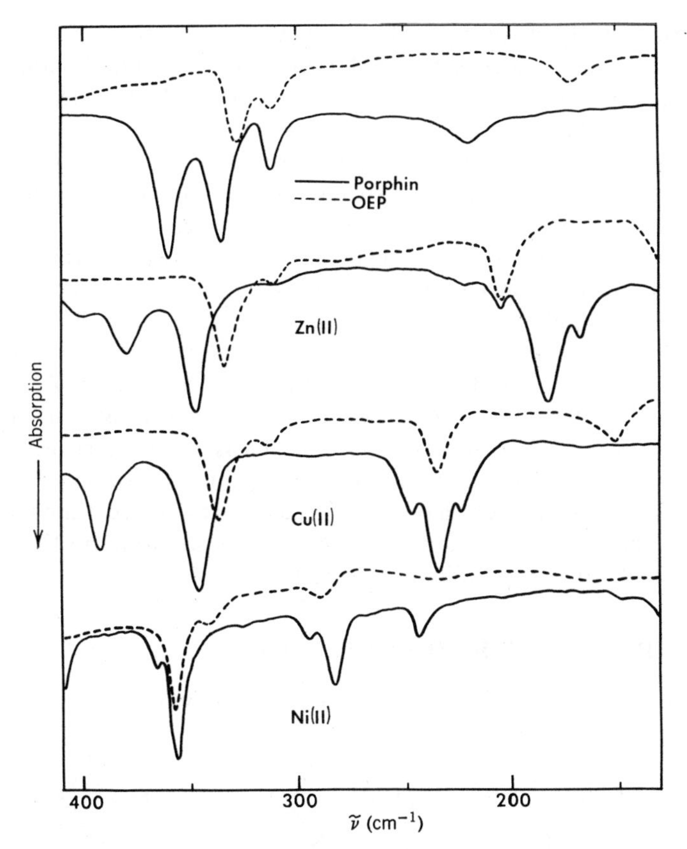


**Fig. III-12.** Far-infrared spectra of metal porphins (solid line) and the corresponding OEP complexes (dotted line).

Fe(OEP)X (X = Cl, Br, etc.) and $[Fe(OEP)L_2]^+$ (L = γ-picoline, imidazole, etc.)[152] metal chelates of tetraphenylporphin,[153] and *trans*-octaethylchlorine.[154] The ν(FeN) of the Fe(OEP)X series (280–250 cm$^{-1}$) are lower than those of the $[Fe(OEP)L_2]^+$ series (320–294 cm$^{-1}$). According to X-ray analysis, the structures of the $FeN_4$ skeletons in these two types are as follows:

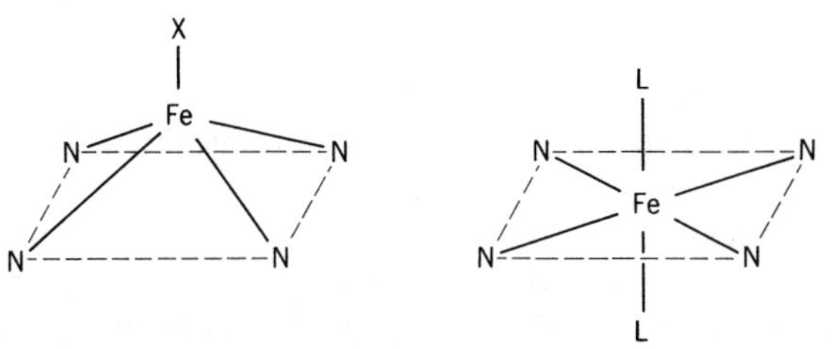

The former is high-spin; the latter, low-spin. In going from high to low-spin complexes, electrons are shifted from the $e_g$ (antibonding) to $t_{2g}$ (bonding) orbitals. The latter forms $\pi$-overlap with the vacant porphin ring orbitals, which strengthens the Fe–N bonds.

As stated in Sec. I-20, some metalloporphyrins are ideal for resonance Raman studies because of their strong absorptions in the visible region. Extensive work by many investigators[155–157] has shown that the spin and oxidation state of Fe atoms in heme proteins can be determined rather simply by the frequencies and intensities of some resonance-enhanced bands in the high frequency region. For example, a band showing anomalous polarization near 1580 $cm^{-1}$ is indicative of the low-spin state. This band shifts to ca. 1553 $cm^{-1}$ in high-spin complexes. A polarized band near 1360 $cm^{-1}$ is indicative of Fe(II), which is shifted to ca. 1375 $cm^{-1}$ in Fe(III) complexes. On the basis of these criteria, it was concluded that the Fe atom in oxyhemoglobin is low-spin Fe(III).[155] Bernstein and co-workers have measured the resonance Raman spectra of protohemin,[158] Ni, Cu, and Co chelates of mesoporphyrin IX dimethyl ester,[159] and Cu tetrapyridinoporphyrazine.[160]

## III-4. NITRO AND NITRITO COMPLEXES

The $NO_2^-$ ion coordinates to a metal in a variety of ways:

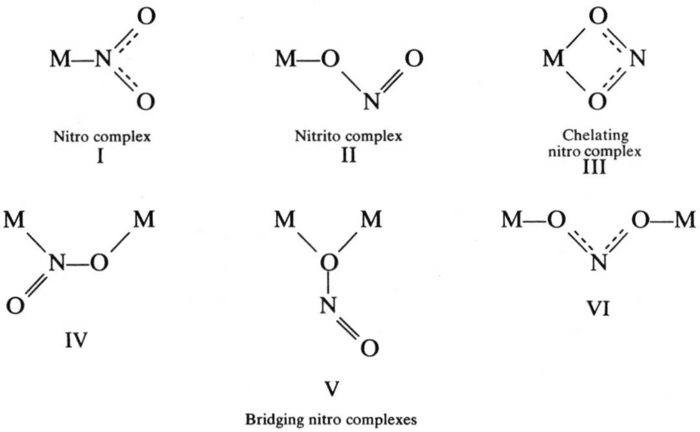

Nitro complex
I

Nitrito complex
II

Chelating
nitro complex
III

IV

V

VI

Bridging nitro complexes

Vibrational spectroscopy is very useful in distinguishing these structures.

### (1) Nitro Complexes

The normal vibrations of the unidentate N-bonded nitro complex may be approximated by those of a planar $ZXY_2$ molecule, as shown in Fig.

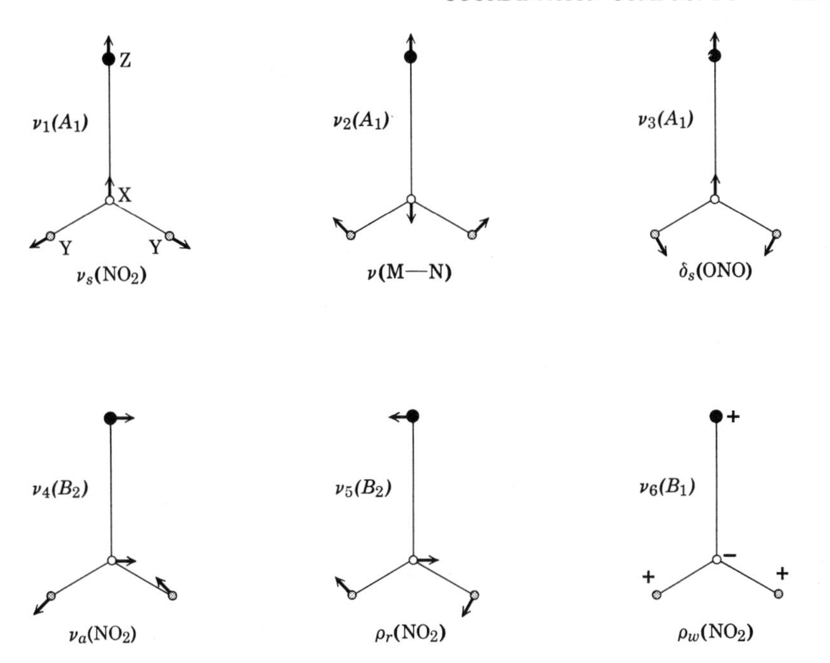


**Fig. III-13.** Normal modes of vibration of planar $ZXY_2$ molecules. (The band assignment is given for an M—$NO_2$ group.)

III-13. In addition to these modes, the $NO_2$ twisting and skeletal modes of the whole complex may appear in the low-frequency region. Table III-9 summarizes the observed frequencies and band assignments for typical nitro complexes. It is seen that these complexes exhibit $\nu_a(NO_2)$ and $\nu_s(NO_2)$ in the 1470–1370 and 1340–1320 cm$^{-1}$ regions, respectively. On the other hand, the free $NO_2^-$ ion exhibits these modes at 1250 and 1335 cm$^{-1}$, respectively. Thus $\nu_a(NO_2)$ shifts markedly to a higher frequency, whereas $\nu_s(NO_2)$ changes very little upon coordination.

Nakagawa and Shimanouchi[30,161] carried out normal coordinate analyses to assign the infrared spectra of crystalline hexanitro cobaltic salts; both internal and lattice modes were assigned completely from factor group analysis. The results indicate that the complex ion takes the $\mathbf{T}_h$ symmetry in K, Rb, and Cs salts but the $\mathbf{S}_6$ symmetry in the Na salt. Figure III-14 illustrates these unusual point groups for the $[Co(NO_2)_6]^{3-}$ ion. The infrared spectra of K and Na salts are compared in Fig. III-15.

There are many nitro complexes containing other ligands such as $NH_3$ and Cl. In these cases, the main interest has been the distinction of stereoisomers by the symmetry selection rules and the differences in

TABLE III-9. OBSERVED INFRARED FREQUENCIES AND BAND ASSIGNMENTS OF NITRO COMPLEXES ($CM^{-1}$)

| Complex | $\nu_a(NO_2)$ | $\nu_s(NO_2)$ | $\delta(ONO)$ | $\rho_w(NO_2)$ | $\nu(MN)$ | $\rho_r(NO_2)^a$ | Refs. |
|---|---|---|---|---|---|---|---|
| $K_3[Co(NO_2)_6]$ | 1386 | 1332 | 827 | 637 | 416 | 293 | 161 |
| $Na_3[Co(NO_2)_6]$ | 1425 | 1333 | 854, 831 | 623 | 449, 372 | 276, 249 | 161 |
| $K_2Ba[Ni(NO_2)_6]$ | 1343 | 1306 | 838 | 433 | 291 | 255 | 161 |
| $K_3[Ir(NO_2)_6]$ | 1395, 1375 | 1330 | 830 | 657 | 390 | 300 | 162 |
| $K_3[Rh(NO_2)_6]$ | 1395 | 1340 | 833 | 627 | 386 | 283 | 162 |
| $K_3[Ir(NO_2)Cl_5]$ | 1374 | 1315 | 835 | 644 | 325 | 288 | 163 |
| $[Pt(NO_2)_6]^{4-}$ | 1488, 1458 | 1328 | 834 | 621 | 368 | 294 | 164 |
| $K_2[Pt(^{15}NO_2)_4]$ | 1466, 1397 | 1343 | 847, 839, 833 | 640, 623 | 421 | | 165, 166 |
| $[Pd(NO_2)_4]^{2-\,b}$ | 1408 | 1364, 1320 | 834, 824 | 440 | 290 | | 166 |
| $K_2[Pt(NO_2)Cl_3]$ | 1401 | 1325 | 844 | 614 | 350 | 304 | 163 |

[a] This mode may couple with other low frequency modes.
[b] Raman data in aqueous solution.

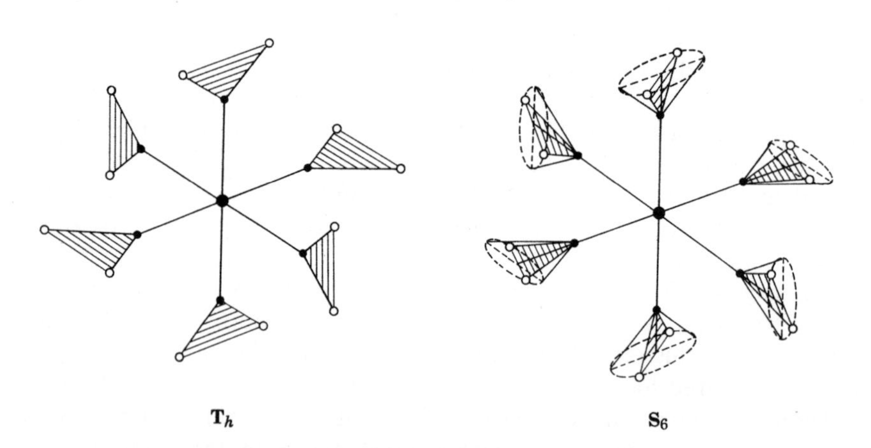


**Fig. III-14.** Possible structures of the $[Co(NO_2)_6]^{3-}$ ion.

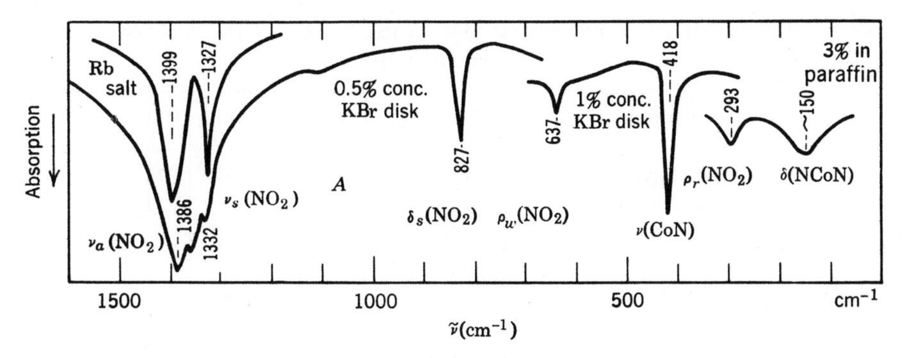


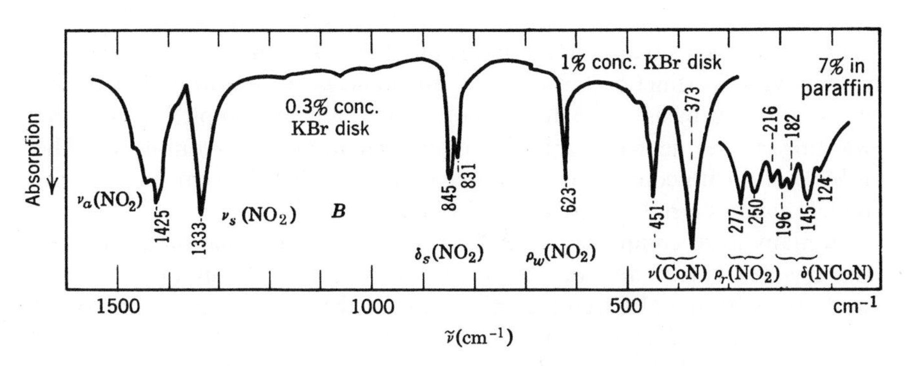


**Fig. III-15.** Infrared spectra of (*A*) $K_3[Co(NO_2)_6]$ and (*B*) $Na_3[Co(NO_2)_6]$.[161]

frequency between isomers. It is possible to distinguish *cis*- and *trans*-$[Co(NH_3)_4(NO_2)_2]^+$,[167] by the rule that the *cis*-isomer exhibits more bands than the *trans*-isomer, and to distinguish *fac*- and *mer*-$[Co(NH_3)_3(NO_2)_3]$[168] by the observation that $\delta(NO_2)$ and $\rho_w(NO_2)$ are higher for the *fac*-isomer ($\mathbf{C}_3$, 832 and 625 $cm^{-1}$, respectively) than for the *mer*-isomer ($\mathbf{C}_{2v}$, 825 and 610 $cm^{-1}$, respectively). Nakagawa and Shimanouchi[169] measured the infrared spectra of the $[Co(NO_2)_n(NH_3)_{6-n}]^{(3-n)+}$ series and carried out normal coordinate analysis on the mono- and dinitro complexes. Nolan and James[170] studied the infrared and Raman spectra of $[Pt(NO_2)_nCl_{6-n}]^{2-}$ type salts in the crystalline state.

### (2) Nitrito Complexes

If the $NO_2$ group is bonded to a metal through one of its O atoms, it is called a nitrito complex. Table III-10 lists the NO stretching frequencies of typical nitrito complexes. The two $\nu(NO_2)$ of nitrito complexes are well

TABLE III-10. VIBRATIONAL FREQUENCIES OF NITRITO COMPLEXES ($cm^{-1}$)

| Complex | ν(N=O) | ν(NO) | δ(ONO) | Ref. |
|---|---|---|---|---|
| $[Co(NH_3)_5(ONO)]Cl_2$ | 1468 | 1065 | 825 | 171 |
| $[Cr(NH_3)_5(ONO)]Cl_2$ | 1460 | 1048 | 839 | 171 |
| $[Rh(NH_3)_5(ONO)]Cl_2$ | 1461, 1445 | 1063 | 830 | 163 |
| $[Ni(py)_4(ONO)_2]$ | 1393 | 1114 | 825 | 172 |
| *trans*-$[Cr(en)_2(ONO)_2]ClO_4$ | 1485, 1430 | — | 835, 825 | 173 |
| $[Co(py)_4(ONO)_2](py)_2$ | 1405 | 1109 | 824 | 174 |

separated, ν(N=O) and ν(NO) being at 1485–1400 and 1110–1050 $cm^{-1}$, respectively. Distinction between the nitro and nitrito coordination can be made on this basis. It is to be noted that nitrito complexes lack the wagging modes near 620 $cm^{-1}$ which appear in all nitro complexes. The ν(MO) of nitrito complexes were assigned in the 360–340 $cm^{-1}$ region for metals such as Cr(III), Rh(III), and Ir(III).[163]

In many nitro complexes several types of nitro coordination are mixed. Goodgame, Hitchman, and their co-workers carried out an extensive study on vibrational spectra of nitro complexes containing various types of coordination. For example, all six nitro groups in $K_4[Ni(NO_2)_6]\cdot H_2O$ are coordinated through the N atom. However, its anhydrous salt exhibits the bands characteristic of nitro as well as nitrito coordination. From UV spectral evidence, Goodgame and Hitchman[175] suggested the structure $K_4[Ni(NO_2)_4(ONO)_2]$ for the anhydrous salt. Table III-11 lists the observed frequencies of two Ni(II) complexes containing both nitro and nitrito groups.

TABLE III-11. VIBRATIONAL FREQUENCIES OF Ni(II) COMPLEXES CONTAINING NITRO AND NITRITO GROUPS ($cm^{-1}$)

| | Nitro Group | | | Nitrito Group | | |
|---|---|---|---|---|---|---|
| Complex | $\nu_a(NO_2)$ | $\nu_s(NO_2)$ | $\rho_w(ONO)$ | ν(N=O) | ν(NO) | Ref. |
| $K_4[Ni(NO_2)_6]H_2O$ | 1346 | 1319 | 427 | — | — | 175 |
| $K_4[Ni(NO_2)_4(ONO)_2]$ | 1347 | 1325 | 423, 414 | 1387 | 1206 | 175 |
| Ni[2-(aminomethyl)-py$]_2(NO_2)(ONO)$ | 1338 | 1318 | — | 1368 | 1251 | 176 |

TABLE III-12. VIBRATIONAL FREQUENCIES OF CHELATING NITRO GROUPS ($cm^{-1}$)

| Complex | $\nu_a(NO_2)$ | $\nu_s(NO_2)$ | $\delta(ONO)$ | Ref. |
|---|---|---|---|---|
| $Co(Ph_3PO)_2(NO_2)_2$ | 1266 | 1199, 1176 | 856 | 180 |
| $Ni(\alpha\text{-pic})_2(NO_2)_2$ | 1272 | 1199 | 866, 862 | 180 |
| $Cs_2[Mn(NO_2)_4]$ | 1302 | 1225 | 841 | 181 |
| $Co(Me_4en)(NO_2)_2$ | 1290 | 1207 | 850 | 174 |
| $Zn(py)_2(NO_2)_2$ | 1351 | 1171 | 850 | 182 |
| $(o\text{-cat})[Co(NO_2)_4]$[a] | 1390 | 1191 | — | 181 |

[a] *o*-cat: [*o*-xylylenebis(triphenylphosphonium)]$^{2+}$ ion.

The red nitritopentammine complex, $[Co(NH_3)_5(ONO)]Cl_2$, is unstable and is gradually converted to the stable yellow nitro complex. The kinetics of this conversion can be studied by observing the disappearance of the nitrito bands.[177,178] Burmeister[179] reviewed the vibrational spectra of these and other linkage isomers.

**(3) Chelating Nitro Groups**

If the nitro group is chelating, both the antisymmetric and the symmetric $NO_2$ stretching frequencies are lower, and the ONO bending frequency is higher, than those of the unidentate N-bonded nitro complexes. Table III-12 lists the vibrational frequencies of chelating nitro groups. It should be noted that $\nu_a(NO_2)$ depends on the degree of asymmetry of the coordinated nitro group; it is lowest when two N–O bonds are equivalent, and becomes higher as the degree of asymmetry increases. The high $\nu_a(NO_2)$ frequencies observed for the last two compounds in Table III-12 may be accounted for on this basis.

**(4) Bridging Nitro Group**

The nitro group is known to form a bridge between two metal atoms. Nakamoto et al.[171] suggested that among the three possible structures, IV, V, and VI, shown before, IV is most probable for

$$\left[(NH_3)_3Co\begin{matrix}\diagup OH \diagdown \\ —OH— \\ \diagdown NO_2 \diagup\end{matrix}Co(NH_3)_3\right]^{3+}$$

since its $NO_2$ stretching frequencies (1516 and 1200 $cm^{-1}$) are markedly different from those of other types discussed thus far. Later, this structure was found by X-ray analysis of

$$[(NH_3)_4Co(\mu\text{-}NH_2)(\mu\text{-}NO_2)Co(NH_3)_4]Cl_4 \cdot 4H_2O.$$ [183]

The $[Co_2\{NO_2(OH)_2\}(NO_2)_6]^{3-}$ ion exhibits the $NO_2$ bands at 1516, 1190, and 860 $cm^{-1}$, indicating the presence of a bridging nitro group:[184]

$$\left[(O_2N)_3Co(\mu\text{-}ONO)(\mu\text{-}OH)_2Co(NO_2)_3\right]^{3-}$$

$[Ni(\beta\text{-pic})_2(NO_2)_2]_3 \cdot C_6H_6$ exhibits a number of bands due to coordinated nitro groups. Goodgame et al.[185] suggested the presence of two different types of bridging nitro groups, IV, V, and III, on the basis of the crystal structure and infrared data for this compound: type IV absorbs at 1412 and 1236, type V at 1460 and 1019, and type III at 1299 and 1236 $cm^{-1}$. Goodgame et al.[186] also studied the infrared spectra of other bridging nitro complexes of Ni(II). For example, they found that $Ni(en)(NO_2)_2$ contains a type IV bridge (1429 and 1241 $cm^{-1}$), while $Ni(py)_2(NO_2)_2\frac{1}{3}C_6H_6$ is similar to that of the analogous $\beta$-picoline complex.

## III-5. LATTICE WATER AND AQUO AND HYDROXO COMPLEXES

Water in inorganic salts may be classified as lattice or coordinated water. There is, however, no definite borderline between the two. The former term denotes water molecules trapped in the crystalline lattice, either by weak hydrogen bonds to the anion or by weak ionic bonds to the metal, or by both:

$$X^- \cdots H\text{—}O(\cdots M)\text{—}H \cdots X^-$$

whereas the latter denotes water molecules bonded to the metal through partially covalent bonds. Although bond distances and angles obtained from X-ray and neutron diffraction data provide direct information about the geometry of the water molecule in the crystal lattice, studies of vibrational spectra are also useful for this purpose. It should be noted, however, that the spectra of water molecules are highly sensitive to their surroundings.

**(1) Lattice Water**

In general, lattice water absorbs at 3550–3200 $cm^{-1}$ (antisymmetric and symmetric OH stretchings) and at 1630–1600 $cm^{-1}$ (HOH bending).[187] If the spectrum is examined under high resolution, the fine structure of these bands is observed. For example, $CaSO_4 \cdot 2H_2O$ exhibits eight peaks in the 3500–3400 $cm^{-1}$ region,[188] and its complete vibrational analysis can be made by factor group analysis (Sec. I-22). In the low-frequency region, lattice water exhibits "librational modes" that are due to rotational oscillations of the water molecule, restricted by interactions with neighboring atoms. These modes have been located by van der Elsken and Robinson[189] in the 600–300 $cm^{-1}$ region for the hydrates of alkali and alkaline earth halides. Ichida et al.[190] measured the far-infrared spectra of $MCl_2 \cdot 2H_2O$ (M = Co, Fe, and Mn) and their deuterated derivatives at liquid nitrogen and room temperatures, and assigned the librational bands to the rocking and wagging motions. For example, they assigned the rocking, wagging, and MO stretching modes of lattice water in $FeCl_2 \cdot 2H_2O$ at 605, 365, and 400 $cm^{-1}$, respectively.

The presence of the hydronium ($H_3O^+$) ion in crystalline acid hydrates is well established, and their spectra were discussed in Sec. II-3. The existence of the $H_5O_2^+$ ion was first detected by X-ray analysis.[191] Pavia and Giguère[192] further confirmed its presence in $HClO_4 \cdot 2H_2O$ (namely, $[H_5O_2]ClO_4$) by the absence of some characteristic bands of the $H_3O^+$ and $H_2O$ species. Its structure is suggested to be centrosymmetric $H_2O$—H—$OH_2$ of approximately $\mathbf{C}_{2h}$ symmetry. Both X-ray[66] and neutron diffraction[193] studies suggest the presence of the $H_5O_2^+$ ion in *trans*-$[Co(en)_2Cl_2]Cl \cdot HCl \cdot 2H_2O$; Thus it should be formulated as *trans*-$[Co(en)_2Cl_2]Cl \cdot [H_5O_2]Cl$. The existence of the $H_7O_3^+$ ion in crystalline $HNO_3 \cdot 3H_2O$ and $HClO_4 \cdot 3H_2O$ was confirmed by infrared studies.[194] The spectra are consistent with a structure in which two of the hydrogens of the $H_3O^+$ ion are bonded to two $H_2O$ molecules through short, asymmetrical hydrogen bonds.

**(2) Aquo ($H_2O$) Complexes**

In addition to the three fundamental modes of the free water molecule, coordinated water exhibits other modes, such as those shown in Fig.

TABLE III-13. OBSERVED FREQUENCIES, BAND ASSIGNMENTS, AND MO STRETCHING FORCE CONSTANTS OF AQUO COMPLEXES[195]

| Compound | $\rho_r(H_2O)$ | $\rho_w(H_2O)$ | $\nu(MO)$ | $K(M—O)^a$ |
|---|---|---|---|---|
| $[Cr(H_2O)_6]Cl_3$ | 800 | 541 | 490 | 1.31 |
| $[Ni(H_2O)_6]SiF_6$ | $(755)^b$ | 645 | 405 | 0.84 |
| $[Ni(D_2O)_6]SiF_6$ | — | 450 | 389 | 0.84 |
| $[Mn(H_2O)_6]SiF_6$ | $(655)^c$ | 560 | 395 | 0.80 |
| $[Fe(H_2O)_6]SiF_6$ | — | 575 | 389 | 0.76 |
| $[Cu(H_2O)_4]SO_4 \cdot H_2O$ | 887, 855 | 535 | 440 | 0.67 |
| $[Zn(H_2O)_6]SO_4 \cdot H_2O$ | — | 541 | 364 | 0.64 |
| $[Zn(D_2O)_6]SO_4 \cdot D_2O$ | 467 | 392 | 358 | 0.64 |
| $[Mg(H_2O)_6]SO_4 \cdot H_2O$ | — | 460 | 310 | 0.32 |
| $[Mg(D_2O)_6]SO_4 \cdot D_2O$ | 474 | 391 | — | 0.32 |

$^a$ UBF field (mdyn/Å).
$^b$ $Ni(H_2O)_4Cl_2$.
$^c$ $Mn(H_2O)_4Cl_2$.

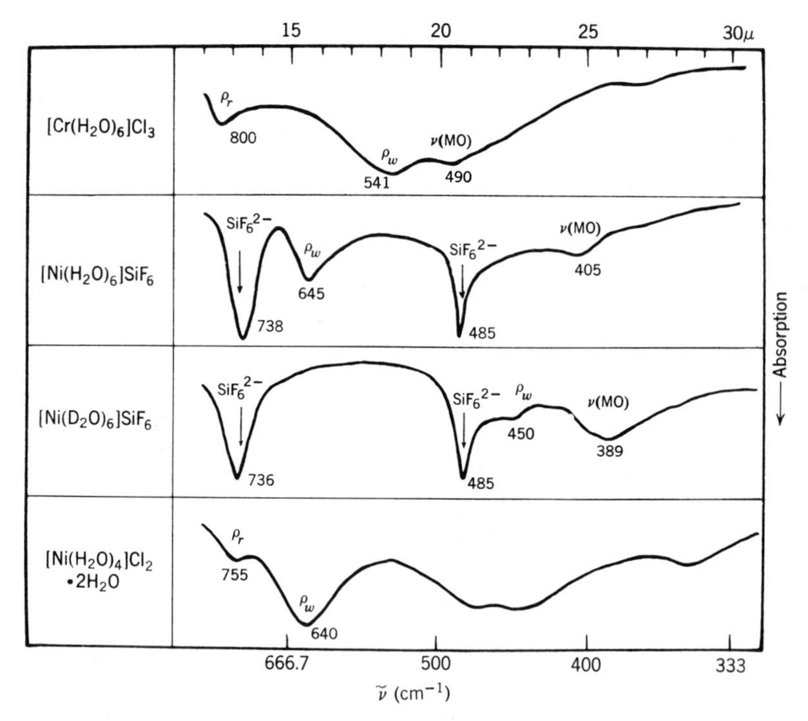


**Fig. III-16.** Infrared spectra of aquo complexes in the low frequency region.[195]

III-13. Nakagawa and Shimanouchi[195] carried out normal coordinate analyses on $[M(H_2O)_6]$ ($\mathbf{T}_h$ symmetry) and $[M(H_2O)_4]$ ($\mathbf{D}_{4h}$ symmetry) type ions to assign these low frequency modes. Table III-13 lists the frequencies and band assignments, and Fig. III-16 illustrates the far-infrared spectra of aquo complexes obtained by these authors. Adams and Lock[196] also studied the infrared and Raman spectra of aquo-halogeno complexes of various types. Their assignments show that the wagging frequencies of these complexes are higher than the rocking frequencies. For example, the wagging, rocking, and MO stretching bands of solid $K_2[FeCl_5(H_2O)]$ have been assigned at 600, 460, and 390 cm$^{-1}$, respectively.

Vibrational spectroscopy is very useful in elucidating the structures of aquo complexes. For example, $TiCl_3 \cdot 6H_2O$ should be formulated as *trans*-$[Ti(H_2O)_4Cl_2]Cl \cdot 2H_2O$ since it exhibits one TiO stretching (500 cm$^{-1}$, $E_u$) and one TiCl stretching (336 cm$^{-1}$, $A_{2u}$) mode.[197] Chang and Irish[198] showed from infrared and Raman studies that the structures of the tetra- and dihydrates resulting from the dehydration of $Mg(NO_3)_2 \cdot 6H_2O$ are as follows:

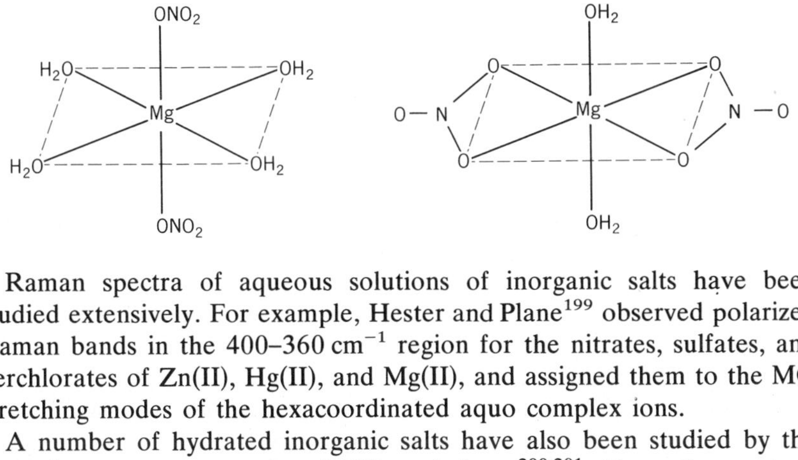

Raman spectra of aqueous solutions of inorganic salts have been studied extensively. For example, Hester and Plane[199] observed polarized Raman bands in the 400–360 cm$^{-1}$ region for the nitrates, sulfates, and perchlorates of Zn(II), Hg(II), and Mg(II), and assigned them to the MO stretching modes of the hexacoordinated aquo complex ions.

A number of hydrated inorganic salts have also been studied by the inelastic neutron scattering (INS) technique.[200,201] Since the proton-scattering cross section is quite large, the INS spectrum reflects mainly the motion of the protons in the crystal. Furthermore, INS spectroscopy has no selection rules involving dipole moments or polarizabilities. Thus it serves as a complementary tool to vibrational spectroscopy in studying the hydrogen vibrations of hydrated salts.

### (3) Hydroxo (OH) Complexes

The spectra of hydroxo complexes are expected to be similar to those of the metal hydroxides discussed in Sec. II-1. The hydroxo group can be distinguished from the aquo group since the former lacks the HOH bending mode near $1600 \text{ cm}^{-1}$. Furthermore, the hydroxo complex exhibits the MOH bending mode below $1200 \text{ cm}^{-1}$. For example, this mode is at $1150 \text{ cm}^{-1}$ for the $[Sn(OH)_6]^{2-}$ ion[202] and at ca. $1065 \text{ cm}^{-1}$ for the $[Pt(OH)_6]^{2-}$ ion.[203] The OH group also forms a bridge between two metals. For example,

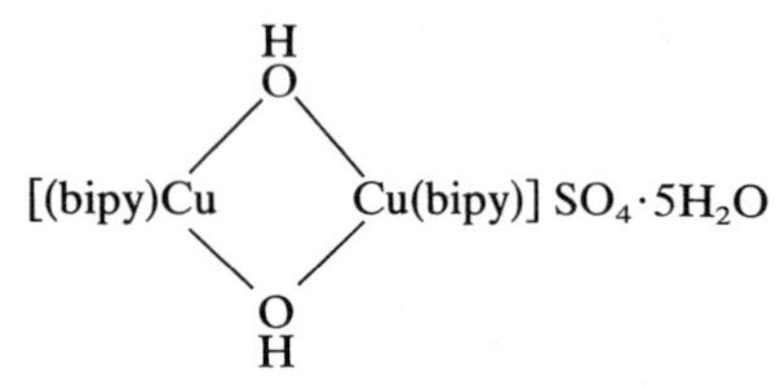


exhibits the bridging OH bending mode at $955 \text{ cm}^{-1}$; this is shifted to $710 \text{ cm}^{-1}$ upon deuteration.[204] For other bridging hydroxo complexes, the following references are given: Cu(II) complexes (204, 205), Cr(III) and Fe(III) complexes (206), Co(III) complexes (207), and Pb(II) complexes (208).

## III-6. COMPLEXES OF ALCOHOLS, ETHERS, KETONES, ALDEHYDES, ESTERS, AND CARBOXYLIC ACIDS

Metal alkoxides, $M(OR)_n$ (R: alkyl), exhibit $\nu(CO)$ at ca. $1000 \text{ cm}^{-1}$ and $\nu(MO)$ at $600–300 \text{ cm}^{-1}$.[209] Infrared spectra have been reported for various alkoxides of Er(III)[210] and isopropoxides of rare-earth metals.[211]

The infrared spectra of alcohol complexes, $[M(EtOH)_6]Y_2$, where M is a divalent metal, Y is $ClO_4^-$, $BF_4^-$, and $NO_3^-$, have been measured by van Leeuwen.[212] As expected, the anions have considerable influence on $\nu(OH)$ and $\delta(MOH)$. In ethylene glycol complexes with $MX_2$ (X = Cl, Br, and I), $\nu(OH)$ are shifted to lower frequencies and $\delta(CCO)$ to higher frequencies relative to those of free ligand. It was shown that ethylene-glycol serves as a bidentate chelating as well as a unidentate ligand, and that the *gauche* form prevails in the complexes.[213]

The vibrational spectra of diethyl ether complexes with $MgBr_2$ and $MgI_2$ have been assigned completely;[214] $\nu(MgO)$ are at $380–300 \text{ cm}^{-1}$. The solid state Raman spectra of 1:1 and 1:2 adducts of 1,4-dioxane with metal halides show that the ligand is bridging between metals in the chair conformation.[215]

There are many coordination compounds with weakly coordinating ligands containing oxygen donors. These include ketones, aldehydes, esters, and some nitro compounds. Driessen and Groeneveld[216–219] prepared metal complexes of these ligands through the reaction

$$MCl_2 + 6L + 2FeCl_3 \xrightarrow[CH_3NO_2]{} [ML_6](FeCl_4)_2$$

in a moisture-free atmosphere; $CH_3NO_2$ was chosen as the solvent because it is the weakest ligand available. In acetone complexes, $\nu$(C=O) are lower, and $\delta$(CO), $\pi$(CO), and $\delta$(CCC) are higher than those of free ligand.[216] Similar results have been obtained for complexes of acetophenone, chloracetone, and butanone.[217] In metal complexes of acetaldehyde, $\nu$(C=O) are lower and $\delta$(CCO) are higher than those of free ligand.[218] In ester complexes,[219] $\nu$(C=O) shifts to lower and $\nu$(C—O) to higher frequency by complex formation. When these shifts are dependent on the metal ions, the magnitudes of the shifts follow the well-known Irving–Williams order: Mn(II) < Fe(II) < Co(II) < Ni(II) < Cu(II) > Zn(II).

TABLE III-14. INFRARED FREQUENCIES AND BAND ASSIGNMENTS FOR FORMATE AND ACETATE IONS ($cm^{-1}$)[220]

| $[HCOO]^-$ | | $[CH_3COO]^-$ | | | |
|---|---|---|---|---|---|
| Na Salt | Aq. Sol'n | Na Salt | Aq. Sol'n | $C_{2v}$ | Band Assignment |
| 2841 | 2803 | 2936 | 2935 | $A_1$ | $\nu$(CH) |
| — | — | — | 1344 | | $\delta(CH_3)$ |
| 1366 | 1351 | 1414 | 1413 | | $\nu$(COO) |
| — | — | 924 | 926 | | $\nu$(CC) |
| 772 | 760 | 646 | 650 | | $\delta$(OCO) |
| — | — | — | — | $A_2$ | $\rho_t(CH_3)$ |
| — | — | 2989 | 3010 or 2981 | $B_1$ | $\nu$(CH) |
| 1567 | 1585 | 1578 | 1556 | | $\nu$(COO) |
| — | — | 1430 | 1429 | | $\delta(CH_3)$ |
| — | — | 1009 | 1020 | | $\rho_r(CH_3)$ |
| 1377 | 1383 | 460 | 471 | | $\delta$(CH) or $\rho_r$(COO) |
| — | — | 2989 | 2981 or 3010 | $B_2$ | $\nu$(CH) |
| — | — | 1443 | 1456 | | $\delta(CH_3)$ |
| — | — | 1042 | 1052 | | $\rho_r(CH_3)$ |
| 1073 | 1069 | 615 | 621 | | $\pi$(CH) or $\pi$(COO) |

Extensive infrared studies have been made on metal complexes of carboxylic acids. Table III-14 gives the infrared frequencies and band assignments for the formate and acetate ions obtained by Itoh and Bernstein.[220] The carboxylate ion may coordinate to a metal in one of the following modes:

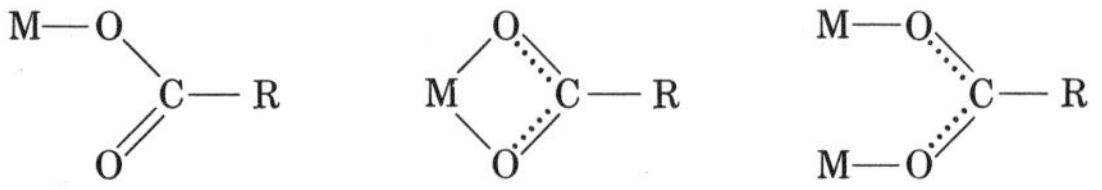

I II III

The $\nu_a(CO_2^-)$ and $\nu_s(CO_2^-)$ of free acetate ion are ca. 1560 and 1416 cm$^{-1}$, respectively. In the unidentate complex (structure I), $\nu$(C═O) is higher than $\nu_a(CO_2^-)$ and $\nu$(C—O) is lower than $\nu_s(CO_2^-)$. As a result, the separation between the two $\nu$(CO) is much larger in unidentate complexes than in the free ion. The opposite trend is observed in the bidentate (chelate) complex (structure II); the separation between the $\nu$(CO) is smaller than that of the free ion in this case. In the bridging complex (structure III), however, two $\nu$(CO) are close to the free ion values. Table III-15 lists the results obtained for some typical compounds. $[Pd(CH_3COO)_2(PPh_3)]_2$ contains one unidentate and one bridging acetate group in one molecule. In a series of unidentate complexes, $\nu$(C═O) increases and $\nu$(C—O) decreases as the M–O bond becomes stronger (see

TABLE III-15. CARBOXYL STRETCHING FREQUENCIES AND STRUCTURES OF ACETATO COMPLEXES

| Compound | $\nu$(C═O)[a] | $\nu$(C—O)[a] | $\Delta$[b] | Structure | Ref. |
|---|---|---|---|---|---|
| $CH_3COO^-$, (ac) | 1560 | 1416 | 144 | — | |
| $Rh(ac)(CO)(PPh_3)_2$ | 1604 | 1376 | 228 | unidentate | 221 |
| $Ru(ac)_2(CO)_2(PPh_3)_2$ | 1613 | 1315 | 289 | unidentate | 221 |
| $RuCl(ac)(CO)(PPh_3)_2$ | 1507 | 1465 | 42 | bidentate | 221 |
| $RuH(ac)(PPh_3)_2$ | 1526 | 1449 | 77 | bidentate | 221 |
| $[Pd(ac)_2(PPh_3)]_2$ | 1629 | 1314 | 315 | unidentate | 222 |
| | 1580 | 1411 | 169 | bridging | 222 |
| $Rh_2(ac)_2(CO)_3(PPh_2)$ | 1580 | 1440 | 140 | bridging | 223 |

[a] These correspond to $\nu_a(COO^-)$ and $\nu_s(COO^-)$ of the symmetrical $COO^-$ group.
[b] Difference between two frequencies.

Sec. III-14). Thus, in a series of monochloroacetates,[224] $\nu(C{=}O)$ increases in this order:

$$\begin{array}{ccccccccc} \text{free ion} & & \text{Ba(II)} & & \text{Sr(II)} & & \text{Ca(II)} & & \text{Cu(II)} \\ 1597 & < & 1600 & < & 1620 & < & 1639 & < & 1660\ \text{cm}^{-1} \end{array}$$

and the separation of the two $\nu(CO)$ increases in the same order as these metals. On the other hand, the order of $\nu(C{=}O)$ in the trichloroacetate series is opposite to the above:

$$\begin{array}{ccccccccc} \text{Cu(II)} & & \text{Ca(II)} & & \text{Sr(II)} & & \text{Ba(II)} & & \text{free ion} \\ 1629 & < & 1649 & < & 1660 & < & 1670 & < & 1677\ \text{cm}^{-1} \end{array}$$

and the separation of the two $\nu(CO)$ increases in the same order.[225] This marked difference between the two series may indicate a difference in structure, which will have to be investigated by X-ray analysis. In the case of bridging complexes, both $\nu(CO)$ bands are shifted to higher frequencies in going from $Cr_2(CH_3COO)_4 \cdot 2H_2O$ to $Cu_2(CH_3COO)_4 \cdot 2H_2O$.[226]

The linkage isomerism involving the acetate group has been reported by Baba and Kawaguchi:[227]

PPh3 / (acac)Pd \ O—C(=O)—CH3          PPh3 / (acac)Pd \ CH2COOH

The O-isomer exhibits $\nu(C{=}O)$ at 1640 cm$^{-1}$, whereas the C-isomer shows $\nu(C{=}O)$ at 1670 and 1650 and $\nu(OH)$ at 2700–2500 cm$^{-1}$. The infrared spectra of metal glycolato complexes have been assigned by Nakamoto et al.[228]

## III-7. COMPLEXES OF OXALIC ACID AND RELATED COMPOUNDS

The oxalato anion ($ox^{2-}$) coordinates to a metal as a unidentate (I) or bidentate (II) ligand:

$[(NH_3)_5Co{-}O{-}C(=O){-}C(=O){-}OH]^{2+}$ (I)          $[(NH_3)_4Co(O_2C{-}CO_2)]^{+}$ (II)

TABLE III-16. FREQUENCIES AND BAND ASSIGNMENTS OF VARIOUS OXALATO COMPLEXES ($CM^{-1}$)[229]

| $K_2[Zn(ox)_2]\cdot 2H_2O$ | $K_2[Cu(ox)_2]\cdot 2H_2O$ | $K_2[Pd(ox)_2]\cdot 2H_2O$ | $K_2[Pt(ox)_2]\cdot 3H_2O$ | $K_3[Fe(ox)_3]\cdot 3H_2O$ | $K_3[V(ox)_3]\cdot 3H_2O$ | $K_3[Cr(ox)_3]\cdot 3H_2O$ | $K_3[Co(ox)_3]\cdot 3H_2O$ | $K_3[Al(ox)_3]\cdot 3H_2O$ | $[Cr(NH_3)_4(ox)]\cdot Cl$ | Band Assignment | |
|---|---|---|---|---|---|---|---|---|---|---|---|
| 1632 | (1720) 1672 | 1698 | 1709 | 1712 | 1708 | 1708 | 1707 | 1722 | 1704 | $\nu_a(C=O)$ | $\nu_7$ |
| — | 1645 | 1675, 1657 | 1674 | 1677, 1649 | 1675, 1642 | 1684, 1660 | 1670 | 1700, 1683 | 1668 | $\nu_a(C=O)$ | $\nu_1$ |
| 1433 | 1411 | 1394 | 1388 | 1390 | 1390 | 1387 | 1398 | 1405 | 1393 | $\nu_s(CO) + \nu(CC)$ | $\nu_2$ |
| 1302 | 1277 | 1245 (1228) | 1236 | 1270, 1255 | 1261 | 1253 | 1254 | 1292, 1269 | 1258 | $\nu_s(CO) + \delta(O-C=O)$ | $\nu_8$ |
| 890 | 886 | 893 | 900 | 885 | 893 | 893 | 900 | 904 | 914, 890 | $\nu_s(CO) + \delta(O-C=O)$ | $\nu_3$ |
| 785 | 795 | 818 | 825 | 797, 785 | 807, 797 | 810, 798 | 822, 803 | 820, 803 | 804 | $\delta(O-C=O) + \nu(MO)$ | $\nu_9$ |
| 622 | 593 | 610 | — | 580 | 581 | 595 | — | — | — | crystal water? | |
| 519 | 541 | 556 | 575, 559 | 528 | 531 | 543 | 565 | 587 | 545 | $\nu(MO) + \nu(CC)$ | $\nu_4$ |
| 519 | 481 | 469 | 469 | 498 | 497 | 485 | 472 | 436 | 486, 469 | ring def. $+ \delta(O-C=O)$ | $\nu_{10}$ |
| 428, 419 | 420 | 417 | 405 | 366 | 368 | 415 | 446 | 485 | 366 | $\nu(MO)$ + ring def. | $\nu_{11}$ |
| 377, 364 | 382, 370 | 368 | 370 | 340 | 336 | 358 | 364 | 364 | 347 | $\delta(O-C=O) + \nu(CC)$ | $\nu_5$ |
| 291 | 339 | 350 | 328 | — | — | 313 | 332 | — | 328 | $\pi$ | |

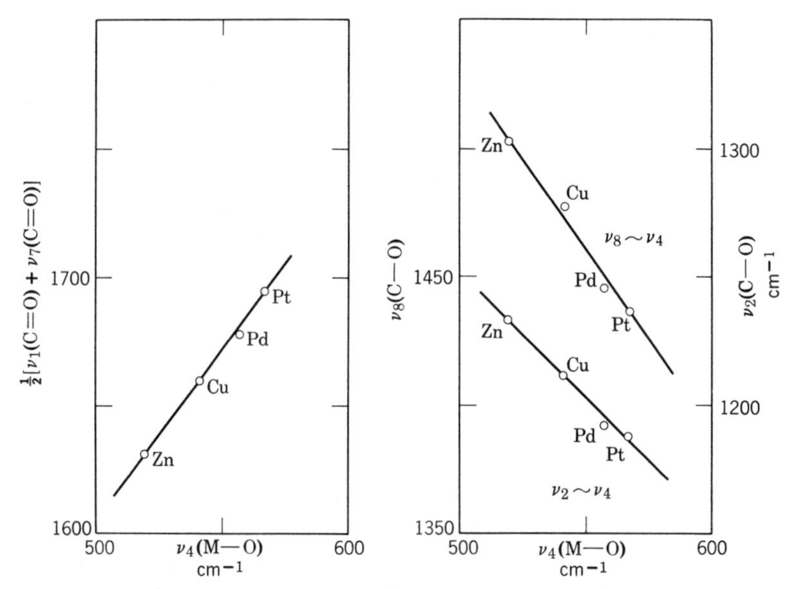


**Fig. III-17.** MO stretching frequency vs. C=O and C—O stretching frequencies in oxalato complexes of divalent metals.

The bidentate chelate structure (II) is most common. Fujita et al.[229] carried out normal coordinate analyses on the 1:1 (metal–ligand) model of the $[M(ox)_2]^{2-}$ and $[M(ox)_3]^{3-}$ series, and obtained the band assignments listed in Table III-16. In the divalent metal series, $\nu$(C=O) (average of $\nu_1$ and $\nu_7$) becomes higher, and $\nu$(C—O) ($\nu_2$ and $\nu_8$) becomes lower, as $\nu_4$(MO) becomes higher in the order Zn(II) < Cu(II) < Pd(II) < Pt(II) (see Fig. III-17). This relation holds in spite of the fact that $\nu_2$, $\nu_4$, and $\nu_8$ are all coupled with other vibrations.

In the trivalent metal series, Hancock and Thornton[230] found that $\nu_{11}$ (MO stretching) follows the same trend as the crystal field stabilization energies (CFSE) of these metals, namely:

| | Sc | | V | | Cr | | Mn | | Fe | | Co | | Ga |
|---|---|---|---|---|---|---|---|---|---|---|---|---|---|
| | $d^0$ | | $d^2$ | | $d^3$ | | $d^4$ | | $d^5$ | | $d^6$ | | $d^{10}$ |
| $\nu$(MO) (cm$^{-1}$) | 340 | < | 367 | < | 416 | > | 372 | > | 354 | < | 446 | > | 368 |
| CFSE ($10^3$ cm$^{-1}$) | 0 | < | 10.2 | < | 21.2 | > | 10.2 | > | 0 | < | 27.0 | > | 0 |

Both quantities are maximized at the $d^3$ and $d^6$ configurations ($d^4$ and $d^5$ ions are in high-spin states).

The oxalato anion may act as a bridging group between metal atoms. According to Scott et al.,[231] the oxalato anion can take the following four bridging structures:

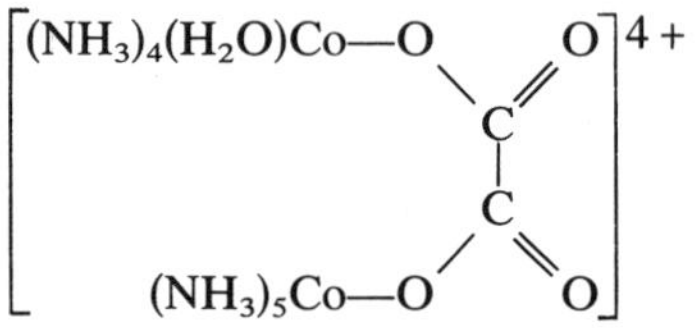

$$\left[\begin{array}{c}(NH_3)_4(H_2O)Co{-}O \quad O \\ \diagdown\;\diagup\!\!\diagup \\ C \\ | \\ C \\ \diagup\;\diagdown\!\!\diagdown \\ (NH_3)_5Co{-}O \quad O\end{array}\right]^{4+}$$

III

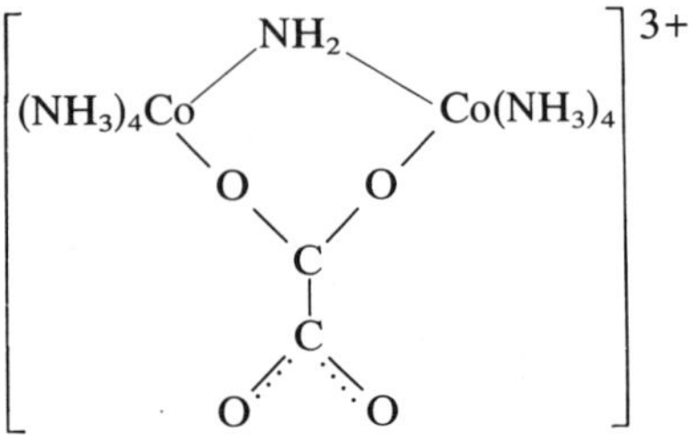

$$\left[\begin{array}{c}NH_2 \\ (NH_3)_4Co \qquad Co(NH_3)_4 \\ O \qquad O \\ C \\ | \\ C \\ O \qquad O\end{array}\right]^{3+}$$

IV

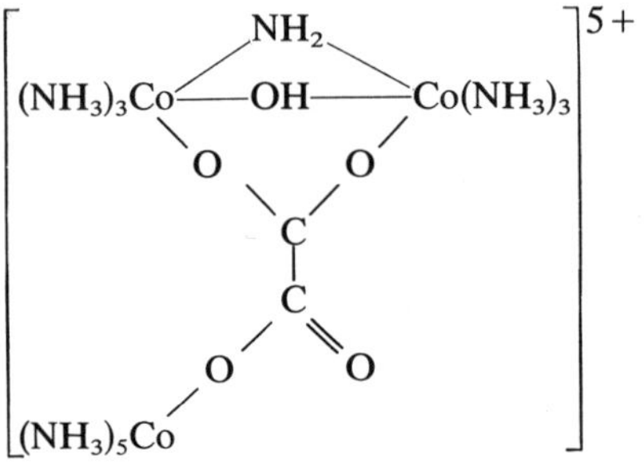

$$\left[\begin{array}{c}NH_2 \\ (NH_3)_3Co{-}OH{-}Co(NH_3)_3 \\ O \qquad O \\ C \\ | \\ C \\ O \qquad O \\ (NH_3)_5Co\end{array}\right]^{5+}$$

V

$$\left[\begin{array}{c}NH_2 \\ (NH_3)_3Co{-}OH{-}Co(NH_3)_3 \\ O \qquad O \\ C \\ | \\ C \\ O \qquad O \\ (NH_3)_3Co{-}OH{-}Co(NH_3)_3 \\ NH_2\end{array}\right]^{6+}$$

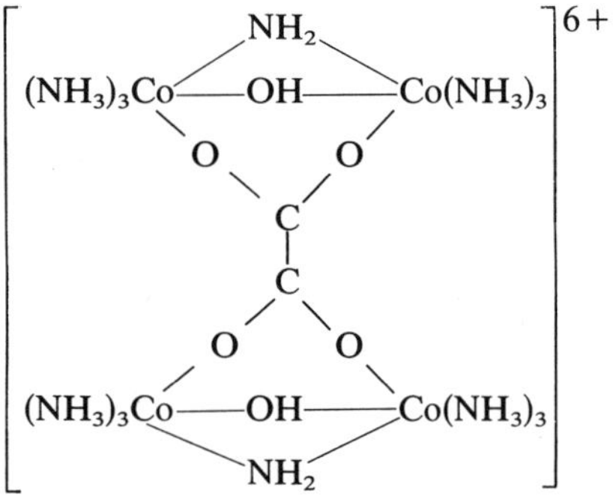

VI

Table III-17 lists the $\nu$(CO) of each type. The spectrum of the tetradentate complex (VI) is the most simple. Because of its high symmetry [$\mathbf{D}_{2h}$ (planar) or $\mathbf{D}_{2d}$ (twisted)], it exhibits only two $\nu$(CO). The spectra of bidentate complexes (III and IV) show four $\nu$(CO), as expected from their $\mathbf{C}_{2v}$ symmetry. The spectrum of the tridentate complex (V) should show four $\nu$(CO), although only three are observed.

The Raman spectra of metal oxalato complexes have also been examined to investigate the solution equilibria and the nature of the M–O bond.[232]

Kuroda et al.[233] interpreted the infrared spectra of metal oxamido complexes on the basis of their structure ($\mathbf{V}_h$ symmetry): Armendarez

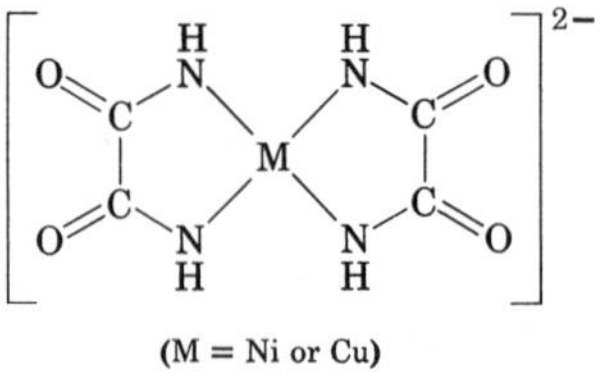

$$\left[\begin{array}{c}O{=}C{-}NH \qquad HN{-}C{=}O \\ | \qquad\; \diagdown M \diagup \qquad\; | \\ O{=}C{-}NH \qquad HN{-}C{=}O\end{array}\right]^{2-}$$

(M = Ni or Cu)

TABLE III-17. CO STRETCHING VIBRATIONS OF Co(III) OXALATO COMPLEXES ($cm^{-1}$)

| Compound | Symmetry[a] | ν(CO) | | | |
|---|---|---|---|---|---|
| I | $\mathbf{C}_s/\mathbf{C}_1$ | 1761 | 1682<br>1665 | 1400 | 1260 |
| II | $\mathbf{C}_{2v}$ | 1696 | 1667 | 1410 | 1268 |
| III | $\mathbf{C}_{2v}/\mathbf{C}_2$ | 1721<br>1701 | 1629<br>1670 | 1439<br>1430 | 1276<br>1250 |
| IV | $\mathbf{C}_{2v}/\mathbf{C}_2$ | 1755 | 1626 | 1320 | 1270 |
| V | $\mathbf{C}_s/\mathbf{C}_1$ | 1650 | 1610 | 1322 | — |
| VI | $\mathbf{D}_{2h}/\mathbf{D}_{2d}$ | — | 1628 | 1345 | — |

[a] The point groups given before and after the slash indicate the symmetry of the oxalato anion for the planar and nonplanar configurations, respectively.

and Nakamoto[234] confirmed this structure from normal coordinate analysis; the NiN stretching force constant (UBF) was estimated to be 0.73 mdyn/Å.

Biuret ($NH_2CONHCONH_2$) is known to form the following two types of chelate rings:

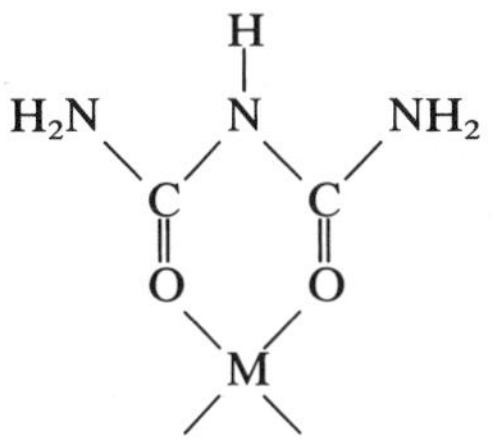

VII 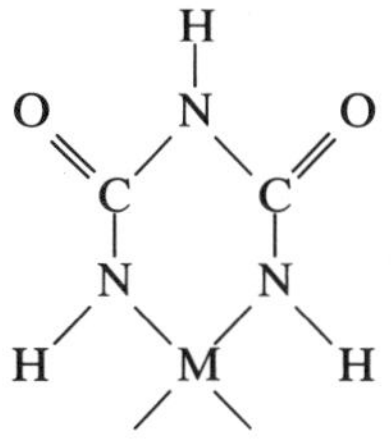 VIII

Violet crystals of composition $K_2[Cu(biureto)_2]\cdot 4H_2O$ are obtained when the Cu(II) ion is added to an alkaline solution of biuret, whereas pale blue–green crystals of composition $[Cu(biuret)_2]Cl_2$ result when the Cu(II) ion is mixed with biuret in neutral (alcoholic) solution. The former contains the N-bonded chelate ring structure (VIII), while the latter consists of the O-bonded chelate rings (VII). Kedzia et al.[235] carried out normal coordinate analyses of both compounds. Figure III-18 shows the infrared spectra and band assignments of these two complexes. It is interesting to note that the ν(C=O) of the N-bonded complex are lower than those of the O-bonded complex. This unexpected result may be

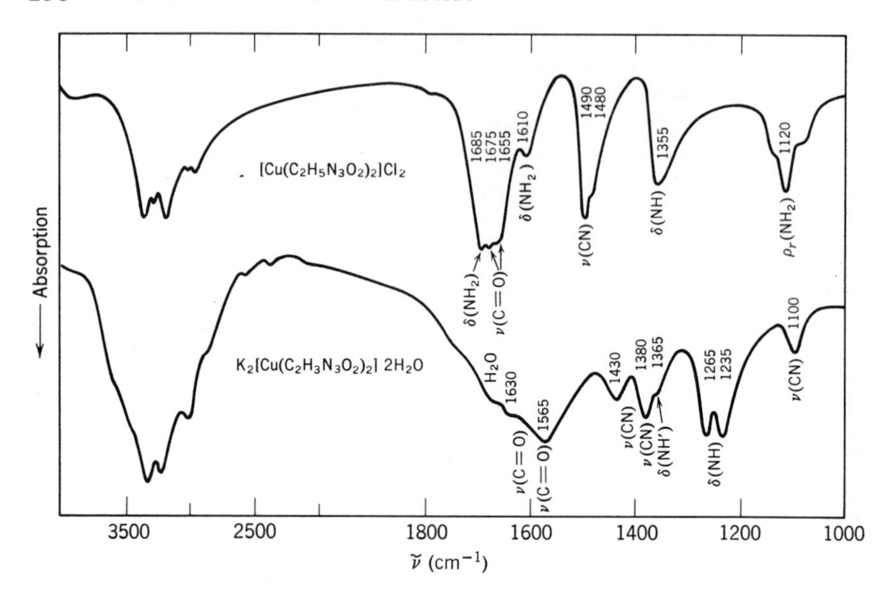


**Fig. III-18.** Infrared spectra of O-bonded and N-bonded biuret complexes of Cu(II).[235]

due to the fact that the C=O groups of the N-bonded complex are strongly hydrogen-bonded to the crystal water. As expected, the infrared spectra of these two complexes are markedly different in the NH and $NH_2$ vibrations. Similar linkage isomers were found in the Ni(II) complexes. The Co(II) complex forms the N-bonded chelate ring, whereas the Zn complex forms the O-bonded ring structure.[235] In $[Cd(biuret)_2]Cl_2$, the biuret molecules are bonded to the metal as follows:[236]

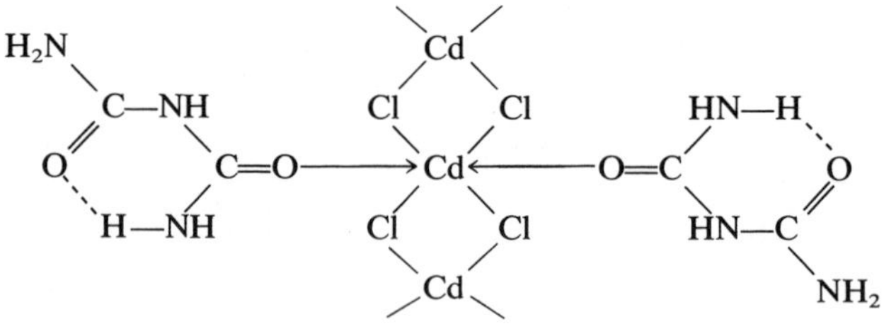

IX

Saito et al.[237] carried out normal coordinate analysis on the ligand portion of the Cd complex.

## III-8. SULFATO, CARBONATO, AND OTHER ACIDO COMPLEXES

When a ligand of relatively high symmetry coordinates to a metal, its symmetry is lowered and marked changes in the spectrum are expected because of changes in the selection rules. This principle has been used extensively to determine whether acido anions such as $SO_4^{2-}$ and $CO_3^{2-}$ coordinate to metals as unidentate, chelating bidentate, or bridging bidentate ligands. Although symmetry lowering is also caused by the crystalline environment, this effect is generally much smaller than the effect of coordination.

### (1) Sulfato ($SO_4$) Complexes

The free sulfate ion belongs to the high symmetry point group $\mathbf{T}_d$. Of the four fundamentals, only $\nu_3$ and $\nu_4$ are infrared active. If the symmetry of the ion is lowered by complex formation, the degenerate vibrations split and Raman active modes appear in the infrared spectrum. The lowering of symmetry caused by coordination is different for the unidentate and bidentate complexes, as shown below:

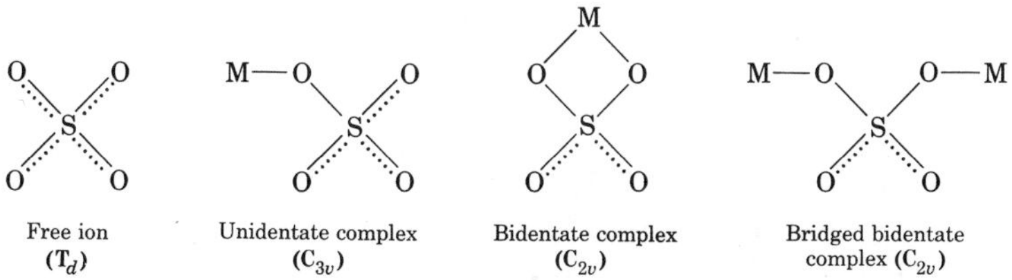

Free ion ($T_d$) | Unidentate complex ($C_{3v}$) | Bidentate complex ($C_{2v}$) | Bridged bidentate complex ($C_{2v}$)

The change in the selection rules caused by the lowering of symmetry was shown in Table II-6*f*. Table III-18 and Fig. III-19 give the frequencies and the spectra of typical Co(III) sulfato complexes obtained by Nakamoto et al.[238] In $[Co(NH_3)_6]_2(SO_4)_3 \cdot 5H_2O$, $\nu_3$ and $\nu_4$ do not split and $\nu_2$ does not appear; although $\nu_1$ is observed, it is very weak. We conclude, therefore, that the symmetry of the $SO_4^{2-}$ ion is approximately $\mathbf{T}_d$. In $[Co(NH_3)_5SO_4]Br$, both $\nu_1$ and $\nu_2$ appear with medium intensity; moreover, $\nu_3$ and $\nu_4$ each splits into two bands. This result can be explained by assuming a lowering of symmetry from $\mathbf{T}_d$ to $\mathbf{C}_{3v}$ (unidentate coordination). In

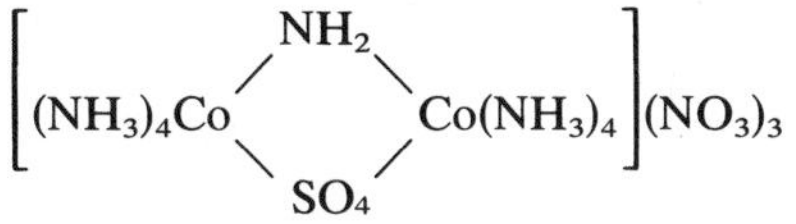

$$\left[(NH_3)_4Co\begin{matrix} NH_2 \\ \\ SO_4 \end{matrix}Co(NH_3)_4\right](NO_3)_3$$

TABLE III-18. VIBRATIONAL FREQUENCIES OF Co(III) SULFATO COMPLEXES ($cm^{-1}$)[238]

| Compound | Symmetry | $\nu_1$ | $\nu_2$ | $\nu_3$ | $\nu_4$ |
|---|---|---|---|---|---|
| Free $SO_4^{2-}$ ion | $\mathbf{T}_d$ | — | — | 1104 (vs)[a] | 613 (s) |
| $[Co(NH_3)_6]_2(SO_4)_3 \cdot 5H_2O$ | $\mathbf{T}_d$ | 973 (vw) | — | 1130–1140 (vs) | 617 (s) |
| $[Co(NH_3)_5SO_4]Br$ | $\mathbf{C}_{3v}$ | 970 (m) | 438 (m) | {1032–1044 (s)<br>1117–1143 (s) | {645 (s)<br>604 (s) |
| $\left[(NH_3)_4Co \begin{smallmatrix} NH_2 \\ SO_4 \end{smallmatrix} Co(NH_3)_4\right][NO_3]_3$ | $\mathbf{C}_{2v}$ | 995 (m) | 462 (m) | {1050–1060 (s)<br>1170 (s)<br>1105 (s) | {641 (s)<br>610 (s)<br>571 (m) |

[a] vs; very strong; s: strong; m: medium; w: weak; vw: very weak.

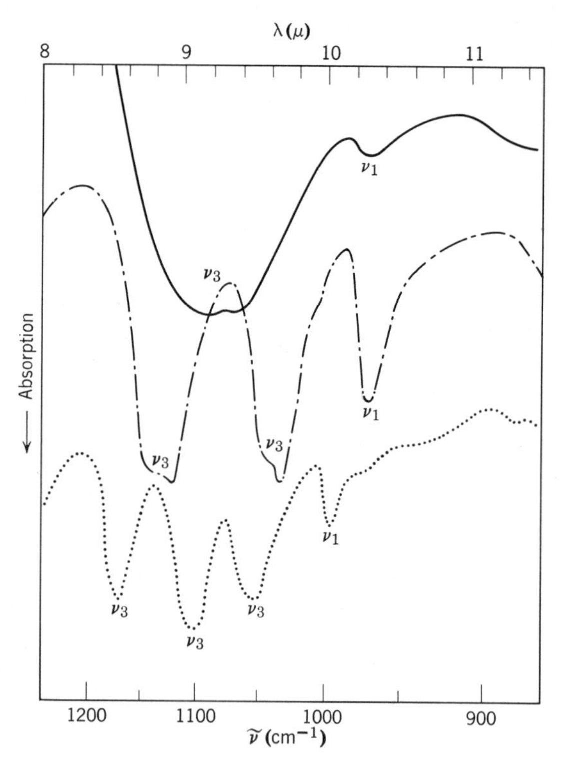


**Fig. III-19.** Infrared spectra of $[Co(NH_3)_6]_2(SO_4)_3 \cdot 5H_2O$ (solid line); $[Co(NH_3)_5SO_4]Br$ (dot-dash line); and $\left[(NH_3)_4Co \begin{smallmatrix} NH_2 \\ SO_4 \end{smallmatrix} Co(NH_3)_4\right](NO_3)_3$ (dotted line).[238]

both $\nu_1$ and $\nu_2$ appear with medium intensity, and $\nu_3$ and $\nu_4$ each splits into three bands. These results suggest that the symmetry is further lowered and probably reduced to $\mathbf{C}_{2v}$, as indicated in Table II-6*f*. Thus the $SO_4^{2-}$ group in this complex is concluded to be a bridging bidentate as depicted in the foregoing diagram.

The chelating bidentate $SO_4^{2-}$ group was discovered by Barraclough and Tobe,[239] who observed three bands (1211, 1176, and 1075 $cm^{-1}$) in the $\nu_3$ region of $[Co(en)_2SO_4]Br$. These frequencies are higher than those of the bridging bidentate complex listed in Table III-18. Eskenazi et al.[240] also found the same trend in Pd(II) sulfato complexes. Thus the distinction between bridging and chelating sulfato complexes can be made on this basis. Table III-19 lists the observed frequencies of the sulfato groups and the modes of coordination as determined from the spectra.

TABLE III-19. VIBRATIONAL FREQUENCIES AND MODES OF COORDINATION OF VARIOUS SULFATO COMPLEXES ($cm^{-1}$)

| Compound | Mode of Coordination | $\nu_1$ | $\nu_2$ | $\nu_3$ | $\nu_4$ | Ref. |
|---|---|---|---|---|---|---|
| $[Cr(H_2O)_5SO_4]Cl\cdot\frac{1}{2}H_2O$ | unidentate | 1002 | — | 1118<br>1068 | — | 241 |
| $[Cu(bipy)SO_4]\cdot 2H_2O$ (polymeric) | bridging bidentate | 971 | — | 1166<br>1096<br>1053–1035 | — | 242 |
| $Ni(morpholine)_2SO_4$ (polymeric) | bridging bidentate | 973 | 493 | 1177<br>1094<br>1042 | 628<br>612<br>593 | 243 |
| $[Co_2\{(SO_4)_2OH\}(NH_3)_6]Cl$ | bridging bidentate | 966 | — | 1180<br>1101<br>1048 | 645<br>598 | 244 |
| $Pd(NH_3)_2SO_4$ | bridging bidentate | 960 | — | 1195<br>1110<br>1030 | — | 240 |
| $Pd(phen)SO_4$ | chelating bidentate | 955 | — | 1240<br>1125<br>1040–1015 | — | 240 |
| $Pd(PPh_3)_2SO_4$ | chelating bidentate | 920 | — | 1265<br>1155<br>1110 | — | 245 |
| $Ir(PPh_3)_2(CO)I(SO_4)$ | chelating bidentate | 856 | 549 | 1296<br>1172<br>880 | 662<br>610 | 246 |

The symmetries of the sulfate and nitrate ions of metal salts at various stages of hydration have been discussed on the basis of their infrared spectra.[247] Normal coordinate analyses on the sulfato, nitrato, and carbonato groups in Co(III) ammine complexes have been carried out by Tanaka et al.[248] Goldsmith et al.[249] carried out normal coordinate analyses on the skeletons of $[Co(NH_3)_5X]^{2+}$ ($X = SO_4^{2-}$ and $CO_3^{2-}$) and *cis*-$[Co(NH_3)_4X_2]^{n+}$ ($X_2 = CO_3^{2-}$ and $PO_4^{3-}$) type complexes.

### (2) Perchlorato ($ClO_4$) Complexes

In general, the perchlorate ($ClO_4^-$) ion coordinates to a metal when its complexes are prepared in nonaqueous solvents. The structure and bonding of metal complexes containing these weakly coordinating ligands have

TABLE III-20. ClO STRETCHING FREQUENCIES OF PERCHLORATO COMPLEXES ($CM^{-1}$)

| Complex | Structure | $\nu_3$ | $\nu_4$ | Ref. |
|---|---|---|---|---|
| $K[ClO_4]$ | ionic | 1170–1050 | (935)[a] | |
| $Cu(ClO_4)_2 \cdot 6H_2O$ | ionic | 1160–1085 | (947)[a] | 251 |
| $Cu(ClO_4)_2 \cdot 2H_2O$ | unidentate | 1158, 1030 | 920 | 251 |
| $Cu(ClO_4)_2$ | bidentate | 1270–1245, 1130, 948–920 | 1030 | 251 |
| $Mn(ClO_4)_2 \cdot 2H_2O$ | bidentate | 1210, 1138, 945 | 1030 | 252 |
| $Co(ClO_4)_2 \cdot 2H_2O$ | bidentate | 1208, 1125, 935 | 1025 | 252 |
| $[Ni(en)_2(ClO_4)_2]$[b] | bidentate | 1130, 1093, 1058 | 962 | 253 |
| $Ni(CH_3CN)_4(ClO_4)_2$ | unidentate | 1135, 1012 | 912 | 254 |
| $Ni(CH_3CN)_2(ClO_4)_2$ | bidentate | 1195, 1106, 1000 | 920 | 254 |
| $[Ni(4\text{-}Me\text{-}py)_4](ClO_4)_2$ | ionic | 1040–1130 | (931)[a] | 255 |
| $Ni(3\text{-}Br\text{-}py)_4(ClO_4)_2$ | unidentate | 1165–1140, 1025 | 920 | 255 |

[a] Weak.
[b] Blue form.

been briefly reviewed by Rosenthal.[250] Infrared spectroscopy has been used extensively to determine the mode of coordination of $ClO_4^-$ group. The structures listed in Table III-20 were determined on the basis of the same symmetry selection rules as are used for sulfato complexes.

### (3) Complexes of Other Tetrahedral Anions

Many other tetrahedral anions coordinate to a metal as unidentate and bidentate ligands, and their modes of coordination have been determined by the same method as is used for $SO_4^{2-}$ and $ClO_4^-$ ions. For example, the $PO_4^{3-}$ ion is a unidentate in $[Co(NH_3)_5PO_4]$ and a bidentate in $[Co(NH_3)_4PO_4]$.[256] Similar pairs have been made with the $AsO_4^{3-}$,[257] $CrO_4^{2-}$, and $MoO_4^{2-}$ ions.[258] The $SeO_4^{2-}$ ion in $[Co(NH_3)_5SeO_4]Cl$ is a unidentate,[259] whereas it is a bridging bidentate in $[Co_2\{(SeO_4)_2OH\}(NH_3)_6]Cl$.[244]

The $S_2O_3^{2-}$ ion can coordinate to a metal in a variety of ways. According to Freedman and Straughan,[260] $\nu_a(SO_3)$ near 1130 $cm^{-1}$ is most useful as a structural diagnosis: >1175 (S-bridging); 1175–1130 (S-coordination); ~1130 (ionic $S_2O_3^{2-}$); <1130 $cm^{-1}$ (O-coordination). On the basis of this criterion, they proposed polymeric structures linked by O-bridges for thiosulfates of $UO_2^{2+}$ and $ZrO_2^{2+}$. The infrared and Raman spectra show that the $SO_3F^-$ ion is a unidentate in $[Sn(SO_3F)_6]^{2-}$.[261]

### (4) Carbonato ($CO_3$) Complexes

The carbonate ion coordinates to the metal in one of two ways:

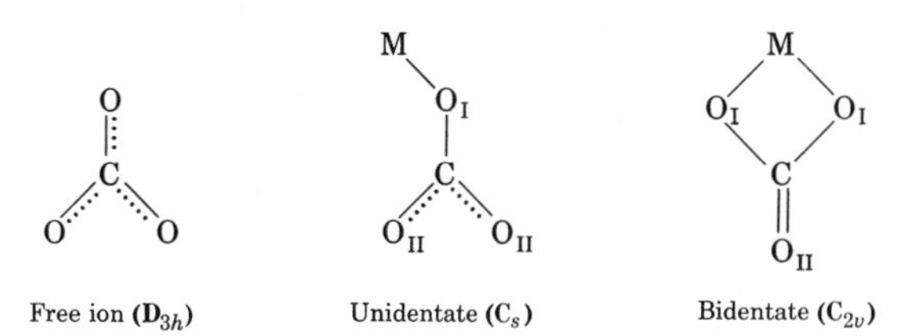


Free ion ($D_{3h}$) Unidentate ($C_s$) Bidentate ($C_{2v}$)

The selection rule changes as shown in Table I-11. In $\mathbf{C}_{2v}$ and $\mathbf{C}_s$,* the $\nu_1$ vibration, which is forbidden in the free ion, becomes infrared active and each of the doubly degenerate vibrations, $\nu_3$ and $\nu_4$, splits into two bands. Although the number of infrared active fundamentals is the same for $\mathbf{C}_{2v}$ and $\mathbf{C}_s$, the splitting of the degenerate vibrations is larger in the bidentate than in the unidentate complex.[238] For example, $[Co(NH_3)_5CO_3]Br$ exhibits two CO stretchings at 1453 and 1373 $cm^{-1}$, whereas $[Co(NH_3)_4CO_3]Cl$ shows them at 1593 and 1265 $cm^{-1}$. In organic carbonates such as dimethyl carbonate, $(CH_3O_I)_2CO_{II}$, this effect is more

* Both unidentate and bidentate carbonato groups have the same $\mathbf{C}_{2v}$ symmetry if the metal atom is ignored.

striking because the $CH_3$—$O_I$ bond is strongly covalent. Thus the $CO_{II}$ stretching is observed at 1870 $cm^{-1}$, whereas the $CO_I$ stretching is at 1260 $cm^{-1}$. Gatehouse and co-workers[262] showed that the separation of the CO stretching bands increases along the following series:

$$\text{basic salt} < \text{carbonato complex} < \text{acid} < \text{organic carbonate}$$

Fujita et al.[263] carried out normal coordinate analyses on unidentate and bidentate carbonato complexes of Co(III). According to their results, the CO stretching force constant, which is 5.46 for the free ion, becomes 6.0 for the C–$O_{II}$ bonds and 5.0 for the C–$O_I$ bond of the unidentate complex, whereas it becomes 8.5 for the C–$O_{II}$ bond and 4.1 for the C–$O_I$ bonds of the bidentate complex (all are UBF force constants in units of mdyn/Å). The observed and calculated frequencies and theoretical band assignments are shown in Table III-21. Normal coordinate analyses on carbonato complexes have also been carried out by Hester and Grossman[264] and Goldsmith and Ross.[265]

As is shown in Table III-21, normal coordinate analysis predicts that the highest frequency CO stretching band belongs to the $B_2$ species in the unidentate and the $A_1$ species in the bidentate complex. Elliott and Hathaway[266] studied the polarized infrared spectra of single crystals of $[Co(NH_3)_4CO_3]Br$ and confirmed these symmetry properties. As will be shown later for nitrato complexes, Raman polarization studies are also useful for this purpose.

### (5) Nitrato ($NO_3$) Complexes

The structures and vibrational spectra of a large number of nitrato complexes have been reviewed by Addison et al.[267] and Rosenthal.[250] X-ray analyses show that the $NO_3^-$ ion coordinates to a metal as a unidentate, symmetric and asymmetric chelating bidentate, and bridging bidentate ligand of various structures. It is rather difficult to differentiate these structures by vibrational spectroscopy since the symmetry of the nitrate ion differs very little among them ($\mathbf{C}_{2v}$ or $\mathbf{C}_s$). Even so, vibrational spectroscopy is still useful in distinguishing unidentate and bidentate ligands.

Originally, Gatehouse et al.[268] noted that the unidentate $NO_3$ group exhibits three NO stretching bands, as expected for its $\mathbf{C}_{2v}$ symmetry. For example, $[Ni(en)_2(NO_3)_2]$ (unidentate) exhibits three bands as follows:

| | | |
|---|---|---|
| $\nu_5(B_2)$ | 1420 $cm^{-1}$ | $\nu_a(NO_2)$ |
| $\nu_1(A_1)$ | 1305 $cm^{-1}$ | $\nu_s(NO_2)$ |
| $\nu_2(A_1)$ | (1008) $cm^{-1}$ | $\nu(NO)$ |

TABLE III-21. CALCULATED AND OBSERVED FREQUENCIES OF UNIDENTATE AND BIDENTATE Co(III) CARBONATO COMPLEXES ($cm^{-1}$)[263]

| Species ($\mathbf{C}_{2v}$)[a] | $\nu_1(A_1)$ | $\nu_2(A_1)$ | $\nu_3(A_1)$ | $\nu_4(A_1)$ | $\nu_5(B_2)$ | $\nu_6(B_2)$ | $\nu_7(B_2)$ | $\nu_8(B_1)$ |
|---|---|---|---|---|---|---|---|---|
| Calc. frequency | 1376 | 1069 | 772 | 303 | 1482 | 676 | 92 | — |
| Assignment | $\nu(CO_{II})$ + $\nu(CO_I)$ | $\nu(CO_I)$ + $\nu(CO_{II})$ | $\delta(O_{II}CO_{II})$ | $\nu(CoO_I)$ | $\nu(CO_{II})$ | $\rho_r(O_{II}CO_{II})$ | $\delta(CoO_IC)$ | $\pi$ |
| $[Co(NH_3)_5CO_3]Br$ | 1373 | 1070 | 756 | 362 | 1453 | 678 | — | 850 |
| $[Co(ND_3)_5CO_3]Br$ | 1369 | 1072 | 751 | 351 | 1471 | 687 | — | 854 |
| $[Co(NH_3)_5CO_3]I$ | 1366 | 1065 | 776 | 360 | 1449 | 679 | — | 850 |
| $[Co(ND_3)_5CO_3]I$ | 1360 | 1063 | 742 | 341 | 1467 | 687 | — | 853 |

| Species ($\mathbf{C}_{2v}$) | $\nu_1(A_1)$ | $\nu_2(A_1)$ | $\nu_3(A_1)$ | $\nu_4(A_1)$ | $\nu_5(B_2)$ | $\nu_6(B_2)$ | $\nu_7(B_2)$ | $\nu_8(B_1)$ |
|---|---|---|---|---|---|---|---|---|
| Calc. frequency | 1595 | 1038 | 771 | 370 | 1282 | 669 | 429 | — |
| Assignment | $\nu(CO_{II})$ | $\nu(CO_I)$ | Ring def. + $\nu(CoO_I)$ | $\nu(CoO_I)$ + ring def. | $\nu(CO_I)$ + $\delta(O_ICO_{II})$ | $\delta(O_ICO_{II})$ + $\nu(CO_I)$ + $\nu(CoO_I)$ | $\nu(CoO_I)$ | $\pi$ |
| $[Co(NH_3)_4CO_3]Cl$ | 1593 | 1030 | 760 | 395 | 1265 | 673 | 430 | 834 |
| $[Co(ND_3)_4CO_3]Cl$ | 1635, 1607 | (1031)[b] | 753 | 378 | 1268 | 672 | 418 | 832 |
| $[Co(NH_3)_4CO_3]ClO_4$ | 1602 | —[c] | 762 | 392 | 1284 | 672 | 428 | 836 |
| $[Co(ND_3)_4CO_3]ClO_4$ | 1603 | —[c] | 765 | 374 | 1292 | 676 | 415 | 835 |

[a] Symmetry assuming a linear Co–O–C bond (see Ref. 263).
[b] Overlapped with $\delta_s(ND_3)$.
[c] Hidden by $[ClO_4]^-$ absorption.

whereas $[Ni(en)_2NO_3]ClO_4$ (chelating bidentate) exhibits three bands at the following:

| | | |
|---|---|---|
| $\nu_1(A_1)$ | 1476 $cm^{-1}$ | $\nu(N{=}O)$ |
| $\nu_5(B_2)$ | 1290 $cm^{-1}$ | $\nu_a(NO_2)$ |
| $\nu_2(A_1)$ | (1025) $cm^{-1}$ | $\nu_s(NO_2)$ |

The separation of the two highest frequency bands is 115 $cm^{-1}$ for the unidentate complex, whereas it is 186 $cm^{-1}$ for the bidentate complex. Thus Curtis and Curtis[269] concluded that $[Ni(dien)(NO_3)_2]$ contains both types, since it exhibits bands due to unidentate (1440 and 1315 $cm^{-1}$) and bidentate (1480 and 1300 $cm^{-1}$) groups. In general, the separation of the two highest frequency bands is larger for bidentate than for unidentate coordination if the complexes are similar. However, this rule does not hold if the complexes are markedly different. This is clearly shown in Table III-22.

Lever et al.[278] proposed using the combination band, $\nu_1 + \nu_4$, of free $NO_3^-$ which appears in the 1800–1700 $cm^{-1}$ region for structural diagnosis. Upon coordination, $\nu_4$ ($E'$, in-plane bending) near 700 $cm^{-1}$ splits

TABLE III-22. NO STRETCHING FREQUENCIES OF UNIDENTATE AND BIDENTATE NITRATO COMPLEXES ($CM^{-1}$)

| Compound | Mode of Coordination | $\nu_5$ | $\nu_1$ | $\nu_2$ | $\nu_5 - \nu_1$ | Ref. |
|---|---|---|---|---|---|---|
| $Re(CO)_5NO_3$ | unidentate | 1497 | 1271 | 992 | 226 | 270 |
| $Cu(py)_4(NO_3)_2$ | unidentate | 1408 | 1306 | 1026 | 102 | 100 |
| $Sn(NO_3)_4$ | chelating bidentate | 1630 | 1255 | 983 | 375 | 271 |
| $K[UO_2(NO_3)_3]$ | chelating bidentate | 1555<br>1521 | 1271 | 1025 | 284<br>250 | 272 |
| $Co(NO_3)_3$ | chelating bidentate | 1619 | 1162 | 963 | 457 | 273 |
| $Na_2[Mn(NO_3)_4]$ | chelating bidentate | 1490 | 1280 | 1041<br>1036 | 210 | 274 |
| $Cu(NO_3)_2MeNO_2$ | bridging bidentate | 1519 | 1291 | 1008 | 228 | 275 |
| $Zn(bt)_2(NO_3)_2$[a] | chelating bidentate | 1485 | 1300 | — | 185 | 276 |
| $Ni(dmpy)_2(NO_3)_2$[b] | chelating bidentate | 1513 | 1270 | 1013 | 243 | 277 |

[a] bt: benzothiazole.
[b] dmpy:2,6-dimethyl-4-pyrone.

into two bands, and the magnitude of this splitting is expected to be larger for bidentate than for unidentate ligands. This should be reflected in the separation of two $(\nu_1+\nu_4)$ bands in the 1800–1700 cm$^{-1}$ region. According to Lever et al., the $NO_3^-$ ion is bidentate if the separation is ca. 66–20 cm$^{-1}$ and is unidentate if it is ca. 26–5 cm$^{-1}$.

As stated previously, the highest frequency CO stretching band of the carbonato complexes belongs to the $A_1$ species in the bidentate and to the $B_2$ species in the unidentate complex. The same holds true for the nitrato complex. Ferraro et al.[279] showed that all the nitrato groups in $Th(NO_3)_4(TBP)_2$ coordinate to the metal as bidentate ligands since the Raman band at 1550 cm$^{-1}$ is polarized. This rule holds very well for other compounds.[280] According to Addison et al.,[267] the intensity pattern of the three NO stretching bands in the Raman spectrum can also be used to distinguish unidentate and symmetrical bidentate $NO_3$ ligands. The middle band is very strong in the former, whereas it is rather weak in the latter.

The use of far-infrared spectra to distinguish unidentate and bidentate nitrato coordination has been controversial. Nuttall and Taylor[281] suggested that unidentate and bidentate complexes exhibit one and two MO stretching bands, respectively, in the 350–250 cm$^{-1}$ region. Bullock and Parrett[282] showed, however, that such a simple rule is not applicable to many known nitrato complexes. Ferraro and Walker[283] assigned the MO stretching bands of anhydrous metal nitrates such as $Cu(NO_3)_2$ and $Pr(NO_3)_3$.

Several workers studied the Raman spectra of metal nitrates in aqueous solution and molten states. For example, Irish and Walrafen[284] found that $E'$ mode degeneracy is removed even in dilute solutions of $Ca(NO_3)_2$. This, combined with the appearance of the $A_1'$ mode in the infrared, suggests $\mathbf{C}_{2v}$ symmetry of the $NO_3^-$ ion. Hester and Krishnan[285] studied the Raman spectra of $Ca(NO_3)_2$ dissolved in molten $KNO_3$ and $NaNO_3$. Their results suggest an asymmetric perturbation of the $NO_3^-$ ion by the $Ca^{2+}$ ion through ion pair formation.

### (6) Sulfito ($SO_3$) and Sulfinato ($RSO_2$) Complexes

The pyramidal sulfite ($SO_3^{2-}$) ion may coordinate to a metal as a unidentate, bidentate, or bridging ligand. The following two structures are probable for unidentate coordination:

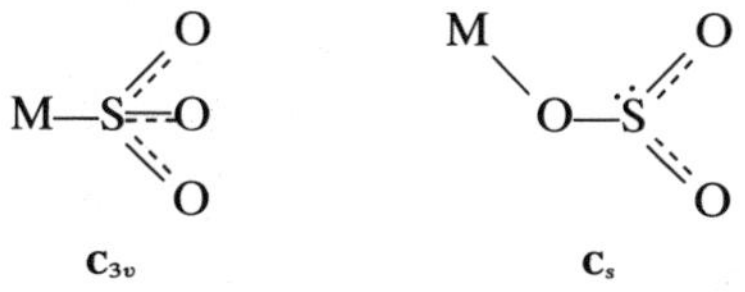

M—S(O)(O)(O) $\mathbf{C}_{3v}$ M—O—S(O)(O) $\mathbf{C}_s$

If coordination occurs through sulfur, the $\mathbf{C}_{3v}$ symmetry of the free ion will be preserved. If coordination occurs through oxygen, the symmetry may be lowered to $\mathbf{C}_s$. In this case, the doubly degenerate vibrations of the free ion will split into two bands. It is anticipated[286] that coordination through sulfur will shift the SO stretching bands to higher frequencies, whereas coordination through oxygen will shift them to lower frequencies, than those of the free ion. On the basis of these criteria, Newman and Powell[287] showed that the sulfito groups in $K_6[Pt(SO_3)_4] \cdot 2H_2O$ and $[Co(NH_3)_5(SO_3)]Cl$ are S-bonded and those in $Tl_2[Cu(SO_3)_2]$ are O-bonded. Baldwin[288] suggested that the sulfito groups in *cis*- and *trans*-$Na[Co(en)_2(SO_3)_2]$ and $[Co(en)_2(SO_3)X]$ (X=Cl or OH) are S-bonded, since they show only two SO stretchings between 1120 and 930 $cm^{-1}$. Table III-23 shows some results obtained by these investigators. According to Nyberg and Larsson,[289] the appearance of a strong SO stretching band above 975 and below 960 $cm^{-1}$ is an indication of S- and O-coordination, respectively.

The structures of complexes containing bidentate sulfito groups are rather difficult to deduce from their infrared spectra. Bidentate sulfito groups may be chelating or bridging through either oxygen or sulfur or both, all resulting in $\mathbf{C}_s$ symmetry. Baldwin[288] prepared a series of complexes of the type $[Co(en)_2(SO_3)]X$ (X=Cl, I, or SCN), which are monomeric in aqueous solution. They show four strong bands in the SO stretching region (one of them may be an overtone or a combination band). She suggests a chelating structure in which two oxygens of the sulfito group coordinate to the Co(III) atom. Newman and Powell[287] obtained the infrared spectra of $K_2[Pt(SO_3)_2] \cdot 2H_2O$, $K_3[Rh(SO_3)_3] \cdot 2H_2O$, and other complexes for which bidentate coordination of the sulfito group is expected. It was not possible, however, to determine their structures from infrared spectra alone.

TABLE III-23. VIBRATIONAL FREQUENCIES OF UNIDENTATE SULFITO COMPLEXES ($cm^{-1}$)

| Complex | Structure | $\nu_3(E)$ | $\nu_1(A_1)$ | $\nu_2(A_1)$ | $\nu_4(E)$ | Ref. |
|---|---|---|---|---|---|---|
| Free $SO_3^{2-}$ | — | 1010 | 961 | 633 | 496 | |
| $K_6[Pt(SO_3)_4] \cdot 2H_2O$ | S-bonded | 1082–1057 | 964 | 660 | 540 | 287 |
| $[Co(NH_3)_5(SO_3)]Cl$ | S-bonded | 1110 | 985 | 633 | 519 | 287 |
| *trans*-$Na[Co(en)_2(SO_3)_2]$ | S-bonded | 1068 | 939 | 630 | — | 288 |
| $[Co(en)_2(SO_3)Cl]$ | S-bonded | 1117–1075 | 984 | 625 | — | 288 |
| $Tl_2[Cu(SO_3)_2]$ | O-bonded | 902, 862 | 989 | 673 | 506, 460 | 287 |

Four types of coordination are probable for sulfinato($RSO_2^-$, R = $CH_3$, $CF_3$, Ph, etc.) groups:

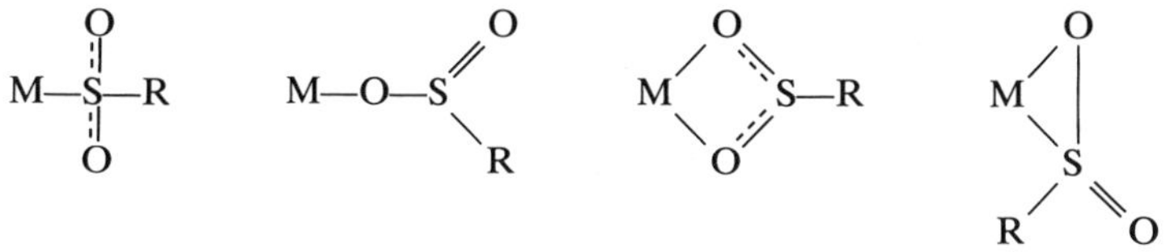

The SO stretching bands at 1200–850 $cm^{-1}$ are useful in distinguishing these structures.[290,290a]

## III-9. COMPLEXES OF β-DIKETONES

### (1) Complexes of Acetylacetonato Ion

A number of β-diketones form metal chelate rings of type A:

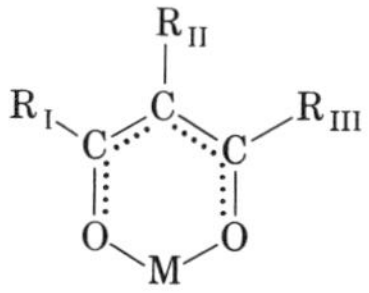

Type A

Among them, acetylacetone (acacH) is most common ($R_I = R_{III} = CH_3$ and $R_{II}$ = H). Infrared spectra of $M(acac)_2$ and $M(acac)_3$ type complexes have been studied extensively. Theoretical band assignments were first made by Nakamoto and Martell,[291] who carried out normal coordinate analysis on the 1:1 model of $Cu(acac)_2$. Mikami et al.[292] performed normal coordinate analyses on the 1:2 (square-planar) and 1:3 (octahedral) models of various acac complexes. Figure III-20 shows the infrared spectra of six acac complexes, and Table III-24 lists the observed frequencies and band assignments for the Cu(II), Pd(II), and Fe(III) complexes obtained by Mikami et al.

The nature of the 1577 and 1529 $cm^{-1}$ bands of $Cu(acac)_2$ has been a subject of controversy. Originally, Nakamoto and Martell[291] assigned the former to a ν(C⋯C) and the latter to a ν(C⋯O). Mikami et al.[292] gave similar assignments, although they showed that these two modes are coupled slightly with each other. However, the results of $^{13}C$ and $^{18}O$ experiments[293,294] suggest that the former should be assigned to a ν(C⋯O) and the latter to a ν(C⋯C).

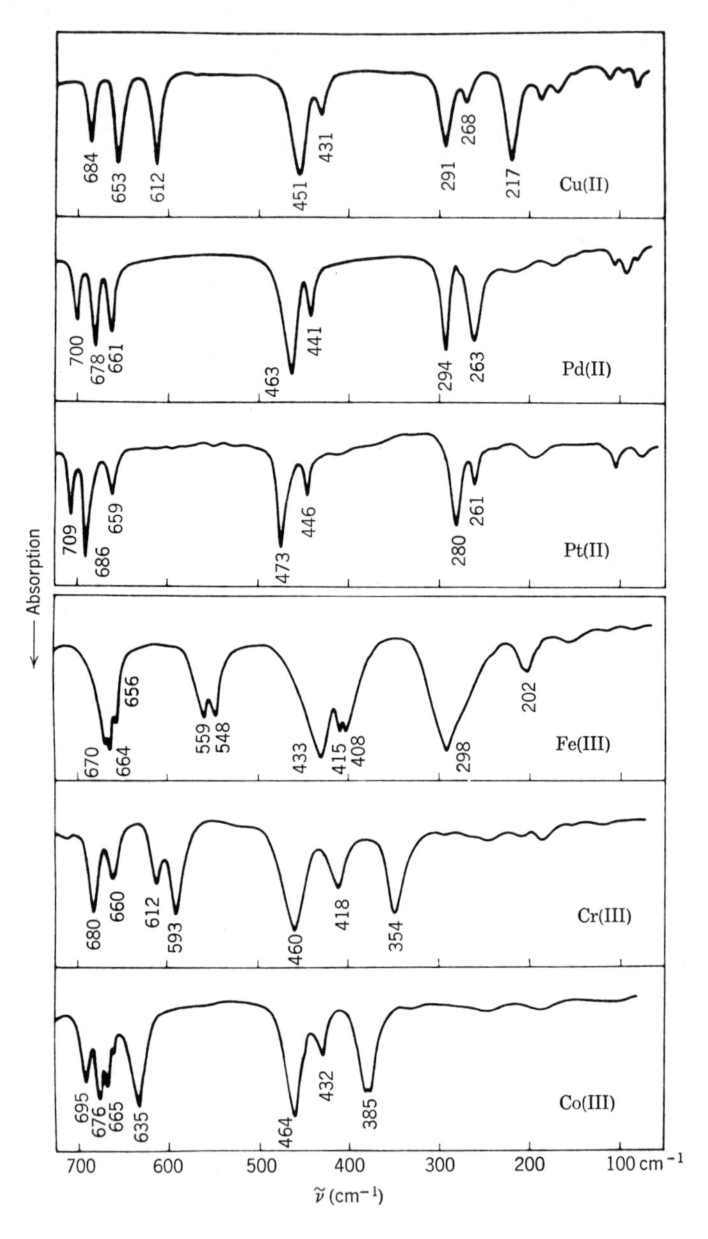


**Fig. III-20.** Infrared spectra of bis- and tris(acetylacetonato) complexes.[292]

TABLE III-24. OBSERVED FREQUENCIES AND BAND ASSIGNMENTS OF ACETYLACETONATO COMPLEXES ($cm^{-1}$)[292]

| $Cu(acac)_2$ | $Pd(acac)_2$ | $Fe(acac)_3$ | Predominant Mode |
|---|---|---|---|
| 3072 | 3070 | 3062 | $\nu(CH)$ |
| 2987<br>2969<br>2920 | 2990<br>2965<br>2920 | 2895<br>2965<br>2920 | $\nu(CH_3)$ |
| 1577 | 1569 | 1570 | $\nu(C \cdots C) + \nu(C \cdots O)$ |
| 1552 | 1549 | — | combination |
| 1529 | 1524 | 1525 | $\nu(C \cdots O) + \nu(C \cdots C)$ |
| 1461 | (1425) | 1445 | $\delta(CH) + \nu(C \cdots C)$ |
| 1413 | 1394 | 1425 | $\delta_d(CH_3)$ |
| 1353 | 1358 | 1385<br>1360 | $\delta_s(CH_3)$ |
| 1274 | 1272 | 1274 | $\nu(C-CH_3) + \nu(C \cdots C)$ |
| 1189 | 1199 | 1188 | $\delta(CH) + \nu(C—CH_3)$ |
| 1019 | 1022 | 1022 | $\rho_r(CH_3)$ |
| 936 | 937 | 930 | $\nu(C \cdots C) + \nu(C \cdots O)$ |
| 780 | 786<br>779 | 801<br>780<br>771 | $\pi(CH)$ |
| 684 | 700 | 670[a]<br>664 | $\nu(C—CH_3)$ + ring def. + $\nu(MO)$ |
| 653 | 678 | 656 | $\pi\left(CH_3—C\begin{smallmatrix}C\\O\end{smallmatrix}\right)$ |
| 612 | 661 | 559[a]<br>548 | ring def. + $\nu(MO)$ |
| 451 | 463 | 433 | $\nu(MO) + \nu(C—CH_3)$ |
| 431 | 441 | 415<br>408 | ring def. |
| 291 | 294 | 298 | $\nu(MO)$ |
| 1.45 | 1.85 | 1.30 | $K(M—O)$ (mdyn/Å)(UBF) |

[a] Pure ring deformation.

The $\nu$(MO) of acac complexes are most interesting since they provide direct information about the M–O bond strength. Using the metal isotope technique, Nakamoto et al.[295] assigned the MO stretching bands of acetylacetonato complexes at the following frequencies ($cm^{-1}$):

| $Cr(acac)_3$ | $Fe(acac)_3$ | $Pd(acac)_2$ | $Cu(acac)_2$ | $Ni(acac)_2(py)_2$ |
|---|---|---|---|---|
| 463.4 | 436.0 | 466.8 | 455.0 | 438.0 |
| 358.4 | 300.5 | 297.1 | 290.5 | 270.8 |
| | | 265.9 | | |

Both normal coordinate calculations and isotope shift studies show that the bands near 450 $cm^{-1}$ are coupled with the C—$CH_3$ bending mode, whereas those in the low frequency region are relatively pure MO stretching vibrations. Figure III-21 shows the actual tracings of the infrared spectra of $^{50}Cr(acac)_3$ and its $^{53}Cr$ analog. It is seen that two bands at 463.4 and 358.4 $cm^{-1}$ of the former give negative shifts of 3.0 and 3.9 $cm^{-1}$, respectively, whereas other bands (ligand vibrations) produce negligible shifts by the $^{50}Cr$–$^{53}Cr$ substitution.

Complexes of the $M(acac)_2X_2$ type may take the *cis* or *trans* structure. Although steric and electrostatic considerations would favor the *trans*-isomer, the greater stability of the *cis*-isomer is expected in terms of metal–ligand $\pi$-bonding. This is the case for $Ti(acac)_2F_2$, which is "*cis*" with two $\nu$(TiF) at 633 and 618 $cm^{-1}$.[296] In the case of $Re(acac)_2Cl_2$, however, both forms can be isolated; the *trans*-isomer exhibits $\nu$(ReO) and $\nu$(ReCl) at 464 and 309 $cm^{-1}$, respectively, while each of these bands

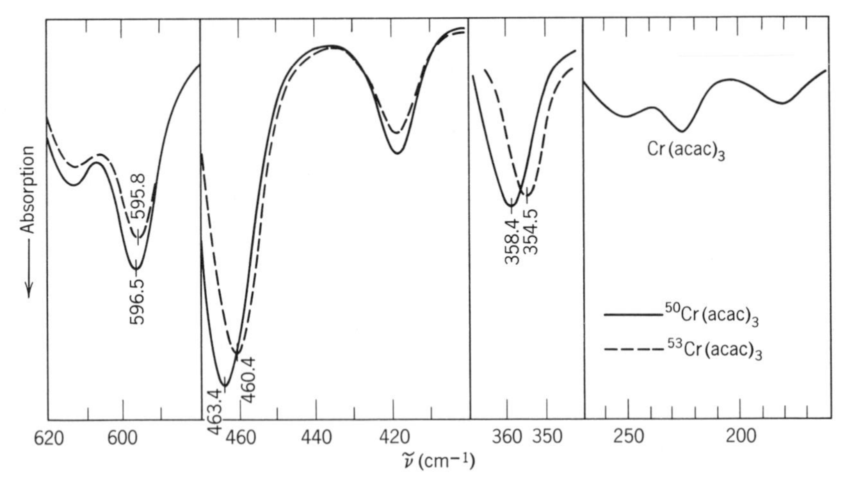


**Fig. III-21.** Infrared spectra of $^{50}Cr(acac)_3$ and its $^{53}Cr$ analog.[295]

splits into two in the *cis*-isomer [472 and 460 $cm^{-1}$ for $\nu$(ReO) and 346 and 333 $cm^{-1}$ for $\nu$(ReCl) in the infrared].[297] For $VO(acac)_2L$, where L is a substituted pyridine, *cis*- and *trans*-isomers are expected. According to Caira et al.,[298] these structures can be distinguished by their infrared spectra. The $\nu$(V=O) and $\nu$(V—O) of the *cis*-isomer are lower than

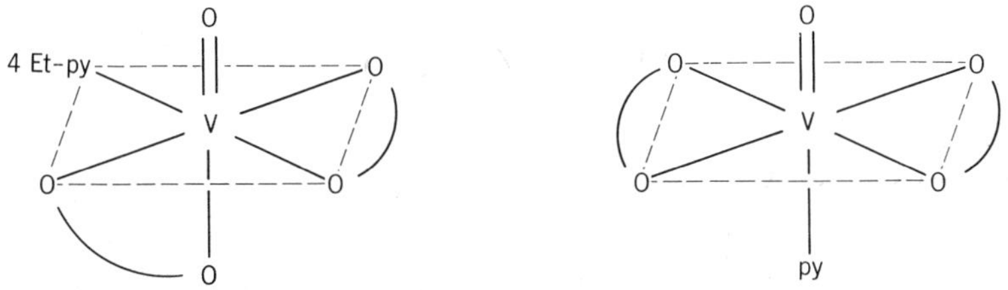

*cis*-isomer

*trans*-isomer

those of the *trans*-isomer. Furthermore, $\nu$(V—O) of the *cis*-isomer splits into two bands. For example, $\nu$(V=O) of $VO(acac)_2$ is 999 $cm^{-1}$, and this band shifts to 959 $cm^{-1}$ for 4-Et-py (*cis*) and to 973 $cm^{-1}$ for py (*trans*).

Vibrational spectra of acetylacetonato complexes have been studied by many other workers. Only references are cited for the following: Raman spectra of tris(acac) complexes,[232] infrared spectra of acac complexes of rare-earth metals,[299] relationships between CH and $CH_3$ stretching frequencies and $^{13}$C—H spin-spin coupling constants,[300] and relationships between $\nu$(C⋯O), $\nu$(C⋯C), and $^{13}$C NMR shifts of CO groups.[301]

According to X-ray analysis,[302] the hexafluoroacetylacetonato ion (hfa) in $[Cu(hfa)_2\{Me_2N\text{-}(CH_2)_2\text{-}NH_2\}_2]$ coordinates to the metal as a unidentate via one of its O atoms. This compound exhibits $\nu$(C=O) at 1675 and 1615 $cm^{-1}$, values slightly higher than those for $Cu(hfa)_2$, in which the hfa ion is chelated to the metal (1644 and 1614 $cm^{-1}$).

The following dimeric bridging structure has been proposed for $[CoBr(acac)]_2$:

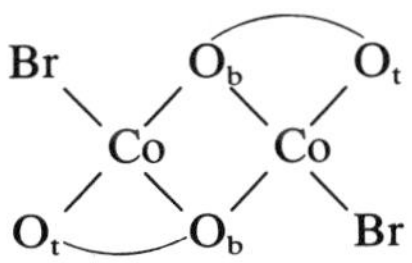

$\nu(CoO_t)$ and $\nu(CoO_b)$ were assigned at 435 and 260 $cm^{-1}$, respectively.[303] In $[Ni(acac)_2]_3$ and $[Co(acac)_2]_4$, the O atoms of the acac ion serve as a bridge between two metal atoms.[304] However, no band assignments are available on these polymeric species.

### (2) Complexes of Neutral Acetylacetone

In some compounds, the keto form of acetylacetone forms a chelate ring of type B:

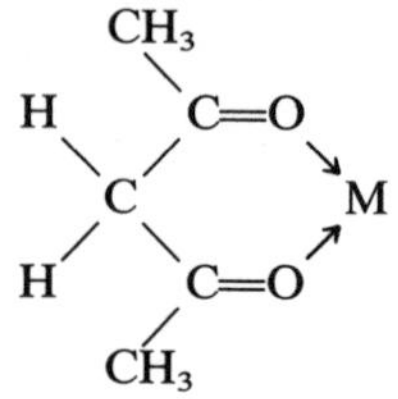

Type B

This type of coordination was found by van Leeuwen[305] in $[Ni(acacH)_3]$-$(ClO_4)_2$ and its derivatives, and by Nakamura and Kawaguchi[306] in $Co(acacH)Br_2$. These compounds were prepared in acidic or netural media, and exhibit strong $\nu(C{=}O)$ bands near 1700 $cm^{-1}$. Similar ketonic coordination was proposed for $Ni(acacH)_2Br_2$[307] and $M(acacH)Cl_2$ (M = Co and Zn).[308]

According to X-ray analysis,[309] the acetylacetone molecule in $Mn(acacH)_2Br_2$ is in the enol form and is bonded to the metal as a unidentate via one of its O atoms:

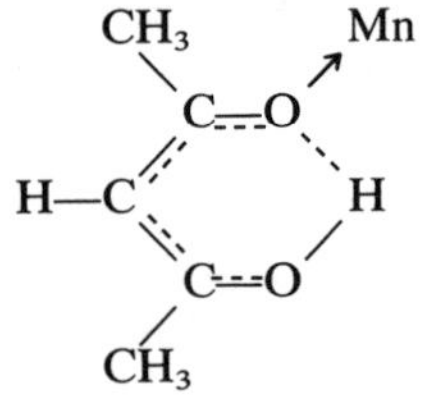

Type C

The C⋯O and C⋯C stretching bands of the enol ring were assigned at 1627 and 1564 $cm^{-1}$, respectively.

### (3) C-bonded Acetylacetonato Complexes

Lewis and co-workers[310] reported the infrared and NMR spectra of a number of Pt(II) complexes in which the metal is bonded to the $\gamma$-carbon atom of the acetylacetonato ion:

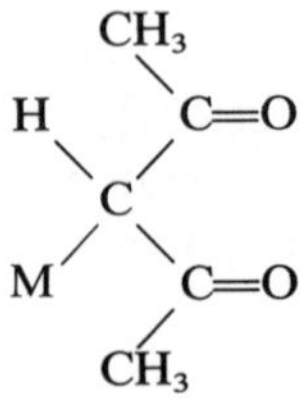

Type D

TABLE III-25. OBSERVED FREQUENCIES, BAND ASSIGNMENTS, AND FORCE CONSTANTS FOR $K[Pt(acac)Cl_2]$ AND $Na_2[Pt(acac)_2Cl_2]\cdot 2H_2O$

| $K[Pt(acac)Cl_2]$ (O-bonded, Type A) | $Na_2[Pt(acac)_2Cl_2]\cdot 2H_2O$ (C-bonded, Type D) | Band Assignment |
|---|---|---|
| — | 1652, 1626 | $\nu$(C═O) |
| 1563, 1380 | — | $\nu$(C┄O) |
| 1538, 1288 | — | $\nu$(C┄C) |
| — | 1352, 1193 | $\nu$(C—C) |
| 1212, 817 | 1193, 852 | $\delta$(CH) or $\pi$(CH) |
| 650, 478 | — | $\nu$(PtO) |
| — | 567 | $\nu$(PtC) |
| $K$(C┄O) = 6.50 | $K$(C═O) = 8.84 | UBF constant (mdyn/Å) |
| $K$(C┄C) = 5.23 | $K$(C—C) = 2.52 | |
| $K(C—CH_3)$ = 3.58 | $K(C—CH_3)$ = 3.85 | |
| $K$(Pt—O) = 2.46 | $K$(Pt—C) = 2.50 | |
| $K$(C—H) = 4.68 | $K$(C—H) = 4.48 | |
| $\rho = 0.43$[a] | | |

[a] The stretching–stretching interaction constant ($\rho$) was used for type A because of the presence of resonance in the chelate ring.

Behnke and Nakamoto carried out normal coordinate analysis on the $[Pt(acac)Cl_2]^-$ ion, in which the acac ion is chelated to the metal (type A),[311] and on the $[Pt(acac)_2Cl_2]^{2-}$ ion, in which the acac ion is C-bonded to the metal (type D).[312] Table III-25 lists the observed frequencies and band assignments for these two types, and Fig. III-22 shows the infrared spectra of these two compounds. The results indicate that (1) two $\nu$(C═O) of type D are higher than those of type A, (2) two $\nu$(C—C) of type D are lower than those of type A, and (3) $\nu$(PtC) of type D is at 567 $cm^{-1}$, while $\nu$(PtO) of type A are at 650 and 478 $cm^{-1}$. Figure III-22 also shows that the structure of $K[Pt(acac)_2Cl]$ is as follows:

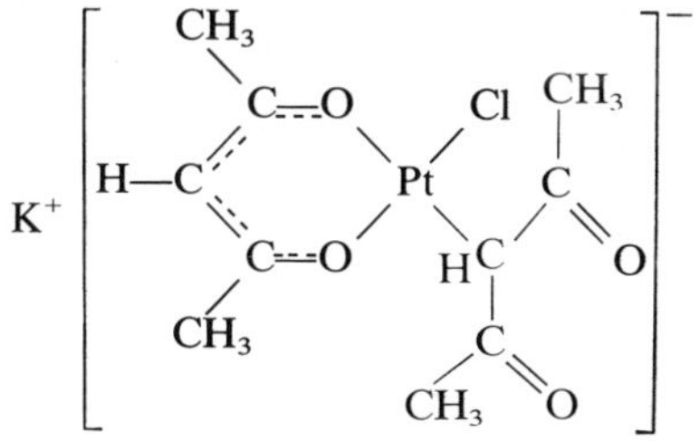

$K^+$ [ $CH_3$–C┄O, H—C, C┄O–$CH_3$ chelated to Pt; Cl; HC(C(═O)$CH_3$)(C(═O)$CH_3$) ]$^-$

since its spectrum is roughly a superposition of those of types A and D. Similarly, the infrared spectrum of $K[Pt(acac)_3]$[310] is interpreted as a

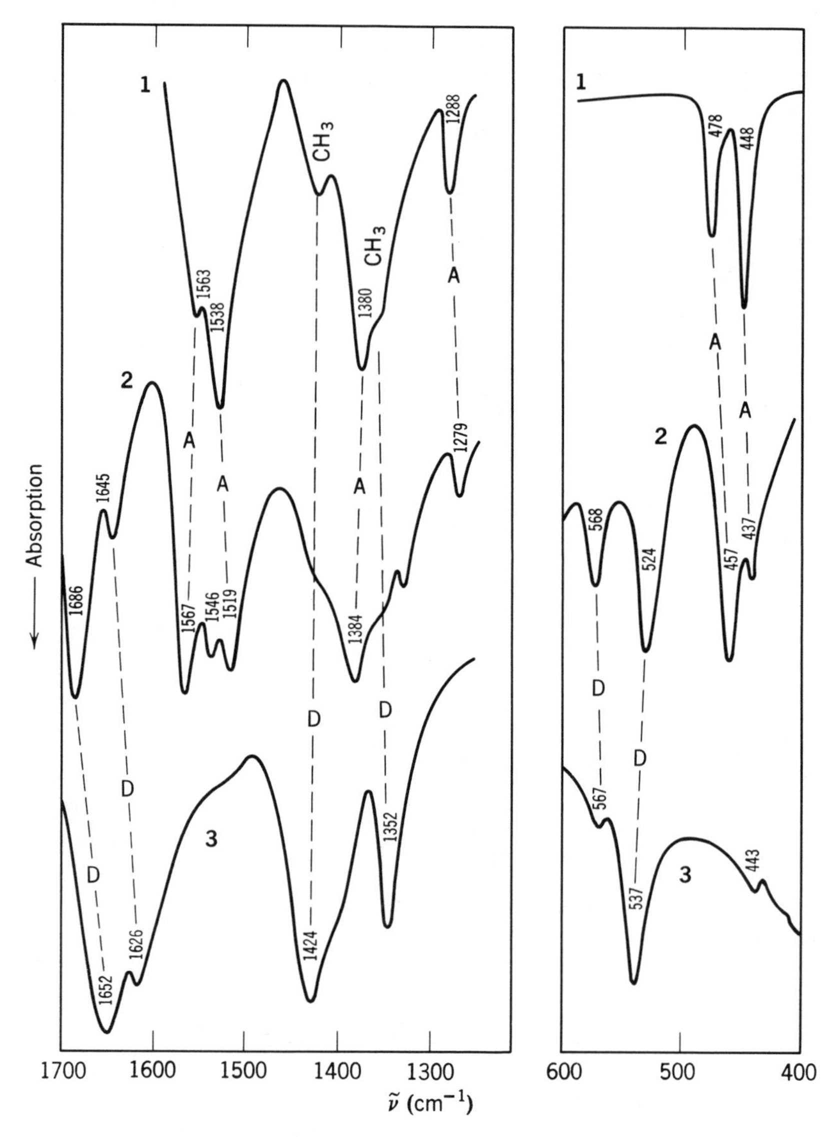


**Fig. III-22.** Infrared spectra of Pt(II) acetylacetonate complexes: (1) $K[Pt(acac)Cl_2]$, (2) $K[Pt(acac)_2Cl]$, and (3) $Na_2[Pt(acac)_2Cl_2]2H_2O$. A and D denote the bands characteristic of types A and D, respectively.

superposition of spectra of types A, D, and D′, in which two C═O bonds are transoid:[313]

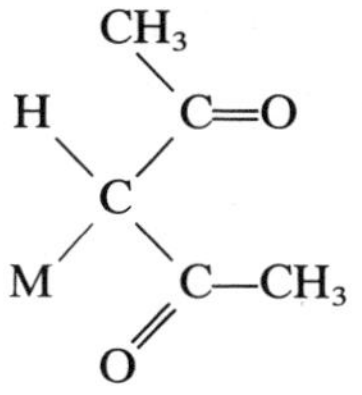

Type D′

The C-bonded acac ion was found in $Hg_2Cl_2(acac)$,[314] $Au(acac)(PPh_3)$,[315] and $Pd(acac)_2(PPh_3)$.[316] In the last compound, one acac group is type A and the other type D. In all these cases, the $\nu$(C═O) of the type D acac groups are at 1700–1630 $cm^{-1}$.

As discussed above, $K[Pt(acac)_2Cl]$ contains one type A and one type D acac group. If a solution of $K[Pt(acac)_2Cl]$ is acidified, its type D acac group is converted into type E:

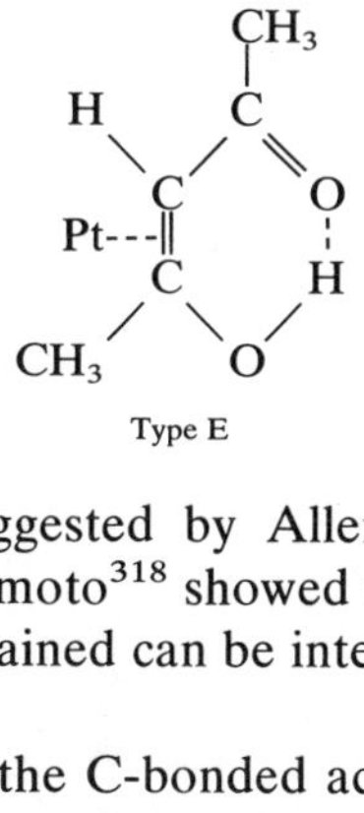

Type E

This structure was first suggested by Allen et al.,[317] based on NMR evidence. Behnke and Nakamoto[318] showed that the infrared spectrum of [Pt(acac)(acacH)Cl] thus obtained can be interpreted as a superposition of spectra of types A and E.

That the two O atoms of the C-bonded acac group (type D) retain the ability to coordinate to a metal was first demonstrated by Lewis and Oldham,[319] who prepared neutral complexes of the following type:

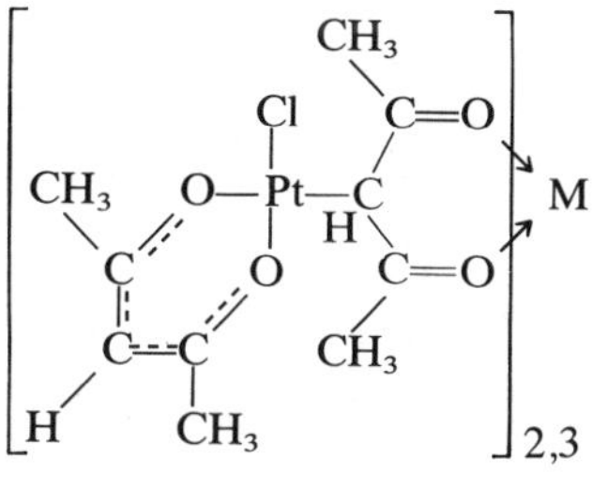

M = Co(II), Ni(II), Fe(III), etc.

Using the metal isotope technique, Nakamura and Nakamoto[320] assigned the $\nu$(NiO) of Ni[Pt(acac)$_2$Cl]$_2$ at 279 and 266 cm$^{-1}$. These values are relatively close to the $\nu$(NiO) of Ni(acacH)$_2$Br$_2$ (264 and 239 cm$^{-1}$), discussed previously. Thus the newly formed Ni-acac ring retains its keto character and is close to type B.

### (4) Complexes of Other β-Diketones

In a series of metal tropolonato complexes, Hulett and Thornton[321] noted a parallel relationship between the $\nu$(MO) and the CFSE energy (see Sec. III-6). However, these workers assigned the $\nu$(MO) of trivalent metal tropolonates in the 660–580 cm$^{-1}$ region, based on the $^{16}$O–$^{18}$O isotope shifts observed for the Cu(II) complex.[322] Using the metal isotope technique, Hutchinson et al.[323] assigned the $\nu$(MO) at the following frequencies (cm$^{-1}$);

| V(III) | | Cr(III) | | Mn(III) | | Fe(III) | | Co(III) |
|---|---|---|---|---|---|---|---|---|
| 377 | ~ | 361 | > | 338 | > | 317 | < | 371 |
| 319 | < | 334 | > | 268 | > | 260 | < | 360 |

It was found that these frequencies still follow the order predicted by the CFSE.

2,4,6-Heptanetrione forms 1 :1 and 1 :2 (metal:ligand) complexes with Cu(II):[324]

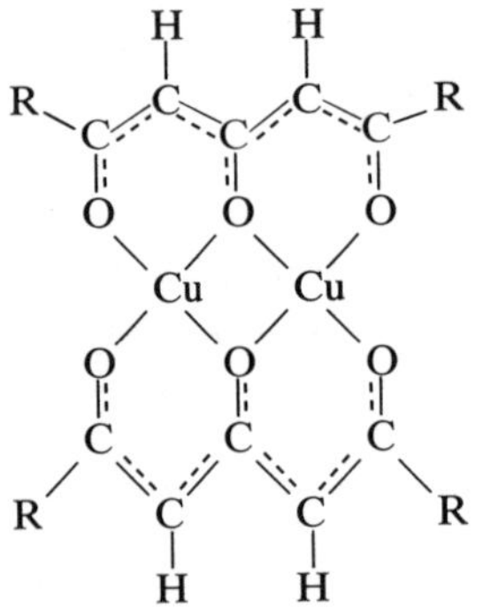

1:1 Complex

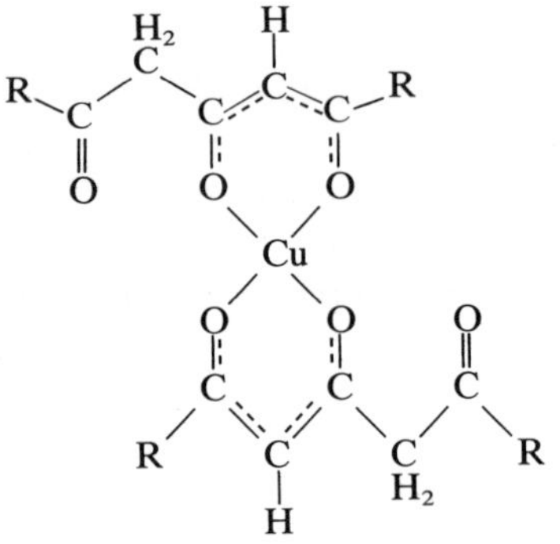

1:2 Complex

Both complexes exhibit multiple bands due to type A rings in the 1600–1500 cm$^{-1}$ region. However, the 1 : 2 complex exhibits $\nu$(C=O) of the uncoordinated C=O groups near 1720 cm$^{-1}$.

## III-10. CYANO AND NITRILE COMPLEXES

### (1) Cyano Complexes

The vibrational spectra of cyano complexes have been studied extensively and these investigations are reviewed by Sharp,[325] Griffith,[326] and Rigo and Turco.[327]

***(a) CN Stretching Bands.*** Cyano complexes can be identified easily since they exhibit sharp $\nu$(CN) at 2200–2000 cm$^{-1}$. The $\nu$(CN) of free $CN^-$ is 2080 cm$^{-1}$ (aqueous solution). Upon coordination to a metal, the $\nu$(CN) shift to higher frequencies, as shown in Table III-26. The $CN^-$ ion acts as a $\sigma$-donor by donating electrons to the metal and also as a $\pi$-acceptor by accepting electrons from the metal. $\sigma$-Donation tends to raise the $\nu$(CN) since electrons are removed from the $5\sigma$ orbital, which is weakly antibonding, while $\pi$-back-bonding tends to decrease the $\nu$(CN) because the electrons enter into the antibonding $2p\pi^*$ orbital. In general, $CN^-$ is a better $\sigma$-donor and a poorer $\pi$-acceptor than CO. Thus the $\nu$(CN) of the complexes are generally higher than the value for free $CN^-$, whereas the opposite prevails for the CO complexes (Sec. III-12).

According to El-Sayed and Sheline,[328] the $\nu$(CN) of cyano complexes are governed by (1) the electronegativity, (2) the oxidation state, and (3)

TABLE III-26. C≡N STRETCHING FREQUENCIES OF CYANO COMPLEXES (cm$^{-1}$)

| Compound | Symmetry | $\nu$(CN) | Ref. |
|---|---|---|---|
| $Tl[Au(CN)_2]$ | $\mathbf{D}_{\infty h}$ | 2164 ($\Sigma_g^+$), 2141 ($\Sigma_u^+$) | 333 |
| $K[Ag(CN)_2]$ | $\mathbf{D}_{\infty h}$ | 2146 ($\Sigma_g^+$), 2140 ($\Sigma_u^+$) | 334 |
| $K_2[Ni(^{12}C^{14}N)_4]$ | $\mathbf{D}_{4h}$ | 2143.5 ($A_{1g}$), 2134.5 ($B_{1g}$), 2123.5 ($E_u$) | 335 |
| $K_2[Pd(^{12}C^{14}N)_4]$ | $\mathbf{D}_{4h}$ | 2160.5 ($A_{1g}$), 2146.4 ($B_{1g}$), 2135.8 ($E_u$) | 335 |
| $K_2[Pt(^{12}C^{14}N)_4]$ | $\mathbf{D}_{4h}$ | 2168.0 ($A_{1g}$), 2148.8 ($B_{1g}$), 2133.4 ($E_u$) | 335 |
| $Na_3[Ni(CN)_5]$ | $\mathbf{C}_{4v}$ | 2130 ($A_1$), 2117 ($B_1$), 2106 ($E$), 2090 ($A_1$) | 336 |
| $Na_3[Co(CN)_5]$ | $\mathbf{C}_{4v}$ | 2115 ($A_1$), 2110 ($B_1$), 2096 ($E$), 2080 ($A_1$) | 336 |
| $K_3[Mn(CN)_6]$ | $\mathbf{O}_h$ | 2129 ($A_{1g}$), 2129 ($E_g$), 2112 ($F_{1u}$) | 337 |
| $K_4[Mn(CN)_6]$ | $\mathbf{O}_h$ | 2082 ($A_{1g}$), 2066 ($E_g$), 2060 ($F_{1u}$) | 337 |
| $K_3[Fe(CN)_6]$ | $\mathbf{O}_h$ | 2135 ($A_{1g}$), 2130 ($E_g$), 2118 ($F_{1u}$) | 337 |
| $K_4[Fe(CN)_6]3H_2O$ | $\mathbf{O}_h$ | 2098 ($A_{1g}$), 2062 ($E_g$), 2044 ($F_{1u}$) | 337 |
| $K_3[Co(CN)_6]$ | $\mathbf{O}_h$ | 2150 ($A_{1g}$), 2137 ($E_g$), 2129 ($F_{1u}$) | 337 |
| $K_4[Ru(CN)_6]3H_2O$ | $\mathbf{O}_h$ | 2111 ($A_{1g}$), 2071 ($E_g$), 2048 ($F_{1u}$) | 337 |
| $K_3[Rh(CN)_6]$ | $\mathbf{O}_h$ | 2166 ($A_{1g}$), 2147 ($E_g$), 2133 ($F_{1u}$) | 337 |
| $K_2[Pd(CN)_6]$ | $\mathbf{O}_h$ | 2185 ($F_{1u}$) | 338 |
| $K_4[Os(CN)_6]3H_2O$ | $\mathbf{O}_h$ | 2109 ($A_{1g}$), 2062 ($E_g$), 2036 ($F_{1u}$) | 337 |
| $K_3[Ir(CN)_6]$ | $\mathbf{O}_h$ | 2167 ($A_{1g}$), 2143 ($E_g$), 2130 ($F_{1u}$) | 337 |

the coordination number of the metal. The effect of electronegativity is seen in the order:

$$\begin{array}{ccccc} [Ni(CN)_4]^{2-} & & [Pd(CN)_4]^{2-} & & [Pt(CN)_4]^{2-} \\ 2128 & < & 2143 & < & 2150 \quad cm^{-1} \end{array}$$

Since the electronegativity of Ni(II) is smallest, the $\sigma$-donation will be the least, and the $\nu$(CN) is expected to be the lowest. The effect of oxidation state is seen in the frequency order:[329]

$$\begin{array}{ccccc} [V(CN)_6]^{5-} & & [V(CN)_6]^{4-} & & [V(CN)_6]^{3-} \\ 1910 & < & 2065 & < & 2077 \quad cm^{-1} \end{array}$$

The higher the oxidation state, the stronger the $\sigma$-bonding, and the higher the $\nu$(CN). The effect of coordination number[330–332] is evident in the frequency order:

$$\begin{array}{ccccc} [Ag(CN)_4]^{3-} & & [Ag(CN)_3]^{2-} & & [Ag(CN)_2]^{-} \\ 2092 & < & 2105 & < & 2135 \quad cm^{-1} \end{array}$$

Here an increase in the coordination number results in a decrease in the positive charge on the metal, which, in turn, weakens the $\sigma$-bonding, thus decreasing the $\nu$(CN).

For $[Mo(CN)_7]^{4-}$, infrared and Raman spectra, together with magnetic susceptibility measurements, suggested a pentagonal-bipyramidal ($\mathbf{D}_{5h}$) structure for which two infrared active ($A_2''$ and $E_1'$) CN stretching modes are observed at 2080 and 2040 $cm^{-1}$.[339] The structure of the $[Mo(CN)_8]^{4-}$ ion in aqueous solution has been controversial. According to X-ray analysis,[340] the $[Mo(CN)_8]^{4-}$ ion in $K_4[Mo(CN)_8] \cdot 2H_2O$ is definitely $\mathbf{D}_{2d}$ (dodecahedron). On the other hand, a Raman study[341] supported the $\mathbf{D}_{4d}$ (archimedean-antiprism) structure of the $[Mo(CN)_8]^{4-}$ ion in aqueous solution. The stereochemical conversion of the $[Mo(CN)_8]^{4-}$ ion from $\mathbf{D}_{2d}$ (solid) to $\mathbf{D}_{4d}$ (solution) symmetry was confirmed by Hartman and Miller[342] and Parish et al.[343] Similar conversions were proposed for the $[W(CN)_8]^{4-}$ [342,343] and $[Nb(CN)_8]^{4-}$ [344] ions. However, Long and Vernon[345] claim that the $\mathbf{D}_{2d}$ geometry is maintained even in aqueous solution.

According to X-ray analysis,[346] the unit cell of $[Cr(en)_3]$-$[Ni(CN)_5] \cdot 1\frac{1}{2}H_2O$ contains both square-pyramidal ($\mathbf{C}_{4v}$) and trigonal-bipyramidal ($\mathbf{D}_{3h}$) structures of the $[Ni(CN)_5]^{3-}$ ion. Terzis et al.[347] showed that the complicated vibrational spectrum of this crystal in the $\nu$(CN) region is simplified dramatically when it is dehydrated. These

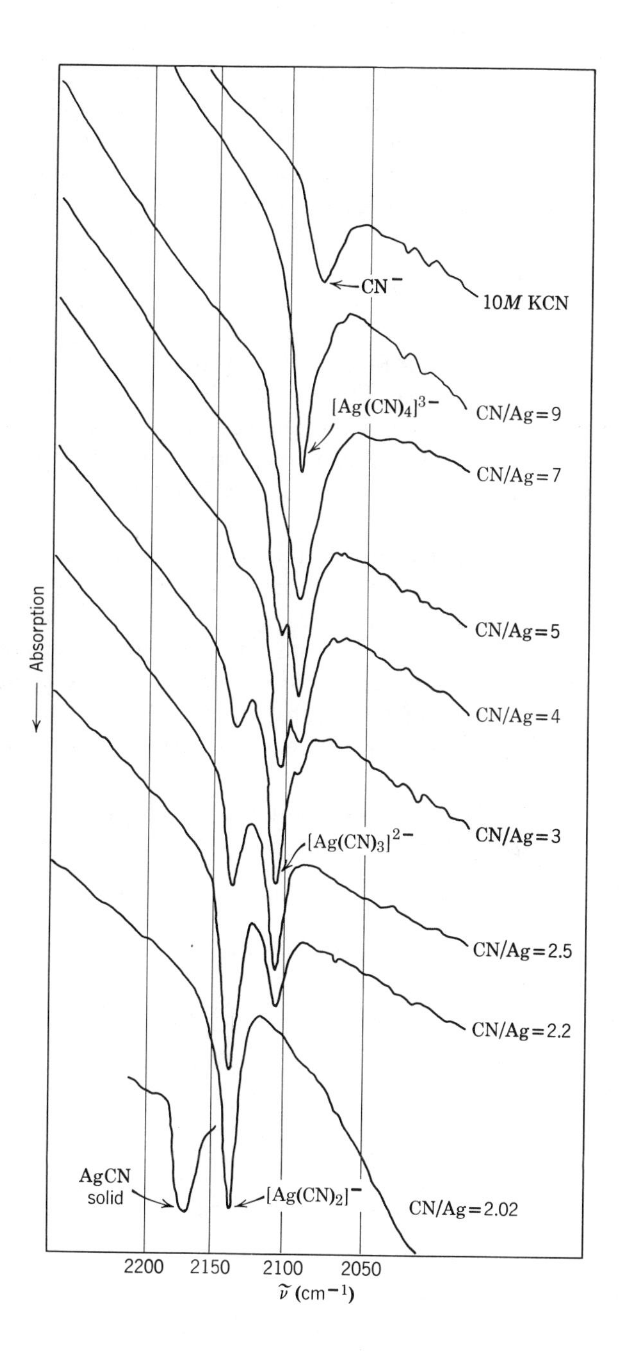


**Fig. III-23.** Infrared spectra of silver cyano complexes in aqueous solution.[330]

spectral changes suggest that the $\mathbf{D}_{3h}$ (somewhat distorted) units have been converted to $\mathbf{C}_{4v}$ geometry upon dehydration. Basile et al.[348] showed that such conversion from $\mathbf{D}_{3h}$ to $\mathbf{C}_{4v}$ also occurs when the crystal is subjected to high pressure. Hellner et al.[349] observed the splitting of the degenerate $\nu$(CN) of $K_2[Zn(CN)_4]$ and a partial reduction of the central metal in $K_3[M(CN)_6]$ [M = Fe(III) and Mn(III)] when these crystals are under high external pressure.

Jones and co-workers[330–332] made an extensive infrared study of the equilibria of cyano complexes in aqueous solution. (For aqueous infrared spectroscopy, see Sec. III-15.) Figure III-23 shows the infrared spectra of aqueous silver cyano complexes obtained by changing the ratio of $Ag^+$ to $CN^-$ ions. Table III-27 lists the frequencies and extinction coefficients from which equilibrium constants can be calculated. Chantry and Plane[351] studied the same equilibria using Raman spectroscopy.

***(b) Lower Frequency Bands.*** In addition to $\nu$(CN), the cyano complexes exhibit $\nu$(MC), $\delta$(MCN), and $\delta$(CMC) bands in the low frequency region. Figure III-24 shows the infrared spectra of $K_3[Co(CN)_6]$ and $K_2[Pt(CN)_4] \cdot 3H_2O$. Normal coordinate analyses have been carried out on various hexacyano complexes to assign these low frequency bands (Table III-28). The results of these calculations indicate that the $\nu$(MC), $\delta$(MCN), and $\delta$(CMC) vibrations appear in the regions 600–350, 500–

TABLE III-27. FREQUENCIES AND MOLECULAR EXTINCTION COEFFICIENTS OF CYANO COMPLEXES IN AQUEOUS SOLUTIONS

| Ion | Frequency ($cm^{-1}$) | Molecular Extinction Coefficient | Ref. |
|---|---|---|---|
| Free $[CN]^-$ | 2080 ± 1 | 29 ± 1 | 330 |
| $[Ag(CN)_2]^-$ | 2135 ± 1 | 264 ± 12 | 330 |
| $[Ag(CN)_3]^{2-}$ | 2105 ± 1 | 379 ± 23 | 330 |
| $[Ag(CN)_4]^{3-}$ | 2092 ± 1 | 556 ± 83 | 330 |
| $[Cu(CN)_2]^-$ | 2125 ± 3 | 165 ± 25 | 331 |
| $[Cu(CN)_3]^{2-}$ | 2094 ± 1 | 1090 ± 10 | 331 |
| $[Cu(CN)_4]^{3-}$ | 2076 ± 1 | 1657 ± 15 | 331 |
| $[Zn(CN)_4]^{2-}$ | 2149 | 113 | 332 |
| $[Cd(CN)_4]^{2-}$ | 2140 | 75 | 332 |
| $Hg(CN)_2$ | 2194 | 3 | 332 |
| $[Hg(CN)_3]^-$ | 2161 | 26 | 332 |
| $[Hg(CN)_4]^{2-}$ | 2143 | 113 | 332 |
| $[Ni(CN)_4]^{2-}$ | 2124 ± 1 | 1068 ± 95 | 350 |
| $[Ni(CN)_5]^{3-}$ | 2102 ± 2 | 1730 ± 230 | 350 |

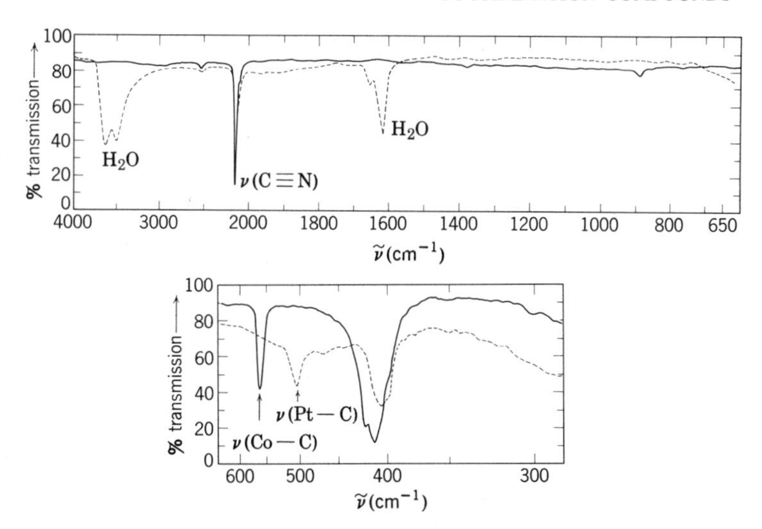


**Fig. III-24.** Infrared spectra of $K_3[Co(CN)_6]$ (solid line) and $K_2[Pt(CN)_4]\cdot 3H_2O$ (broken line).

350, and 130–60 $cm^{-1}$, respectively. The MC and C≡N stretching force constants obtained are also given in Table III-28.

Nakagawa and Shimanouchi[356] noted that the MC stretching force constant increases in the order Fe(III) < Co(III) < Fe(II) < Ru(II) < Os(II), and the C≡N stretching force constant decreases in the same order of metals. This result was interpreted as indicating that the M–C $\pi$-bonding is increasing in the above order. The degree of M–C $\pi$-bonding may be proportional to the number of $d$-electrons in the $F_{2g}$ electronic level. According to Jones,[357] the integrated absorption coefficient of the C≡N stretching band ($F_{1u}$) becomes larger as the number of $d$-electrons in the $F_{2g}$ level increases. Thus the results shown in Table III-29 suggest that the M–C $\pi$-bonding increases in the order Cr(III) < Mn(III) < Fe(III) < Co(III). The order of $\nu$(MC) shown in the same table confirms this conclusion. Griffith and Turner[337] found a similar trend in the Fe(II) < Ru(II) < Os(II) series. Nakagawa and Shimanouchi[358] carried out complete normal coordinate analyses on $K_3[M(CN)_6]$ [M = Fe(III) and Cr(III)] crystals, including all lattice modes. Jones et al.[358a] also performed complete normal coordinate analyses on crystalline $Cs_2Li[Fe(CN)_6]$, including its $^{13}C$, $^{15}N$, and $^6Li$ analogs.

Normal coordinate analyses have been made on tetrahedral, square-planar, and linear cyano complexes of various metals; Table III-30 gives the results of these studies. Far-infrared spectra of various cyano complexes have been measured.[370] Jones and co-workers[371] carried out an

TABLE III-28. VIBRATIONAL FREQUENCIES AND BAND ASSIGNMENTS OF HEXACYANO COMPLEXES ($cm^{-1}$)

| | | $[Cr(CN)_6]^{3-}$ | $[Co(CN)_6]^{3-}$ | $[Ir(CN)_6]^{3-}$ | $[Rh(CN)_6]^{3-}$ | $[Co(CN)_6]^{3-}$ | $[Fe(CN)_6]^{4-}$ | $[Fe(CN)_6]^{3-}$ | $[Ru(CN)_6]^{4-}$ | $[Os(CN)_6]^{4-}$ |
|---|---|---|---|---|---|---|---|---|---|---|
| $A_{1g}$ | $\nu$(MC) | 374 | 408 | 469 | 445 | 406 | (410) | (390) | (460) | (480) |
| $E_g$ | $\nu$(MC) | 336 | (391) | 450 | 435 | (375) | (390) | — | (410) | (450) |
| $F_{1g}$ | $\delta$(MCN) | 536 | (358) | (415) | (380) | (380) | (350) | — | (340) | (360) |
| $F_{1u}$ | $\nu$(MC) | 457 | 564 | 520 | 520 | 565 | 585 | 511 | 550 | 554 |
| | $\delta$(MCN) | 694 | 416 | 398 | 386 | 414 | 414 | 387 | 376 | 392 |
| | $\delta$(CMC) | 124 | (84) | (82) | (88) | — | — | 89 | — | — |
| $F_{2g}$ | $\delta$(MCN) | 536 | (480) | 483 | (475) | — | (420) | — | (400) | (430) |
| | $\delta$(CMC) | 106 | 98 | 95 | 94 | 98 | — | 99 | — | — |
| $F_{2u}$ | $\delta$(MCN) | 496 | (440) | 445 | — | (395) | — | — | (365) | (390) |
| | $\delta$(CMC) | 95 | (72) | (69) | — | — | — | 70 | — | — |
| Force field | | GVF | GVF | GVF | GVF | UBF | UBF | UBF | UBF | UBF |
| K(M—C) (mdyn/Å) | | 1.928 | 2.063 | 2.704 | 2.366 | 2.308 | 2.428 | 1.728 | 2.793 | 3.343 |
| K(C≡N) (mdyn/Å) | | 16.422 | 16.767 | 16.678 | 16.831 | 16.5 | 15.1 | 17.0 | 15.3 | 14.9 |
| References | | 352 | 353<br>354<br>355 | 354 | 354 | 356 | 356 | 356<br>355 | 356 | 356 |

TABLE III-29. RELATION BETWEEN INFRARED SPECTRUM AND ELECTRONIC STRUCTURE IN HEXACYANO COMPLEXES[357]

| Compound | Number of *d*-Electrons in $F_{2g}$ Level | $\nu$(CN) ($cm^{-1}$) | $\nu$(MC) ($cm^{-1}$) | Integrated Absorption Coefficient ($mole^{-1}\ cm^{-2}$) |
|---|---|---|---|---|
| $K_3[Cr(CN)_6]$ | 3 | 2128 | 339 | 2,100 |
| $K_3[Mn(CN)_6]$ | 4 | 2112 | 361 | 8,200 |
| $K_3[Fe(CN)_6]$ | 5 | 2118 | 389 | 12,300 |
| $K_3[Co(CN)_6]$ | 6 | 2129 | 416 | 18,300 |

TABLE III-30. FREQUENCIES AND BAND ASSIGNMENTS OF THE LOWER FREQUENCY BANDS OF CYANO COMPLEXES ($CM^{-1}$)

| Ion | Symmetry | $\nu$(MC) | $\delta$(MCN) | $\delta$(CMC) | Force constant[a] K(M—C) | K(C≡N) | Ref. |
|---|---|---|---|---|---|---|---|
| $[Cu(CN)_4]^{3-}$ | $\mathbf{T}_d$ | 364(IR) | 324(R) | (74) | 1.25– | 16.10– | 359 |
| | | 288(R) | 306(IR) | (63) | 1.30 | 16.31 | 360 |
| $[Zn(CN)_4]^{2-}$ | $\mathbf{T}_d$ | 359(IR)[b] | 315(IR)[b] | 71(R) | 1.30 | 17.22 | 361 |
| | | 342(R) | 230(R) | | | | |
| $[Cd(CN)_4]^{2-}$ | $\mathbf{T}_d$ | 316(IR)[b] | 250(R)[b] | 61(R) | 1.28 | 17.13 | 361 |
| | | 324(R) | 194(R) | | | | |
| $[Hg(CN)_4]^{2-}$ | $\mathbf{T}_d$ | 330(IR)[b] | 235(R)[b] | 54(R) | 1.53 | 17.08 | 361 |
| | | 335(R) | 180(R) | | | | |
| $[Pt(CN)_4]^{2-}$ | $\mathbf{D}_{4h}$ | 505(IR) | 318(R) | 95(R) | 3.425 | 16.823 | 362 |
| | | 465(R) | 300(IR) | | | | 363 |
| | | 455(R) | | | | | |
| $[Ni(CN)_4]^{2-}$ | $\mathbf{D}_{4h}$ | 543(IR) | 433(IR) | | 2.6 | 16.67 | 364 |
| | | (419) | 421(IR) | (54) | | | |
| | | (405) | 488(IR) | | | | |
| | | | (325) | | | | |
| $[Au(CN)_4]^{-}$ | $\mathbf{D}_{4h}$ | 462(IR) | 415(IR) | 110(R) | 3.28– | 17.40– | 365 |
| | | 459(R) | | | 3.42 | 17.44 | |
| | | 450(R) | | | | | |
| $[Hg(CN)_2]$ | $\mathbf{D}_{\infty h}$ | 442(IR) | 341(IR) | (100) | 2.607 | 17.62 | 366 |
| | | 412(R) | 276(R) | | | | 367 |
| $[Ag(CN)_2]^{-}$ | $\mathbf{D}_{\infty h}$ | 390(IR) | (310) | (107) | 1.826 | 17.04 | 368 |
| | | (360) | 250(R) | | | | |
| $[Au(CN)_2]^{-}$ | $\mathbf{D}_{\infty h}$ | 427(IR) | (368) | (100) | 2.745 | 17.17 | 369 |
| | | 445(R) | 305(R) | | | | |

[a] Force constants (mdyn/Å) were obtained by using the GVF field for all ions except the $[Pt(CN)_4]^{2-}$ ion, for which the UBF field was used.
[b] Coupled vibrations between $\nu$(MC) and $\delta$(MCN).

extensive study on mixed cyano–halide complexes such as $[Au(CN)_2X_2]^-$, where X is $Cl^-$, $Br^-$, or $I^-$. An ultraviolet and infrared study[372] showed that the $[Ni(CN)_4]^{2-}$ and $[Ni(CN)_5]^{3-}$ ions are in equilibrium in a solution containing $Na_2[Ni(CN)_4]$, KCN, and KF. The integrated absorption coefficient of the C≡N stretching band increases in the order Hg(II) < Ag(I) < Au(I) in linear dicyano complexes, indicating that the M–C $\pi$-bonding increases in the same order.[357] From the measurements of infrared dichroism, Jones determined the orientation of $[Ag(CN)_2]^-$ and $[Au(CN)_2]^-$ ions in their potassium salts.[368,369] His results are in good agreement with those of X-ray analysis.

***(c) Cyano Complexes Containing NO and Halogens.*** The infrared spectra of nitroprusside salts have been reviewed briefly.[373] Khanna et al.[374] assigned the infrared and Raman spectra of the $Na_2[Fe(CN)_5NO] \cdot 2H_2O$ crystal and its deuterated analog. On the basis of a comparison of $\nu$(CN), $\delta$(MCN), and $\nu$(MC) between Fe(II) and Fe(III) complexes of $[Fe(CN)_5X]^{n-}$ type ions, Brown suggests that the Fe–NO bonding of the $[Fe(CN)_5NO]^{2-}$ ion be formulated as Fe(III)—NO and not as Fe(II)—$NO^+$.[375] The infrared spectra of $K_3[Mn(CN)_5NO]$ and its $^{15}NO$ analog have been reported.[376,377] Tosi and Danon[378] studied the infrared spectra of $[Fe(CN)_5X]^{n-}$ ions (X = $H_2O$, $NH_3$, $NO_2^-$, $NO^-$, and $SO_3^{2-}$). The $\nu$(C≡N) of the $[Fe(CN)_5NO]^{2-}$ ion (2170, 2160, and 2148 $cm^{-1}$) are unusually high in this series because the M–C $\pi$-bonding in this ion is much less than in other compounds due to extensive M–NO $\pi$-bonding. The infrared spectra of $K_2[Pt(CN)_4X_2]$[379] and $K_2[Pt(CN)_5X]$[380] (X = $Cl^-$, $Br^-$, $I^-$, etc) have been reported.

***(d) Bridged Cyano Complexes.*** If the M—C≡N group forms a M—C≡N—M′ type bridge, $\nu$(C≡N) shifts to a higher, and $\nu$(MC) to a lower, frequency. The higher frequency shift of $\nu$(C≡N) should be noted since the opposite trends are observed for bridging carbonyl and halogeno complexes. Shriver[381] observed that $\nu$(C≡N) of $K_2[Ni(CN)_4]$ at 2130 $cm^{-1}$ shifts to 2250 $cm^{-1}$ in $K_2[Ni(CN)_4] \cdot 4BF_3$ because of the formation of the Ni—C≡N—$BF_3$ type bridge. They[382] also found that, for $KFeCr(CN)_6$, the green isomer containing the Fe(II)—C≡N—Cr(III) bridges exhibits $\nu$(C≡N) at 2092 $cm^{-1}$, while the red isomer containing the Cr(III)—C≡N—Fe(II) bridges shows $\nu$(C≡N) at 2168 and 2114 $cm^{-1}$. Brown et al.[383] studied the mechanism of conversion from green to red isomer by combining infrared and Mossbauer spectroscopy with other techniques. The infrared and Mossbauer spectra of $K_4[Fe(CN—SbX_3)_6]$ (X = F and Cl) and $K_4[Fe(CN—SbX_3)_4(CN)_2]$ (X = Cl and Br) have been studied.[384] As expected, the infrared spectrum of

Prussian blue is identical to that of Turnbull's blue.[385] Originally, the $[Ni_2(CN)_6]^{4-}$ ion was thought to have structure I below.[386] However,

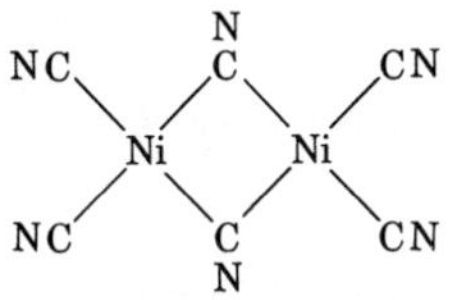

I 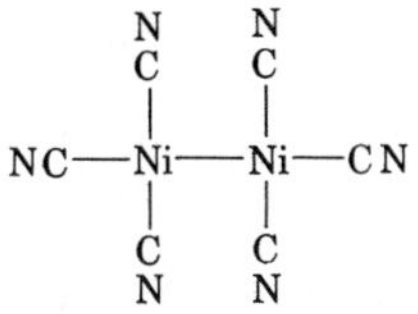 II

recent X-ray[387] and vibrational studies[388] have confirmed structure II containing the Ni–Ni bonding.

**(2) Nitrile and Isonitrile Complexes**

Nitriles (R—C≡N, R = alkyl or phenyl) form a number of metal complexes by coordination through their N atoms. Again, $\nu$(CN) becomes higher upon complex formation. For example, Walton[389] measured the infrared spectra of $MX_2(RCN)_2$ type compounds, where M is Pt(II) and Pd(II) and X is $Cl^-$ and $Br^-$. When R is phenyl, the $\nu$(CN) are near 2285 $cm^{-1}$, which is higher than the value for the $\nu$(CN) of free benzonitrile (2231 $cm^{-1}$). It was noted that the $\nu$(CN) of benzonitrile (2231 $cm^{-1}$) shifts to a higher frequency (2267 $cm^{-1}$) when it coordinates to the pentammine Ru(III) species but to a lower frequency (2188 $cm^{-1}$) when coordinated to the pentammine Ru(II) species. This result may indicate that the latter species has unusually strong $\pi$-back-bonding ability.[390] A strong band at 174 $cm^{-1}$ of $ZnCl_2(CH_3CN)_2$ was suggested to be $\nu$(ZnN).[391] The $\nu$(MN) bands of other acetonitrile complexes have been assigned in the 450–160 $cm^{-1}$ region.[392]

Farona and Kraus[393] observed $\nu$(CN) of $Mn(CO)_3(NC—CH_2—CH_2—CN)Cl$ at 2068 $cm^{-1}$, although $\nu$(CN) of free succinonitrile (sn) is at 2257 $cm^{-1}$. This large shift to a lower frequency was attributed to the chelating bidentate coordination through its CN triple bonds:

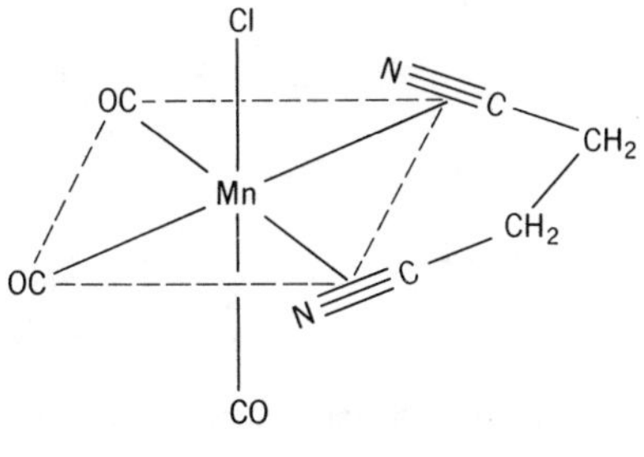

According to X-ray analysis,[394] the complex ion in $[Cu(sn)_2]NO_3$ takes a polymeric chain structure in which the ligand is in the *gauche* conformation:

```
    NC—CH2—CH2—CN       NC—CH2—CH2—CN
 \ /             \ /                 \ /
 Cu               Cu                  Cu
 / \             / \                 / \
    NC—CH2—CH2—CN       NC—CH2—CH2—CN
```

In these dinitrile complexes, $\nu$(CN) are shifted to higher frequencies upon coordination. As in the case of ethylenediamine complexes (Sec. III-2), infrared spectroscopy has been used to determine the conformation of the ligand in metal complexes. The Cu(I) complex, which is known to contain the *gauche* conformation, exhibits two $CH_2$ rocking modes at 966 and 835 $cm^{-1}$, whereas the Ag(I) complex, $Ag(sn)_2BF_4$, shows a single $CH_2$ rocking mode at 770 $cm^{-1}$, which is characteristic of the *trans* conformation.[395]

There are four rotational isomers for glutaronitrile (gn), NC—$CH_2$—$CH_2$—$CH_2$—CN, which are spectroscopically distinguishable. Figure III-25 shows the conformation, the symmetry, and the number of infrared active $CH_2$ rocking vibrations for each isomer. According to X-ray analysis on $Cu(gn)_2NO_3$,[396] the ligand in this complex is in the *gg* conformation. The infrared spectrum of this complex is very similar to that of solid glutaronitrile in the stable form. Matsubara[397] therefore concluded that the latter also takes the *gg* conformation. However, the spectrum of solid glutaronitrile in the metastable form (produced by rapid cooling) is different from that of the *gg* conformation and it could have been *tt*, *tg*, or *gg′*. The *tt* conformation was excluded because of the absence of the 730 $cm^{-1}$ band characteristic of the *trans*-planar

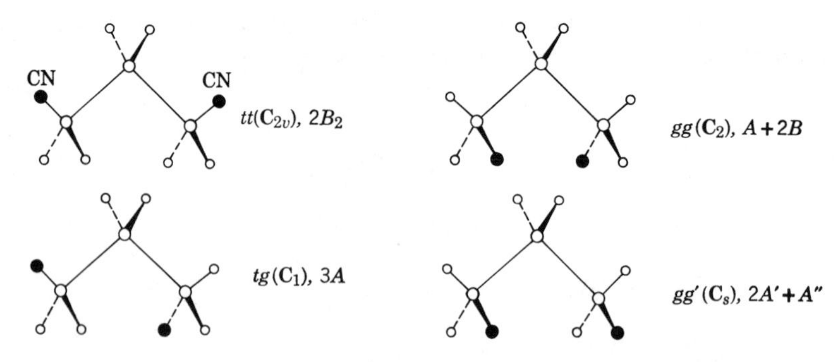


**Fig. III-25.** Rotational isomers of glutaronitrile.

methylene chain,[398] and the *gg′* conformation was considered to be improbable because of steric repulsion between two CN groups. This left only the *tg* conformation for the metastable solid. The complicated spectrum of liquid glutaronitrile was accounted for by assuming that it is a mixture of the *tg*, *gg*, and *tt* conformations. Kubota and Johnston,[399] using these results, have been able to show that the glutaronitrile molecules in $Ag(gn)_2ClO_4$ and $Cu(gn)_2ClO_4$ are in the *gg* conformation, while those in $TiCl_4gn$ and $SnCl_4gn$ have the *tt* conformation. Table III-31 summarizes the $CH_2$ rocking frequencies of glutaronitrile and its metal complexes. An infrared study similar to the above has been extended to adiponitrile ($NC—(CH_2)_4—CN$) and its Cu(I) complex.[400]

Cotton and Zingales[401] studied the N≡C stretching bands of isonitrile complexes. When isonitriles are coordinated to zero-valent metals such as Cr(O), back donation of electrons from the metal to the ligand is extensive and the N≡C stretching band is shifted to a lower frequency. For mono- and dipositive metal ions, little or no back donation occurs and the N≡C stretching band is shifted to a higher frequency as a result of the inductive effect of the metal ion. Sacco and Cotton[402] obtained the infrared spectra of $Co(CH_3NC)_4X_2$ and $[Co(CH_3NC)_4][CoX_4]$ type compounds (X = Cl, Br, etc.). Dart et al.[403] report the $\nu$(NC) of bis(phosphine) tris(isonitrile) complexes of Co(I). Boorman et al.[404] made rather complete assignments of vibrational spectra of some isonitrile complexes of Co(I) and Co(II) in the 4000–33 $cm^{-1}$ region.

TABLE III-31. INFRARED ACTIVE $CH_2$ ROCKING FREQUENCIES OF GLUTARONITRILE AND ITS METAL COMPLEXES ($CM^{-1}$)

| | | | | | |
|---|---|---|---|---|---|
| Liquid[a] | 945 (*tg*) | 904 (*gg*) | 835 (*tg*, *gg*) | 757 (*tg*, *gg*) | 737(*tt*)[b] |
| Solid[a] (metastable) | 943 (*tg*) | — | 839 (*tg*) | 757 (*tg*) | — |
| Solid[a] (stable) | — | 903 (*gg*) | 837 (*gg*) | 768 (*gg*) | — |
| $Cu(gn)_2NO_3$[a] | — | 913 (*gg*) | 830 (*gg*)[c] | 778 (*gg*) | — |
| $Cu(gn)_2ClO_4$[d] | — | 908 (*gg*) | 875 (*gg*) | 767 (*gg*) | — |
| $Ag(gn)_2ClO_4$[d] | — | 904 (*gg*) | 872 (*gg*) | 772 (*gg*) | — |
| $SnCl_4(gn)$[d] | — | — | — | — | 733 (*tt*) |
| $TiCl_4(gn)$[d] | — | — | — | — | 730 (*tt*) |

[a] Ref. 397.
[b] The *tt* form should exhibit two infrared active $CH_2$ rocking vibrations. The other one is not known, however.
[c] Overlapped with a $NO_3^-$ absorption.
[d] Ref. 399.

## III-11. THIOCYANATO AND OTHER PSEUDOHALOGENO COMPLEXES

The $CN^-$, $OCN^-$, $SCN^-$, $SeCN^-$, $CNO^-$, and $N_3^-$ ions are called pseudohalide ions, since they resemble halide ions in their chemical properties. These ions may coordinate to a metal through either one of the end atoms. As a result, the following linkage isomers are possible:

| | |
|---|---|
| M—CN, cyano complex | M—NC, isocyano complex |
| M—OCN, cyanato complex | M—NCO, isocyanato complex |
| M—SCN, thiocyanato complex | M—NCS, isothiocyanato complex |
| M—SeCN, selenocyanato complex | M—NCSe, isoselenocyanato complex |
| M—CNO, fulminato complex | M—ONC, isofulminato complex |

Two compounds are called true linkage isomers if they have exactly the same composition and two of the different linkages mentioned above. A well-known example is nitro (and nitrito) pentammine Co(III) chloride, discussed in Sec. III-4. A pair of true linkage isomers is difficult to obtain since, in general, one form is much more stable than the other. As will be shown later, a number of new linkage isomers have been isolated, and infrared spectroscopy has proved to be very useful in distinguishing them. Burmeister[405] reviewed linkage isomerism in metal complexes. Bailey et al.[406] and Norbury[407] reviewed the infrared spectra of SCN, SeCN, NCO, and CNO complexes and their linkage isomers in detail.

### (1) Thiocyanato (SCN) Complexes

The SCN group may coordinate to a metal through the nitrogen or the sulfur or both (M—NCS—M′). In general, Class A metals (first transition series, such as Cr, Mn, Fe, Co, Ni, Cu, and Zn) form M–N bonds, whereas Class B metals (second half of the second and third transition series, such as Rh, Pd, Ag, Cd, Ir, Pt, Au, and Hg) form M–S bonds.[408] However, other factors, such as the oxidation state of the metal, the nature of other ligands in a complex, and steric consideration, also influence the mode of coordination.

Several empirical criteria have been developed to determine the bonding type of the NCS group in metal complexes.

1. The CN stretching frequencies are generally lower in N-bonded complexes (near and below 2050 $cm^{-1}$) than in S-bonded complexes (near 2100 $cm^{-1}$).[409] The bridging (M—NCS—M′) complexes exhibit $\nu$(CN) well above 2100 $cm^{-1}$. However, this rule must be applied with caution since $\nu$(CN) are affected by many other factors.[406]

2. Several workers[410–412] considered $\nu$(CS) as a structural diagnosis: 860–780 $cm^{-1}$ for N-bonded, and 720–690 $cm^{-1}$ for S-bonded, complexes. However, this band is rather weak and is often obscured by the presence of other bands in the same region.

3. It was suggested[411,412] that the N-bonded complex exhibits a single sharp $\delta$(NCS) near 480 $cm^{-1}$, whereas the S-bonded complex shows several bands of low intensity near 420 $cm^{-1}$. However, these bands are also weak and tend to be obscured by other bands.

4. Several workers[413–416] used the integrated intensity of $\nu$(CN) as a criterion; it is larger than $9 \times 10^4\ M^{-1}\ cm^{-2}$ per $NCS^-$ for N-bonded complexes, and close to or smaller than $2 \times 10^4\ M^{-1}\ cm^{-2}$ for S-bonded complexes. However, this rule is also difficult to apply when the spectrum consists of multiple components or when the dissociation occurs in solution.

5. Some workers[417,418] proposed using $\nu$(MN) and $\nu$(MS) in the far-infrared region as a criterion; in general, $\nu$(MN) is higher than $\nu$(MS). However, these frequencies are very sensitive to the overall structure of the complex and the nature of the central metal. Thus extreme caution must be taken in applying this criterion.

It is clear that only a combination of these five criteria would provide reliable structural diagnosis. Table III-32 lists the vibrational frequencies of typical isothiocyanato and thiocyanato complexes.

Clark and Williams[417] measured the infrared spectra of tetrahedral $M(NCS)_2L_2$, monomeric octahedral $M(NCS)_2L_4$, and polymeric octahedral $M(NCS)_2L_2$ type complexes (M = Fe, Co, Ni, etc.; L = py, $\alpha$-pic, etc), and studied the relationship between the spectra and stereochemistry. They found that $\nu$(CS) are higher by 40–50 $cm^{-1}$ for tetrahedral than for octahedral complexes for the same metal, although $\nu$(CN) are very similar for both.

The *cis*- and *trans*-isomers of $[Co(en)_2(NCS)_2]Cl \cdot H_2O$, for example, can be distinguished by infrared spectra in the $\nu$(CN) region: *trans*, 2136 $cm^{-1}$; *cis*, 2122 and 2110 $cm^{-1}$.[422] Lever et al.[423] have found, however, that no splittings of $\nu$(CN) are observed at room temperature for *cis*-octahedral $ML_2(NCS)_2$, where M is Co(II) and Ni(II) and L is 1,2-bis-(2′-imidazolin-2′-yl)benzene. The splitting of $\nu$(CN) of this complex was observed only at liquid nitrogen temperature.

Turco and Pecile[410] noted that the presence of other ligands in a complex influences the mode of the N–C–S bonding. For example, in $Pt(NCS)_2L_2$, the NCS ligand is N-bonded if L is a phosphine ($\pi$-acceptor), and is S-bonded if L is an amine. For $[Cr(NCS)_4L_2]^{n-}$ ions, Contreras and Schmidt[424] proposed, based on the $\nu$(CN) and $\nu$(CS) of

TABLE III-32. VIBRATIONAL FREQUENCIES OF ISOTHIOCYANATO AND THIOCYANATO COMPLEXES ($cm^{-1}$)[a]

| Compound | $\nu$(CN) | $\nu$(CS) | $\delta$(NCS) | Ref. |
|---|---|---|---|---|
| K[NCS] | 2053 | 748 | 486, 471 | II-226 |
| $(NEt_4)_2[Co(—NCS)_4]$ | 2062 (s) | 837 (w) | 481 (m) | 406 |
| $K_3[Cr(—NCS)_6]$ | 2098 (vs) | 820 (vw) | 474 (s) | 419 |
| | 2058 (vs) | | | |
| $(NEt_4)_2[Cu(—NCS)_4]$ | 2074 (s) | 835 (w) | — | 420 |
| $(NEt_4)_3[Fe(—NCS)_6]$ | 2098 (sh) | 822 (w) | 479 (m) | 406 |
| | 2052 (s) | | | |
| $(NEt_4)_4[Ni(—NCS)_6]$ | 2109 sh | 818 (w) | 469 (m) | 406 |
| | 2102 (s) | | | |
| $(NEt_4)_2[Zn(—NCS)_4]$ | 2074 (s) | 832 (w) | 480 (m) | 406 |
| $(NH_4)[Ag(—SCN)_2]$ | 2101 (s) | 718 (w) | 453 (m) | 406 |
| | 2086 (s) | | | |
| $K[Au(—SCN)_4]$ | 2130 (s) | 700 (w) | 458 (w) | 406 |
| | | | 413 (s) | |
| $K_2[Hg(—SCN)_4]$ | 2134 (m) | 716 (m) | 461 (m) | 406 |
| | 2122 (sh) | 709 (sh) | 448 (m) | |
| | 2109 (s) | 703 (sh) | 432 (sh) | |
| | | | 419 (m) | |
| $(NBu_4)_3[Ir(—SCN)_6]$ | 2127 (m) | 822 (m) | 430 (w) | 421 |
| | 2098 (s) | 693 (w) | | |
| $K_2[Pd(—SCN)_4]$ | 2125 (s) | 703 (w) | 474 (w) | 406 |
| | 2095 (s) | 697 (sh) | 467 (w) | |
| | | | 442 (m) | |
| | | | 432 (m) | |
| $K_2[Pt(—SCN)_4]$ | 2128 (s) | 696 (w) | 477 (w) | 406 |
| | 2099 (s) | | 469 (w) | |
| | 2077 (sh) | | 437 (m) | |
| | | | 426 (m) | |

[a] vs: very strong; s: strong; m: medium; w: weak; sh: shoulder.

these ions, N-bonding for L = urea, glycinate ion, and so on, and S-bonding for L = thiourea, acetamide, and so on. These results have been explained in terms of the steric and electronic effects of L.

Since 1963,[425] a variety of true linkage isomers involving the NCS group have been prepared. Table III-33 lists the $\nu$(CN) and $\nu$(CS) of typical pairs of these linkage isomers. Epps and Marzilli[432] isolated three

TABLE III-33. VIBRATIONAL FREQUENCIES OF TRUE LINKAGE ISOMERS INVOLVING THE NCS GROUP ($cm^{-1}$)

| Compound | Type | $\nu$(CN) | $\nu$(CS) | Refs. |
|---|---|---|---|---|
| *trans*-$[Pd(AsPh_3)_2(NCS)_2]$ | N-bonded | 2089 | 854 | 425, 427, 428 |
| | S-bonded | 2119 | — | |
| $Pd(bipy)(NCS)_2$ | N-bonded | 2100 | 842 | 426, 427 |
| | S-bonded | 2117 | 700 | |
| | | 2108 | | |
| ($\pi$-Cp)$Mo(CO)_3(NCS)$ | N-bonded | 2099 | — | 429 |
| | S-bonded | 2114 | 699 | |
| $K_3[Co(CN)_5(NCS)]$ | N-bonded | 2065 | 810 | 430 |
| | S-bonded | 2110 | 718 | |
| *trans*-$[Co(DMG)_2(py)(NCS)]$ | N-bonded | 2128 | — | 431 |
| | S-bonded | 2118 | — | |

linkage isomers of $AsPh_4[Co(DMG)_2(NCS)_2]$:

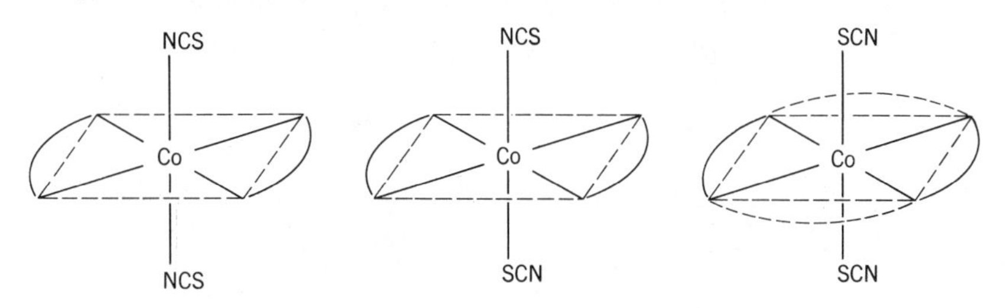


Although all these isomers exhibit $\nu$(CN) at 2110 $cm^{-1}$, they can be distinguished by the differences in the intensity of the $\nu$(CN) band; the (NCS, NCS) isomer is the strongest, the (SCN, SCN) isomer is the weakest, and the (NCS, SCN) isomer is in between.

Both N-bonded and S-bonded NCS groups have been found in [Pd(4,4′-dimethyl-bipy)(NCS)(SCN)][433] and $[Pd\{Ph_2P(CH_2)_3NMe_2\}$-(NCS)(SCN)].[434] Similar mixed NCS–SCN bonding was found for [PdL(NCS)(SCN)], where L is $Ph_2P(o\text{-}C_6H_4)AsPh_2$ and $Ph_2P(CH_2)_2NMe_2$.[435] These bidentate ligands contain two different donor atoms which give different electronic effects on the NCS groups *trans* to them. Thus the *trans* effect, together with the steric effect of these ligands, may be responsible for the mixing of the N- and S-bonding. In the case of $[Ni(DPEA)(NCS)_2]_2$[DPEA: di(2-pyridyl-$\beta$-ethyl)amine], infrared spectra suggest that terminal N-bonded and bridging NCS groups are mixed [$\nu$(CN): 2094 and 2128 $cm^{-1}$, respectively].[436]

In $CH_3CN$ solution, the infrared and Raman spectra of hexathiocyanato complexes indicate the formation of mixed $CH_3CN$–NCS complexes.[437] Burmeister et al.[438] found that in the $ML_2X_2$ type complexes [M = Pd(II), Pt(II); L = a neutral ligand; X = SCN, SeCN, NCO, etc.], the mode of bonding of X to the metal is determined by the nature of the solvent. For example, $Pd(AsPh_3)_2(NCS)_2$ is N-bonded in pyridine and acetone solution, whereas it is S-bonded in DMF and DMSO solution. However, the bonding of the NCO group is insensitive to the nature of the solvent.

The NCS group also forms a bridge between two metal atoms. The CN stretching frequency of a bridging group is generally higher than that of a terminal group. For example, $HgCo(NCS)_4$(Co—NCS—Hg) absorbs at 2137 $cm^{-1}$, whereas $(NEt_4)_2[Co(—NCS)_4]$ absorbs at 2065 $cm^{-1}$. According to Chatt and Duncanson,[439] the CN stretching frequencies of Pt(II) complexes are 2182–2150 $cm^{-1}$ for the bridging and 2120–2100 $cm^{-1}$ for the terminal NCS group. $[(P(n\text{-}Pr)_3)_2Pt_2(SCN)_2Cl_2]$ (compound I) exhibits one bridging CN stretching, whereas $[(P(n\text{-}Pr)_3)_2Pt_2(SCN)_4]$ (compound II) exhibits both bridging and terminal CN stretching bands. Thus the infrared spectra suggest that the structure of each compound is as follows:

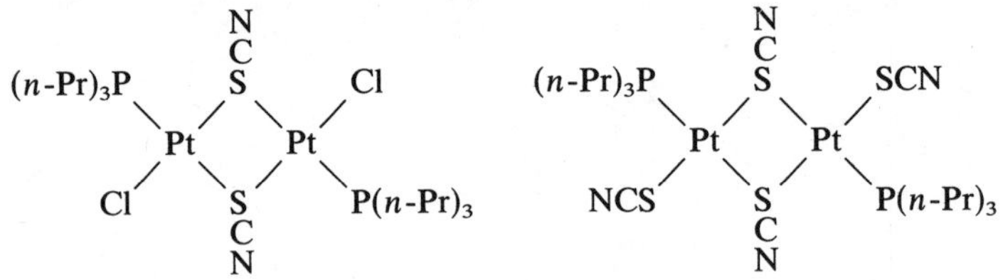

```
              N                                  N
              C                                  C
(n-Pr)3P      S       Cl          (n-Pr)3P       S       SCN
        \   /   \   /                     \   /   \   /
         Pt       Pt                        Pt      Pt
        /   \   /   \                     /   \   /   \
      Cl      S       P(n-Pr)3         NCS      S       P(n-Pr)3
              C                                 C
              N                                 N

              I          n-Pr: n-Propyl         II
```

Compound I, however, exists as two isomers, $\alpha$ and $\beta$, which absorb at 2162 and 2169 $cm^{-1}$, respectively. Chatt and Duncanson[439] originally suggested a geometrical isomerism in which two SCN groups were in a *cis* or *trans* position with respect to the central ring. Later,[440–442] "bridge isomerism" such as the following was demonstrated by X-ray analysis:

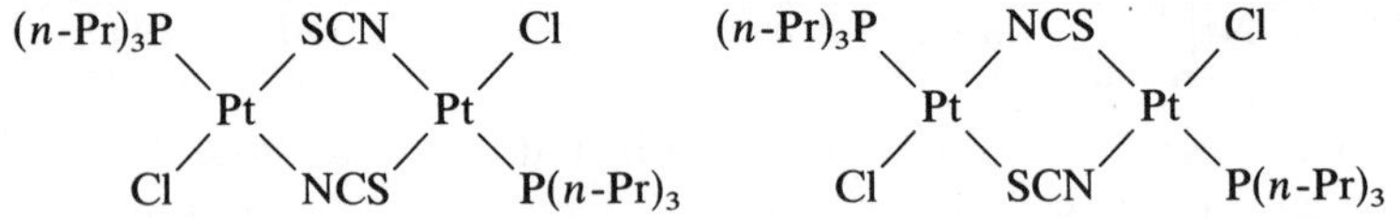

```
(n-Pr)3P     SCN      Cl          (n-Pr)3P     NCS      Cl
        \   /   \   /                     \   /   \   /
         Pt       Pt                        Pt      Pt
        /   \   /   \                     /   \   /   \
      Cl     NCS      P(n-Pr)3          Cl     SCN      P(n-Pr)3

              α                                 β
```

The infrared spectra of metal complexes containing bridging NCS groups have been reported for $Sn(NCS)_2$,[443] $M(py)_2(NCS)_2$,[444] [M = Mn(II), Co(II), and Ni(II)] $[Me_3Pt(NCS)]_4$,[445] and $M[Pt(SCN)_6]$ [M = Co(II), Ni(II), Fe(II), etc.].[446]

**(2) Selenocyanato (SeCN) Complexes**

The SeCN group also coordinates to a metal through the nitrogen (M—NCSe) or the selenium (M—SeCN) or both (M—NCSe-M′). Again, Class A metals tend to form M–N bonds, while Class B metals prefer to form M–Se bonds. Although the number of SeCN complexes studied is much smaller than that of SCN complexes, these studies suggest the following trends:

1. $\nu$(CN) is below 2080 cm$^{-1}$ for N-bonded, but higher for Se-bonded, complexes. The $\nu$(CN) of a bridged complex [HgCo(NCSe)$_4$] is at 2146 cm$^{-1}$.[447]
2. The $\nu$(CSe) is at 700–620 cm$^{-1}$ for N-bonded and 550–500 cm$^{-1}$ for Se-bonded complexes.
3. The $\delta$(NCSe) of N-bonded complexes are above 400 cm$^{-1}$, whereas Se-bonded complexes show at least one component of $\delta$(NCSe) below 400 cm$^{-1}$.
4. The integrated intensity of $\nu$(CN) is larger for the N-bonded than for the Se-bonded group.[448]

Table III-34 lists the observed frequencies of typical N-bonded and Se-bonded complexes.

TABLE III-34. VIBRATIONAL FREQUENCIES OF ISOSELENOCYANATO AND SELENOCYANATO COMPLEXES (CM$^{-1}$)

| Compound[a] | $\nu$(CN) | $\nu$(CSe) | $\delta$(NCSe) | Ref. |
|---|---|---|---|---|
| K[NCSe] | 2070 | 558 | 424, 416 | II-227 |
| $R_4$[Mn(—NCSe)$_6$] | 2079, 2082<br>2070 | 640<br>617 | 424 | 448 |
| $R_2$[Fe(—NCSe)$_4$] | 2067, 2055 | 673, 666 | 432 | 448 |
| $R_4$[Ni(—NCSe)$_6$] | 2118, 2102 | 625 | 430 | 448 |
| [Ni(pn)$_2$(—NCSe)$_2$] | 2096, 2083 | 692 | — | 449 |
| $R'_2$[Co(—NCSe)$_4$] | 2053 | 672 | 433, 417 | 450 |
| [Co(NH$_3$)$_5$(—NCSe)](NO$_3$)$_2$ | 2116 | 624 | — | 451 |
| $R_2$[Zn(—NCSe)$_4$] | 2087 | 661 | 429 | 448 |
| [Cu(pn)$_2$(—SeCN)$_2$] | 2053, 2028 | — | — | 449 |
| $R_3$[Rh(—SeCN)$_6$] | 2104, 2071 | 515 | — | 448 |
| $R''_2$[Pd(—SeCN)$_4$] | 2114, 2105 | 521 | 410, 374 | 452 |
| $R_2$[Pt(—SeCN)$_4$] | 2105, 2060 | 516 | — | 448 |
| [Pt(bipy)(—SeCN)$_2$] | 2135, 2125 | 532, 527 | — | 451 |
| $K_2$[Pt(—SeCN)$_6$] | 2130 | 519 | 390, 379<br>367 | 450 |

[a] R: [N($n$-$C_4H_9$)$_4$]$^+$; R′: [N($C_2H_5$)$_4$]$^+$; R″: [N($CH_3$)$_4$]$^+$; pn: propylenediamine; bipy: 2,2′-bipyridine.

Burmeister and Gysling[453] observed that in $[PdL_2(SeCN)_2]$ type compounds the effect of changing the $\pi$-bonding ability and basicity of L on the Pd–SeCN bonding is negligible in contrast to the analogous SCN complexes. A pair of true linkage isomers has been isolated and characterized by infrared spectra for $[(\pi\text{-Cp})Fe(CO)(PPh_3)(SeCN)]$[454] and $[Pd(Et_4dien)(SeCN)]BPh_4$,[455] where $Et_4dien$ is 1,1,7,7-tetraethyldiethylenetriamine.

### (3) Cyanato (OCN) Complexes

The OCN group may coordinate to a metal through the nitrogen (M—NCO) or the oxygen (M—OCN) or both. Thus far, the majority of complexes are reported to be N-bonded. Table III-35 lists the observed frequencies of N-bonded NCO groups in typical complexes; $\nu_a$(NCO) and $\nu_s$(NCO) denote vibrations consisting mainly of $\nu$(CN) and $\nu$(CO), respectively. For $ML_2(NCO)_2$ (M = Pd or Pt; L = $NH_3$, py, etc.) and $In(NCO)_3L_3$ (L = py, DMSO, etc.), see Refs. 462 and 463, respectively.

Thus far, O-bonded structures have been suggested for $[M(OCN)_6]^{n-}$ [M = Mo(III), Re(IV), and Re(V)].[461] Anderson and Norbury[464] prepared the first example of linkage isomers: yellow $Rh(PPh_3)_3(NCO)$ and orange

TABLE III-35. VIBRATIONAL FREQUENCIES OF ISOCYANATO COMPLEXES ($cm^{-1}$)

| Compound | $\nu_a$(NCO) | $\nu_s$(NCO) | $\delta$(NCO) | Refs. |
|---|---|---|---|---|
| K[NCO] | 2155 | 1282, 1202 | 630 | II-224 |
| $Si(NCO)_4$ | 2284 | 1482 | 608, 546 | 456 |
| $Ge(NCO)_4$ | 2247 | 1426 | 608, 528 | 456 |
| $[Zn(NCO)_4]^{2-}$ | 2208 | 1326 | 624 | 456a |
| $[Mn(NCO)_4]^{2-}$ | 2222 | 1335 | 623 | 457, 458 |
| $[Fe(NCO)_4]^{2-}$ | 2182 | 1337 | 619 | 457, 458 |
| $[Co(NCO)_4]^{2-}$ | 2217, 2179 | 1325 | 620, 617 | 457, 458 |
| $[Ni(NCO)_4]^{2-}$ | 2237, 2186 | 1330 | 619, 617 | 457 |
| $[Fe(NCO)_4]^{-}$ | 2208, 2171 | 1370 | 626, 619 | 457 |
| $[Pd(NCO)_4]^{2-}$ | 2200–2190 | 1319 | 613 604, 594 | 459 |
| $[Sn(NCO)_6]^{2-}$ | 2270, 2188 | 1307 | 667, 622 | 459 |
| $[Zr(NCO)_6]^{2-}$ | 2205 | 1340 | 628 | 460 |
| $[Mo(OCN)_6]^{3-}$ | 2205 | 1296, 1140 | 595 | 461 |

$Rh(PPh_3)_3(OCN)$. The integrated $\nu$(CN) intensity of the former is smaller than that of the free ion, whereas the intensity of the latter is larger than that of the free ion. Also, the latter exhibits two $\delta$(OCN) at 607 and 590 $cm^{-1}$, whereas the former shows only one band at 592 $cm^{-1}$.

A bridging NCO group of the type:

$$\begin{array}{l} M \\ \quad \searrow \\ \quad\quad N{=}C{=}O \\ \quad \nearrow \\ M \end{array}$$

has been proposed for $ML_2(NCO)_2$ (M = Mn, Fe, Co, and Ni; L = 3- or 4-CN-py)[465] and $Re_2(CO)_8(NCO)_2$.[466] Forster and Horrocks[456a] carried out normal coordinate analyses on $[Zn(NCX)_4]^{2-}$ anions, where X is O, S, or Se.

### (4) Fulminato (CNO) Complexes

The fulminate ($CNO^-$) ion may coordinate to a metal through the carbon (M—CNO) or the oxygen (M—ONC) or both. So far, all the complexes containing the CNO group are presumed to be C-bonded. Table III-36 lists the observed CNO frequencies of typical fulminato complexes. Thus far, not much work has been done on fulminato complexes.

TABLE III-36. OBSERVED FREQUENCIES OF TYPICAL FULMINATO COMPLEXES ($cm^{-1}$)

| Ion | $\nu$(CN) | $\nu$(NO) | $\delta$(CNO) | Ref. |
|---|---|---|---|---|
| $[CNO]^-$ | 2052 | 1057 | 471 | 467 |
| $[Ag(CNO)_2]^-$ | 2119 | 1144 | — | 468 |
| $[Fe(CNO)_6]^{4-}$ | 2187 | 1040 | 514 | 468 |
| | | | 466 | |
| $[Hg(CNO)_4]^{2-}$ | 2130 | 1143 | — | 469 |
| $[Ni(CNO)_4]^{2-}$ | 2184 | 1122 | 479 | 470 |
| | | | 470 | |
| $[Zn(CNO)_4]^{2-,a}$ | 2146 ($A_1$) | 1177 ($A_1$) | 498 ($E$) | 471 |
| | 2130 ($F_2$) | 1154 ($F_2$) | 475 ($F_2$) | |
| $[Pt(CNO)_4]^{2-,b}$ | 2194 ($A_{1g}$) | 1174 ($A_{1g}$) | 476 ($B_{2g}$) | 471 |
| | 2189 ($B_{1g}$) | 1140 ($B_{1g}$) | 453 ($E_g$) | |

[a] $\mathbf{T}_d$ symmetry.
[b] $\mathbf{D}_{4h}$ symmetry.

TABLE III-37. VIBRATIONAL FREQUENCIES OF AZIDO COMPLEXES ($cm^{-1}$)

| Compound | $\nu_a$(NNN) | $\nu_s$(NNN) | $\delta$(NNN) | $\nu$(MN) | Refs. |
|---|---|---|---|---|---|
| $K[N_3]$ | 2041 | 1344 | 645 | — | II-225 |
| $R_2[Pt(N_3)_4]$ | 2075, 2060<br>2024, 2029 | 1276 | 582 | 394 | 472 |
| $R[Au(N_3)_4]$ | 2030, 2034 | 1261<br>1251 | 578 | 432 | 472 |
| $R''_2[Zn(N_3)_4]$ | 2097, 2058 | 1330<br>1282 | — | — | 472 |
| $R_2[VO(N_3)_4]$ | 2088, 2051<br>2092, 2060<br>2005 | 1340 | 652 | 442<br>405 | 472 |
| $R_2[Pb(N_3)_6]$ | 2045, 2056<br>2037 | 1262<br>1253 | 640<br>597 | 327<br>313 | 472 |
| $R_2[Pt(N_3)_6]$ | 2022, 2028 | 1275<br>1262<br>1253 | 578 | 402<br>397<br>320 | 472 |
| $R'_2[Co(N_3)_4]$ | 2089, 2050 | 1338<br>1280 | 642<br>610 | 368 | 473 |
| $R'_2[Zn(N_3)_4]$ | 2098, 2055 | 1342<br>1290 | 649<br>615 | 351<br>295 | 473 |
| $R'_2[Sn(N_3)_4]$ | 2115, 2080 | 1340 | 659<br>601 | 390<br>330 | 473 |

R: $[As(Ph)_4]^+$; R′: $[N(C_2H_5)_4]^+$; R″: $[P(Ph)_4]^+$.

### (5) Azido ($N_3$) Complexes

Table III-37 lists the observed frequencies of typical azido complexes. The two $N_3$ groups around the Hg atom in $Hg_2(N_3)_2$ are in the *trans* position ($\mathbf{C}_{2h}$), whereas they are in a twisted configuration ($\mathbf{C}_2$) in $Hg(N_3)_2$. The former exhibits one $\nu_a(N_3)$ at 2080 $cm^{-1}$, whereas the latter shows two $\nu_a(N_3)$ at 2090 and 2045 $cm^{-1}$.[474] For Co(III) azido ammine complexes and $[M(N_3)_2(py)_2]$(M = Cu, Zn, and Cd), see Refs. 475 and 476, respectively. The bridging azido groups are found in

$$[(PPh_3)(N_3)Pd\underset{N_3}{\overset{N_3}{\diamond}}Pd(N_3)(PPh_3)]^{477}$$

and

$$[(acac)_2Co\underset{N_3}{\overset{N_3}{\diamond}}Co(acac)_2].^{478}$$

However, it is not possible to determine the structures of these bridges from infrared spectra. Forster and Horrocks[473] made complete assignments of vibrational spectra of the $[Co(N_3)_4]^{2-}$ and $[Zn(N_3)_4]^{2-}$ ($\mathbf{D}_{2d}$) and $[Sn(N_3)_6]^{2-}$ ($\mathbf{D}_{3d}$) ions. The spectra suggest that the M–NNN bonds in these anions are not linear.

## III-12. CARBONYL AND NITROSYL COMPLEXES

In the last few decades, a large number of carbonyl complexes have been synthesized, and their spectra and structures have been studied exhaustively. This section describes only typical results obtained from these investigations. For more comprehensive information, several review articles[479–482] should be consulted.

Most carbonyl complexes exhibit strong and sharp $\nu$(CO) bands at ca. 2100–1800 cm$^{-1}$. Since $\nu$(CO) is generally free from coupling with other modes and is not obscured by the presence of other vibrations, studies of $\nu$(CO) alone often provide valuable information about the structure and bonding of carbonyl complexes. In the majority of compounds, $\nu$(CO) of free CO (2155 cm$^{-1}$) is shifted to lower frequencies. In terms of simple M.O. theory, this observation has been explained as follows. First, the $\sigma$-bond is formed by donating $5\sigma$ electrons of CO to the empty orbital of the metal (see Fig. III-26). This tends to raise $\nu$(CO), since the $5\sigma$ orbital is slightly antibonding. Second, the $\pi$-bond is formed by back-donating the $d\pi$-electrons of the metal to an empty antibonding orbital, the $2p\pi^*$ orbital of CO. This tends to lower $\nu$(CO). Although these two components of bonding are synergic, the net result is a drift of electrons from the metal to CO when the metal is in a relatively low oxidation state. Thus the $\nu$(CO) of metal carbonyl complexes are generally lower than the

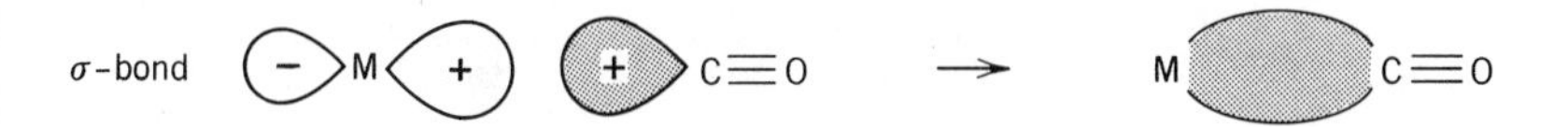


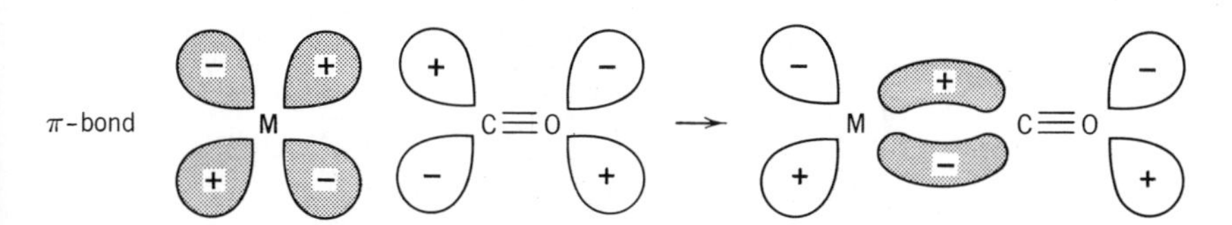


**Fig. III-26.** The $\sigma$- and $\pi$-bonding in metal carbonyls.

value for free CO. The opposite trend is observed, however, when CO is complexed with metal halides in which the metals are in a relatively higher oxidation state [see Sec. III-12(6)].

If CO forms a bridge between two metals, its $\nu$(CO)(1900–1800 cm$^{-1}$) is much lower than that of the terminal CO group (2100–2000 cm$^{-1}$). An extremely low $\nu$(CO)(ca. 1300 cm$^{-1}$) is observed when the bridging CO group forms an adduct via its O atom (see below).[483]

### (1) Mononuclear Carbonyls

Table III-38 lists the observed frequencies and band assignments of mononuclear carbonyls of tetrahedral ($\mathbf{T}_d$), trigonal-bipyramidal ($\mathbf{D}_{3h}$), and octahedral ($\mathbf{O}_h$) structures. Complete normal coordinate analyses have been made on most of these carbonyls. Jones and co-workers[488,489] carried out extensive vibrational studies on $Fe(CO)_5$ and $M(CO)_6$ (M = Cr, Mo, and W), including their $^{13}C$ and $^{18}O$ species. In the $\mathbf{T}_d$ series,

TABLE III-38. VIBRATIONAL FREQUENCIES AND BAND ASSIGNMENTS OF MONONUCLEAR METAL CARBONYLS (CM$^{-1}$)[a]

| Compound | Symmetry | State | $\nu$(CO) | $\nu$(MC) | $\delta$(MCO) | $\delta$(CMC) | Ref. |
|---|---|---|---|---|---|---|---|
| $Ni(CO)_4$ | $\mathbf{T}_d$ | gas | 2131 ($A_1$)<br>2057.6 ($F_2$) | 367.5 ($A_1$)<br>421 ($F_2$) | 380 ($E$)<br>458.8 ($F_2$)<br>300 ($F_1$) | 64 ($E$)<br>80 ($F_2$) | 484 |
| $[Co(CO)_4]^-$ | $\mathbf{T}_d$ | DMF sol'n | 2002 ($A_1$)<br>1890 ($F_2$) | 431 ($A_1$)<br>556 ($F_2$) | 523 ($F_2$) | 91 ($E$) | 485 |
| $[Fe(CO)_4]^{2-}$ | $\mathbf{T}_d$ | aq. sol'n | 1788 ($A_1$)<br>1788 ($F_2$) | 464 ($A_1$)<br>644 ($F_2$) | 550 ($F_2$)<br>785 ($E$) | 100–85<br>($E$, $F_2$) | 486 |
| $Fe(CO)_5$ | $\mathbf{D}_{3h}$ | liquid | 2116 ($A_1'$)<br>2030 ($A_1'$)<br>1989 ($E'$) | 418 ($A_1'$)<br>381 ($A_1'$)<br>482 ($E'$) | 278 ($A_2'$)<br>653 ($E'$)<br>559 ($E'$)<br>491 ($E''$)<br>448 ($E''$) | 107 ($E'$)<br>64 ($E'$) | 487<br>488 |
| $Cr(CO)_6$ | $\mathbf{O}_h$ | gas | 2118.7 ($A_{1g}$)<br>2026.7 ($E_g$)<br>2000.4 ($F_{1u}$) | 379.2 ($A_{1g}$)<br>390.6 ($E_g$)<br>440.5 ($F_{1u}$) | 364.1 ($F_{1g}$)<br>668.1 ($F_{1u}$)<br>532.1 ($F_{2g}$)<br>510.9 ($F_{2u}$) | 97.2 ($F_{1u}$)<br>89.7 ($F_{2g}$)<br>67.9 ($F_{2u}$) | 489 |
| $Mo(CO)_6$ | $\mathbf{O}_h$ | gas | 2120.7 ($A_{1g}$)<br>2024.8 ($E_g$)<br>2000.3 ($F_{1u}$) | 391.2 ($A_{1g}$)<br>381 ($E_g$)<br>367.2 ($F_{1u}$) | 341.6 ($F_{1g}$)<br>595.6 ($F_{1u}$)<br>477.4 ($F_{2g}$)<br>507.2 ($F_{2u}$) | 81.6 ($F_{1u}$)<br>79.2 ($F_{2g}$)<br>60 ($F_{2u}$) | 489 |
| $W(CO)_6$ | $\mathbf{O}_h$ | gas | 2126.2 ($A_{1g}$)<br>2021.1 ($E_g$)<br>1997.6 ($F_{1u}$) | 426 ($A_{1g}$)<br>410 ($E_g$)<br>374.4 ($F_{1u}$) | 361.6 ($F_{1g}$)<br>586.6 ($F_{1u}$)<br>482.0 ($F_{2g}$)<br>521.3 ($F_{2u}$) | 82.0 ($F_{1u}$)<br>81.4 ($F_{2g}$)<br>61.4 ($F_{2u}$) | 489 |
| $[V(CO)_6]^-$ | $\mathbf{O}_h$ | $CH_3CN$ sol'n | 2020 ($A_{1g}$)<br>1894 ($E_g$)<br>1858 ($F_{1u}$) | 374 ($A_{1g}$)<br>393 ($E_g$)<br>460 ($F_{1u}$) | 356 ($F_{1g}$)<br>650 ($F_{1u}$)<br>517 ($F_{2g}$)<br>506 ($F_{2u}$) | 92 ($F_{1u}$)<br>84 ($F_{2g}$) | 490 |
| $[Re(CO)_6]^+$ | $\mathbf{O}_h$ | $CH_3CN$ sol'n | 2197 ($A_{1g}$)<br>2122 ($E_g$)<br>2085 ($F_{1u}$) | 441 ($A_{1g}$)<br>426 ($E_g$)<br>356 ($F_{1u}$) | 354 ($F_{1g}$)<br>584 ($F_{1u}$)<br>486 ($F_{2g}$)<br>522 ($F_{2u}$) | 82 ($F_{1u}$)<br>82 ($F_{2g}$) | 490 |
| $[Mn(CO)_6]^+$ | $\mathbf{O}_h$ | $CH_3CN$<br>sol'n (IR)<br>solid (R) | 2192 ($A_{1g}$)<br>2125 ($E_g$)<br>2095 ($F_{1u}$) | 384 ($A_{1g}$)<br>390 ($E_g$)<br>412 ($F_{1u}$) | 347 ($F_{1g}$)<br>636 ($F_{1u}$)<br>500 ($F_{2g}$)<br>500 ($F_{2u}$) | 101 ($F_{1u}$)<br>101 ($F_{2g}$) | 491 |

[a] The three low frequency vibrations may be coupled.

$\nu$(CO) decreases and $\nu$(MC) increases in going from $Ni(CO)_4$ to $[Co(CO)_4]^-$ to $[Fe(CO)_4]^{2-}$. This result indicates that the M→CO $\pi$-back-donation is increasing in the order Ni(0) < Co(−I) < Fe(−II). The existence of $Ru(CO)_5$ and $Os(CO)_5$ has been confirmed by their infrared spectra in heptane solution. Both compounds seem to be trigonal-bipyramidal.[492]

In the $\mathbf{O}_h$ series, Jones et al.[489] obtained the following $F_{1u}$ force constants (mdyn/Å) from gas-phase spectra:

| | $Cr(CO)_6$ | $Mo(CO)_6$ | $W(CO)_6$ | |
|---|---|---|---|---|
| $F$(C≡O) | 17.22 | 17.39 | 17.21 | (GVF) |
| $F$(M—C) | 1.64 | 1.43 | 1.80 | |

This result indicates that the M–C bond strength increases in the order Mo < Cr < W, an order also supported by a Raman intensity study of these compounds.[493] On the other hand, Hendra and Qurashi[494] related the Raman intensity ratio of two $A_{1g}$ modes, $I(\nu_1$, CO stretching)/$I(\nu_2$, MC stretching), to the $\pi$-character of the M–C bond, and concluded that the M–C bond strength increases in the order Cr < W < MO < Re(I). The infrared intensity of the CO stretching bands has been reviewed by Kettle and Paul.[495]

Solvent effects on the infrared spectra of $M(CO)_6$ (M = Cr, Mo, and W) have been studied by Clark and Crociani.[496] Edgell and co-workers[497] attributed the band at 413 cm$^{-1}$ of $Li[Co(CO)_4]$ in a THF solution to the vibrations of the alkali ions, which form ion pairs with $[Co(CO)_4]^-$: For sodium and potassium salts, the corresponding bands are observed at 192 and 142 cm$^{-1}$, respectively. One of the probable structures of these compounds may be structure I of Fig. III-27. Pribula and Brown[498] proposed the $\mathbf{C}_{3v}$ structure (II) for ion pairs $M[Mn(CO)_5]$($M = Li^+$ and $Na^+$) in ether solution.

### (2) Polynuclear Carbonyls

Since polynuclear carbonyls take a variety of structures, elucidation of their structures by vibrational spectroscopy has been a subject of considerable interest in the past. The principles involved in these structure determinations were described in Sec. I–10. However, the structures of some polynuclear complexes are too complicated to allow elucidation by simple application of selection rules based on symmetry. Thus the results are often ambiguous. In these cases, one must resort to X-ray analysis to obtain definitive and accurate structural information. However, vibrational spectroscopy is still useful in elucidating the structures of metal carbonyls in solution.

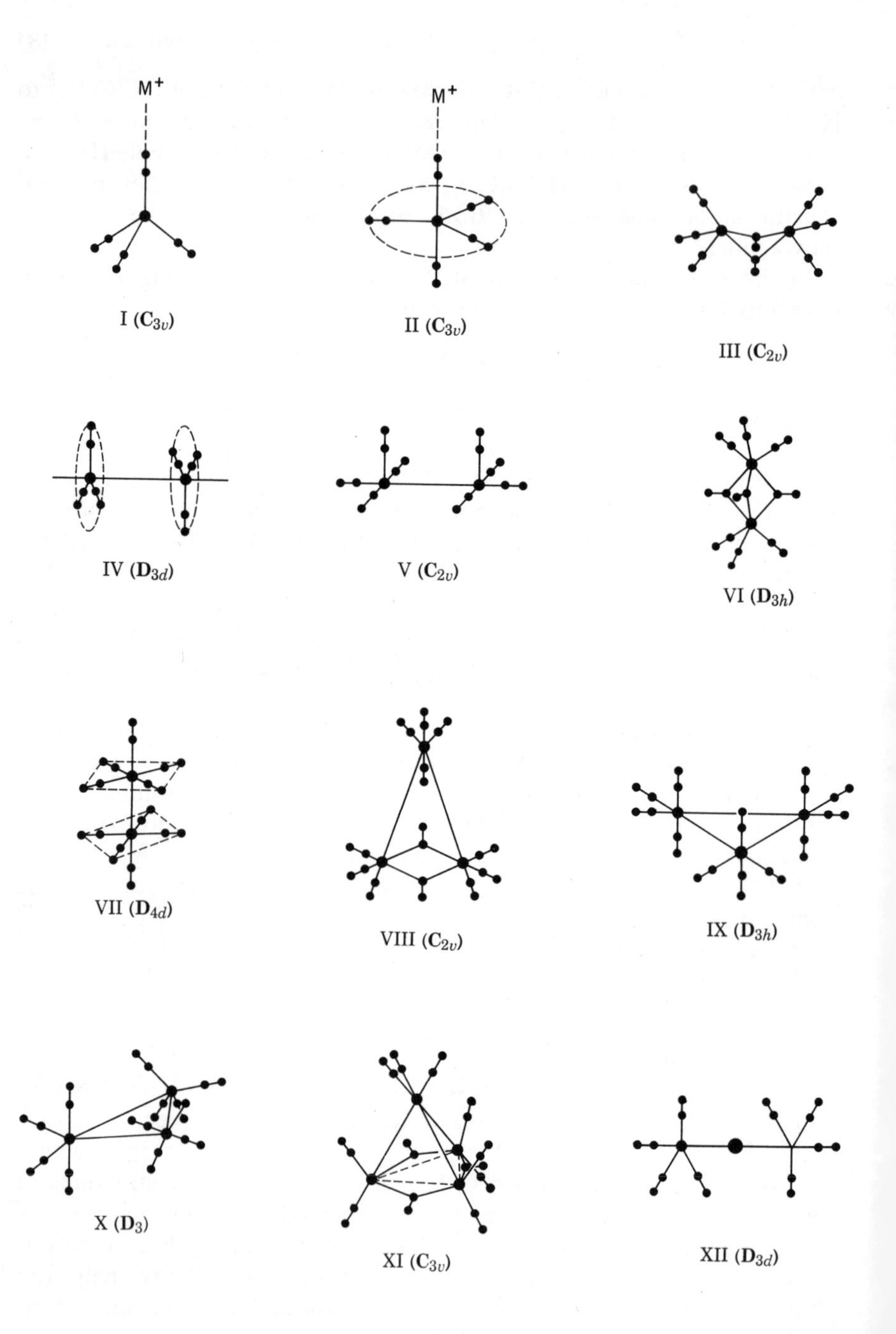


**Fig. III-27.** Structures of metal carbonyls.

According to X-ray analysis,[499] $Co_2(CO)_8$ takes structure III of Fig. III-27. For this $\mathbf{C}_{2v}$ structure, five terminal and two bridging $\nu(CO)$ are expected to be infrared active; the former are observed at 2075, 2064, 2047, 2035, and 2028 cm$^{-1}$, and the latter are located at 1867 and 1859 cm$^{-1}$.[500] In solution, $Co_2(CO)_8$ exists as an equilibrium mixture of three isomers; one isomer has structure III, while the other has structure IV according to Noack[501] and structure V according to Bor.[502] Their joint work,[503] however, revealed that structure V must be abandoned. Furthermore, they suggested the presence of the third isomer of unknown structure.

The infrared spectrum of $Fe_2(CO)_9$ was first obtained by Sheline and Pitzer,[504] who observed two terminal (2080 and 2034 cm$^{-1}$) and one bridging (1828 cm$^{-1}$) CO stretching bands. This result agrees with that expected from structure VI, determined by X-ray analysis.[505] Again according to X-ray analysis,[506] the structure of $Mn_2(CO)_{10}$ is that shown by VII of Fig. III-27. This $\mathbf{D}_{4d}$ structure predicts four Raman and three infrared active CO stretching bands. Adams et al.[507] observed the former at 2116 ($A_1$), 1997 ($A_1$), 2024 ($E_2$), and 1981 ($E_3$) cm$^{-1}$, and Bor[508] observed the latter at 2046 ($B_2$), 1984 ($B_2$), and 2015 ($E_1$) cm$^{-1}$ in solution. Levenson et al.[509] confirmed these infrared assignments by studying the polarization properties of the three bands in a nematic liquid crystal. The structures of $Re_2(CO)_{10}$ and $Tc_2(CO)_{10}$ are similar to that of $Mn_2(CO)_{10}$. The Raman[507] and infrared[510,511] spectra of $Re_2(CO)_{10}$ and the infrared spectrum of $Tc_2(CO)_{10}$[511] have been assigned.

The structure of $Fe_3(CO)_{12}$ had been controversial. Dahl et al.[512] finally confirmed structure VIII of Fig. III-27 by X-ray analysis. This structure can account for Mossbauer[513] and solid state infrared spectra. In solution, the infrared spectrum does not agree with that expected for structure VIII; the bridging $\nu(CO)$ is very weak and the terminal $\nu(CO)$ region is broad without resolution. Cotton and Hunter[514] suggest that a whole range of structures varying from $\mathbf{D}_{3h}$ (structure IX) to $\mathbf{C}_{2v}$ (structure VIII) are in equilibrium, the majority being close to $\mathbf{D}_{3h}$. Johnson suggests the presence of a new isomer of $\mathbf{D}_3$ symmetry, shown by structure X.[515] According to X-ray analysis,[516] $Os_3(CO)_{12}$ takes the $\mathbf{D}_{3h}$ structure(IX), for which four terminal $\nu(CO)$ should be infrared active. Huggins et al.[517] assigned them at 2068 ($E'$), 2035 ($A_2''$), 2014 ($E'$), and 2002 ($E'$) cm$^{-1}$. Quicksall and Spiro[518] assigned the Raman spectra of $Os_3(CO)_{12}$ and analogous $Ru_3(CO)_{12}$, for which six $\nu(CO)$ are expected in the Raman spectrum. For $Os_3(CO)_{12}$, they are observed at 2130 ($A_1'$), 2028 ($E''$), 2019 ($E'$), 2006 ($A_1'$), 2000 ($E'$), and 1989 ($E'$) cm$^{-1}$.

According to X-ray analysis,[519] $Co_4(CO)_{12}$ takes the $\mathbf{C}_{3v}$ structure (XI of Fig. III-27), for which six terminal and two bridging $\nu(CO)$ are expected to be infrared active. Bor[502] found that the observed spectrum is in good agreement with this prediction. Stammreich et al.[520] proposed structure XII of $\mathbf{D}_{3d}$ symmetry from a Raman study of $M[Co(CO)_4]_2$ (M = Cd or Hg). For this structure, three $\nu$(CO) are Raman active and the other three are infrared active. The former were observed at 2107 ($A_{1g}$), 2030 ($A_{1g}$), and 1990 ($E_g$) $cm^{-1}$.[520] and the latter were located at 2072 ($A_{2u}$), 2022 ($A_{2u}$), and 2007 ($E_u$) $cm^{-1}$.[521] Ziegler et al.[522] made complete vibrational assignments of the $M[Co(CO)_4]_2$ series, where M is Zn, Cd, and Hg.

As stated before, Shriver et al.[523] discovered that the O atom of the bridging CO group can form a bond with a Lewis acid such as $AlEt_3$. Kristoff and Shriver[483] found that $Co_2(CO)_8$ forms an adduct of the following type:

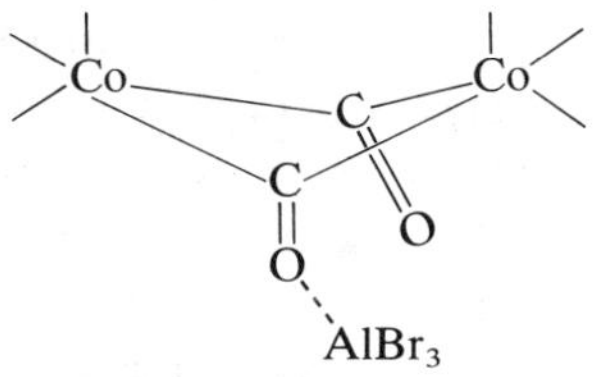

Co Co C C O O $AlBr_3$

As expected from this structure, the adduct exhibits two bridging $\nu$(CO) in the infrared: one at 1867 $cm^{-1}$, which is 15 $cm^{-1}$ higher, and the other at 1600 $cm^{-1}$, which is 232 $cm^{-1}$ lower, than that of the parent compound. In the case of $Fe_2(CO)_9AlBr_3$, only one bringing $\nu$(CO) is observed at 1557 $cm^{-1}$. This suggests the following structure, which resulted from rearrangement of the CO groups of the parent compound:

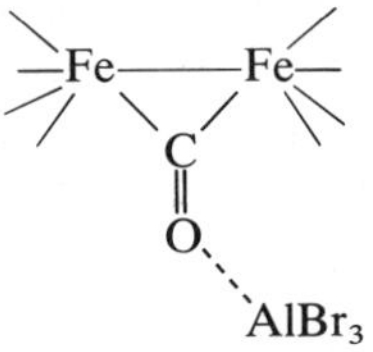

Fe Fe C O $AlBr_3$

### (3) Metal Carbonyls Containing Other Ligands

Carbonyl halides exhibit bands characteristic of both M—CO and M—X (X: a halogen) groups. The MX vibrations will be discussed in Sec. III-16. If the CO group is substituted by a halogen, $\nu$(CO) tends to shift to a higher frequency, since the M–CO $\pi$-back-bonding decreases as the metal becomes more electropositive by forming a M–X bond. In a series of halogenocarbonyls of the same type, $\nu$(CO) is highest for the chloro compound and lowest for the iodo compound, the bromo compound being between the two, as expected from their electronegativities. Table III-39 lists the observed frequencies for $\nu$(CO) and $\nu$(MX). Complete assignments have been made for all these compounds.

Noack[528] studied by infrared spectroscopy the reaction of $Fe(CO)_5$ with $Br_2$ below −80°C, and inferred the presence of two unstable products, $[Fe(CO)_5Br]Br$ and $Br—CO—Fe(CO)_4Br$, in $CHCl_3$-$CH_2Cl_2$ solutions. Farona and Camp[529] studied the same system at −70° to −30°C in pure $CH_2Cl_2$. They proposed a seven-coordinate pentagonal-bipyramidal structure, in which two halide atoms occupy *cis*-equatorial positions, or a capped octahedral structure. Anderson and Brown[530] predicted the infrared intensity of $\nu$(CO) bands in $M(CO)_5L$ and *fac*-$M(CO)_3L_3$ type compounds from M.O. calculations.

El-Sayed and Kaesz[531] studied the $\nu$(CO) of $M_2(CO)_8X_2$ (M = Mn, Tc, and Re; X = Cl, Br, and I), and proposed the halogen-bridging structure shown in Fig. III-28(I). Four infrared active $\nu$(CO) have been observed in accordance with this structure. Garland and Wilt[532] interpreted the infrared spectrum of $Rh_2(CO)_4X_2$ (X = Cl and Br) on the basis of the $\mathbf{C}_{2v}$

TABLE III-39. VIBRATIONAL FREQUENCIES OF METAL CARBONYL HALIDES ($CM^{-1}$)

| Compound | IR or Raman and Symmetry | $\nu$(CO) | $\nu$(MX) | Ref. |
|---|---|---|---|---|
| $Mn(CO)_5Cl$ | IR ($\mathbf{C}_{4v}$) | 2138 ($A_1$)<br>2056 ($E$)<br>2000 ($A_1$) | 291 ($A_1$) | 524 |
| $Mn(CO)_5Br$ | IR ($\mathbf{C}_{4v}$) | 2138 ($A_1$)<br>2052 ($E$)<br>2007 ($A_1$) | 222 ($A_1$)[a] | 525 |
| *fac*-$[Os(CO)_3Cl_3]^-$ | Raman ($\mathbf{C}_{3v}$) | 2125 ($A_1$)<br>2022 ($E$)<br>2033 ($E$) | 321 ($A_1$)<br>287 ($E$) | 526 |
| *cis*-$[Os(CO)_2Cl_4]^{2-}$ | Raman ($\mathbf{C}_{2v}$) | 2016 ($A_1$)<br>1910 ($B_2$) | 316 ($A_1$)<br>281 ($A_1$)<br>308 ($B_2$) | 526 |
| $[Os(CO)Cl_5]^{3-}$ | IR ($\mathbf{C}_{4v}$) | 1968 ($A_1$) | 332 ($A_1$)[a]<br>316 ($A_1$)[a]<br>306 ($E$) | 526 |
| $[Pt(CO)Cl_3]^-$ | IR ($\mathbf{C}_{2v}$) | 2120 ($A_1$) | 331 ($A_1$)<br>310 ($A_1$) | 526<br>527 |

[a] Raman frequency.

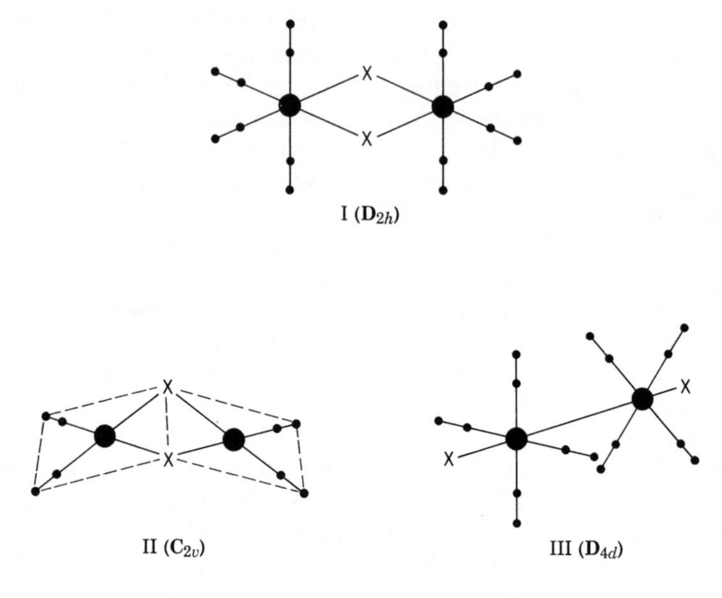


**Fig. III-28.** Structures of metal carbonyl halides.

structure [Fig. III-28(II)] found by X-ray analysis.[533] As predicted, three infrared active $\nu$(CO) have been observed for this compound. Johnson et al.[534] studied the exchange of $C^{18}O$ with CO groups of $Rh_2(CO)_4X_2$ (X = Cl, Br, I, etc.) with time by following the variation of infrared spectra in the $\nu$(CO) region. Cotton and Johnson[535] proposed the staggered structure (III) for $Fe_2(CO)_8I_2$, since only two $\nu$(CO) were observed in the infrared.

If CO is replaced by a phosphine, $\nu$(CO) decreases since the latter is a strong $\sigma$-donor but a weak $\pi$-acceptor. Ligands such as arsines, amines, and isonitriles give similar trends. Vibrational assignments have been made for $\nu$(CO) for most of these carbonyl derivatives. Table III-40 lists the observed $\nu$(CO) of typical compounds; the $\nu$(CO) of the last compound are very low (1778 and 1719 $cm^{-1}$).[540] Low frequency infrared studies have been reported for $M(CO)_{6-n}(PR_3)_n$ (M = Cr, Mo, and W),[541] $M(CO)_{6-n}(CH_3CN)_n$ (M = Cr and W: $n = 1$ and 2),[542] and $Fe(CO)_4L$ (L = $PPh_3$, $AsPh_3$, and $SbPh_3$).[543] For phosphine and arsine complexes, see Sec. III-18.

### (4) Normal Coordinate Calculations

Normal coordinate analyses on metal carbonyl compounds have been carried out by many investigators. Among them, Jones and co-workers

TABLE III-40. CO STRETCHING FREQUENCIES OF METAL CARBONYLS CONTAINING OTHER LIGANDS ($cm^{-1}$)

| Compound | IR or Raman and Symmetry | $\nu$(CO) | Ref. |
|---|---|---|---|
| $Ni(CO)_3(PMe_3)$ | Raman ($\mathbf{C}_{3v}$) | 2069 ($A_1$), 1980 ($E$) | 536 |
| $Fe(CO)_4(PMe_3)$ | Raman ($\mathbf{C}_{3v}$) | 2051 ($A_1$), 1967 ($A_1$) 1911 ($E$) | 537 |
| $Fe(CO)_4(AsMe_3)$ | Raman ($\mathbf{C}_{3v}$) | 2050 ($A_1$), 1964 ($A_1$) 1911 ($E$) | 537 |
| $Co(CO)_5(PEt_3)$ | IR ($\mathbf{C}_{4v}$) | 2060 ($A_1$), 1973 ($B_1$) 1943 ($A_1$), 1935 ($E$) | 538 |
| $W(CO)_5(NMe_3)$ | IR ($\mathbf{C}_{4v}$) | 2073 ($A_1$), 1932 ($E$) 1920 ($A_1$) | 539 |
| $W(CO)_4(bipy)$ | IR ($\mathbf{C}_{2v}$) | 2010 ($A_1$), 1900 ($B_1$) 1874 ($A_1$), 1832 ($B_2$) | 540 |
| $W(CO)_2(bipy)_2$ | IR ($\mathbf{C}_2$) | 1778 ($A_1$), 1719 ($B_2$) | 540 |

have made the most extensive study in this field. For example, they performed rigorous calculations on the $M(CO)_6$ (M = Cr, Mo, and W) series,[489] $Fe(CO)_5$,[488] and $Mn(CO)_5Br$,[525] including their $^{13}C$ and $^{18}O$ analogs. For the last compound, 5 stretching, 16 stretching–stretching interaction, and 33 bending–bending interaction constants (GVF) were used to calculate its 30 normal vibrations.

On the other hand, Cotton and Kraihanzel[544] developed an approximation (C–K) method for calculating the CO stretching and CO—CO stretching interaction constants, while neglecting all other low frequency modes. For $Mn(CO)_5Br$, they used only the five force constants[545] shown in Fig. III-29. Since only four CO stretching bands are observed for this type of compound, it was assumed that $\frac{1}{2}k_t = k_c = k_d$ holds. This was justified on the basis of the symmetry properties of the metal $d\pi$ orbitals involved. This C–K method has since been applied to many other carbonyls in making band assignments, in interpreting intensity data, and in discussing the bonding schemes of metal carbonyls.[481] It is clear that the choice of a rigorous approach (Jones) or a simplified method (C—K) depends on the availability of observed data and the purpose of the investigation. Jones[546] and Cotton[547] discuss the merits of their respective approaches relative to the alternative.

As mentioned earlier, $\nu$(CO) of metal carbonyls are determined by two factors: (1) donation of the $5\sigma$-electrons to the empty metal orbital tends to raise the $\nu$(CO) since the $5\sigma$ orbital is slightly antibonding, and (2)

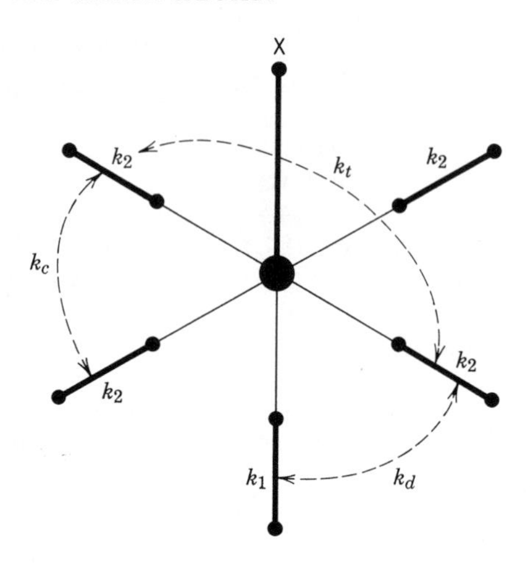


**Fig. III-29.** Definition of force constants for $M(CO)_5X$.

back donation of metal $d\pi$-electrons to the $2p\pi$ orbitals of CO tends to lower $\nu$(CO) since the $2p\pi$ orbitals are antibonding. Vibrational spectroscopy does not allow observation of these two effects separately since the observed $\nu$(CO) and the corresponding force constant reflect only the net result of the two counteracting components. It is possible, however, to correlate the CO stretching force constants (C–K) with the occupancies of the $5\sigma$ and $2p\pi$ orbitals as calculated by M.O. theory. Table III-41 lists the results obtained for $d^6$ carbonyl halides and dihalides by Hall and Fenske.[548] It is interesting that the *trans* CO in $Fe(CO)_4I_2$ and the *cis* CO in $Mn(CO)_5Cl$ have almost the same force constants since the $5\sigma$ occupancy of the former is smaller by 0.102 than that of the latter, while the $2p\pi$ occupancy of the former is larger by 0.108 than that of the latter. It is also noteworthy that the *trans*-CO in $Fe(CO)_4I_2$ and the *cis*-CO in $Cr(CO)_5Cl^-$ have identical $2p\pi$ occupancies (0.537) but substantially different force constants (17.43 and 15.58 mdyn/Å, respectively). In this case, the difference in force constants originates in the difference in the $5\sigma$ occupancies (1.293 vs. 1.457). Hall and Fenske[548] found a linear relationship between the C–K CO stretching force constants and the occupancies of the $5\sigma$ and $2p\pi$ levels:

$$k = -11.73[2\pi_x + 2\pi_y + (0.810)5\sigma] + 35.81$$

A similar attempt has been made for a series of Mn carbonyls containing isocyanide groups.[549]

TABLE III-41. CARBONYL ORBITAL OCCUPANCIES[a] AND FORCE CONSTANTS

| Compound | Structure | $5\sigma$ | $2\pi_x$ | $2\pi_y$ | $k$(mdyn/Å)[b] |
|---|---|---|---|---|---|
| $Cr(CO)_5Cl^-$ | *trans* | 1.407 | 0.355 | 0.355 | 14.07 |
| $Cr(CO)_5Br^-$ | *trans* | 1.405 | 0.353 | 0.353 | 14.10 |
| $Mn(CO)_4I_2^-$ | *trans* | 1.354 | 0.302 | 0.330 | 15.48 |
| $Mn(CO)_4IBr^-$ | *trans* | 1.355 | 0.302 | 0.327 | 15.48 |
| $Mn(CO)_4Br_2^-$ | *trans* | 1.357 | 0.302 | 0.325 | 15.50 |
| $Cr(CO)_5Br^-$ | *cis* | 1.456 | 0.261 | 0.282 | 15.56 |
| $Cr(CO)_5Cl^-$ | *cis* | 1.457 | 0.261 | 0.276 | 15.58 |
| $Mn(CO)_5Cl$ | *trans* | 1.352 | 0.286 | 0.286 | 16.28 |
| $Mn(CO)_5Br$ | *trans* | 1.350 | 0.286 | 0.286 | 16.32 |
| $Mn(CO)_5I$ | *trans* | 1.349 | 0.286 | 0.286 | 16.37 |
| $Mn(CO)_4I_2^-$ | *cis* | 1.402 | 0.251 | 0.251 | 16.75 |
| $Mn(CO)_4IBr^-$ | *cis* | 1.404 | 0.241 | 0.252 | 16.77 |
| $Mn(CO)_4Br_2^-$ | *cis* | 1.406 | 0.242 | 0.242 | 16.91 |
| $Mn(CO)_5I$ | *cis* | 1.394 | 0.213 | 0.240 | 17.29 |
| $Mn(CO)_5Br$ | *cis* | 1.394 | 0.212 | 0.228 | 17.39 |
| $Fe(CO)_4I_2$ | *trans* | 1.293 | 0.252 | 0.285 | 17.43 |
| $Mn(CO)_5Cl$ | *cis* | 1.395 | 0.211 | 0.218 | 17.46 |
| $Fe(CO)_4Br_2$ | *trans* | 1.295 | 0.250 | 0.272 | 17.53 |
| $Fe(CO)_5Br^+$ | *trans* | 1.287 | 0.233 | 0.233 | 17.93 |
| $Fe(CO)_5Cl^+$ | *trans* | 1.289 | 0.233 | 0.233 | 17.95 |
| $Fe(CO)_4I_2$ | *cis* | 1.337 | 0.221 | 0.221 | 17.95 |
| $Fe(CO)_4Br_2$ | *cis* | 1.338 | 0.205 | 0.205 | 18.26 |
| $Fe(CO)_5Cl^+$ | *cis* | 1.325 | 0.171 | 0.177 | 18.99 |
| $Fe(CO)_5Br^+$ | *cis* | 1.325 | 0.171 | 0.193 | 19.00 |

[a] The *cis* and *trans* designations of the CO groups are made with respect to the position of the halogen or halogens,
[b] C–K force constants (see Ref. 548).

**(5) Hydrocarbonyls**

Hydrocarbonyls exhibit bands characteristic of both M—H and M—CO groups. Kaesz and Saillant[550] reviewed the vibrational spectra of metal carbonyls containing the hydrido group. Vibrational spectra of hydrido complexes containing other groups will be discussed in Sec. III-13. In general, the terminal M—H group exhibits a relatively sharp and medium intensity $\nu$(MH) band in the 2200–1800 $cm^{-1}$ region. The MH stretching band can be distinguished easily from the CO stretching band by the deuteration experiment.

Edgell and co-workers[551] assigned the infrared bands at 1934 and 704 $cm^{-1}$ of $HCo(CO)_4$ to $\nu$(CoH) and $\delta$(CoH), respectively, and pro-

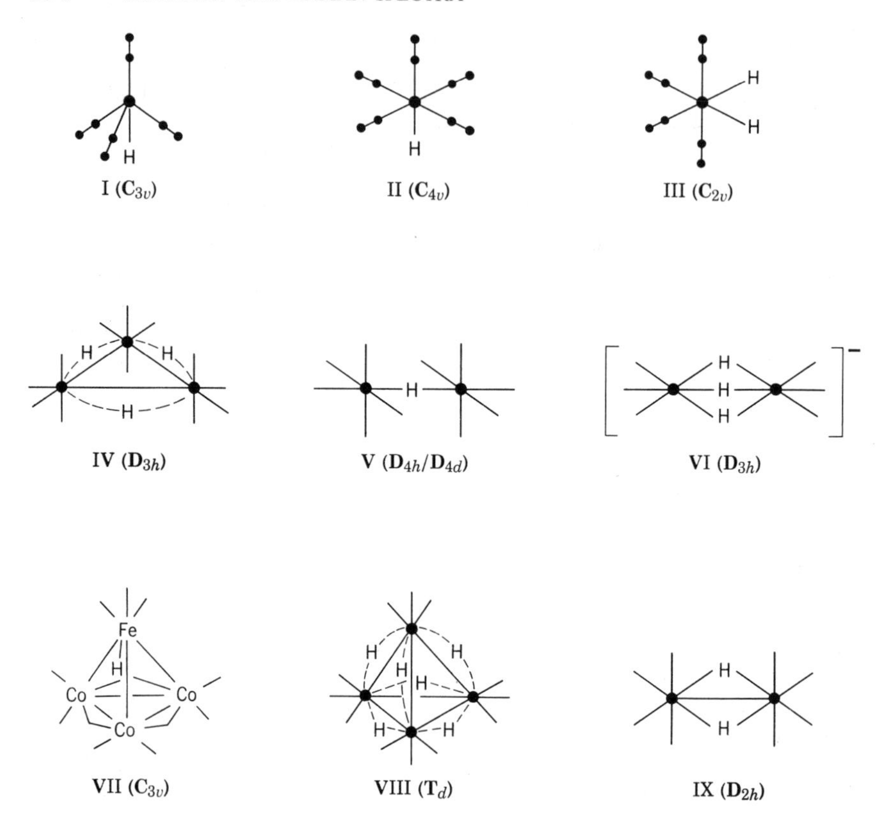


**Fig. III-30.** Structures of hydrocarbonyls.

posed structure I of Fig. III-30, in which the H atom is on the $C_3$ axis. Stammreich et al.[486] reported the Raman spectrum of $HFe(CO)_4^-$, which is expected to have a structure similar to that of $HCo(CO)_4$. According to X-ray analysis,[552] the $Mn(CO)_5$ skeleton of $HMn(CO)_5$ takes the $\mathbf{C}_{4v}$ structure shown in Fig. III-30(II). Kaesz and co-workers[553,554] assigned the infrared spectrum of $HMn(CO)_5$ in the $\nu(CO)$ region on the basis of this structure. The Raman spectra of $HMn(CO)_5$ and $HRe(CO)_5$ exhibit their $\nu(MH)$ at 1780 and 1824 $cm^{-1}$, respectively.[555] A complete vibrational assignment of gaseous $HMn(CO)_5$ has been made by Edgell et al.[556] The infrared spectrum of $H_2Fe(CO)_4$ in hexane at $-78°C$ exhibits three or more $\nu(CO)$ above 2000 $cm^{-1}$ and a weak, broad $\nu(FeH)$ at 1887 $cm^{-1}$. Thus Farmery and Kilner[557] suggested structure III of Fig. III-30. Table III-42 lists the observed frequencies of other hydrocarbonyl compounds.

It is rather difficult to locate the bridging $\nu(MH)$ in polynuclear

TABLE III-42. VIBRATIONAL FREQUENCIES OF METAL HYDROCARBONYL COMPOUNDS ($cm^{-1}$)[a]

| Compound | $\nu$(CO) | $\nu$(MH) | $\delta$(MH) | Ref. |
|---|---|---|---|---|
| $RhH(CO)(PPh_3)_3$ | 1926 | 2004 | 784 | 558 |
| $IrH(CO)(PPh_3)_3$ | 1930 | 2068 | 822 | 558 |
| $IrHCl_2(CO)(PEt_2Ph)_2$ | 2101 | 2008 | — | 559 |
| $IrHBr_2(CO)(PEt_2Ph)_2$ | 2035 | 2232 | — | 559 |
| $IrHCl_2(CO)(PPh_3)_2$ | 2027 | 2240 | — | 560 |
| $OsHCl(CO)(PPh_3)_3$ | 1912 | 2097 | — | 560 |
| $OsH_2(CO)_2(PPh_3)_2$ | 2014 | 1928 | — | 561 |
| | 1990 | 1873 | | |

[a] For the configurations of these molecules, see the original references.

hydrocarbonyls. These vibrations appear at ca. 1100 $cm^{-1}$ and are extremely broad since the H atom with its large vibrational amplitude interacts strongly with its environment. In some cases, the bridging $\nu$(MH) can be observed more easily in the Raman than in the infrared. Huggins et al.[562] were the first to suggest the presence of bridging hydrogens in $Re_3H_3(CO)_{12}$ (structure IV of Fig. III-30) since no terminal $\nu$(ReH) bands were observed. Smith et al.[563] observed a very weak and broad band at 1100 $cm^{-1}$ in the Raman spectrum of $Re_3H_3(CO)_{12}$ and assigned it to the bridging $\nu$(ReH) since it shifted to 787 $cm^{-1}$ upon deuteration. Hayter[564] proposed structure V for $[M_2H(CO)_{10}]^-$ (M = Cr, Mo, and W), and Ginsberg and Hawkes[565] suggested structure VI for $[Re_2H_3(CO)_6]^-$ since they could not observe any terminal $\nu$(ReH) vibrations.

The bridging $\nu$(FeH) band of $FeHCo_3(CO)_{12}$ in the infrared was finally located at 1114 $cm^{-1}$ by Mays and Simpson,[566] using a highly concentrated KBr pellet. This band shifts to 813 $cm^{-1}$ upon deuteration. On the basis of mass spectroscopic and infrared evidence, they proposed structure VII, in which the H atom is located inside the metal atom cage. From the spectra in the $\nu$(CO) region, together with X-ray evidence, Kaesz et al.[567] proposed the $\mathbf{T}_d$ skeleton (structure VIII) for $[Re_4H_6(CO)_{12}]^{2-}$. It showed no terminal $\nu$(ReH), but a broad bridging $\nu$(ReH) centered at 1165 $cm^{-1}$ was observed in its Raman spectrum. This band shifts to 832 $cm^{-1}$ with less broadening upon deuteration. Bennett et al.[568] found no terminal $\nu$(ReH) in the infrared spectrum of $Re_2H_2(CO)_8$. However, its Raman spectrum exhibits bands at 1382 and 1272 $cm^{-1}$, which shift to 974 and 924 $cm^{-1}$, respectively, upon deuteration. The $\mathbf{D}_{2h}$ structure (IX) was proposed for this compound.

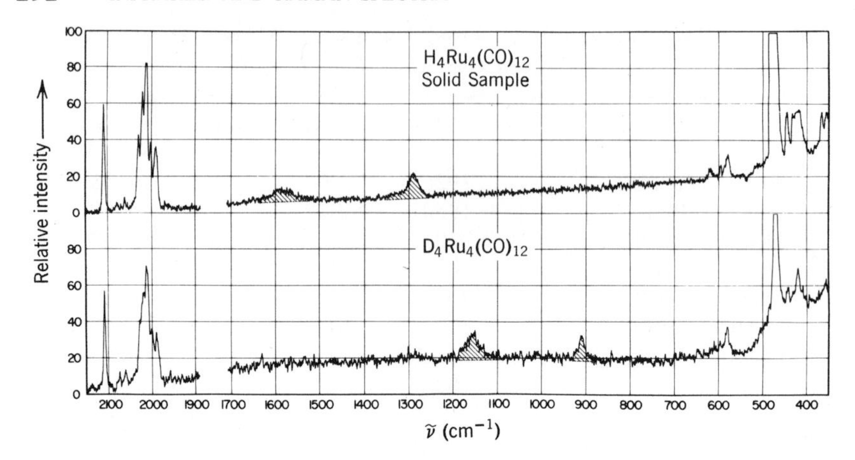


**Fig. III-31.** Raman spectra of $Ru_4H_4(CO)_{12}$ and its deuterated analog.[569]

Figure III-31 shows the Raman spectra of $Ru_4H_4(CO)_{12}$ and $Ru_4D_4(CO)_{12}$ obtained by Knox et al.[569] Two $\nu$(RuH) bands at 1585 and 1290 cm$^{-1}$ of the former compound are shifted to 1153 and 909 cm$^{-1}$, respectively, upon deuteration. Its infrared spectrum exhibits five $\nu$(CO) instead of the two expected for $\mathbf{T}_d$ symmetry. Thus a structure of $\mathbf{D}_{2d}$ symmetry, which lacks two H atoms from structure VIII, was proposed.[570]

As stated in Sec. III-5 (aquo complexes), the inelastic neutron scattering (INS) technique is very effective in locating hydrogen vibrations. White and Wright[571] found two hydrogen vibrations at 608 and 312 cm$^{-1}$ in the INS spectrum of $Mn_3H_3(CO)_{12}$. However, the nature of these vibrations is not clear.

## (6) Metal Carbonyls in Inert Gas Matrices

A number of unstable and transient metal carbonyls have been synthesized and their structures determined by vibrational spectroscopy in inert gas matrices. In most cases, only $\nu$(CO) vibrations have been measured to determine the structures of these compounds since it is rather difficult to observe low frequency modes in inert gas matrices.

These transient carbonyls can be prepared by two methods. The first involves direct reaction of metal vapor with CO diluted in inert gas matrices: $M + xCO \rightarrow M(CO)_x$. As discussed in Sec. I-21, DeKock[572] first succeeded in preparing the $Ni(CO)_x$ ($x = 1$–$3$) series by this method. Although the method yields a mixture of carbonyls of various stoichiometries, bands characteristic of each species can be determined in several ways: warm-up experiments, concentration dependence studies, isotope substitutions, and so on.[573] Table III-43 lists the number of infrared active $\nu$(CO) predicted for possible structures. The structures of $M(CO)_2$, $M(CO)_3$, and $M(CO)_4$ (M = Ni,[572] Pd,[574,575] and Pt[576]) have been found to be linear, trigonal-planar, and tetrahedral, respectively, since all of these compounds exhibit only one $\nu$(CO). In the case of the $M(CO)_{1-6}$ series (M = Ta.[572] U,[577] Pr, etc.[578]), it was more difficult to determine the structures of the transient species because the spectra were more complicated.

The second method utilizes *in situ* photolysis of stable metal carbonyls in inert gas matrices. For example, Poliakoff and Turner[579] carried out UV photolysis of $^{13}$CO-enriched $Fe(CO)_5$ in $SF_6$ and Ar matrices [$Fe(CO)_5 \xrightarrow{h\nu} Fe(CO)_4 + CO$], and concluded that the structure of $Fe(CO)_4$ is $\mathbf{C}_{2v}$ since it exhibits four $\nu$(CO) (two $A_1 + B_1 + B_2$) in the infrared spectrum. Graham et al.[580] proposed the $\mathbf{C}_{4v}$ structure for $Cr(CO)_5$ produced by the photolysis of $Cr(CO)_6$ in inert gas matrices. On the other hand, Kündig and Ozin[581] proposed the $\mathbf{D}_{3h}$ structure for $Cr(CO)_5$ prepared by cocondensation of Cr atoms with CO in inert gas matrices. They derived a general rule that $M(CO)_5$ species take the $\mathbf{D}_{3h}$ structure when the number of valence shell electrons is even [Cr (16), Fe (18)], and the $\mathbf{C}_{4v}$ structure when it is odd [V (15), Mn (17)]. Other examples of *in situ* photolysis are the following: $Co(CO)_4(NO) \xrightarrow{h\nu} Co(CO)_4(\mathbf{C}_{3v}) + NO$,[582] and $Fe_2(CO)_9 \xrightarrow{h\nu} Fe_2(CO)_8$ (bridging + nonbridging isomers) + CO.[583]

Carbonyl complexes of the type $MX_2CO$ are formed by reacting metal halide vapor directly with CO in inert gas matrices.[584,585] In this case, $\nu$(CO) shifts to higher frequencies by complexation, since the bonding is

TABLE III-43. NUMBER OF INFRARED ACTIVE CO STRETCHING VIBRATIONS FOR $M(CO)_x$

| Molecule | Symmetry and Structure | | IR Active $\nu$(CO) |
|---|---|---|---|
| M(CO) | $\mathbf{C}_{\infty v}$ | linear | $\Sigma^+$ |
| $M(CO)_2$ | $\mathbf{D}_{\infty h}$ | linear | $\Sigma_u^+$ |
| $M(CO)_2$ | $\mathbf{C}_{2v}$ | bent | $A_1 + B_2$ |
| $M(CO)_3$ | $\mathbf{D}_{3h}$ | trigonal-planar | $E'$ |
| $M(CO)_3$ | $\mathbf{C}_{3v}$ | trigonal-pyramidal | $A_1 + E$ |
| $M(CO)_4$ | $\mathbf{T}_d$ | tetrahedral | $F_2$ |
| $M(CO)_4$ | $\mathbf{D}_{4h}$ | tetragonal-pyramidal | $E_u$ |
| $M(CO)_5$ | $\mathbf{C}_{4v}$ | tetragonal-pyramidal | $2A_1 + E$ |
| $M(CO)_5$ | $\mathbf{D}_{3h}$ | trigonal-bipyramidal | $A_2'' + E'$ |
| $M(CO)_6$ | $\mathbf{O}_h$ | octahedral | $F_{1u}$ |

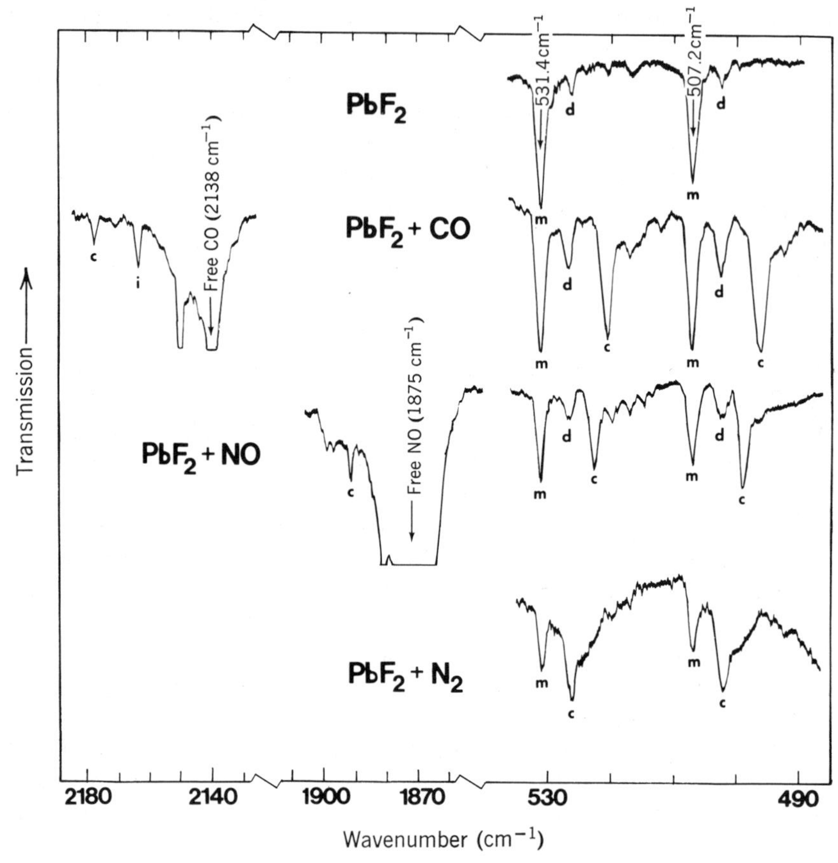


**Fig. III-32.** Infrared spectra of $PbF_2$, $PbF_2CO$, $PbF_2NO$, and $PbF_2N_2$ in argon matrices: m, monomeric $PbF_2$; d, dimeric $PbF_2$; c, complex; i, impurity(HF—CO).

dominated by the donation of $\sigma$-electrons to the metal. On the other hand, $\nu(MX)$ shifts to lower frequencies because the oxidation state of the metal is lowered by accepting $\sigma$-electrons from CO. Figure III-32 shows infrared spectra of the $PbF_2$—L system (L = CO, NO, and $N_2$) in Ar matrices obtained by Tevault and Nakamoto.[585] In this series, the magnitudes of the shifts of the $PbF_2$ and L stretching bands ($cm^{-1}$) relative to the free state are as follows:

| | $PbF_2CO$ | $PbF_2NO$ | $PbF_2N_2$ |
|---|---|---|---|
| $\nu_s(PbF_2)$ | −10.8 | −8.8 | −5.8 |
| $\nu_a(PbF_2)$ | −10.9 | −8.5 | −5.0 |
| $\nu(L)$ | +38.4 | +16.4 | — |

This result definitely indicates that CO is the best, NO is the next best, and $N_2$ is the poorest $\sigma$-donor.

Other work involves the direct deposition of stable carbonyls in inert gas matrices, mainly to study the effect of matrix environments on the structure. Both $Fe(CO)_5$[586] and $M_3(CO)_{12}$ (M = Ru and Os)[587] were found to be distorted from $\mathbf{D}_{3h}$ symmetry in inert gas matrices. If a thick deposit is made on a cryogenic window while maintaining a relatively high sample/inert gas dilution ratio, it is possible to observe low frequency modes such as $\nu$(MC) and $\delta$(MCO). It was found that these bands show splittings due to the mixing of metal isotopes. For example, the $F_{1u}$ $\nu$(CrC) of $Cr(CO)_6$ in a $N_2$ matrix exhibits four bands due to $^{50}Cr$, $^{52}Cr$, $^{53}Cr$, and $^{54}Cr$ (see Fig. I-20). The magnitude of these isotope splittings may be used to estimate the degree of the $\nu$(MC)–$\delta$(MCO) mixing in the low frequency vibrations.[588]

## (7) Nitrosyl (NO) Complexes

Many review articles[589–594] are available for nitrosyl complexes. Like CO, NO acts as a $\sigma$-donor and a $\pi$-acceptor. NO contains one more electron than CO, and this electron is in the $2p\pi^*$ orbital. The loss of this electron gives the nitrosonium ion ($NO^+$), which is much more stable than NO. Thus the $\nu$(NO) of the nitrosonium ion (ca. 2200 $cm^{-1}$) is much higher than that of the latter (1876 $cm^{-1}$). In nitrosyl complexes, $\nu$(NO) ranges from 1900 to 1500 $cm^{-1}$. Recent X-ray studies on nitrosyl complexes have revealed the presence of linear and bent M—NO groups:

M—N≡O: (I) M—N̈=Ö (II)

In the valence-bond theory, the hybridizations of the N atom in (I) and (II) are *sp* and $sp^2$, respectively. If the pair of electrons forming the M–N bond is counted as the ligand electrons, the nitrosyl groups in (I) and (II) are regarded as $NO^+$ and $NO^-$, respectively. Thus, one is tempted to correlate $\nu$(NO) with the charge on NO and the MNO angle. It was not possible, however, to find simple relationships between them since $\nu$(NO) is governed by several other factors (electronic effects of other ligands, nature of the metal, structure and charge of the whole complex etc.)[593] According to X-ray analysis, $RuCl(NO)_2(PPh_3)_2PF_6$ contains one linear and one bent M—NO group which absorb at 1845 and 1687 $cm^{-1}$,

TABLE II-44. VIBRATIONAL FREQUENCIES OF NITROSYL COMPLEXES ($cm^{-1}$)

| Compound | $\nu$(NO) | $\nu$(MN) | $\delta$(MNO) | Ref. |
|---|---|---|---|---|
| $Cr(NO)_4$ | 1721 | 650 | 496 | 597 |
| $Co(NO)_3$ | 1860, 1795 | — | — | 598 |
| $Co(CO)_3(NO)$ | 1822 | 609 | 566 | 599 |
| $Mn(CO)_4(NO)$ | 1781 | 524 | 657 | 600 |
| $Mn(PF_3)(NO)_3$ | 1836, 1744 | — | — | 601 |
| *cis*-$[MoCl_4(NO)_2]^{2-}$ | 1720, 1600 | — | — | 602 |
| $NiCl_2(NO)_2$ | 1872, 1842 | — | — | 603 |
| $[RuCl_5(NO)]^{2-}$ | 1904 | 606 | 588 | 604 |
| $[RuBr_5(NO)]^{2-}$ | 1870 | 572 | 300 | 605 |

respectively.[595] $CoCl_2(NO)L_2$ [L = $P(CH_3)Ph_2$] exists in two isomeric forms:

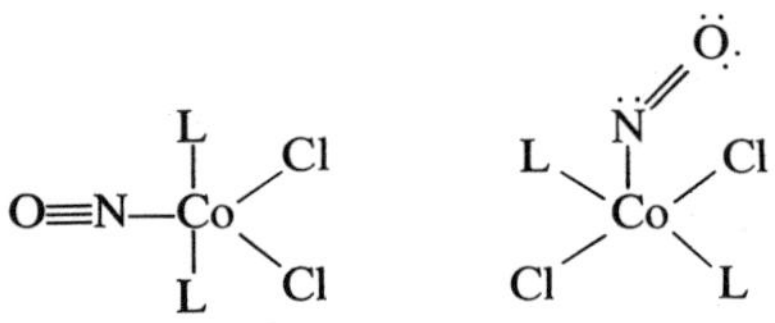

The $\nu$(NO) of the former is at 1750 $cm^{-1}$, whereas that of the latter is at 1650 $cm^{-1}$ [596]

Table III-44 lists the vibrational frequencies of typical nitrosyl complexes. Although the M—NO group is expected to show $\nu$(NO), $\nu$(MN), and $\delta$(MNO), only $\nu$(NO) have been observed in most cases. The latter two modes are often coupled since their frequencies are close to each other. Jones et al.[599] carried out a complete analysis of the vibrational spectra of $Co(CO)_3(NO)$ and its $^{13}C$, $^{18}O$, and $^{15}N$ analogs. For $M(CN)_5(NO)$ type compounds, see Sec. III-10.

It has been known for many years that the complex of composition $[Co(NH_3)_5NO]X_2$ exists in two isomeric forms: a black salt ($X = Cl^-$) and a red salt ($X = NO_3^-$ or $Br^-$). According to X-ray analysis,[606–608] the black salt, $[Co(NH_3)_5NO]Cl_2$, is monomeric, and the red salt is dimeric with a hyponitrite bridge:

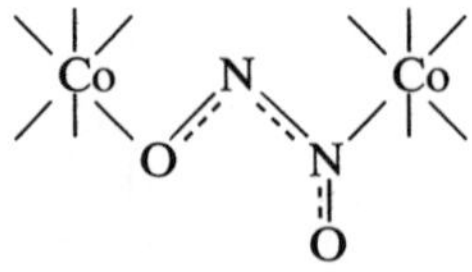

Mercer et al.[609] assigned the 1610 $cm^{-1}$ band of the black salt to $\nu(NO)$, and two bands at 1046 and 932 $cm^{-1}$ of the red salt to $\nu_a(NO)$ and $\nu_s$ (NO), respectively. For the infrared spectra of $[Ru(NH_3)_4(NO)X]X_2$ (X = Cl, Br, and I), see Ref. 610. Freshly prepared $[Fe(NH_3)_5NO]Cl_2$ exhibits $\nu(NO)$ at 1600 $cm^{-1}$, indicating $Fe^{3+}$—$NO^-$ type bonding, but after an hour $\nu(NO)$ appears at 1750 $cm^{-1}$, as is characteristic of $NO^+$ in a complex.[611] Mercer et al.[612] noted that, in a series of *trans*-$[Ru(NH_3)_4(NO)X]^{n+}$, $\nu(RuNO)$ decreases as the *trans*-influence of X becomes stronger:

$$X = Cl^- < NH_3 < Br^- < I^-$$
$$\nu(RuNO)(cm^{-1}),\ 608 > 602 > 591 > 572$$

According to X-ray analysis, the structure of $M_3(CO)_{10}(NO)_2$ (M = Ru and Os) resembles that of $Fe_3(CO)_{12}$ [Fig. III-27(VIII)] with double nitrosyl bridges in place of the double carbonyl bridges in the latter. As expected, $\nu(NO)$ of these nitrosyl groups are very low: 1517 and 1500 $cm^{-1}$ for the Ru compound, and 1503 and 1484 $cm^{-1}$ for the Os compound.[613]

## III-13. COMPLEXES OF MOLECULAR OXYGEN AND NITROGEN, AND NITRIDO AND HYDRIDO COMPLEXES

### (1) Molecular Oxygen ($O_2$) Complexes

A number of molecular oxygen complexes have been synthesized, and these are reviewed by several authors.[593,614,615] The $\nu(OO)$ of neutral $O_2$, $O_2^-$ (superoxide ion), and $O_2^{2-}$ (peroxide ion) are 1555, ca. 1143, and ca. 1060 $cm^{-1}$, respectively. Since simple M.O. theory predicts the bond orders of these species to be 2, $1\frac{1}{2}$, and 1, respectively, the vibrational frequencies are not directly proportional to these bond orders.[616] However, they are very useful in determining the oxidation state of $O_2$ in the complex irrespective of the mode of bonding. Thus superoxo complexes exhibit $\nu(OO)$ near 1100 $cm^{-1}$, whereas the majority of complexes that would be formulated as peroxo complexes show $\nu(OO)$ in the 900–800 $cm^{-1}$ region.

Structurally, molecular oxygen complexes are classified into three types:

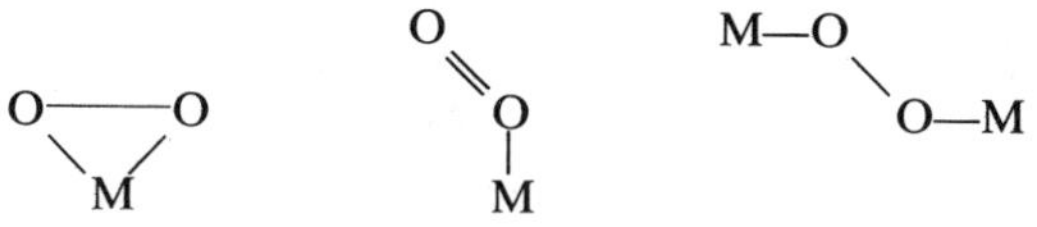

O—O, M (Symmetric) | O=O—M (Asymmetric) | M—O—O—M (Bridging)

Symmetric Asymmetric Bridging

TABLE III-45. OBSERVED FREQUENCIES OF MOLECULAR OXYGEN COMPOUNDS ($cm^{-1}$)

| Compound | $\nu(OO)$ | $\nu(MO)$ | Ref. |
|---|---|---|---|
| $Pt(O_2)(PPh_3)_2$ | 828 | 472 | 617 |
| $Ni(O_2)(t\text{-}BuNC)_2$ | 898 | 552 | 617 |
| $Pd(O_2)(t\text{-}BuNC)_2$ | 893 | 484 | 617 |
| $Rh(O_2)Cl(PPh_3)_2(t\text{-}BuNC)$ | 892 | 576 | 617 |
| $[VO(O_2)(H_2O)(bipicoline)]^-$ | 839 | 610–570 | 618 |
| $Ir(O_2)F(CO)(PPh_3)_2$ | 850 | — | 619 |
| $Ir(O_2)N_3(CO)(PPh_3)_2$ | 855 | — | 620 |

Complexes containing the symmetric type $O_2$ show $\nu(OO)$ in the 900–800 $cm^{-1}$ region (peroxo type). Several $^{18}O_2$ isotope studies show that $\nu(OO)$ couples strongly with $\nu_a(MO)$ and $\nu_s(MO)$ in the 600–400 $cm^{-1}$ region. Table III-45 lists the observed frequencies of typical complexes belonging to this class. Nakamura et al.[617] carried out normal coordinate analyses on several symmetric $O_2$ complexes, including their $^{18}O_2$ analogs. Dunn et al.[621] showed from the resonance Raman spectrum of oxyhemerythrin that the bound oxygen is in a peroxo type electronic state (844 $cm^{-1}$). A similar conclusion was drawn from a resonance Raman study of the hemocyanin–$O_2$ complex, which exhibits $\nu(OO)$ at 742 $cm^{-1}$.[622]

Originally, the $O_2$ coordination in $Co(O_2)(py)$(3-methoxysalen) was thought to be symmetrical, although it exhibits $\nu(OO)$ at 1140 $cm^{-1}$.[623] However, Diemente et al.[624] showed from an ESR study that its $O_2$ coordinates to the metal asymmetrically.

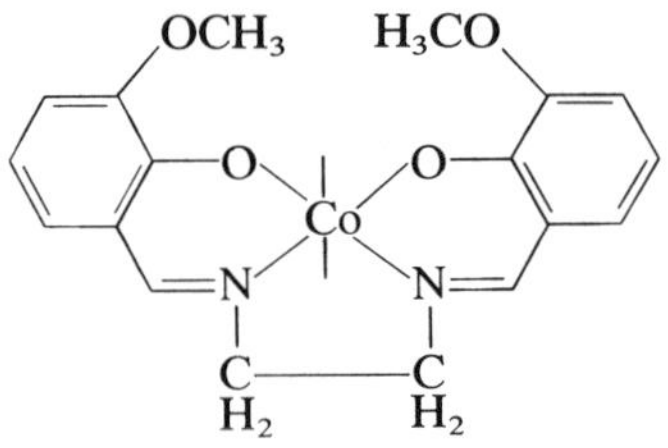

Co(3-methoxysalen)

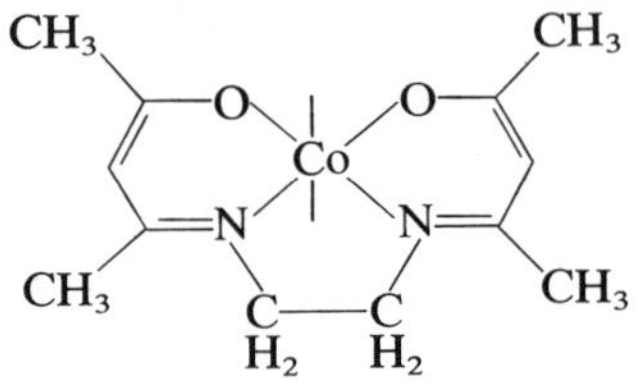

Co(acac-en)

Crumbliss and Basolo[625] observed strong bands at 1140–1120 $cm^{-1}$ [superoxo $\nu(OO)$] in a series of $Co(O_2)L$(acac-en) type complexes, where L is py, 4-methyl-py, and so on. Hoffman et al.[625a] showed from ESR studies that the $O_2$ molecules in these complexes also coordinate to the Co atom asymmetrically.

Dinuclear bridging complexes do not show $\nu(OO)$ in the infrared if it is centrosymmetric. If the molecule loses its center of symmetry at the O–O bond by distortion or for other reasons, $\nu(OO)$ may appear weakly. For example, the infrared spectra of $[(NH_3)_5Co(O_2)Co(NH_3)_5]^{n+}$ ($n = 4$ and 5) ions show no bands in the 1200–800 $cm^{-1}$ region.[626,627] On the other hand,

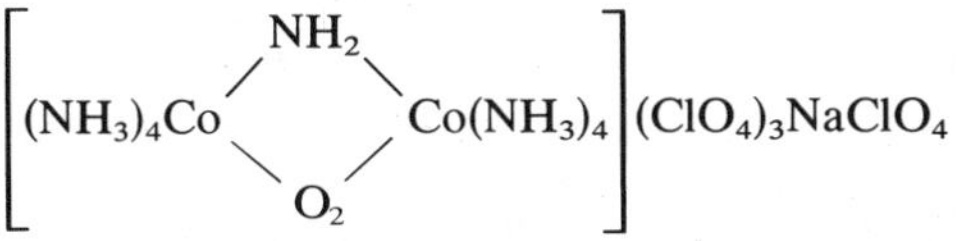

$$\left[(NH_3)_4Co\underset{O_2}{\overset{NH_2}{\diamond}}Co(NH_3)_4\right](ClO_4)_3NaClO_4$$

exhibits a weak band at 830 $cm^{-1}$, which was assigned to $\nu(OO)$.[628] However, $\nu(OO)$ is expected to appear strongly in the Raman. For example, a strong $\nu(OO)$ band was observed at 1122 $cm^{-1}$ in the Raman spectrum of paramagnetic $[(NH_3)_5Co(O_2)Co(NH_3)_5]Cl_5 \cdot 4H_2O$ (superoxo complex).[629] This band is shifted to 800 $cm^{-1}$ in the resonance Raman spectrum of diamagnetic $[(NH_3)_5Co(O_2)Co(NH_3)_5](NO_3)_4$ (peroxo complex).[630] Jere and Gupta[631] proposed the following structure for $Zr_2(O_2)_3SO_4 \cdot 10H_2O$:

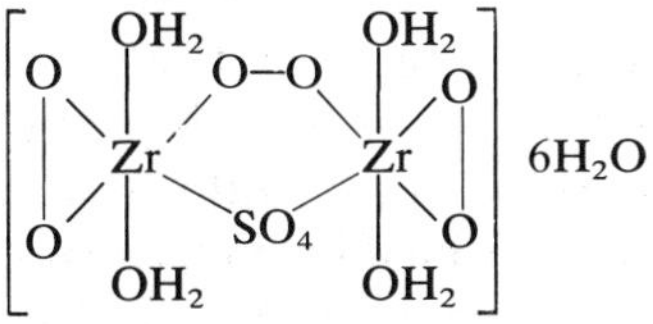

OH₂, O–O, OH₂, O, O, Zr, Zr, SO₄, O, O, OH₂, OH₂ — 6H₂O

$$\left[(O_2)(H_2O)_2Zr(\mu\text{-}O_2)(\mu\text{-}SO_4)Zr(H_2O)_2(O_2)\right]6H_2O$$

This compound exhibits three Raman bands at 872, 840, and 790 $cm^{-1}$ which are due to symmetrical and bridging $O_2$ groups.

### (2) Molecular Nitrogen ($N_2$) and Nitrido (N) Complexes

Since Allen and Senoff[632] prepared the first stable molecular nitrogen compounds, $[Ru(N_2)(NH_3)_5]X_2$ ($X = Br^-, I^-, BF_4^-$, etc.), a large number of molecular nitrogen compounds have been synthesized. The chemistry and spectroscopy of these compounds have been reviewed extensively.[594,633–636] The structures of molecular nitrogen compounds are classified into three types:

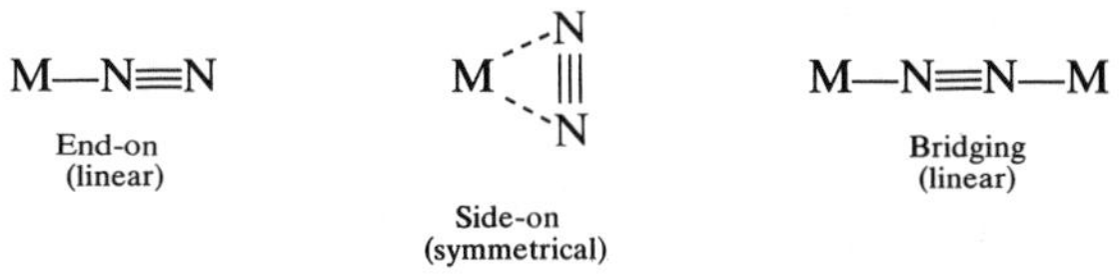

| M—N≡N | M(N≡N) | M—N≡N—M |
|---|---|---|
| End-on (linear) | Side-on (symmetrical) | Bridging (linear) |

The terminal end-on coordination is most common. The M–$N_2$ bonding is interpreted in terms of the $\sigma$-donation and $\pi$-back-bonding, which were discussed in Secs. III-10 and III-11. Since $N_2$ is a weaker Lewis base than CO, $\pi$-back-bonding may be more important in nitrogen complexes than in CO complexes.[637] Free $N_2$ exhibits $\nu(N{\equiv}N)$ at 2331 $cm^{-1}$, and this band shifts to 2220–1850 $cm^{+1}$ upon coordination to the metal. Table III-46 lists the $\nu(N{\equiv}N)$ of typical complexes. Very little information is available for $\nu(M{-}N_2)$ and $\delta(M{-}N{\equiv}N)$ in the low frequency region. Allen et al.[637] assigned $\nu(Ru{-}N_2)$ of $[Ru(N_2)(NH_3)_5]^{2+}$ type compounds in the 508–474 $cm^{-1}$ region, whereas other workers[638,639] attributed these bands to $\delta(Ru{-}N{\equiv}N)$. Figure III-33 shows the infrared spectrum of $[Ru(NH_3)_5N_2]Br_2$ obtained by Allen et al.[637]

According to Srivastava and Bigorgne,[647] $Co(N_2)H(PPh_3)_3$ exists in two forms in the solid state; one form exhibits $\nu(N{\equiv}N)$ at ca. 2087 $cm^{-1}$, and the other shows two bands of equal intensity at 2101 and 2085 $cm^{-1}$. However, their structural differences are unknown. Darensbourg[648] obtained a linear relationship between $\nu(N{\equiv}N)$ and the absolute integrated intensity in a series of molecular nitrogen compounds.

Armor and Taube[649] postulated the occurrence of the side-on structure as a possible transition state in linkage isomerization: $[(NH_3)Ru{-}^{14}N{\equiv}{}^{15}N]Br_2 \leftrightarrow [(NH_3)Ru{-}^{15}N{\equiv}{}^{14}N]Br_2$. Ozin and Vander Voet[650] confirmed the side-on symmetrical coordination in $Co(N_2)$ by the $^{14}N_2$–$^{15}N_2$ isotope scrambling experiment [see Sec. III-13(3)]. Jonas et al.[651] carried out X-ray analysis on $[\{(C_6H_5Li)_3Ni\}_2(N_2)\{(C_2H_5)_2O\}_2]_2$ and confirmed the presence of the side-on coordination in this compound; the N—N distance was found to be extremely long (1.35 Å). Thus far, no $\nu(N{\equiv}N)$

TABLE III-46. OBSERVED N≡N STRETCHING FREQUENCIES ($CM^{-1}$)

| Complex | $\nu(N{\equiv}N)$ | Ref. |
|---|---|---|
| $[Ru(N_2)(NH_3)_5]Br_2$ | 2105 | 638 |
| $[Ru(N_2)(NH_3)_5]I_2$ | 2124 | 639 |
| $[Os(N_2)(NH_3)_5]Cl_2$ | 2022, 2010 | 640 |
| $Co(N_2)(PPh_3)_3$ | 2093 | 641 |
| $Co(N_2)H(PPh_3)_3$ | 2105 | 642 |
| $Ir(N_2)Cl(PPh_3)_2$ | 2105 | 643 |
| *trans*-$Mo(N_2)_2(DPE)_2$ | 1970, (2020) | 644 |
| *cis*-$W(N_2)_2(PMe_2Ph)_4$ | 1998, 1931 | 645 |
| $Co(N_2)(PR_3)(PR_2)_2^{2-}$ [a] | 1904 ~ 1864 | 646 |

[a] R: *n*-Bu.

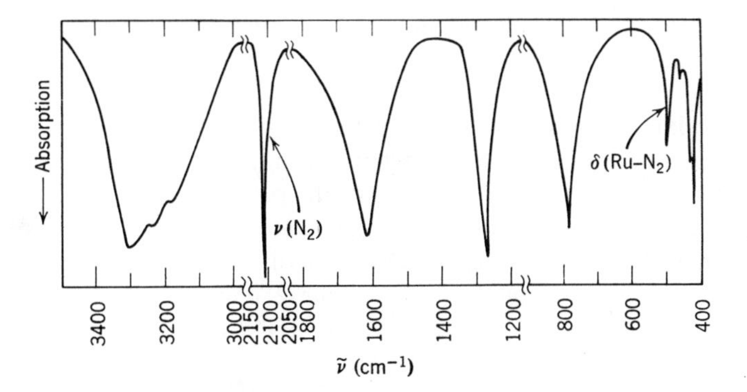


**Fig. III-33.** Infrared spectrum of $[Ru(NH_3)_5N_2]Br_2$.[637]

have been reported for complexes of the side-on structure except for $Co(N_2)$ (2101 cm$^{-1}$), which was prepared in an Ar matrix at ~10°K.

The bridging M—N≡N—M type complex should not show ν(N≡N) if it is linear. However, it may show a strong ν(N≡N) in the Raman spectrum. Thus $[\{Ru(NH_3)_5\}_2(N_2)]^{4+}$ shows no infrared bands in the 2220–1920 cm$^{-1}$ region, whereas a strong ν(N≡N) band appears at 2100 cm$^{-1}$ in the Raman.[652] If $N_2$ forms a bridge between two different metals, ν(N≡N) is observed in the infrared. For example, ν(N≡N) is at 1875 cm$^{-1}$ in the infrared spectrum of $[(PMe_2Ph)_4ClRe—N_2—CrCl_3(THF)_2]$.[653] According to X-ray analysis,[654] an analogous compound, $[(PMe_2Ph)_4ClRe—N_2—MoCl_4(OMe)]$, has a N≡N distance of 1.21 Å, and its ν(N≡N) is at 1660 cm$^{-1}$. As expected, the ν(N═N) of $[(CO)_5Cr—NH═NH—Cr(CO)_5]$ is very low (1415 cm$^{-1}$).[655]

If the $N^{3-}$ ion coordinates to a metal, it is called a nitrido complex. Nitrido complexes of transition metals can be prepared by several methods, and their preparations, structures, and spectra have been reviewed by Griffith.[656] The M≡N triple bonds are formed as a result of the strong π-donating property of the $N^{3-}$ ion. The ν(M≡N) of non-bridging nitrido complexes are in the 1200–950 cm$^{-1}$ region.[657] For example, ν(M≡N) of $[Mo(N)Cl_5]^{2-}$ is at 1023 cm$^{-1}$,[658] and those of $[M(N)X_5]^{2-}$ [M = Ru(VI) and Os(VI); X = $Cl^-$ and $Br^-$] are at 1120–1000 cm$^{-1}$.[659]

Dinuclear nitrido complexes of the types $[M_2(N)X_8(H_2O)_2]^{3-}$ and $[M_2(N)(NH_3)_8Y_2]^{3+}$ (M = Ru and Os; X = $Cl^-$ and $Br^-$; Y = $Cl^-$, $Br^-$, $NCS^-$, and $N_3^-$) contain the linear M—N—M bridging group, which exhibits $\nu_a(NM_2)$ and $\nu_s(NM_2)$ at 1120–1050 and 350–280 cm$^{-1}$, respectively. In trinuclear nitrido complexes containing the trigonal-planar $NIr_3$ unit, $\nu_a(NIr_3)$ and $\nu_s(NIr_3)$ are at 800–700 and ca. 230 cm$^{-1}$, respectively.

In both cases, the symmetric modes have been observed only in the Raman spectra.[657]

### (3) Molecular Oxygen and Nitrogen Compounds in Inert Gas Matrices

In addition to the molecular oxygen and nitrogen complexes described in the preceding sections, it is possible to prepare simple $M(O_2)_{1,2}$ and $M(N_2)_{1-4}$ type complexes by reacting metal atoms directly with the respective ligand in inert gas matrices. Ozin, Moskovits, and their co-workers have carried out extensive studies on these compounds. For details of their work, see Refs. 660 and 573. Table III-47 lists some representative compounds prepared by this method.

The $O_2$ and $N_2$ ligands may coordinate to a metal in the end-on or side-on fashion. These two structures can be distinguished by using the isotope scrambling technique. Andrews[661] first applied this method to the structure determination of the ion-pair complex $Li^+O_2^-$; a mixture of $^{16}O_2$, $^{16}O^{18}O$, and $^{18}O_2$ was prepared by Tesla coil discharge of a $^{16}O_2$–$^{18}O_2$ mixture, and reacted with Li vapor in an Ar matrix. Three $\nu(OO)$ were observed in the Raman spectrum:

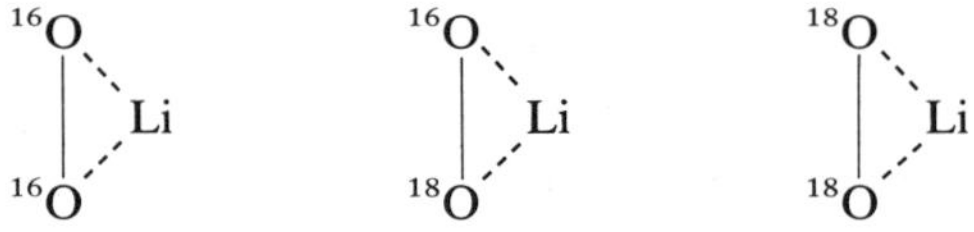

| $^{16}O$–$^{16}O$ (side-on to Li) | $^{16}O$–$^{18}O$ (side-on to Li) | $^{18}O$–$^{18}O$ (side-on to Li) |
|---|---|---|
| 1096.1 cm$^{-1}$ | 1065.7 cm$^{-1}$ | 1034.6 cm$^{-1}$ |

This result clearly indicates side-on coordination since four bands are expected for end-on coordination (see below). Using the same technique, Ozin and co-workers[662,663] showed that, in all cases they studied, $O_2$ coordinates to a metal in the side-on fashion and that, in $M(O_2)_2$ (M = Ni, Pd, and Pt), the complexes take the spiro $\mathbf{D}_{2d}$ structure.

TABLE III-47. TYPICAL COMPLEXES PREPARED BY METAL ATOM REACTION IN INERT GAS MATRICES

| Complex | Ref. | Complex | Refs. |
|---|---|---|---|
| $M(O_2)$, M = Li, Na, K | 661 | $Ni(N_2)_{1-4}$ | 664, 665 |
| $Ni(O_2)_{1,2}$ | 662 | $Pd(N_2)_{1-3}$ | 664, 666 |
| $Pd(O_2)_{1,2}$ | 662 | $Pt(N_2)_{1-3}$ | 667 |
| $Pt(O_2)_{1,2}$ | 662 | $Co(N_2)$ | 668 |
| $Cu(O_2)_2$ | 663 | $Rh(N_2)_{1-4}$ | 669 |
| $Ag(O_2)_2$ | 663 | $Ni(N_2)_m(CO)_{4-m}$, $m = 1$–3 | 670, 671 |
| $Au(O_2)$ | 663 | | |

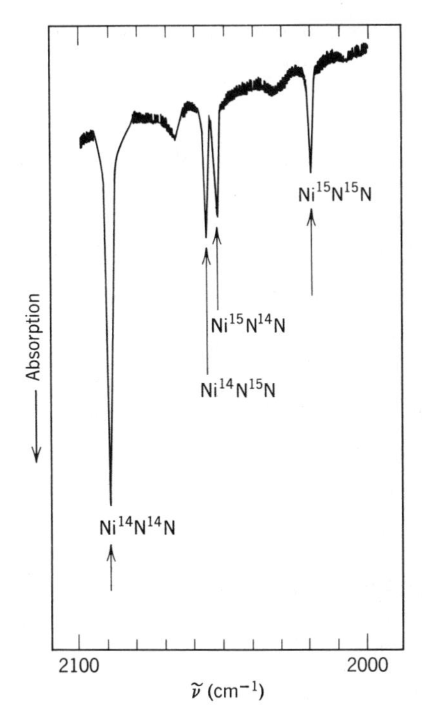


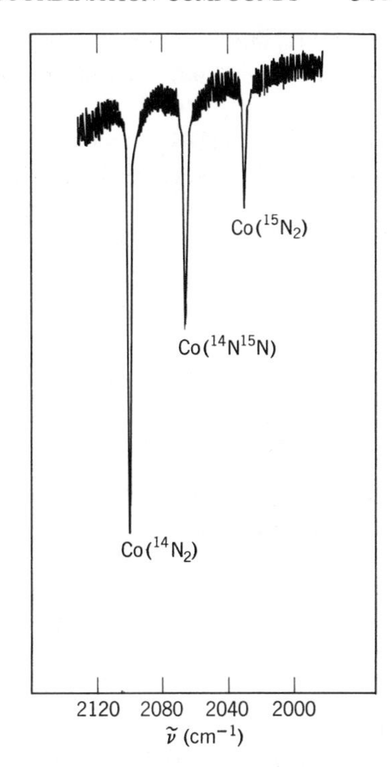


**Fig. III-34.** The matrix infrared spectra of the products of the condensation reactions of Ni and Co atoms with $^{14}N_2/^{14}N^{15}N/^{15}N_2/Ar$ at 10°K.[664,668]

Ozin and co-workers also applied the isotope scrambling technique to molecular nitrogen complexes. A mixture of $^{14}N_2$, $^{14}N^{15}N$, and $^{15}N_2$ prepared by Tesla coil discharge of a $^{14}N_2$–$^{15}N_2$ mixture was reacted with metal vapor. In all cases but $Co(N_2)$ they observed the four-peak pattern expected for end-on coordination. Figure III-34 shows the infrared spectra of $Ni(N_2)$ (end-on) and $Co(N_2)$ (side-on). The observed frequencies ($cm^{-1}$) and assignments of the four bands of the former are as follows:

| $Ni—^{14}N{\equiv}^{14}N$ | $Ni—^{14}N{\equiv}^{15}N$ | $Ni—^{15}N{\equiv}^{14}N$ | $Ni—^{15}N{\equiv}^{15}N$ |
|---|---|---|---|
| 2089.9 | 2057.4 | 2053.6 | 2020.6 |

Like analogous carbonyl complexes [Sec. III-12(6)], the structures of $M(N_2)_4$, $M(N_2)_3$, and $M(N_2)_2$ are tetrahedral, trigonal-planar, and linear, respectively, although slight distortion from these ideal symmetries occurs because of the matrix effect. Finally, it should be mentioned that most of these studies have been made in the high-frequency region [$\nu$(OO) and

$\nu$(N≡N)] since low-frequency vibrations are generally weak and difficult to measure in inert gas matrices.

**(4) Hydrido (H) Complexes**

As stated in Sec. III-12(5) (hydrocarbonyls), the terminal M—H group exhibits a relatively sharp band of medium intensity in the 2250–1700 cm$^{-1}$ region.[550] In addition, it shows $\delta$(MH) in the 800–600 cm$^{-1}$ region. These assignments can be confirmed by deuteration experiments. Table III-48 lists the observed M–H group frequencies of typical hydrido complexes.

The $\nu$(MH) is sensitive to other ligands, particularly those in the *trans*-position in square-planar Pt(II) complexes. Thus Chatt et al.[681] found that the order of $\nu$(PtH) in *trans*-[Pt(H)X($PEt_3$)$_2$] is as follows:

| X = | $NO_3^-$ | < | $Cl^-$ | < | $Br^-$ | < | $I^-$ | < | $NO_2^-$ | < | $SCN^-$ | < | $CN^-$ |
|---|---|---|---|---|---|---|---|---|---|---|---|---|---|
| $\nu$(PtH)(cm$^{-1}$) | 2242 | > | 2183 | > | 2178 | > | 2156 | > | 2150 | > | 2112 | > | 2041 |

This is the increasing order of *trans*-influence. Church and Mays[682] found that the NMR Pt—H coupling constant and $\nu$(PtH) decrease in the same order in the *trans*-[Pt(H)L($PEt_3$)$_2$]$^+$ series:

| L = | py | < | CO | < | P(Ph)$_3$ | < | P(OPh)$_3$ | < | P(OMe)$_3$ | < | $PEt_3$ |
|---|---|---|---|---|---|---|---|---|---|---|---|
| *J*(PtH)(Hz) | 1106 | > | 967 | > | 890 | > | 872 | > | 846 | > | 790 |
| $\nu$(PtH)(cm$^{-1}$) | 2216 | > | 2167 | > | 2100 | > | 2090 | > | 2067 | < | 2090 |

In the above series, the $\sigma$-donor strength of L increases as the *J*(PtH) value decreases and $\nu$(PtH) shifts to a lower frequency. Atkins et al.[683]

TABLE III-48. M—H FREQUENCIES OF TYPICAL HYDRIDO COMPLEXES (CN$^{-1}$)

| Complex | $\nu$(MH) | $\delta$(MH) | Ref. |
|---|---|---|---|
| Co(H)$_2$(PPh$_3$)$_3$ | 1755 | — | 672 |
| *mer*-Co(H)$_3$(PPh$_3$)$_3$ | 1933, 1745 | — | 673 |
| [Co(H)(CN)$_5$]$^{3-}$ | 1840 | 774 | 674 |
| [Ir(H)(CN)$_5$]$^{3-}$ | 2043 | 810 | 675 |
| *trans*-[Fe(H)(Cl){$C_2H_4$(PEt$_2$)$_2$}$_2$] | 1849 | 656 | 676 |
| *trans*-[Fe(H)$_2${o-$C_6H_4$(PEt$_2$)$_2$}$_2$] | 1726 | 716 | 676 |
| Ir(H)(Cl)$_2$(PPh$_3$)$_3$ (isomer I) | 2197 | 840, 804 | 677 |
| *cis*-[Ir(H)$_2$(CO)(PPh$_3$)$_3$] | 2160, 2107 | — | 678 |
| *cis*-[Ir(H)$_2$(Ph$_2$P-(CH$_2$)$_2$-PPh$_2$)$_2$]$^-$ | 2091, 2080 | — | 679 |
| *trans*-[Os(H)$_2${$C_2H_4$(PEt$_2$)$_2$}$_2$] | 1721 | — | 680 |

found linear relationships between the chemical shift of the hydride, the Pt—H coupling constant, $\nu$(PtH), and the p$K_a$ value of the parent carboxylic acid in a series of *trans*-[Pt(H)L($PEt_3)_2$], where L is a carboxylate ligand.

## III-14. COMPLEXES OF AMINO ACIDS, EDTA, AND RELATED COMPOUNDS

Amino acids exist as zwitterions in the crystalline state. Table III-49 gives band assignments made for the zwitterions of glycine[684] and $\alpha$-alanine.[685] According to X-ray analysis, two glycino anions (gly) in [Ni(gly)$_2$]·2$H_2O$,[686] for example, coordinate to the metal by forming a

TABLE III-49. INFRARED FREQUENCIES AND BAND ASSIGNMENTS OF GLYCINE AND $\alpha$-ALANINE IN THE CRYSTALLINE STATE ($cm^{-1}$)[684,685]

| Glycine | $\alpha$-Alanine | Band Assignment |
|---|---|---|
| 1610 | 1597 | $\nu_a(COO^-)$ |
| 1585 | 1623 | $\delta_d(NH_3^+)$ |
| 1492 | 1534 | $\delta_s(NH_3^+)$ |
| — | 1455 | $\delta_d(CH_3)$ |
| 1445 | — | $\delta(CH_2)$ |
| 1413 | 1412 | $\nu_s(COO^-)$ |
| — | 1355 | $\delta_s(CH_3)$ |
| 1333 | — | $\rho_w(CH_2)$ |
| — | 1308 | $\delta(CH)$ |
| 1240(R) | — | $\rho_t(CH_2)$ |
| 1131, 1110 | 1237, 1113 | $\rho_r(NH_3^+)$[a] |
| 1033 | 1148 | $\nu_a(CCN)$[a] |
| — | 1026, 1015 | $\rho_r(CH_3)$[a] |
| 910 | — | $\rho_r(CH_2)$ |
| 893 | 918, 852 | $\nu_s(CCN)$[a] |
| 694 | 648 | $\rho_w(COO^-)$ |
| 607 | 771 | $\delta(COO^-)$ |
| 516 | 492 | $\rho_t(NH_3^+)$ |
| 504 | 540 | $\rho_r(COO^-)$ |

[a] These bands are coupled with other modes in $\alpha$-alanine.

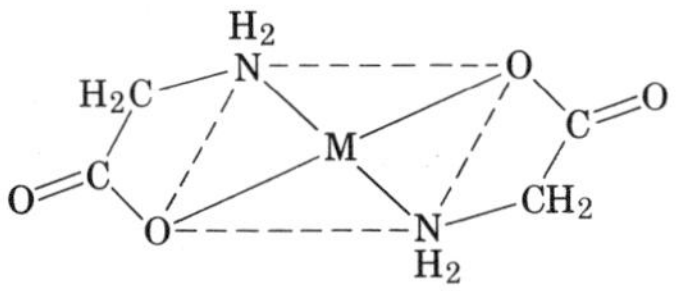

Fig. III-35. Structure of bis(glycino) complex.

*trans*-planar structure (Fig. III-35), and the noncoordinating C=O groups are hydrogen-bonded to the neighboring molecule or water of crystallization, or weakly bonded to the metal of the neighboring complex. Thus $\nu(CO_2)$ of amino acid complexes are affected by coordination as well as by intermolecular interactions.

To examine the effects of coordination and hydrogen bonding, Nakamoto et al.[687] made extensive measurements of the COO stretching frequencies of various metal complexes of amino acids in $D_2O$ solution, in the hydrated crystalline state, and in the anhydrous crystalline state. The results showed that, in any one physical state, the same frequency order is found for a series of metals, regardless of the nature of the ligand. The antisymmetric frequencies increase, the symmetric frequencies decrease, and the separation between the two frequencies increases in the following order of metals:

$$\text{Ni(II)} < \text{Zn(II)} < \text{Cu(II)} < \text{Co(II)} < \text{Pd(II)} \approx \text{Pt(II)} < \text{Cr(III)}$$

Although there are several exceptions to this order, these results generally indicate that the effect of coordination is still the major factor in determining the frequency order in a given physical state. The above frequency order indicates the increasing order of the metal–oxygen interaction since the COO group becomes more asymmetrical as the metal-oxygen interaction becomes stronger. (Sec. III-6). It is rather difficult, however, to discuss the strength of the metal–nitrogen bond from the frequency order of the $NH_2$ vibration since it is very sensitive to the effect of intermolecular interaction. The $\nu(MN)$ and $\nu(MO)$ vibrations that appear in the low frequency region would be ideal for this purpose. Unfortunately, it is not simple to assign these vibrations empirically. Use of the metal isotope effect combined with normal coordinate analysis seems to give the most reliable results, as will be discussed below.

To give theoretical band assignments on metal glycino complexes, Condrate and Nakamoto[688] carried out a normal coordinate analysis on the metal-glycino chelate ring. Figure III-36 shows the infrared spectra of bis(glycino) complexes of Pt(II), Pd(II), Cu(II), and Ni(II). Table III-50 lists the observed frequencies and theoretical band assignments. The $CH_2$ group frequencies are not listed, since they are not metal sensitive. It is

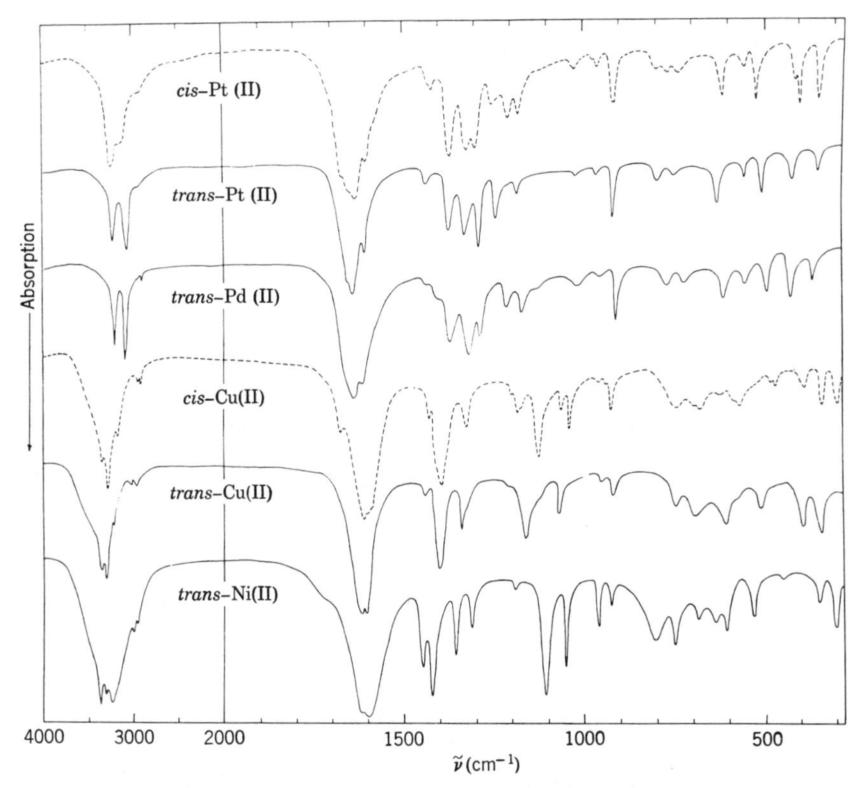


**Fig. III-36.** Infrared spectra of bis(glycino) complexes of divalent metals.[688]

seen that the C═O stretching, $NH_2$ rocking, and MN and MO stretching bands are metal sensitive and are shifted progressively to higher frequencies as the metal is changed in the order Ni(II) < Cu(II) < Pd(II) < Pt(II). Table III-50 shows that both the MN and MO stretching force constants also increase in the same order of the metals. These results provide further support to the preceding discussion of the M–O bonds of glycino complexes.

Normal coordinate analyses on metal complexes of amino acids have also been made by Lane et al.[689] and Walter and co-workers.[690] These workers, however, assigned the MO stretching modes below 200 $cm^{-1}$. Thus their MO stretching force constants are much smaller than the MN stretching force constants. Rayner-Canham and Lever[691] assigned the CuN and CuO stretching bands of *cis*-$[Cu(gly)_2]H_2O$ at 379 and 334 $cm^{-1}$, respectively, based on the $^{NA}Cu$–$^{65}Cu$ (NA: natural abundance) shift data. To give definitive band assignments in the low frequency region of *bis*(glycino) complexes of Ni(II), Cu(II), and Co(II),

TABLE III-50. OBSERVED FREQUENCIES AND BAND ASSIGNMENTS OF BIS(GLYCINO) COMPLEXES ($cm^{-1}$)[688]

| *trans*-[Pt(gly)$_2$] | *trans*-[Pd(gly)$_2$] | *trans*-[Cu(gly)$_2$] | *trans*-[Ni(gly)$_2$] | Band Assignment |
|---|---|---|---|---|
| 3230<br>3090 | 3230<br>3120 | 3320<br>3260 | 3340<br>3280 | $\nu(NH_2)$ |
| 1643 | 1642 | 1593 | 1589 | $\nu(C{=}O)$ |
| 1610 | 1616 | 1608 | 1610 | $\delta(NH_2)$ |
| 1374 | 1374 | 1392 | 1411 | $\nu(C{-}O)$ |
| 1245 | 1218 | 1151 | 1095 | $\rho_t(NH_2)$ |
| 1023 | 1025 | 1058 | 1038 | $\rho_w(NH_2)$ |
| 792 | 771 | 644 | 630 | $\rho_r(NH_2)$ |
| 745 | 727 | 736 | 737 | $\delta(C{=}O)$ |
| 620 | 610 | 592 | 596 | $\pi(C{=}O)$ |
| 549 | 550 | 439 | 439 | $\nu(MN)$ |
| 415 | 420 | 360 | 290 | $\nu(MO)$ |
| 2.10 | 2.00 | 0.90 | 0.70 | $K$(M—N), (mdyn/Å)[a] |
| 2.10 | 2.00 | 0.90 | 0.70 | $K$(M—O), (mdyn/Å)[a] |

[a] UBF.

Kincaid and Nakamoto[692] carried out H–D, $^{14}N$–$^{15}N$, $^{58}Ni$–$^{62}Ni$ and $^{63}Cu$–$^{65}Cu$ substitutions, and performed normal coordinate analyses on the skeletal modes of bis(glycino) complexes. Their results show that, in *trans*-[M(gly)$_2$]2$H_2O$, the infrared active $\nu$(MN) and $\nu$(MO) are at 483 and 337 $cm^{-1}$, respectively, for the Cu(II) complex, and at 442 and 289 $cm^{-1}$, respectively, for the Ni(II) complex. Both modes are coupled strongly with other skeletal modes, however.

A Raman study on metal glycino complexes was made by Krishnan and Plane, who assigned $\nu$(ZnN) and $\nu$(ZnO) at 470 and 395 $cm^{-1}$, respectively.[693] Similarly, Long and Yoshida[694] assigned $\nu$(HgN) and $\nu$(HgO) of the Hg–glycino complex at 464 and 385 $cm^{-1}$, respectively, in its aqueous Raman spectrum.

Square-planar bis(glycino) complexes can take the *cis* or the *trans* configuration. As expected from symmetry consideration, the *cis*-isomer exhibits more bands in infrared spectra than does the *trans*-isomer (see Fig. III-36). In the low frequency region, the *cis*-isomer exhibits two $\nu$(MN) and two $\nu$(MO), whereas the *trans*-isomer exhibits only one for each of these modes.[688] This criterion has been used by Herlinger et al. to assign the geometry of a series of bis(amino acidato)Cu(II) complexes.[695,696] Octahedral tris(glycino) complexes may take the *fac* and

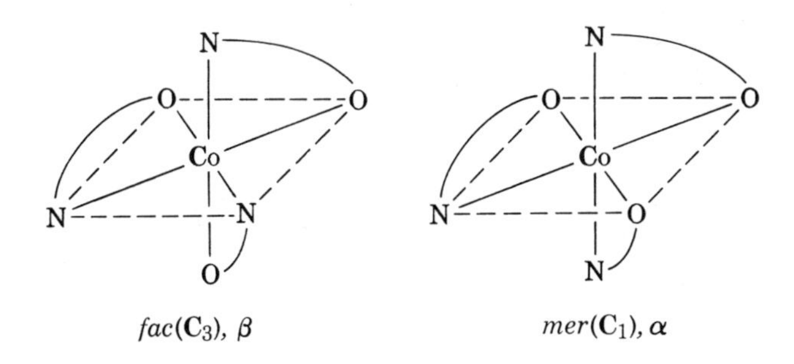


**Fig. III-37.** Structures of tris(glycino) complex.

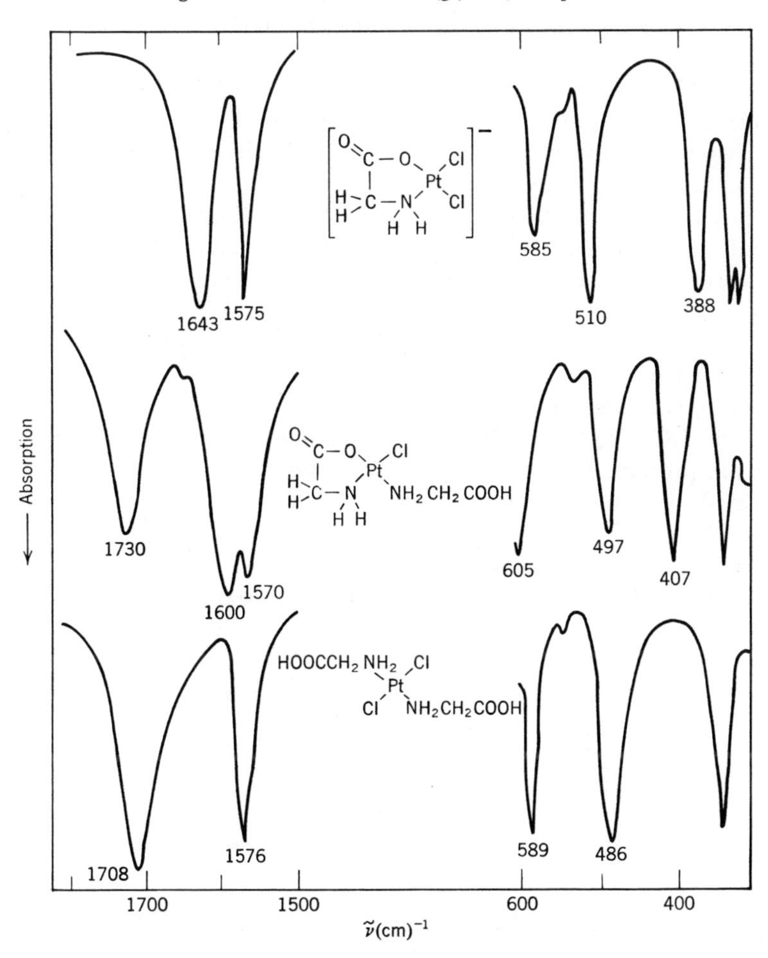


**Fig. III-38.** Infrared spectra of K[Pt(gly)$Cl_2$], [Pt(gly)(glyH)Cl] and *trans*- [Pt$(glyH)_2Cl_2$].[698]

*mer* configurations shown in Fig. III-37. For example, $[Co(gly)_3]$ exists in two forms: purple crystals (dihydrate, $\alpha$-form), and red crystals (monohydrate, $\beta$-form). The $\alpha$-form is assigned to the *mer* configuration since it exhibits more infrared bands than does the $\beta$-form (*fac* configuration).[697]

Glycine also coordinates to the Pt(II) atom as a unidentate ligand:

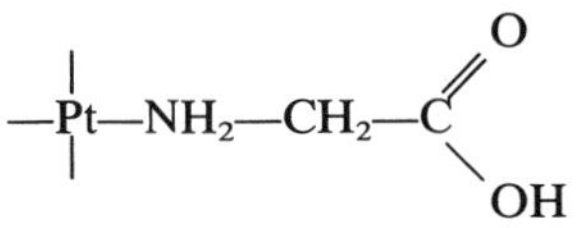

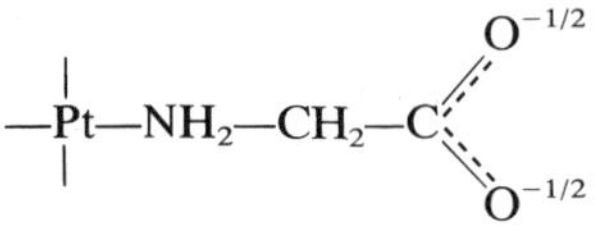

$$-Pt-NH_2-CH_2-C(=O)OH \qquad -Pt-NH_2-CH_2-C(O^{-1/2})(O^{-1/2})$$

The carboxyl group is not ionized in *trans*-$[Pt(glyH)_2X_2]$ (X: a halogen), whereas it is ionized in *trans*-$[Pt(gly)_2(NH_3)_2]$. The former exhibits the un-ionized COO stretching band near 1710 $cm^{-1}$, while the latter shows the ionized COO stretching band near 1610 $cm^{-1}$.[698]

The distinction between uni- and bidentate glycino complexes of Pt(II) can be made readily from their infrared spectra. Figure III-38 illustrates the infrared spectra of *trans*-$[Pt(glyH)_2Cl_2]$ and $K[Pt(gly)Cl_2]$ in the COO stretching and PtO stretching regions. The bidentate (chelated) glycino group absorbs at 1643 $cm^{-1}$, unlike either the ionized unidentate group (1610 $cm^{-1}$) or the unionized unidentate group (1710 $cm^{-1}$). Furthermore, the bidentate glycino group exhibits the PtO stretching band at 388 $cm^{-1}$, whereas the unidentate glycino group has no absorption between 470 and 350 $cm^{-1}$. Figure III-38 also shows the spectrum of [Pt(gly)(glyH)Cl], in which both the unidentate and bidentate glycino groups are present. It is seen that the spectrum of this compound can be interpreted as a superposition of the spectra of the former two compounds.[698]

Normal coordinate analysis has been carried out on bis(glycinamido)Cu(II) monohydrate.[699] The infrared spectra of

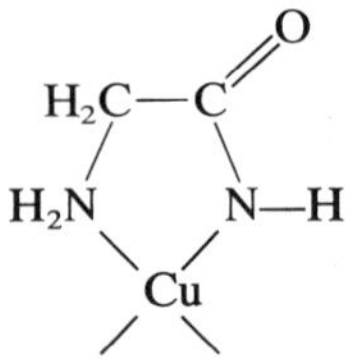

Glycinamido–Cu(II) complex

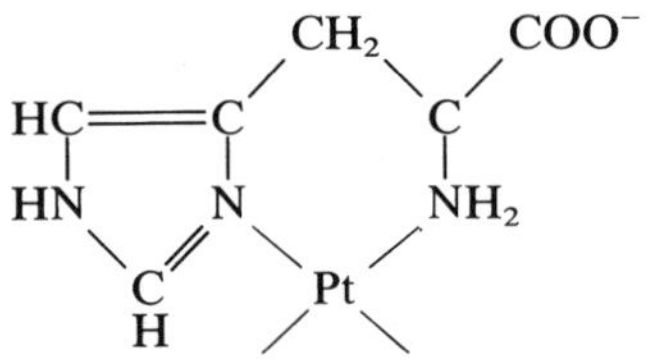

Histidino–Pt(II) complex

$[Pt(hist)_2]$ and $[Pt(hist-H)_2]^{2+}$ show that both compounds take the *trans*-planar structure, in which the Pt atom is chelated by two nitrogens. The carboxyl group in the former is ionized [$\nu(CO_2^-)$, 1625 and

1400 $cm^{-1}$], whereas it is not ionized in the latter [$\nu$(C=O), 1740 $cm^{-1}$].[700] In these compounds, the O atoms are not involved in coordination. Watt and Knifton[701] observed that $\nu$(MN) of the coordinated $NH_2$ group in metal complexes of glycine and iminodiacetic acid are shifted to higher frequencies by proton abstraction. This observation suggests that the M–N bond becomes stronger as a result of deprotonation.

From the infrared spectra observed in the solid state, Busch and co-workers[702] determined the coordination numbers of the metals in metal chelate compounds of EDTA and its derivatives:

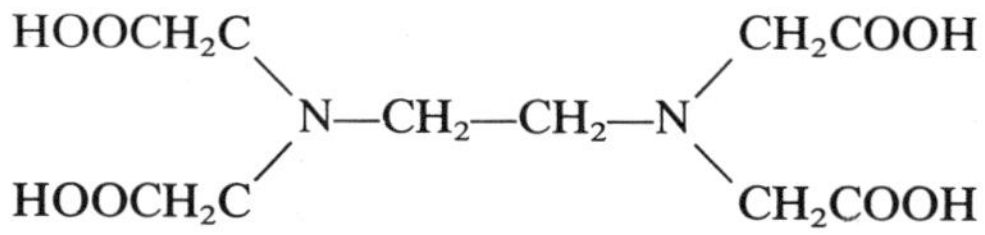

$$\begin{array}{ccc} HOOCH_2C & & CH_2COOH \\ & \diagdown\ \ \ \ \ \ \ \ \ \diagup & \\ & N{-}CH_2{-}CH_2{-}N & \\ & \diagup\ \ \ \ \ \ \ \ \ \diagdown & \\ HOOCH_2C & & CH_2COOH \end{array}$$

Ethylenediaminetetraacetic acid
(EDTA) or ($H_4Y$)

The method is based on the simple rule that the un-ionized and uncoordinated COO stretching band occurs at 1750–1700 $cm^{-1}$, whereas the ionized and coordinated COO stretching band is at 1650–1590 $cm^{-1}$. The latter frequency depends on the nature of the metal: 1650–1620 $cm^{-1}$ for metals such as Cr(III) and Co(III), and 1610–1590 $cm^{-1}$ for metals such as Cu(II) and Zn(II). Since the free ionized $COO^-$ stretching band is at 1630–1575 $cm^{-1}$, it is also possible to distinguish the coordinated and free $COO^-$ stretching bands if a metal such as Co(III) is chosen for complex formation. Table III-51 shows the results obtained by Busch et al.

Tomita and Ueno[704] studied the infrared spectra of metal complexes of NTA, using the method described above. They concluded that NTA:

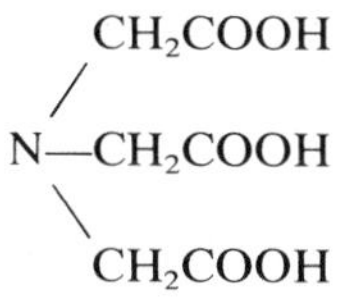

$$\begin{array}{l} \ \ \ \ CH_2COOH \\ \ \ \diagup \\ N{-}CH_2COOH \\ \ \ \diagdown \\ \ \ \ \ CH_2COOH \end{array}$$

Nitrilotriacetic acid (NTA)

acts as a quadridentate ligand in complexes of Cu(II), Ni(II), Co(II), Zn(II), Cd(II), and Pb(II), and as a tridentate in complexes of Ca(II), Mg(II), Sr(II), and Ba(II).

Krishnan and Plane[705] studied the Raman spectra of EDTA and its metal complexes in aqueous solution. They noted that $\nu$(MN) appears strongly in the 500–400 $cm^{-1}$ region for Cu(II), Zn(II), Cd(II), Hg(II), and so on, and that its frequency decreases with an increasing radius of the metal ion, independently of the solubility of the metal complex.

TABLE III-51. ANTISYMMETRIC COO STRETCHING FREQUENCIES AND NUMBER OF FUNCTIONAL GROUPS USED FOR COORDINATION IN EDTA COMPLEXES ($cm^{-1}$)[702]

| Compound[a] | Un-ionized COOH | Coordinated $COO^- \cdots M$ | Free $COO^-$ | Number of Coordinated Groups |
|---|---|---|---|---|
| $H_4[Y]$ | 1697[b] | — | — | |
| $Na_2[H_2Y]$ | 1668[b] | — | 1637[b] | |
| $Na_4[Y]$ | — | — | 1597[b] | |
| $Ba[Co(Y)]_2 \cdot 4H_2O$ | — | 1638 | — | 6 |
| $Na_2[Co(Y)Cl]$ | — | 1648 | 1600 | 5 |
| $Na_2[Co(Y)NO_2]$ | — | 1650 | 1604 | 5 |
| $Na[Co(HY)Cl] \cdot \frac{1}{2}H_2O$ | 1750 | 1650 | — | 5 |
| $Na[Co(HY)NO_2] \cdot H_2O$ | 1745 | 1650 | — | 5 |
| $Ba[Co(HY)Br] \cdot 9H_2O$ | 1723 | 1628 | — | 5 |
| $Na[Co(YOH)Cl] \cdot \frac{3}{2}H_2O$ | — | 1658 | — | 5 |
| $Na[Co(YOH)Br] \cdot H_2O$ | — | 1654 | — | 5 |
| $Na[Co(YOH)NO_2]$ | — | 1652 | — | 5 |
| $[Pd(H_2Y)] \cdot 3H_2O$ | 1740 | 1625 | — | 4 |
| $[Pt(H_2Y)] \cdot 3H_2O$ | 1730 | 1635 | — | 4 |
| $[Pd(H_4Y)Cl_2] \cdot 5H_2O$ | 1707, 1730 | — | — | 2 |
| $[Pt(H_4Y)Cl_2] \cdot 5H_2O$ | 1715, 1730 | — | — | 2 |

[a] Y: tetranegative ion; HY: trinegative ion; $H_2Y$: dinegative ion; $H_4Y$: neutral species of EDTA; YOH: trinegative ion of HEDTA (hydroxyethylenediamine-triacetic acid).
[b] Reference 703.

Infrared spectroscopy has been used to deduce the structures of Cu(II) complexes with mono-, di-, and triethanolamines in the solid state.[706] For the infrared spectra of amino acids, EDTA, and related compounds in aqueous solution, see Sec. III-15.

## III-15. INFRARED SPECTRA OF AQUEOUS SOLUTIONS

Infrared studies of aqueous solutions provide a valuable tool for elucidating the structures of complex ions in equilibria. To observe the infrared spectrum of an aqueous solution, it is necessary to use window materials such as AgCl and $BaF_2$, which are insoluble in water, and a thin spacer (0.02–0.01 mm) to reduce the strong absorption of water. The latter condition necessitates a solution of relatively high concentration. Even if these conditions are met, it is still difficult to measure the spectrum of a solute in the regions at 3700–2800, 1800–1600, and below

1000 $cm^{-1}$, where water ($H_2O$) absorbs strongly. The C≡N stretching band (2200–2000 $cm^{-1}$) can be measured in aqueous solution since it is outside of these regions. Thus the solution equilibria of cyano complexes have been studied extensively by using aqueous infrared spectroscopy (Sec. III-10). Fronaeus and Larsson[707] extended similar studies to thiocyanato complexes that exhibit the C≡N stretching bands in the same region. They[708] also studied the solution equilibria of oxalato complexes in the 1500–1200 $cm^{-1}$ region, where the CO stretching bands of the coordinated oxalato group appear. Larsson[709] studied the infrared spectra of metal glycolato complexes in aqueous solution. In this case, the C—OH stretching band near 1060 $cm^{-1}$ was used to elucidate the structures of the complex ions in equilibria.

If $D_2O$ is used instead of $H_2O$, it is possible to observe infrared spectra in the regions 4000–2900, 2000–1300, and 1100–900 $cm^{-1}$. The COO stretching bands of NTA, EDTA, and their metal complexes appear between 1750 and 1550 $cm^{-1}$ (Sec. III-14). Nakamoto and co-workers,[710] therefore, studied the solution ($D_2O$) equilibria of NTA, EDTA, and related ligands in this frequency region. By combining the results of potentiometric studies with the spectra obtained as a function of the pH(pD) of the solution, it was possible to establish the following COO stretching frequencies:

Type A, un-ionized carboxyl ($R_2N$—$CH_2COOH$), 1730–1700 $cm^{-1}$
Type B, $\alpha$-ammonium carboxylate ($R_2N^+H$—$CH_2COO^-$), 1630–1620 $cm^{-1}$
Type C, $\alpha$-aminocarboxylate ($R_2N$—$CH_2COO^-$), 1595–1575 $cm^{-1}$

As stated in Sec. III-14, the coordinated (ionized) COO group absorbs at 1650–1620 $cm^{-1}$ for Cr(III) and Co(III), and at 1610–1590 $cm^{-1}$ for Cu(II) and Zn(II). Thus it is possible to distinguish the coordinated COO group from those of types B and C if a proper metal ion is selected.

Tomita et al.[711] studied the complex formation of NTA with Mg(II) by aqueous infrared spectroscopy. Figure III-39 shows the infrared spectra of equimolar mixtures of NTA and $MgCl_2$ at concentrations about 5–10% by weight. The spectra of the mixture from pD 3.2 to 4.2 exhibit a single band at 1625 $cm^{-1}$, which is identical to that of the free $NTA^{3-}$ ion in the same pD range.[712] This result indicates that no complex formation occurs in this pD range, and that the 1625 $cm^{-1}$ band is due to the $NTA^{3-}$ ion (type B). If the pD is raised to 4.2, a new band appears at 1610 $cm^{-1}$, which is not observed for the free NTA solution over the entire pD range investigated. Figure III-39 shows that this 1610 $cm^{-1}$ band becomes stronger, and the 1625 $cm^{-1}$ band becomes weaker, as the pD increases.

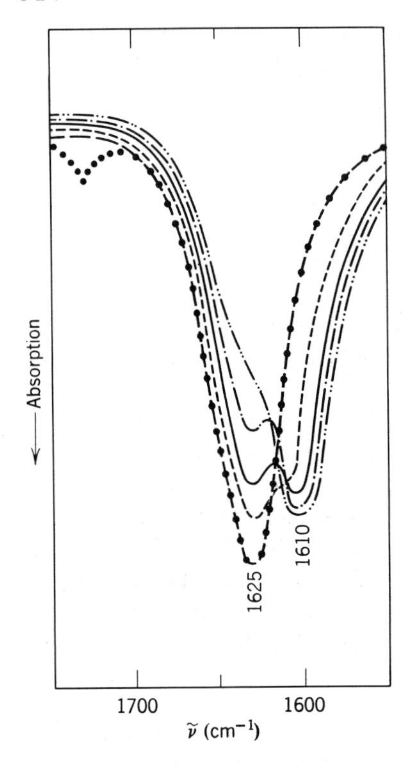


**Fig. III-39.** Infrared spectra of Mg—NTA complex in $D_2O$ solutions: . . . . . , pD 3.2; — — —, pD 4.2; - - - - -, pD 5.5; ——, pD 6.8; —·—·—, pD 10.0; · ·—· ·—, pD 11.6.[711]

It was concluded that this change is due mainly to a shift of the following equilibrium in the direction of complex formation:

$$\overset{+}{HN}(CH_2COO^-)_3 + Mg^{2+} \rightleftharpoons \left[ N(CH_2COO^-)_3 \cdots Mg \right]^- + H^+$$

1625 cm$^{-1}$ (Type B) — 1610 cm$^{-1}$

By plotting the intensity of these two bands as a function of pD, the stability constant of the complex ion was calculated to be 5.24. This value is in good agreement with that obtained from potentiometric titration (5.41).

Martell and Kim[713] carried out an extensive study on solution equilibria involving the formation of Cu(II) complexes with various polypeptides. As an example, the glycylglycino–Cu(II) system is discussed below.[714] Figure III-40 illustrates the infrared spectra of free glycylglycine in $D_2O$

solution as a function of pD. The observed spectral changes were interpreted in terms of the solution equilibria shown below:

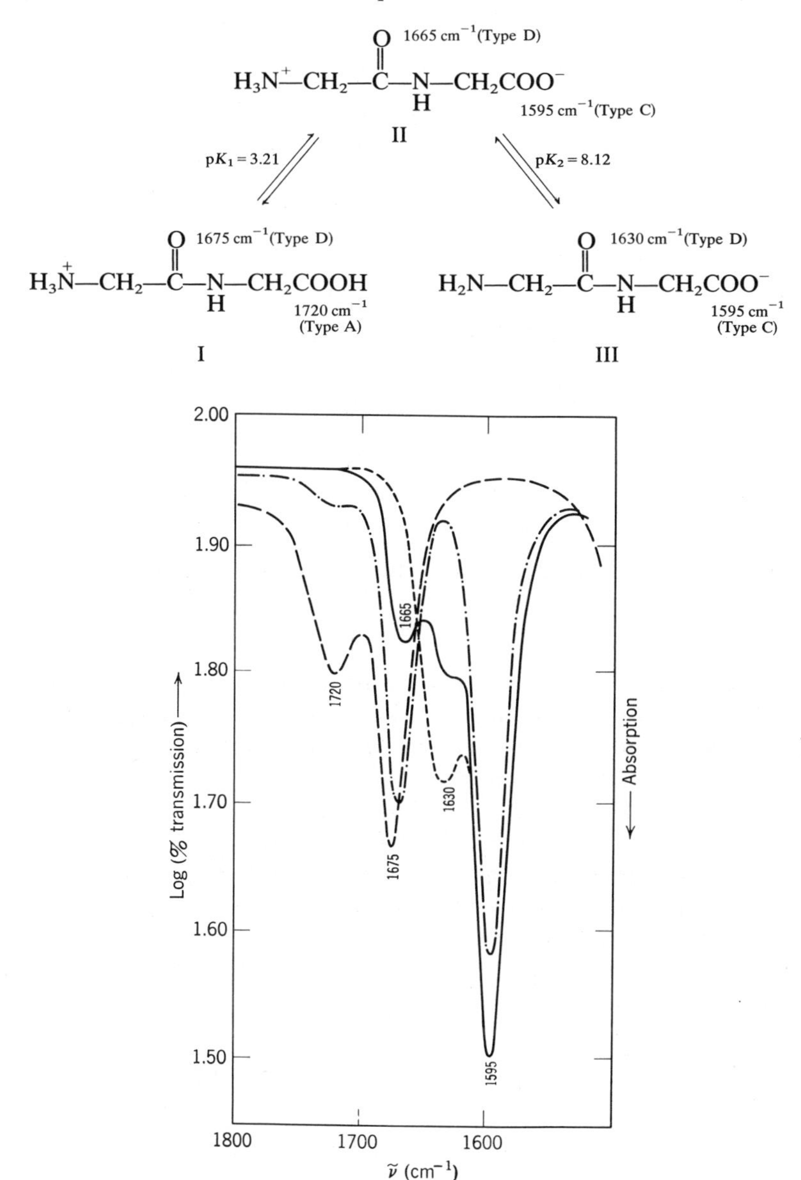


**Fig. III-40.** Infrared absorption spectra of glycylglycine in $D_2O$ solution at 0.288*M* concentration and ionic strength 1.0 adjusted with KCl; ———, pD 1.75; -·-·-, pD 4.31; ——, pD 8.77; -----, pD 10.29.[714]

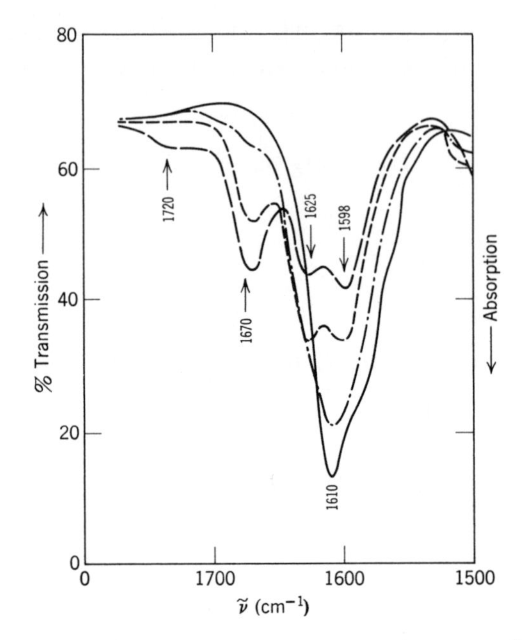


**Fig. III-41.** Infrared spectra of Cu(II)–glycylglycine complexes in aqueous ($D_2O$) solutions: — — —, pD 3.58; - - - - -, pD 4.24; —·—·—, pD 5.18; ——, pD 10.65. Total concentration of ligand and metal is 0.2333*M*, and ionic strength is 1.0 adjusted with KCl.[715]

Band assignments have been made by using the criteria given previously. In addition, type D frequency (1680–1610 cm$^{-1}$) was introduced to denote the peptide carbonyl group. The exact frequency of this group depends on the nature of the neighboring groups.

Figure III-41 shows the infrared spectra of glycylglycine mixed with copper chloride at equimolar ratio in $D_2O$ solution.[715] At pD = 3.58, the ligand exhibits three bands at 1720, 1675, and 1595 cm$^{-1}$ (Fig. III-40). This result indicates that I and II are in equilibrium. At the same pD value, however, the mixture exhibits one extra band at 1625 cm$^{-1}$. This band was attributed to the metal complex (IV), which was formed by the following reaction:

$$\mathrm{I,II} + Cu^{2+} \rightarrow \left[ \begin{array}{l} \qquad\qquad\qquad \mathrm{NH{-}CH_2COO^-} \\ \mathrm{H_2C{-}C} \qquad\qquad\quad 1598\ \mathrm{cm^{-1}} \\ \mathrm{H_2N} \qquad \mathrm{O}\ \ 1625\ \mathrm{cm^{-1}} \\ \qquad\ \mathrm{Cu} \\ \mathrm{H_2O} \qquad \mathrm{OH_2} \end{array} \right]^{+} + xH^{+}$$

IV

At pD = 5.18, the solution exhibits one broad band at about 1610 $cm^{-1}$. This result was interpreted as an indication that the following equilibrium was shifted almost completely to the right-hand side, and that the 1610 $cm^{-1}$ band is an overlap of two bands at 1610 and 1598 $cm^{-1}$:

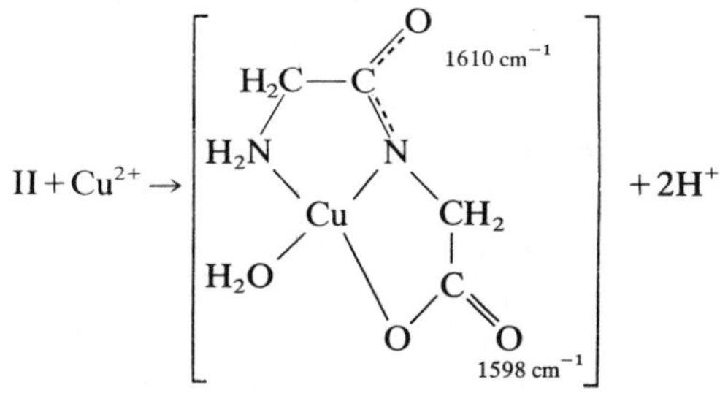

$$II + Cu^{2+} \rightarrow [\text{V}] + 2H^{+}$$

(Structure V: $H_2C$—C(=O, 1610 $cm^{-1}$)—N; $H_2N$, N, $H_2O$ and O coordinated to Cu; N—$CH_2$—C(=O, 1598 $cm^{-1}$)—O)

V

The shift of the peptide carbonyl stretching band from 1625 (IV) to 1610 (V) $cm^{-1}$ may indicate the ionization of the peptide NH hydrogen, since such an ionization results in the resonance of the O—C—N system, as indicated by the dotted line in structure V. Kim and Martell[716] also studied the triglycine and tetraglycine Cu(II) systems.

## III-16. HALOGENO COMPLEXES

Halogens (X) are the most common ligands in coordination chemistry. Several review articles[717–719] summarize the results of extensive infrared studies on halogeno complexes. Part II of this book lists the vibrational frequencies of many halogeno complexes. Here the vibrational spectra of halogeno complexes containing other ligands are discussed. In most cases, $\nu$(MX) can readily be assigned by halogen or metal (isotope) substitution.

### (1) Terminal Metal–Halogen Bond

Terminal MX stretching bands appear in the regions of 750–500 $cm^{-1}$ for MF, 400–200 $cm^{-1}$ for MCl, 300–200 $cm^{-1}$ for MBr, and 200–100 $cm^{-1}$ for MI. According to Clark and Williams,[97] the $\nu$(MBr)/$\nu$(MCl) and $\nu$(MI)/$\nu$(MCl) ratios are 0.77–0.74 and 0.65, respectively. Several factors govern $\nu$(MX).[720] If other conditions are equal, $\nu$(MX) is higher as the oxidation state of the metal is higher. Examples have already been given for tetrahedral $MX_4$ and octahedral $MX_6$ type compounds, discussed in Part II. It is interesting to note, however, that in the $[M(dias)_2Cl_2]^{n+}$ series* $\nu$(MCl) changes rather drastically in going from

---

* dias: *o*-phenylenebis(dimethylarsine).

Ni(III) to Ni(IV) (Fig. III-51), while very little change is observed between Fe(III) and Fe(IV):

| | $d^4$ Fe(IV) | $d^5$ Fe(III) | $d^6$ Ni(IV) | $d^7$ Ni(III) |
|---|---|---|---|---|
| $\nu$(MCl), ($cm^{-1}$) | 390 | 384 | 421 | 240 |

This was attributed to the presence of one electron in the antibonding $e_g^*$ orbital in the Ni(III) complex.[721]

If other conditions are equal, $\nu$(MX) is higher as the coordination number of the metal is smaller. Table III-52 indicates the structure dependence of $\nu$(NiX), obtained by Saito et al.[99] According to Wharf and Shriver,[726] the SnX stretching force constants of halogenotin compounds are approximately proportional to the oxidation number of the metal divided by the coordination number of the complex.

It is interesting to note that the $\nu$(SnCl) of free $SnCl_3^-$ ion [289 ($A_1$) and 252 ($E$) $cm^{-1}$] are shifted to higher frequencies upon coordination to a metal. Thus $\nu$(SnCl) of $[Rh_2Cl_2(SnCl_3)_4]^{2-}$ are at 339 and 323 $cm^{-1}$. According to Shriver and Johnson,[727] the L-X force constant of the $LX_n$ type ligand will increase upon coordination to a metal if X is significantly more electronegative than L. In the above example, chlorine is more

TABLE III-52. STRUCTURAL DEPENDENCE OF NiX STRETCHING FREQUENCIES ($CM^{-1}$)[a]

| Stretching Frequency | Linear Triatomic | *trans*-Planar | *cis*-Planar | Tetrahedral | *trans*-Octahedral |
|---|---|---|---|---|---|
| $\nu$(NiCl) | $NiCl_2$[b] 521 | $Ni(PEt_3)_2Cl_2$[c] 403 | $Ni(DPE)Cl_2$[d] 341, 328 | $Ni(PPh_3)_2Cl_2$[c] 341, 305 | $Ni(py)_4Cl_2$ 207 |
| $\nu$(NiBr) | $NiBr_2$[b] 414 | $Ni(PEt_3)_2Br_2$[c] 338 | $Ni(DPE)Br_2$[d] 290, 266 | $Ni(PPh_3)_2Br_2$[e] 265, 232 | $Ni(py)_4Br_2$ 140 |
| $\nu$(NiI) | | | $Ni(DPE)I_2$[d] 260, 212 | $Ni(PPh_3)_2I_2$[e] 215 | $Ni(py)_4I_2$ 105 |
| $\frac{\nu(NiBr)}{\nu(NiCl)}$ | 0.80 | 0.84 | 0.83[f] | 0.77[f] | 0.68 |
| $\frac{\nu(NiI)}{\nu(NiCl)}$ | | | 0.70[f] | 0.67[f] | 0.51 |

[a] DPE: 1,2-bis(diphenylphosphino)ethane.
[b] Ref. 722.
[c] Ref. 723.
[d] Ref. 724.
[e] Ref. 725.
[f] This value was calculated by using average frequencies of two bands.

electronegative than tin. In metal amine complexes (Sec. III-1), $\nu$(NH) shifts to lower frequencies because nitrogen is more electronegative than hydrogen. As expected, the $\nu$(GeCl) of free $GeCl_3^-$ ion [303 ($A_1$) and 285 ($E$) cm$^{-1}$] are also shifted to higher frequencies in [Pd(PhNC)($PPh_3$)-($GeCl_3$)Cl] (384 and 360 cm$^{-1}$).[728]

The MX vibrations are very useful in determining the stereochemistry of the complex. Appendix III tabulates the number of infrared and Raman active vibrations of various $MX_nY_m$ type compounds. Using these tables, it is possible to determine the stereochemistry of a halogeno complex simply by counting the number of $\nu$(MX) fundamentals observed. Examples of this method will be given in the following sections.

***(a) Square-Planar Complexes.*** Vibrational spectra of planar $M(NH_3)_2X_2$ [M = Pt(II) and Pd(II)] were discussed in Sec. III-1. The *trans*-isomer ($\mathbf{C}_{2h}$) exhibits one $\nu$(MX) ($B_u$), whereas the *cis*-isomer ($\mathbf{C}_{2v}$) exhibits two $\nu$(MX) ($A_1$ and $B_2$) bands in the infrared. The infrared spectra of *cis*- and *trans*-[$Pd(NH_3)_2Cl_2$] were shown in Fig. III-4. Similar results have been obtained for a pair of *cis*- and *trans*- [$Pt(py)_2Cl_2$][729] and $PtL_2X_2$, where L is one of a variety of neutral ligands.[730]

In planar Pt(II) and Pd(II) complexes, $\nu$(MX) is sensitive to the ligand *trans* to the M–X bond. Thus the effect of "*trans*-influence"[731] has been studied extensively by using infrared spectroscopy. In the [$PtCl_3L$]$^-$ series,[732] $\nu$($PtCl_{trans}$) follows the order:

| L = | CO | | $SMe_2$ | | $C_2H_4$ | | $SEt_2$ | | $AsEt_3$ | | $PPh_3$ | | $PMe_3$ | | $AsMe_3$ | | $PEt_3$ |
|---|---|---|---|---|---|---|---|---|---|---|---|---|---|---|---|---|---|
| $\nu$(PtCl) (cm$^{-1}$) | 322 | > | 310 | ~ | 309 | ~ | 307 | > | 280 | ~ | 279 | ~ | 275 | ~ | 272 | ~ | 271 |

Their order represents an increasing degree of *trans*-influence, since $\nu$(PtCl) becomes lower as a ligand of stronger *trans*-influence is introduced *trans* to the Pt–Cl bond. It was found that $\nu$($PtCl_{cis}$) is insensitive to the change in L. An order of *trans*-influence such as

$$Cl^- < Br^- < I^- \sim CO < CH_3 < PR_3 \sim AsR_3 < H$$

was noted from the order of $\nu$(M–$Cl_{trans}$) in a series of octahedral Rh(III) and Os(III) complexes.[733]

Fujita et al.[734] prepared two isomers of PtCl($C_2H_4$)(L-ala), where L-ala is L-alanino anion:

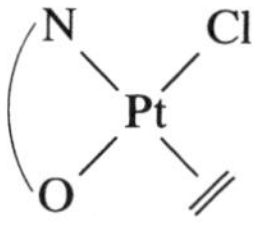

N Cl Pt O (N-isomer) I

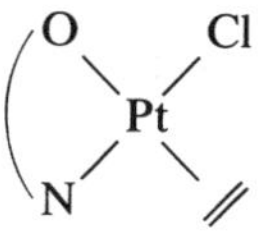

O Cl Pt N (O-isomer) II

Isomers I and II exhibit their $\nu$(PtCl) at 360 and 340 cm$^{-1}$, respectively. Since the *trans*-influence of the N-donor is expected to be stronger than that of the O-donor, the structures of these two isomers have been assigned as shown above.

Complexes of the type $Ni(PPh_2R)_2Br_2$ (R = alkyl) exist in two isomeric forms: tetrahedral (green) and *trans*-planar (brown). Distinction between these two can be made easily since the numbers and frequencies of infrared active $\nu$(NiBr) and $\nu$(NiP) are different for each isomer. Wang et al.[735] studied the infrared spectra of a series of compounds of this type, and confirmed that $\nu$(NiBr) and $\nu$(NiP) are at ca. 330 and 260 cm$^{-1}$, respectively, for the planar form and at ca. 270–230 and 200–160 cm$^{-1}$, respectively, for the tetrahedral form. The presence or absence of the 330 cm$^{-1}$ band is particularly useful in distinguishing these two isomers. According to X-ray analysis,[736] the green form of $Ni(PPh_2Bz)_2Br_2$ (Bz = benzyl) is a mixture of the planar and tetrahedral molecules in a 1:2 ratio. Ferraro et al.[737] studied the effect of high pressure on the infrared spectra of this compound, and found that all the bands characteristic of the tetrahedral form disappear as the pressure is increased to ca. 20,000 atm. This result indicates that the tetrahedral molecule can be converted to the planar form under high pressure if the energy difference between the two is relatively small. This conversion is completely reversible; the

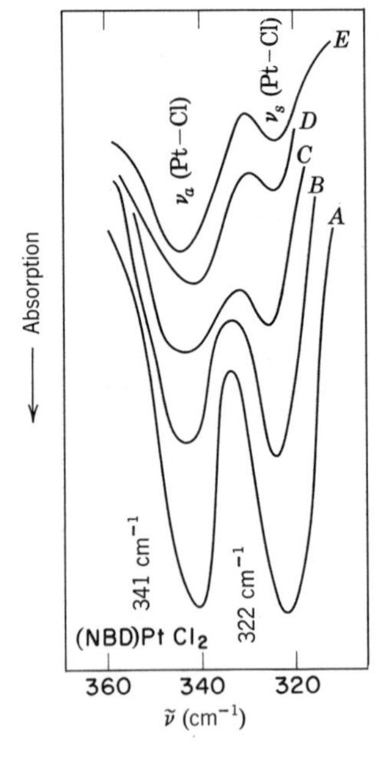


**Fig. III-42.** Effect of pressure on PtCl stretching bands: *A*, 1 atm; *B*, 6,000 atm; *C*, 12,000 atm; *D*, 18,000 atm; and *E*, 24,000 atm.

original green form is recovered as the pressure is reduced. High pressure infrared spectrosocpy has also been used to distinguish symmetric and antisymmetric MX stretching vibrations. For example, Fig. III-42 shows the effect of pressure on $\nu_a$(PtCl) and $\nu_s$(PtCl) of Pt(NBD)$Cl_2$ (NBD: norbornadiene).[738] It is seen that by increasing pressure the intensity of $\nu_s$(PtCl) is suppressed to a greater degree than that of $\nu_a$(PtCl). For high pressure vibrational spectroscopy, see a review by Ferraro and Long.[738a]

***(b) Octahedral Complexes.*** *cis*-$MX_2L_4$ ($\mathbf{C}_{2v}$) should exhibit two $\nu$(MX), while *trans*-$MX_2L_4$ ($\mathbf{D}_{4h}$) should give only one $\nu$(MX) in the infrared. Thus *cis*-[$IrCl_2(py)_4$]Cl shows two $\nu$(IrCl) at 333 and 327 $cm^{-1}$, while *trans*-[$IrCl_2(py)_4$] Cl exhibits only one $\nu$(IrCl) at 335 $cm^{-1}$.[97] If $MX_3L_3$ is *fac* ($\mathbf{C}_{3v}$), two $\nu$(MX) are expected in the infrared. If it is *mer* ($\mathbf{C}_{2v}$), three $\nu$(MX) should be infrared active. As is shown in Fig. III-43, *fac*-[$RhCl_3(py)_3$] gives two bands at 341 and325 $cm^{-1}$ and *mer*-[$RhCl_3(py)_3$] shows three bands at 355, 322, and 295 $cm^{-1}$.

In $MX_4L_2$ type compounds, group theory predicts one $\nu$(MX) for the *trans*-isomer ($\mathbf{D}_{4h}$) and four $\nu$(MX) for the *cis*-isomer ($\mathbf{C}_{2v}$). For example, *trans*-[$PtCl_4(NH_3)_2$] exhibits one $\nu$(PtCl) at 352 $cm^{-1}$ (with a shoulder at 346 $cm^{-1}$), whereas *cis*-[$PtCl_4(NH_3)_2$] exhibits four $\nu$(PtCl) at 353,

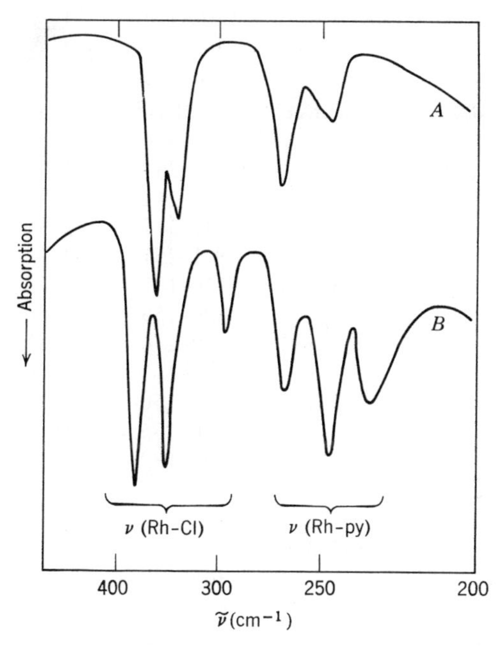


**Fig. III-43.** Far-infrared spectra of (*A*) *fac*- and (*B*) *mer*- $Rh(py)_3Cl_3$.[97]

344, 330, and 206 cm$^{-1}$.[739] Using Sn isotopes, Ohkaku and Nakamoto[740] confirmed that *trans*-[$SnCl_4L_2$] (L = py, THF, etc.) exhibits one $\nu$(SnCl) in the 342–370 cm$^{-1}$ region, while *cis*-[$SnCl_4$(L—L)](L—L = bipy, phen, etc.) shows four $\nu$(SnCl) in the 340–280 cm$^{-1}$ region. For $MX_5L(\mathbf{C}_{4v})$, one expects three $\nu$(MX) in the infrared. The $\nu$(InCl) of $[InCl_5(H_2O)]^{2-}$ were observed at 280, 271, and 256 cm$^{-1}$.[741]

### (2) Bridging Metal–Halogen Bond

Halogens tend to form bridges between two metal atoms. In general, bridging MX stretching frequencies [$\nu_b$(MX)] are lower than terminal MX stretching frequencies [$\nu_t$(MX)]. Vibrational spectra of simple $M_2X_6$ type ions having bridging halogens were discussed in Sec. II-10. Table III-53 lists the $\nu_t$(MX) and $\nu_b$(MX) of bridging halogeno complexes containing other ligands.

The *trans*-planar $M_2X_4L_2$ type compounds ($\mathbf{C}_{2h}$) exhibit three infrared active ($B_u$) $\nu$(MX) modes: one $\nu(MX_t)$ and two $\nu(MX_b)$. For the latter two, the higher frequency band corresponds to $\nu(MX_b)$ *trans* to *X*, whereas the lower frequency mode is assigned to $\nu(MX_b)$ *trans* to L since it is sensitive to the nature of L.[742] Strong coupling is expected, however, among these modes since they belong to the same symmetry species.

TABLE III-53. TERMINAL AND BRIDGING METAL—HALOGEN STRETCHING FREQUENCIES (CM$^{-1}$)

| Compound[a] | $\nu_t$(MX) | $\nu_b$(MX) | $\nu_b/\nu_t$[b] | Ref. |
|---|---|---|---|---|
| *trans*-$Pd_2Cl_4L_2$ | 360–339 | 308–294 | 0.86 | 742 |
| | | 283–241 | 0.75 | |
| *trans*-$Pt_2Cl_4L_2$ | 368–347 | 331–317 | 0.91 | 742 |
| | | 301–257 | 0.78 | |
| $Pd_2Br_4L_2$ | 285–265 | 220–185 | 0.74 | 743 |
| | | 200–165 | 0.66 | |
| $Pt_2Br_4L_2$ | 260–235 | 230–210 | 0.89 | 743 |
| | | 190–175 | 0.74 | |
| $Pt_2I_4L_2$ | 200–170 | 190–150 | 0.92 | 743 |
| | | 150–135 | 0.77 | |
| $Ni(py)_2Cl_2$ | — | 193, 182 | — | 744 |
| $Ni(py)_2Br_2$ | — | 147 | — | 744 |
| $Co(py)_2Cl_2$ | | | | |
| Monomeric | 347, 306 | — | | 102 |
| Polymeric | — | 186, 174 | | 102 |

[a] L = $PMe_3$, $PEt_3$, $PPh_3$, etc.

[b] These values were calculated using average frequencies.

$Co(py)_2Cl_2$ is known to exist in two forms: the monomeric tetrahedral (blue) and the polymeric octahedral (lilac). The latter is an infinite chain polymer bonded through chlorine bridges:

Figure III-44 shows the infrared spectra of both forms. The $\nu(CoCl_b)$ of the polymer is very low relative to $\nu(CoCl_t)$ of the monomer because of an increase in coordination number and the effect of bridging.[102] Polymeric $Ni(py)_2X_2$ also exhibits $\nu(NiX_b)$ below 200 $cm^{-1}$.[744]

Figure III-45*a* shows the structure of the metal cluster ion such as $[(Mo_6X_8)Y_6]^{2-}$. Because of its high symmetry ($\mathbf{O}_h$) only five $F_{1u}$ modes are infrared active. For $[Mo_6Cl_8)Cl_6]^{2-}$, Cotton et al.[745] proposed assigning two $\nu[MoCl(X)]$, $\nu[MoCl(Y)]$, and $\delta(XMoY)$ at 350–310, 246, and 110 $cm^{-1}$, respectively; $\nu(MoMo)$, which will be discussed in Sec. III-17,

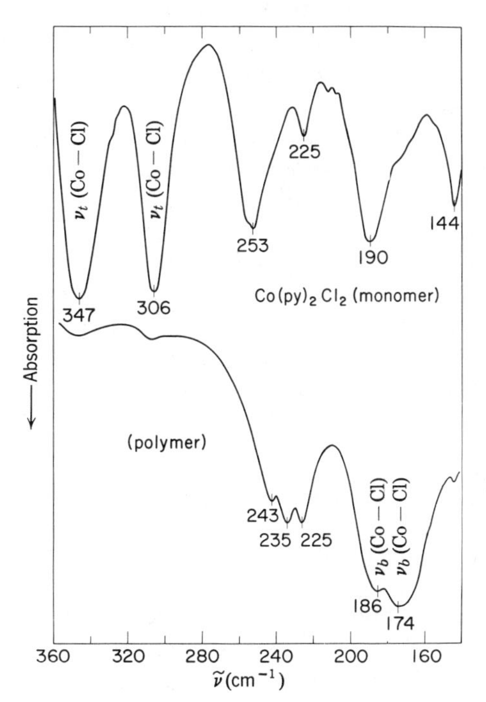


**Fig. III-44.** Infrared spectra of monomeric and polymeric forms of $Co(py)_2Cl_2$.

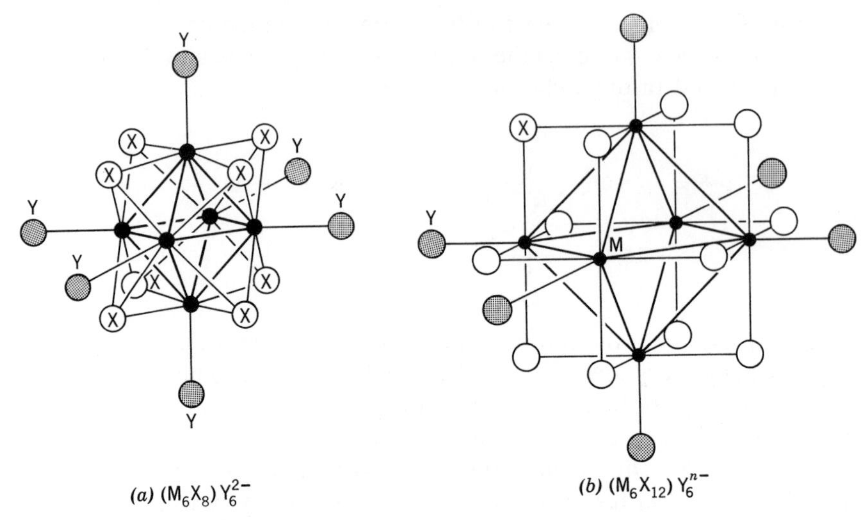


**Fig. III-45.** Structures of metal cluster compounds.

was assigned at 220 $cm^{-1}$. On the other hand, Hogue and McCarley[746] assigned five bands of $[(W_6X_8)Y_n]$ type compounds to the following five modes: three $W_6X_8$ unit vibrations, $\nu$(WY), and $\delta$(XWY). For X = Y = Cl, these bands are at 318, 284, 225, 305, and 105 $cm^{-1}$, respectively. The $W_6X_8$ unit vibrations were not assigned to individual internal modes because of the strong coupling between them.

Figure III-45*b* shows the structure of the $[(M_6X_{12})Y_6]^{n-}$ type metal cluster. The infrared spectra of the $M_6X_{12}$ unit (M = Nb and Ta; X = Cl, Br)[747,748] have been reported. Under $\mathbf{O}_h$ symmetry, the $M_6X_{12}$ unit is expected to show four infrared active fundamentals: three $\nu$(MX) and one $\nu$(MM). For the $Nb_6Cl_{12}$ unit, these were assigned at 340, 280, 232, and 145 [$\nu$(NbNb)] $cm^{-1}$, respectively.[748] For the $[(Nb_6Cl_{12})Cl_6]^{3-}$ ion, two more vibrations, $\nu$[NbCl(Y)] and $\delta$(XNbY), are expected to be infrared active. The former was assigned at 200 $cm^{-1}$, but the latter was not located. Fleming et al.[749] concluded that $\nu$(MM) cannot be identified in these compounds because of strong coupling among individual modes.

## III-17. COMPLEXES CONTAINING METAL–METAL BONDS

A large number of complexes containing metal–metal bonds are known, and their vibrational spectra have been reviewed extensively.[750–753] Vibrational spectra of polynuclear carbonyls were reviewed in Sec. III-12 and will also be disscussed in Sec. IV-7. Vibrational spectra of metal clusters containing halogen bridges were reviewed in Sec. III-16. In this section, only metal–metal stretching, $\nu$(MM), of these and other compounds are discussed. In general, $\nu$(MM) appear in the low-frequency region (250–100 $cm^{-1}$) because the M–M bonds are relatively weak and the masses of metals are relatively large. If the complex is perfectly centrosymmetric with respect to the M–M bond at the center, $\nu$(MM) is forbidden in the infrared. If it is not, $\nu$(MM) may appear weakly in the infrared. The $\nu$(M—M′) vibration of a heteronuclear complex is expected to be stronger because of the presence of a dipole moment along the M–M′ bond. In contrast, Raman spectroscopy has distinct advantages in that both $\nu$(MM) and $\nu$(MM′) appear strongly since large changes in polarizabilities are expected as a result of stretching long, covalent M–M(M′) bonds.

**(1) Polynuclear Carbonyls**

The $\nu$(MM) of polynuclear carbonyls have been assigned for a number of compounds, and the MM stretching force constants obtained from normal coordinate analysis have been used to discuss the nature of the M–M bond. Figure III-46 shows the Raman spectra of $Mn_2(CO)_{10}$,

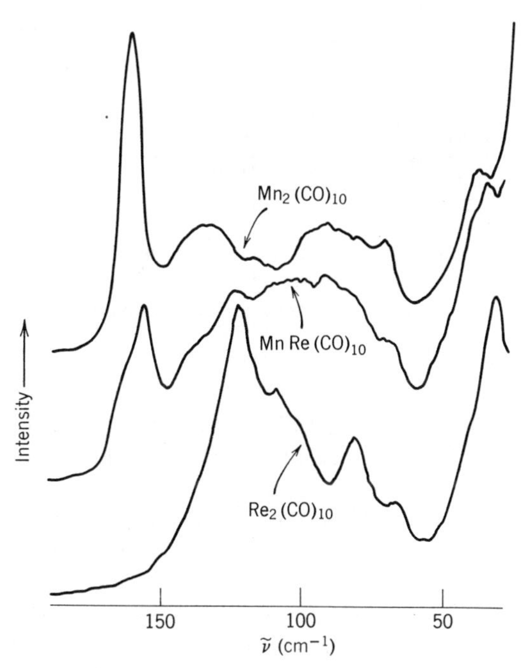


**Fig. III-46.** Low frequency Raman spectra of polycrystalline $Mn_2(CO)_{10}$, $MnRe(CO)_{10}$ and $Re_2(CO)_{10}$ (He-Ne 632.8 nm excitation).[754]

TABLE III-54. METAL—METAL STRETCHING FREQUENCIES ($cm^{-1}$) AND FORCE CONSTANTS

| Compound | $\nu$(MM) | Force Constant (mdyn/Å) | | Ref. |
|---|---|---|---|---|
| | | Rigorous Calculation | Approx. Calculation[a] | |
| $(CO)_5Mn—Mn(CO)_5$ | 160 | 0.59 | 0.41 | 754 |
| $(CO)_5Tc—Tc(CO)_5$ | 148 | 0.72 | 0.63 | 754 |
| $(CO)_5Re—Re(CO)_5$ | 122 | 0.82 | 0.82 | 754 |
| $(CO)_5Re—Mn(CO)_5$ | 157 | 0.81 | 0.62 | 754 |
| $(CO)_5Mn—W(CO)_5^-$ | 153 | 0.71 | 0.55 | 755 |
| $(CO)_5Mn—Mo(CO)_5^-$ | 150 | 0.60 | 0.47 | 755 |
| $(CO)_5Mn—Cr(CO)_5^-$ | 149 | 0.50 | 0.37 | 755 |
| $Cl_3Sn—Co(CO)_4$ | 204 | 1.23 | 0.97 | 756 |
| $Cl_3Ge—Co(CO)_4$ | 240 | 1.05 | 1.11 | 756 |
| $Cl_3Si—Co(CO)_4$ | 309 | 1.32 | 1.07 | 756 |
| $Br_3Ge—Co(CO)_4$ | 200 | 0.96 | — | 757 |
| $I_3Ge—Co(CO)_4$ | 161 | 0.52 | — | 757 |
| $Br_3Sn—Co(CO)_4$ | 182 | 1.05 | — | 757 |
| $I_3Sn—Co(CO)_4$ | 156 | 0.64 | — | 757 |
| $H_3Ge—Re(CO)_5$ | 209 | — | 1.34 | 758 |
| $H_3Ge—Mn(CO)_5$ | 219 | — | 0.88 | 758 |
| $H_3Ge—Co(CO)_4$ | 228 | — | 1.00 | 758 |
| $(CO)_4Co—Zn—Co(CO)_4$ | 170, 284[b] | 1.30 | — | 522 |
| $(CO)_4Co—Cd—Co(CO)_4$ | 163, 218[b] | 1.28 | — | 522 |
| $(CO)_4Co—Hg—Co(CO)_4$ | 163, 195[b] | 1.26 | — | 522 |

[a] Calculations considering only metal atom skeletons.
[b] Under $\mathbf{D}_{3d}$ symmetry, these frequencies correspond to the $A_{1g}$ (symmetric) and $A_{2u}$ (antisymmetric) MCo stretching modes, respectively.

$MnRe(CO)_{10}$, and $Re(CO)_{10}$ obtained by Quicksall and Spiro.[754] Risen and co-workers[755–757,522] carried out normal coordinate analyses on many dinuclear and trinuclear metal carbonyls. Table III-54 lists the observed $\nu$(MM) and the corresponding force constants obtained by these and other workers. It is noted that the MM stretching force constants obtained by rigorous calculations are surprisingly close to those obtained by approximate calculations considering only metal atoms. It is also noted that the MM stretching force constants are in the range of 1.3–0.5 mdyn/Å, which is close to the metal–nitrogen stretching force constants of typical ammine complexes, for example.

In Sec. III-12(3), we discussed the spectra of $M_2(CO)_4X_2$ type compounds (M = Mn, Tc, Re, Rh, etc.) in which the metals are bonded through halogen (X) bridges. Goggin and Goodfellow[759] concluded, however, that the $[Pt_2(CO)_2X_4]^{2-}$ ion (X = Cl and Br) contains the direct Pt–Pt bond:

```
  O   O
  C   C
  |   |
X—Pt—Pt—X
  |   |
  X   X
```

They isolated two isomers of $[N(n-Pr)_4]_2[Pt_2(CO)_2Cl_4]$ which differ only in the angle of rotation about the Pt–Pt bond. Both isomers exhibit $\nu$(PtPt) at ~170 cm$^{-1}$.

Trinuclear complexes such as $Ru_3(CO)_{12}$ and $Os_3(CO)_{12}$ contain a triangular $M_3$ skeleton for which two $\nu$(MM) are expected under $\mathbf{D}_{3h}$ symmetry. Quicksall and Spiro[518] assigned the Raman bands at 185 and 149 cm$^{-1}$ of the Ru complex to $\nu$(RuRu) of the $A_1'$ and $E'$ species, respectively. The latter is coupled with other modes. The corresponding RuRu stretching force constant is 0.82 mdyn/Å. The same workers[760] assigned the Raman spectrum of $Ir_4(CO)_{12}$, which consists of a tetrahedral $Ir_4$ skeleton; three $\nu$(IrIr) bands were assigned at 207 ($A_1$), 161 ($F_2$), and 131 ($E$) cm$^{-1}$. The ratio of these three frequencies, 2:1.56:1.27, is far from that predicted by a "simple cluster model" $(2:\sqrt{2}:1)$,[761] indicating the substantial coupling between the individual stretching modes. Their rigorous calculations gave $K$(Ir–Ir) of 1.69 mdyn/Å, together with interaction constants of −0.13 and +0.13 mdyn/Å for the adjacent and opposite Ir–Ir bonds, respectively.

$Fe_2(CO)_9$ and $Fe_3(CO)_{12}$ exhibit very strong Raman bands at 225 and 219 cm$^{-1}$, respectively. San Filippo and Sniadoch[762] assigned them to $\nu$(FeFe). Later studies[763] showed, however, that these bands are due to decomposition products resulting from strong laser irradiation. Thus the appearance of strong Raman bands in the low frequency region does not necessarily mean that they are due to $\nu$(MM). It was also noted that $Re_2(CO)_8X_2$ (X = Cl and Br), which do not contain Re–Re bonds, show strong Raman bands at 125 cm$^{-1}$ where $\nu$(ReRe) of $Re_2(CO)_{10}$ appears.[763]

### (2) Metal Cluster Compounds

As shown in Fig. III-45, metal atoms in compounds such as $[(Mo_6Cl_8)Cl_6]^{2-}$ and $[(Nb_6Cl_{12})Cl_6]^{3-}$ are connected by halogens. Vibrational spectroscopy alone cannot determine whether these metal atoms are bonded to each other directly or through halogen bridges. As stated in

Sec. III-16(2), Cotton et al.[745] assigned $\nu$(MoMo) of $[(Mo_6Cl_8)Cl_6]^{2-}$ at 220 $cm^{-1}$. Hogue and McCarley[746] could not assign $\nu$(MoMo) empirically since strong coupling between $\nu$(MoCl) and $\nu$(MoMo) of the $Mo_6Cl_8$ unit was expected. Although Mattes[764] carried out normal coordinate analysis on the $(Mo_6X_8)Y_4$ cluster (M = Mo, W; X, Y = Cl, Br), he did not calculate the potential energy distribution, which would indicate the degree of coupling between individual modes.

The infrared spectra of $[(M_6X_{12})Y_6]^{2-}$ type clusters have been reported by several investigators.[747] Mackay and Schneider[748] assigned the $\nu$(NbNb) of the $Nb_6Cl_{12}$ cluster at 140 $cm^{-1}$. Again, Fleming et al.[749] pointed out that $\nu$(MM) of such a cluster cannot be assigned because of strong coupling with other modes. Although Mattes[765] carried out normal coordinate analysis on the $(M_6X_{12})Y_n$ system (M = Nb, Ta; $n$ = 2–4), the potential energy distribution in individual modes was not calculated. He concluded that $\nu$(MM) are too low (below 100 $cm^{-1}$) to be observed.

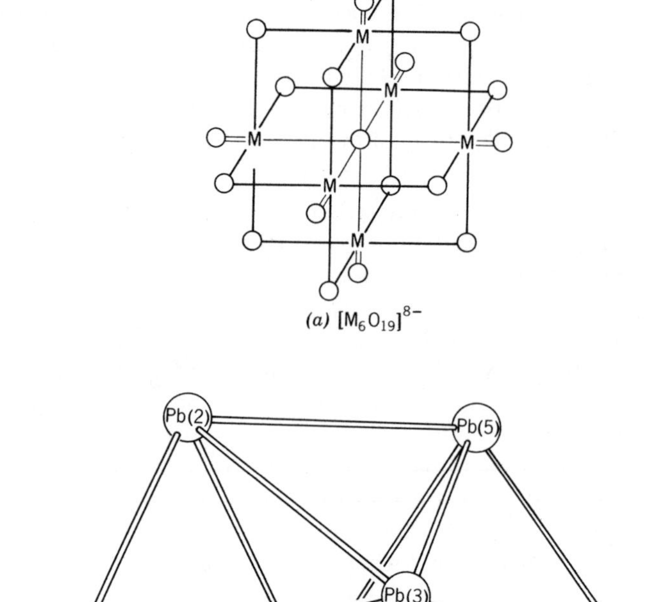


(*a*) $[M_6O_{19}]^{8-}$

(*b*) $[Pb_6O(OH)_6]^{4+}$

**Fig. III-47.** Structures of metal cluster compounds.

Figure III-47*a* shows the structure of $[M_6O_{19}]^{8-}$ ions (M = Nb, Ta). Farrell et al.[766] obtained the following set of force constants for the $[Nb_6O_{19}]^{8-}$ ion:

| K(Nb—O) | | | K(Nb—Nb) |
|---|---|---|---|
| 5.66 | 2.92 | 0.91 | 1.01 mdyn/Å |
| terminal | bridging | central | |

The ratio of the first three force constants is about 6:3:1. Although the NbNb stretching constant was estimated to be ca. 1 mdyn/Å, this value does not represent the strength of this bond, since such a value can be obtained without any M–M interaction.[766] Mattes[767] carried out normal coordinate analysis on the same system, and obtained a ratio of 8:4:1 for the three MO stretching force constants mentioned above.

The metal skeleton of the $[Pb_6O(OH)_6]^{4+}$ ion takes the very unusual structure shown in Fig. III-47*b*.[768] A band at 150 $cm^{-1}$ was assigned to $\nu$(PbPb), which corresponds to the shortest Pb–Pb bond ($Pb_3$—$Pb_4$ of Fig. III-47*b*).[769]

### (3) Other Compounds Containing Metal–Metal Bonds

Many other compounds containing metal–metal bonds do not belong to the groups already discussed. Their $\nu$(MM) have been assigned on the premise that $\nu$(MM) appears strongly in the Raman spectrum. Table III-55 lists $\nu$(MM) of some typical compounds. These frequencies scatter over a wide region, depending on the strength of the M–M bond. San Filippo and

TABLE III-55. METAL—METAL STRETCHING FREQUENCIES ($cm^{-1}$) AND BOND ORDERS

| Compound | $\nu$(MM) | Bond Order | Ref. |
|---|---|---|---|
| $Re_2Cl_8^{2-}$ | 275 | 4 | 762, 771 |
| $Re_2Br_8^{2-}$ | 278 | 4 | 762 |
| $Re_2Cl_6(PPh_3)_2$ | 278 | 4 | 771 |
| $Re_2Cl_5(dth)_2^a$ | 267 | 3 | 762 |
| $Re_2OCl_5(O_2CCH_2CH_3)$-$(PPh_3)_2$ | 216 | 2 | 762 |
| $Re_2(CO)_{10}$ | 122 | 1 | 762 |
| $Mo_2(OAc)_4$ | 406 | 4 | 770 |
| $Mo_2Cl_8^{4-}$ | 349 | 3 | 770 |
| $Rh_2(OAc)_4$ | 351 | 3 | 770 |

[a] dth: 2,5-dithiahexane.

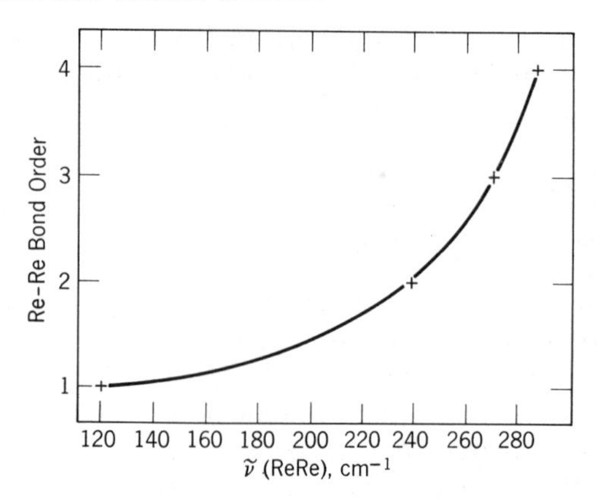


**Fig. III-48.** $\nu$(ReRe) vs. Re–Re bond order.

Sniadoch[762] plotted $\nu$(ReRe) against the Re–Re bond order[772] and obtained a monotonic curve such as shown in Fig. III-48. Ketteringham and Oldham[770] assigned $\nu$(MoMo) of $Mo_2(CH_3COO)_4$ and $[Mo_2Cl_8]^{4-}$ at 406 and 349 $cm^{-1}$, respectively. The Mo–Mo bond order in $[(\pi\text{–Cp})Mo(CO)_3]_2$ is one, and its $\nu$(MoMo) is at 193 $cm^{-1}$. Using simple diatomic harmonic approximation, they estimated the Mo–Mo bond orders of the former two compounds to be four and three, respectively. Angell et al.[773] noted that, among a series of compounds containing Mo–Mo bonds, $\nu$(MoMo) correlates only crudely with the Mo—Mo distance obtained by X-ray analysis. Thus $\nu$(MoMo) serves only as a rough measure of the bond strength, probably because it couples with other modes. Because of its strong absorption in the visible region, it was possible to measure the resonance Raman spectrum of the $[Mo_2Cl_8]^{4-}$ ion with 514.5 nm excitation. As is shown in Fig. III-49, it gave a series of overtones, $n\nu_1$(MoMo), up to $n = 11$.[774] A series of combination bands, $n\nu_1 + \nu_4$, was also observed up to $n = 4$, where $\nu_4$ is probably $\nu$(MoCl), $A_{1g}$ mode. The same figure shows similar results obtained for the $[Mo_2Cl_9]^{5-}$ ion.

Raman intensity measurements of the MM stretching modes can also provide information about the M–M bond order. For example, Quicksall and Spiro[775] estimate that the M–M bond orders are 0.52 and 0.95, respectively, for $Re_2(CO)_{10}$ and $Mn_2(CO)_{10}$ but only 0.061 and 0.023, respectively, for $[Bi_6(OH)_{12}]^{4+}$ and $[Pb_6(OH)_4]^{4+}$.

## III-18. COMPLEXES OF PHOSPHINES AND ARSINES

Ligands such as phosphines ($PR_3$) and arsines ($AsR_3$) (R = alkyl, aryl, halogen, etc.) form complexes with a variety of metals in various oxida-

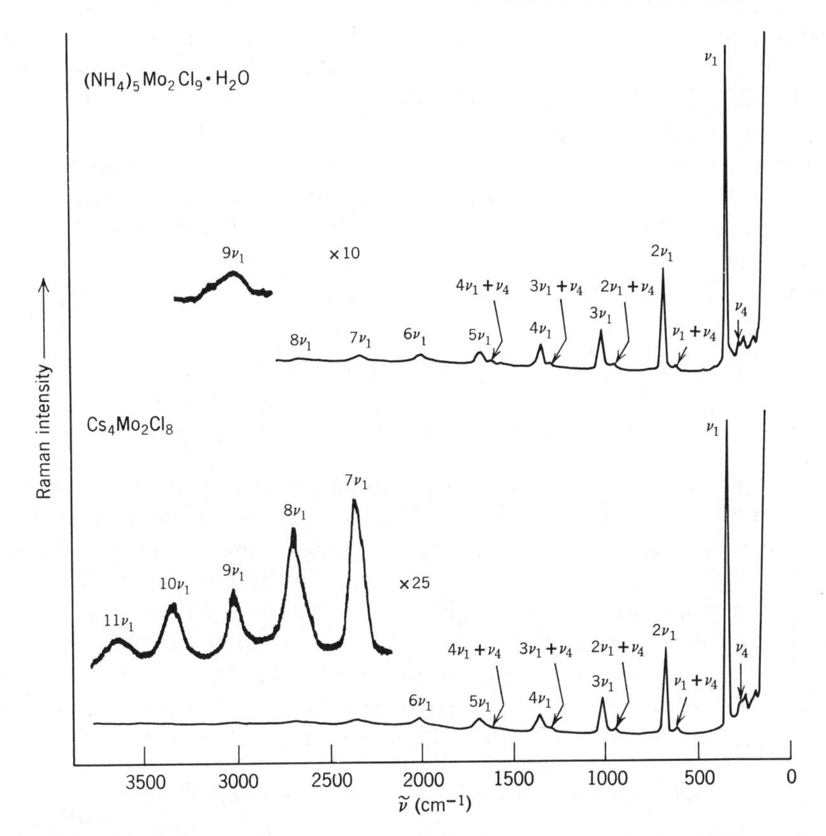


**Fig. III-49.** Resonance Raman spectra of $(NH_4)_5Mo_2Cl_9 \cdot H_2O$ and $Cs_4Mo_2Cl_8$ obtained with $Ar^+$ 514.5 nm excitation.

tion states. Vibrational spectroscopy has been used extensively to determine the structures of these compounds and to discuss the nature of the metal–phosphorus (M–P) bonding. Verkade[776] reviewed spectroscopic studies of M–P bonding with emphasis on cyclic phosphine ligands.

The most simple phosphine ligand is $PH_3$. The vibrational spectra of $Ni(PH_3)_4$,[777] $Ni(PH_3)(CO)_3$,[778] and $Ni(PH_3)(PF_3)_3$[779] have been reported by Bigorgne and co-workers. All these compounds exhibit $\nu$(PH), $\delta(PH_3)$, and $\nu$(NiP) at 2370–2300, 1120–1000, and 340–295 $cm^{-1}$, respectively. Trifluorophosphine ($PF_3$) forms a variety of complexes with transition metals. According to Kruck,[780] the $\nu$(PF) of free $PF_3$ (892, 860 $cm^{-1}$) are shifted slightly to higher frequencies (960–850 $cm^{-1}$) in $M(PF_3)_n$ ($n$ = 4, 5, and 6) and $HM(PF_3)_4$ and to lower frequencies (850–750 $cm^{-1}$) in $[M(PF_3)_4]^-$ (M = Co, Rh, and Ir). These results have been explained by assuming that the P–F bond possesses a partial double bond character which is governed by the oxidation state of the metal. For individual compounds, only references are given: $M(PF_3)_4$ (M = Ni, Pd, and Pt) (781), $M(PF_3)_5$ (782), and *cis*-$H_2M(PF_3)_4$ (783) (M = Fe, Ru, and Os). These compounds exhibit $\delta(PF_3)$ and $\nu(MP)$ at 590–280 and 250–180 $cm^{-1}$, respectively. Bénazeth et al.[784] showed that the skeletal symmetry of $HCo(PF_3)_4$ is $\mathbf{C}_{3v}$, while that of $[Co(PF_3)_4]^-$ is $\mathbf{T}_d$. The $\nu$(CoP) of these compounds are at 250–210 $cm^{-1}$. Woodward and co-workers[785] carried out complete vibrational analyses of the $M(PF_3)_4$ (M = Ni, Pd, and Pt) series. Their results give the following:

| | | Ni | Pd | Pt | |
|---|---|---|---|---|---|
| $\nu(MP)(cm^{-1})$ | $A_1$ | 195 | 204 | 213 | |
| | $F_2$ | 219 | 222 | 219 | |
| $K$(M—P)(mdyn/Å) | | 2.71 | 3.17 | 3.82 | (GVF) |

The infrared spectra of trialkyl phosphine halogeno complexes have been studied by many investigators. The main interest in these investigations has been to determine the stereochemistry and the nature of the M–P bond from the vibrational spectra. Although metal—halogen stretching vibrations can be assigned easily (see Sec. III-16), it is much more difficult to assign the metal—phosphorus (MP) stretching bands because they appear in the region where alkylphosphines exhibit many bands. To distinguish these two types of vibration, Nakamoto et al.[723] developed the metal isotope technique (Sec. I-15). Figure III-50 shows

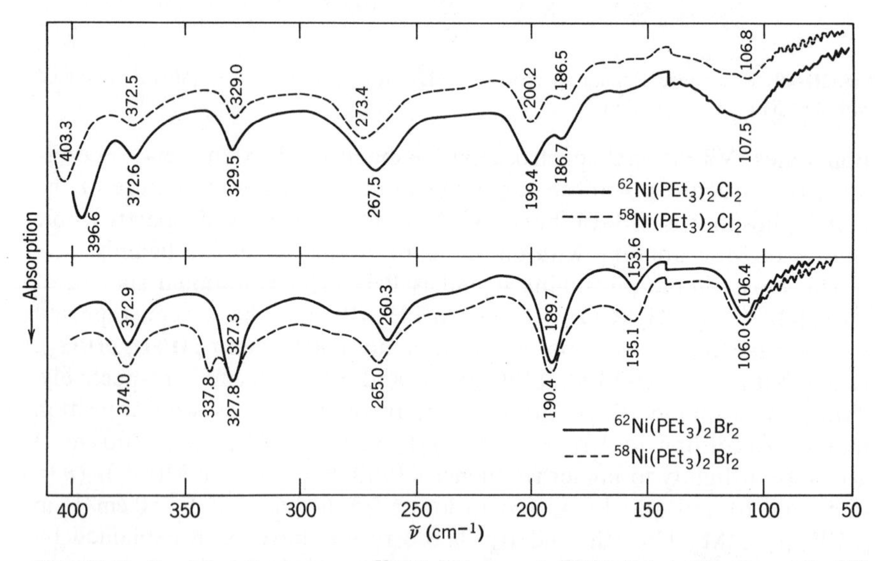


**Fig. III-50.** Far-infrared spectra of $^{58}NiX_2(PEt_3)_2$ and $^{62}NiX_2(PEt_3)_2$(X = Cl and Br.)[723]

the infrared spectra of *trans*-[$^{58,62}$ $Ni(PEt_3)_2X_2$] (X = Cl and Br), and Table III-56 lists the observed frequencies, metal isotope shifts, and band assignments. It is clear that the $\nu$(NiP) of these complexes must be assigned near 270 $cm^{-1}$, in contrast to previous investigations, which placed these vibrations near 450–410 $cm^{-1}$.[743,786–788]

Triphenylphosphine ($PPh_3$) is most common among phosphine ligands. It is not simple, however, to assign the $\nu$(MP) of $PPh_3$ complexes since $PPh_3$ exhibits a number of ligand vibrations in the low frequency region.[790] Using the metal isotope technique, Nakamoto et al.[723] showed that tetrahedral $Ni(PPh_3)_2Cl_2$, for example, exhibits two $\nu$(NiP) at 189.6 and 164.0 $cm^{-1}$, in agreement with the result of previous workers.[791]

As stated in Sec. III-16, complexes of the type $Ni(PPh_2R)_2Br_2$ (R = alkyl) exist in two forms (tetrahedral and square-planar) which can be distinguished by the $\nu$(NiBr) and $\nu$(NiP).[735] For R = Et, the $\nu$(NiP) of the planar complex is at 243 $cm^{-1}$, whereas these vibrations are at 195 and 182 $cm^{-1}$ in the tetrahedral complex.

Udovich et al.[724] studied the infrared spectra of $Ni(DPE)X_2$, where DPE is 1,2-bis(diphenylphosphino)ethane and X is Cl, Br, and I, by using the metal isotope technique. It was found that the $\nu$(NiX) are always lower and $\nu$(NiP) are always higher in the *cis*-$Ni(DPE)X_2$ than in the corresponding *trans*-$Ni(PEt_3)_2X_2$. This difference has been attributed to the strong *trans*-influence of phosphine ligands.

TABLE III-56. INFRARED FREQUENCIES, ISOTOPIC SHIFTS, AND BAND ASSIGNMENTS OF $NiX_2(PEt_3)_2$ (X = Cl AND Br)($cm^{-1}$)[723]

| $PEt_3$ | $^{58}NiCl_2(PEt_3)_2$ | | $^{58}NiBr_2(PEt_3)_2$ | | |
|---|---|---|---|---|---|
| $\nu$ | $\nu$ | $\Delta\nu^a$ | $\nu$ | $\Delta\nu^a$ | Assignment$^b$ |
| 408 | 416.7 | 0.0 | 413.6 | 1.2 | $\delta$(CCP) |
| — | 403.3 | 6.7 | 337.8 | 10.5$^c$ | $\nu$(NiX) |
| 365 | 372.5 | −0.1 | 374.0 | 1.1 | $\delta$(CCP) |
| 330 | 329.0 | −0.5 | 327.8 | 0.5$^c$ | $\delta$(CCP) |
| — | 273.4 | 5.9 | 265.0 | 4.7 | $\nu$(NiP) |
| 245 | (hidden) | | (hidden) | | $\delta$(CCP) |
| | 200.2 | 0.8 | 190.4 | 0.7 | $\delta$(CPC) |
| | 186.5 | −0.2 | 155.1 | 1.5 | $\delta$(NiX) |
| | 161.5 | −0.5 | (hidden) | | $\delta$(NiP) |

$^a$ $\Delta\nu$ indicates metal isotope shift, $\nu(^{58}Ni) - \nu(^{62}Ni)$.

$^b$ Ligand vibrations were assigned according to Ref. 789.

$^c$ Since these two bands are overlapped (Fig. III-50), $\Delta\nu$ values are only approximate.

A number of investigators have discussed the nature of the M–P bonding, based on electronic, vibrational, and NMR spectra,[776] and controversy has arisen about the degree of $\pi$-back-bonding in the M—P bond. For example, Park and Hendra[792] suggest the presence of a considerable degree of $\pi$-bonding in square-planar Pd(II) and Pt(II) complexes of $PMe_3$ and $AsMe_3$. On the other hand, Venanzi[793] claims from NMR evidence that the Pt–P $\pi$-bonding is much less than originally predicted.[794] It is rather difficult, however, to discuss the degree of $\pi$-bonding from vibrational spectra alone since the MP stretching frequency and force constant are determined by the net effect, which involves both $\sigma$- and $\pi$-bonding.

Complexes of the type $M(CO)_5L$, where L is arsine ($AsH_3$) and stibine ($SbH_3$) and M is Cr, Mo, and W, have been prepared by Fischer et al.[795] $\nu$(AsH) and $\delta$ ($AsH_3$) are near 2200 and 900 $cm^{-1}$, respectively. Complexes of trimethylarsine ($AsMe_3$) have been studied by several investigators. Goodfellow et al.[796] and Park and Hendra[792] measured the infrared spectra of $M(AsMe_3)_2X_2$ (M = Pt and Pd; X = Cl, Br, and I) type

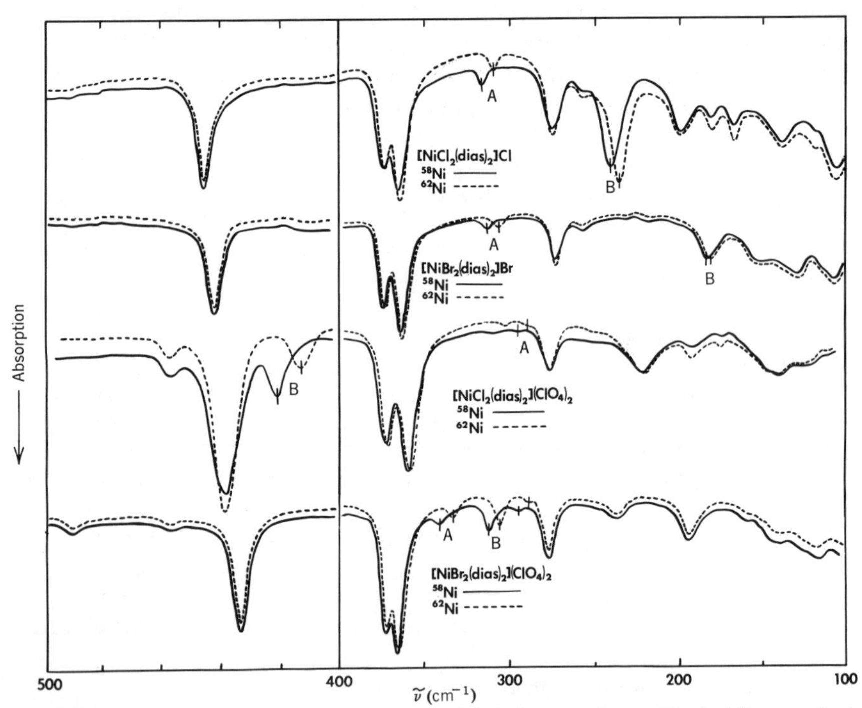


**Fig. III-51.** Far-infrared spectra of octahedral nickel dias complexes. Vertical lines marked by A and B indicate Ni-As and Ni-X stretching modes, respectively.

complexes and assigned $\nu$(MAs) in the 300–260 cm$^{-1}$ region. The latter workers assigned $\nu$(MSb) of analogous alkylstibine complexes at ca. 200 cm$^{-1}$. Konya and Nakamoto[721] assigned $\nu$(MAs) and $\nu$(MX) of $[M(dias)_2]^{2+}$ and $[M(dias)_2X_2]Y_n$ type complexes by using the metal isotope technique. Figure III–51 shows the infrared spectra of $[^{58}Ni(dias)_2X_2]X$ and $[^{58}Ni(dias)_2X_2](ClO_4)_2$ (X = Cl and Br) and their $^{62}$Ni analogs. Their results show that the $\nu$(MAs) are very weak and appear at 325–295 cm$^{-1}$ for the Ni, Co, and Fe complexes and at 270–210 cm$^{-1}$ for the Pd and Pt complexes. For the $\nu$(MX) of these complexes, see Sec. III-16.

Tertiary phosphine oxides and arsine oxides coordinate to a metal through their O atoms. The $\nu$(P=O) of triphenylphosphine oxide (TPPO) at 1193 cm$^{-1}$ is shifted by ca. 35 cm$^{-1}$ to a lower frequency when it coordinates to Zn(II).[797] The shift is much larger in $MX_4(TPPO)_2$ (160–120 cm$^{-1}$), where $MX_4$ is a tetrahalide of Pa, Np, and Pu.[798] A similar observation has been made for $\nu$ (As=O) of arsine oxide complexes. Exceptions to this rule are found in $MnX_2(Ph_3AsO)_2$ (X = Cl and Br); their $\nu$(As=O) are higher by 30–20 cm$^{-1}$ than the frequency of the free ligand (880 cm$^{-1}$).[799] Rodley et al.[800] have assigned the $\nu$(MO) of tertiary arsine oxide complexes at 440–370 cm$^{-1}$.

## III-19. COMPLEXES OF SULFUR, SELENIUM, AND TELLURIUM-CONTAINING LIGANDS

A large number of metal complexes of ligands containing sulfur, selenium, and tellurium are known. Here the vibrational spectra of typical compounds will be reviewed briefly. For $SO_3$ and thiourea complexes that form metal–sulfur bonds, see Sec. III-8 and III-20, respectively.

### (1) Complexes of Relatively Simple Ligands

Like molecular oxygen, $S_2$ and $Se_2$ form $\pi$-complexes with transition metals. The $\nu$(SS) of $[Rh(S_2)(DPE)_2]Cl$* and the $\nu$(SeSe) of $[Ir(Se_2)(DME)_2]Cl$† are at 525 and 310 cm$^{-1}$, respectively.[801] These values are much lower than those of free $S_2$ (726 cm$^{-1}$) and $Se_2$(392 cm$^{-1}$). The $[Pt(S_5)_3]^{2-}$ and $[Pt(S_5)_2]^{2-}$ ions contain three and two $S_5$ chelate rings, respectively. The $\nu$(PtS) of the former was assigned at 294 cm$^{-1}$.[802]

Allkins and Hendra[803] carried out an extensive vibrational study on *cis*- and *trans*-$[MX_2Y_2]$ and their halogen-bridged dimers, where M is Pd(II) and Pt(II), X is Cl, Br, and I, and Y is $(CH_3)_2S$, $(CH_3)_2Se$, and $(CH_3)_2Te$. The $\nu$(MS), $\nu$(MSe), and $\nu$(MTe) were assigned in the ranges

---

*DPE: $Ph_2P—CH_2—CH_2—PPh_2$.
†DME: $(CH_3)_2P—CH_2—CH_2—P(CH_3)_2$.

350–300, 240–170, and 230–165 cm$^{-1}$, respectively. The vibrational spectra of $PtX_2L_2$,[804] $PdX_2L_2$,[805] $AuX_3L$, and $AuXL$,[806] where X is a halogen and L is a dialkylsulfide, have been assigned. Aires et al.[807] reported the infrared spectra of $MX_3L_3$ type compounds, where M is Ru(III), Os(III), Rh(III), and Ir(III), X is Cl or Br, and L is $Et_2S$ and $Et_2Se$. The $\nu$(MS) and $\nu$(MSe) of these compounds were assigned at 325–290 and 225–200 cm$^{-1}$, respectively, based on the *fac*-structure. On the other hand, Allen and Wilkinson[808] proposed the *mer*-structure for these compounds, based on far-infrared and other evidence.

According to X-ray analysis, $Pd_2Br_4(Me_2S)_2$ is bridged via Br atoms, whereas $Pt_2Br_4(Et_2S)_2$ is bridged via S atoms:[809]

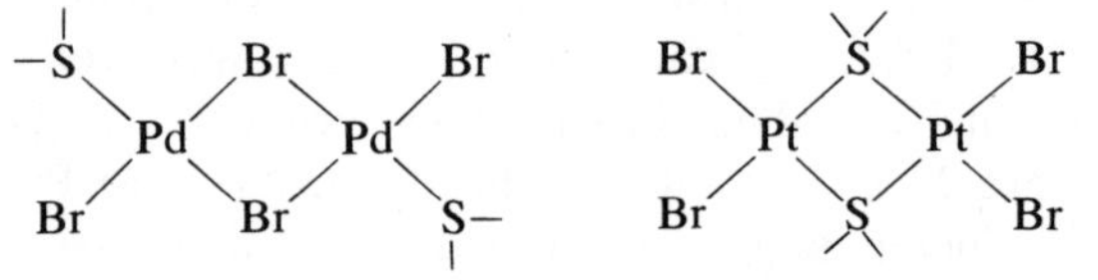

Adams and Chandler[810] have shown that halogen-bridged $Pd_2Cl_4(SEt_2)_2$ exhibits $\nu(PdCl_t)$, $\nu(PdCl_b)$, and $\nu$(PdS) at 366, 266, and 358 cm$^{-1}$, respectively, whereas sulfur-bridged $Pt_2Cl_4(SEt_2)_2$ exhibits $\nu(PtCl_t)$ and $\nu$(PtS) at 365–325 and 422–401 cm$^{-1}$, respectively. In $Fe_2S_2(CO)_6$, the two Fe atoms are bridged by S atoms:

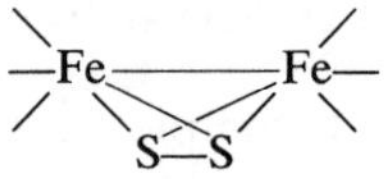

Its $\nu$(FeS) and $\nu$(FeFe) are reported to be at 329 and 191 cm$^{-1}$, respectively. However, the former is only 50% pure Fe—S vibration.[811]

The vibrational spectra of $CS_2$ and CS, which form complexes with a variety of transition metals, have been reviewed by Butler and Fenster.[812] According to Wilkinson et al.,[813,814] $CS_2$ coordinates to the metal in four ways:

I

II

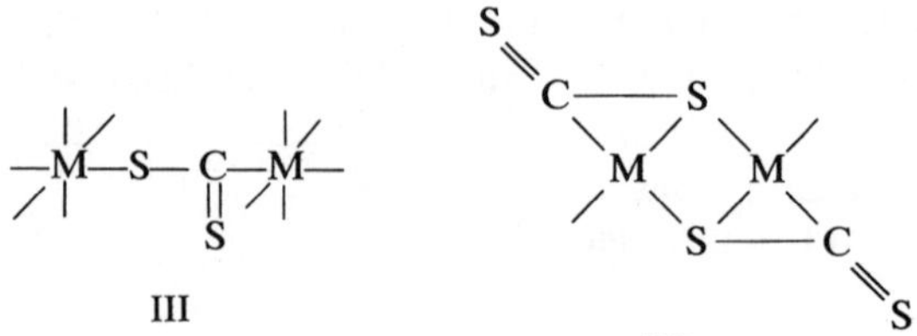

III

IV

Free $CS_2$ exhibits $\nu(CS_2)$ at 1533 $cm^{-1}$. The $\nu(CS_2)$ of structure I ($\pi$-bonded) and structure II ($\sigma$-bonded) are at ca. 1100 and 1510 $cm^{-1}$, respectively. The $\nu$(CS) of the bridging CS groups are at ca. 980–840 $cm^{-1}$ for structure III,[814] and 1120 $cm^{-1}$ for structure IV.[813] For the infrared spectra of other $CS_2$ complexes, see Refs. 814–816. From infrared and other evidence, $[Ir(CS_2)(CO)(PPh_3)_3]BPh_4$ was originally thought to be a six-coordinate complex with a $\pi$-bonded $CS_2$.[816] However, recent X-ray analysis[817] revealed an unexpected structure; a five-coordinate complex of the type $[Ir(CO)(PPh_3)_2(S_2C{-}PPh_3)]BPh_4$.

Like CO, thiocarbonyl (CS) forms complexes with transition metals, Thus far, the following three structures have been found or suggested:

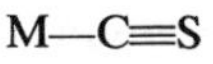

M—C≡S (V)

M(μ-C=S)₂M (VI)

M—C≡S—M′ (VII)

The $\nu$(CS) of free CS is at 1275 $cm^{-1}$. The $\nu$(CS) of the C-bonded terminal CS group (Structure V) is higher than that of free CS (1360–1290 $cm^{-1}$).[814,815] The bridging C-bonded structure (VI) was suggested for $Mn_2(\pi{-}Cp)_2(NO)_2(CS)_2$.[818] The bridging C and S-bonded structure (VII) was suggested for $(DPE)_2(CO)W(CS)W(CO)_5$ since its $\nu$(CS) is very low (1161 $cm^{-1}$).[819] In $[Ir(CS)(CS_2)(PPh_3)_2]^+$, $\nu$(CS) is at 1305 $cm^{-1}$, and the $\nu(CS_2)$ of the $\pi$-bonded ligand are found at 1106 and 1009 $cm^{-1}$.[816]

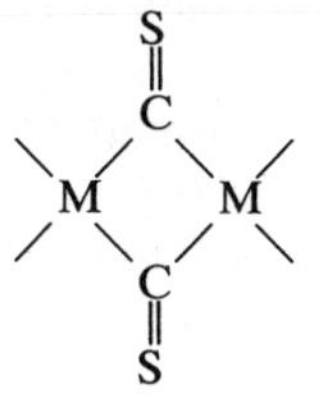

The infrared spectra of thiocarbonato complexes of the type $[M(CS_3)_2]^{2-}$ [M = Ni(II), Pd(II), and Pt(II)] have been studied by Burke and Fackler[820] and Cormier et al.[821] The latter workers carried out normal coordinate analyses on the $[^{58}Ni(CS_3)_2]^{2-}$ ion and its $^{62}Ni$ analog. The $\nu$(NiS) were assigned at 385 and 366 $cm^{-1}$, with a corresponding force constant of 1.41 mdyn/Å (UBF).

Sulfur dioxide ($SO_2$) may take one of the following structures when it coordinates to a metal:

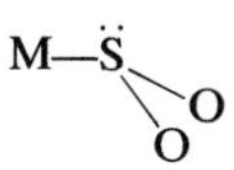

VIII: M—S̈(O)O (pyramidal)

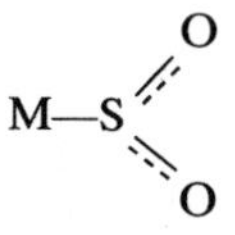

IX: M—S(O)₂ (planar)

X: M—O—S=O

XI: M—S(=O)₂—M

XII: M—S(=O) with M—O—S ring

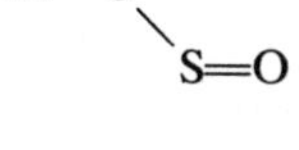

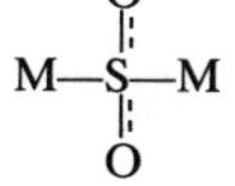

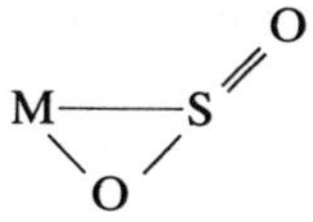

Free $SO_2$ exhibits $\nu_a(SO_2)$ and $\nu_s(SO_2)$ at 1362 and 1151 $cm^{-1}$, respectively. In $IrCl(CO)(SO_2)(PPh_3)_2$, $SO_2$ is bonded to the metal through the S atom by the formation of a pyramidal structure (VIII), and its $SO_2$ stretching frequencies are at 1198–1185 ($\nu_a$) and 1048 $cm^{-1}$ ($\nu_s$).[822] In $[Ru(NH_3)_4(SO_2)Cl]Cl$, the $SO_2$ molecule is bonded to the metal via the S atom by the formation of a planar structure (IX), and its $\nu_a(SO_2)$ and $\nu_s(SO_2)$ are at 1301–1278 and 1100 $cm^{-1}$, respectively.[823] Although the $SO_2$ stretching frequencies of the latter are higher than those of the former, they are lower than those of free $SO_2$ in both cases. In $[SbF_5(SO_2)]$, $SO_2$ is bonded to the metal via the O atom, as shown in structure X,[824] and its $\nu(SO_2)$ are observed at 1327 ($\nu_a$) and 1102 $cm^{-1}$ ($\nu_s$).[825] According to Byler and Shriver,[825] $SO_2$ in a complex is O-bonded if $(\nu_a - \nu_s)$ is larger than 190 $cm^{-1}$, and is S-bonded if it is smaller than 190 $cm^{-1}$. According to X-ray analysis,[826] $SO_2$ in $[(\pi\text{-Cp})Fe(CO)_2]_2(SO_2)$ takes the bridging structure XI. As expected, the $\nu(SO_2)$ of this compound (1135 and 993 $cm^{-1}$) are lower than those of the other types discussed above. Recently, a chelate structure (XII) was found from X-ray analysis of $[Rh(SO_2)(NO)(PPh_3)_2]$.[826a]

Müller and co-workers[827] carried out an extensive vibrational study on $MS_4^{n-}$, $MS_3O^{n-}$, and $MS_2O_2^{n-}$ type ions (Part II) and their metal complexes. Complete normal coordinate analyses have been performed on $[^{58}Ni(^{92}MoS_4)_2]^{2-}$ and its $^{62}Ni$ and $^{100}Mo$ analogs,[828] and empirical assignments have been made for $[Ni(WS_3O)_2]^{2-}$,[829] and $[Ni(WS_2O_2)_2]^{2-}$.[830] Structures such as the following have been deduced from the infrared spectra:

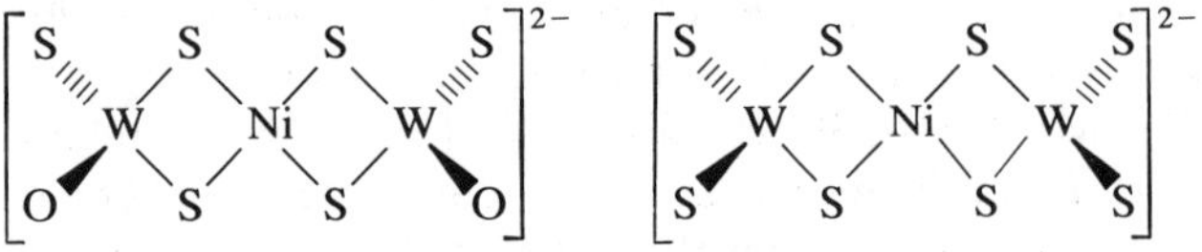

$C_{2h}$ or $C_{2v}$ $D_{2h}$

The infrared spectra of $M(S_2PR_2)_2$, where M is Ni(II), Pd(II), and Pt(II), and R is $CH_3$, $C_6H_5$, and so on, have been reported.[831]

### (2) Complexes of Relatively Large Ligands

2,5-Dithiahexane (dth) forms metal complexes such as $[ReCl_3(dth)]_n$ and $Re_3Cl_9(dth)_{1.5}$. Cotton et al.[832] showed from infrared spectra that dth of the former forms a chelate ring in the *gauche* conformation, whereas that of the latter forms a bridge between two metals by taking the *trans* conformation (see Sec. III-2). Infrared spectra have been used to show that ethanedithiol forms a chelate ring of the *gauche* conformation in $Bi(S_2C_2H_4)X$, where X is Cl and Br.[833] Schläpfer et al.[834] assigned the $\nu$(NiS) and $\nu$(NiN) of dth, ete [2-(ethylthio)ethylamine], and mea [mercaptoethylamine] complexes with the metal isotope technique:

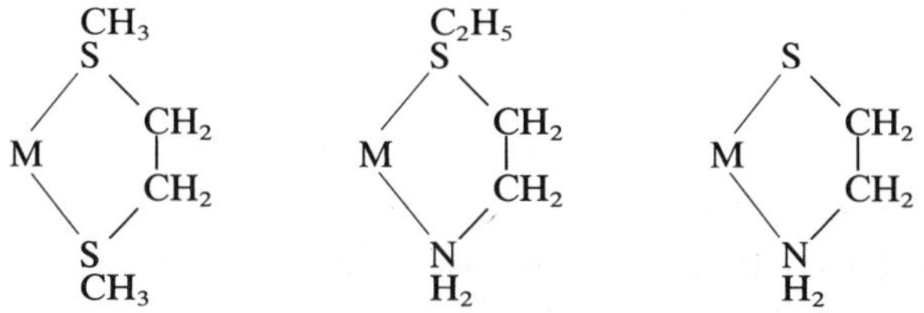

dth ete mea

The infrared spectra of *N,N*-dialkyldithiocarbamato complexes have been studied extensively. All these compounds exhibit strong $\nu$(C═N) bands in the 1600–1450 $cm^{-1}$ region. These compounds are roughly classified into two types:

Bidentate coordination

Unidentate coordination

The former exhibits $\nu$(CS) near 1000 $cm^{-1}$ as a single band, whereas the latter shows a doublet in the same region.[835] Also, the $\nu$(C═N) of the former (above 1485 $cm^{-1}$) is higher than that of the latter (below 1485 $cm^{-1}$).[836] The $\nu$(MS) of the bidentate complexes are observed at 400–300 $cm^{-1}$.[837] The infrared spectra of diselenocarbamato complexes have been reported,[838] and assigned on the basis of normal coordinate analysis.[839] In the Ni(II) complex, $\nu$(NiSe) is assigned at 298 $cm^{-1}$, which is lower by 85 $cm^{-1}$ than the $\nu$(NiS) of the corresponding dithiocarbamato complex. The infrared spectra of xanthato complexes:

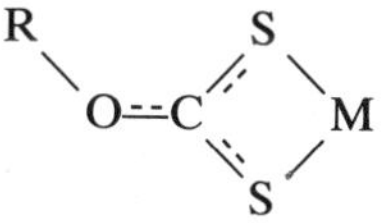

have been studied by Watt and McCormick.[840] The $\nu$(CO), $\nu$(CS), and $\nu$(MS) were assigned at 1325–1250, 760–540, and 380–340 $cm^{-1}$, respectively.

Savant et al.[841] roughly classified monothiobenzoato complexes into three categories:

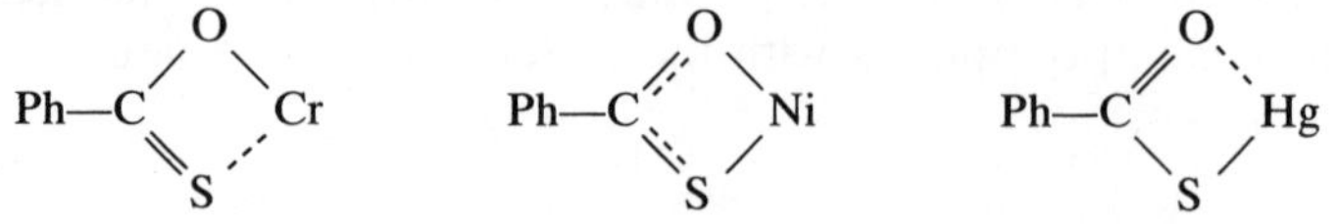

| | Ph—C(O)(S)Cr | Ph—C(O)(S)Ni | Ph—C(O)(S)Hg |
|---|---|---|---|
| $\nu$(C=O)(cm$^{-1}$) | 1465 | 1508 | 1630 |
| $\nu$(CS)(cm$^{-1}$) | 982 | 958 | 912 |

In the Hg(II), Cu(I) and Ag(I) complexes, coordination occurs mainly via sulfur. In the Cr(III) complex, however, the Cr–O bond is stronger than the Cr—S bond. The Ni(II) complex is between these two cases and is close to symmetrical coordination. This is reflected in the frequency trends (cm$^{-1}$) shown above.

In thiocarboxylato complexes of the type Ni(R—COS)$_2\cdot\frac{1}{2}$(EtOH) (R = $CH_3$, $C_2H_5$, and Ph), the ligand serves as a bridge between two metals as shown:

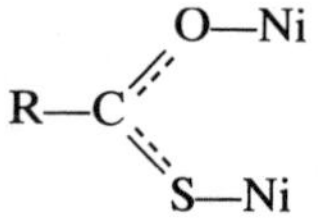

R—C(=O—Ni)(=S—Ni)

and their $\nu$(CO) are reported to be at 1580–1520 cm$^{-1}$.[842]

The infrared spectra of metal complexes of thiosemicarbazides

$NH_2$—C(=S)—NH—$NH_2$

and thiosemicarbazones

$NH_2$—C(=S)—NH—N=C$R_1R_2$; ($R_1$ and $R_2$ are alkyl, phenyl, etc.)

have been reviewed by Campbell.[843] For metal complexes of dithiocarbazic acid, infrared spectra support the N,S-chelated structure (shown below) rather than the S,S-chelated' structure normally found for dithiocarbamato complexes.[844] In 1:1 complexes of N,N-monosubstituted dithiooxamides, Desseyn et al.[845] concluded from infrared spectra that metals such as Ni(II) and Cu(II) are primarily bonded to the N, whereas metals such as Hg(II), Pb(II), and Pd(II) are bonded to the S, atom.

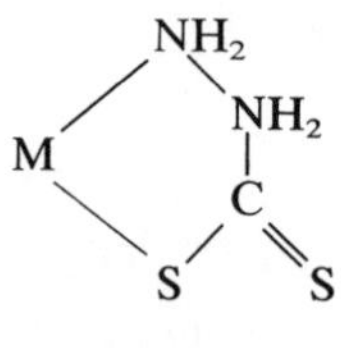

Dithiocarbazato complex

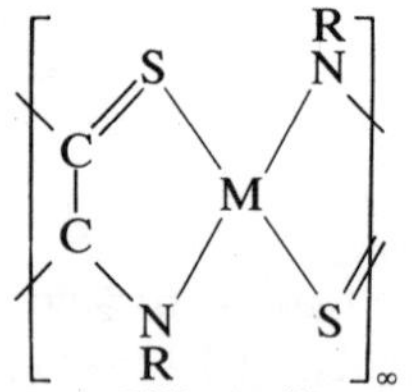

Dithiooxamido complex

Coucouvanis et al.[846] synthesized novel tin halide adducts of Ni(II) and Pd(II) dithiooxalato (DTO) complexes:

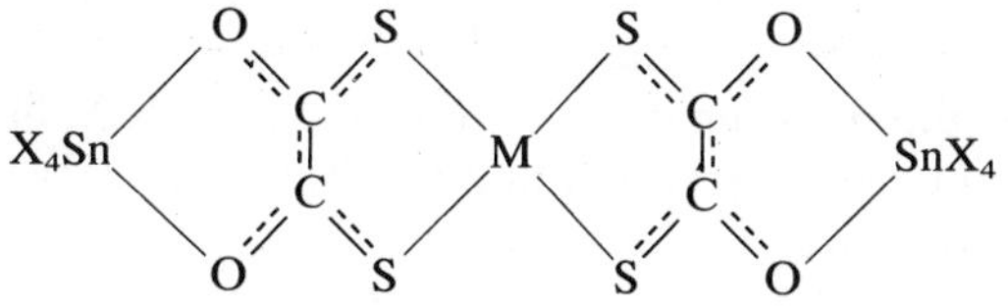

The $\nu$(NiS) of the $[Ni(DTO)_2]^{2-}$ ion is at 349 cm$^{-1}$. In $SnX_4$ (X = Cl, Br, and I) adducts, this band shifts to 385–375 cm$^{-1}$, indicating a strengthening of the Ni–S bond by complexation. It was found that $Cr(DTO)_3[Cu(PPh_3)_2]_3$ exists in two isomeric forms;[847] one in which the Cr atom is bonded to sulfur, and another in which it is bonded to oxygen of the DTO ion. As expected, $\nu$(C=O), $\nu$(CS), and $\nu$[CrO(S)] are markedly different between these two isomers.

Metal complexes of 1,2-dithiolates (or dithienes) have been of great interest to inorganic chemists because of their redox properties.[848] Schläpfer and Nakamoto[849] prepared a series of complexes of the type $[Ni(S_2C_2R_2)_2]^n$, where R is H, Ph, $CF_3$, and CN and $n$ is 0, −1, or −2, and carried out normal coordinate analysis to obtain rough estimates of the charge distribution based on the calculated force constants.

Infrared spectra of metal complexes with thio-$\beta$-diketones have been reviewed briefly.[850,851] Siimann and Fresco[852] and Martin et al.[853] carried out normal coordinate analyses on metal complexes of dithioacetylacetone, monothioacetylacetone, and related ligands. For dithioacetylacetonato complexes, two $\nu$(MS) have been assigned at 390–340 and 300–260 cm$^{-1}$.

L-Cysteine has three potential coordination sites (S, N, and O), and infrared spectra have been used to determine the structures of its metal complexes. For example, the Zn(II) complex shows no $\nu$(SH), and its carboxylate frequency indicates the presence of a free $COO^-$ group. Thus the following structure was proposed:[854]

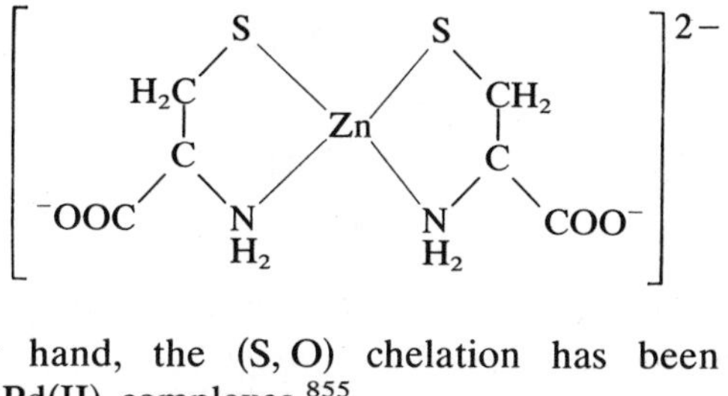

On the other hand, the (S, O) chelation has been suggested for the Pt(II) and Pd(II) complexes.[855]

McAuliffe et al.[856] studied the infrared spectra of metal complexes of methionine [$CH_3$—S—$CH_2$—$CH_2$—CH($NH_2$)—COOH]. They found that most of the metals they studied [Except Ag(I)] coordinate through the $NH_2$ and $COO^-$ groups, and that the $CH_3S$ groups of these complexes are available for further coordination to other metals. McAuliffe[857] suggested that complexes of the type M(methionine)$Cl_2$ [M = Pd(II) and Pt(II)] take the polymeric structure:

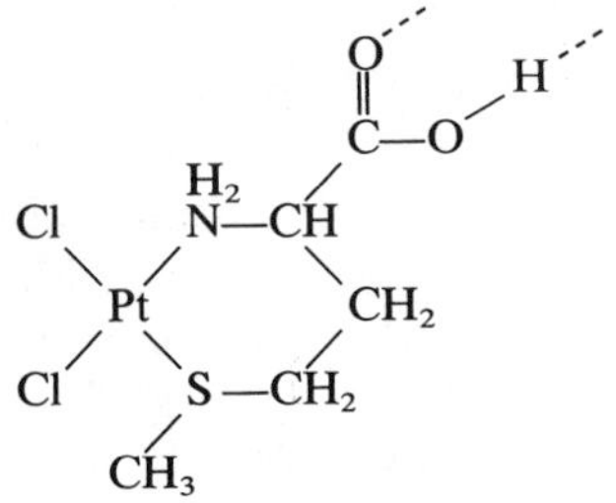

```
                           O···
                           ‖        H···
                           C—O ⁄
              H2         ⁄
    Cl        N—CH
      \     /      \
        Pt          CH2
      /     \      /
    Cl        S —CH2
            /
         CH3
```

Infrared spectra of metal complexes of sulfur-containing ligands are reported for 2-methylthioaniline,[858] 8-mercaptoquinoline,[859] cyclic thioethers,[860] and *N*-alkylthiopicolinamides.[861]

## III-20. COMPLEXES OF UREA, SULFOXIDES, AND RELATED COMPOUNDS

Linkage isomerism involving $NO_2$ (Sec. III-4), NCS, and so forth (Sec. III-11) has already been discussed. Here the vibrational spectra of linkage isomers involving larger ligands are reviewed.

### (1) Complexes of Urea and Related Ligands

Penland et al.[862] first studied the infrared spectra of urea complexes to determine whether coordination occurs through nitrogen or oxygen. The electronic structure of urea may be represented by a resonance hybrid of structures I, II, and III, with each contributing roughly an equal amount:

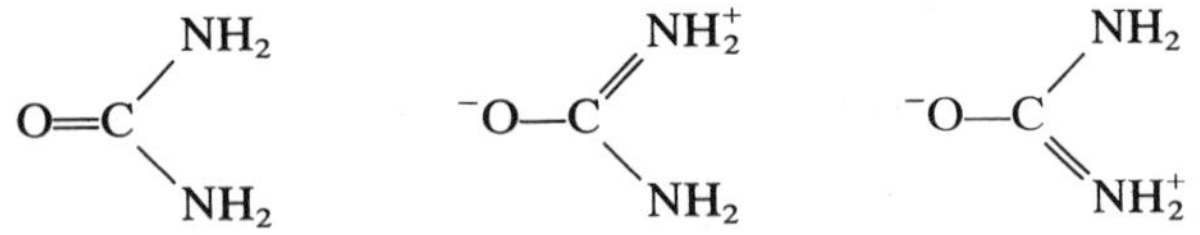

```
         NH2              NH2+              NH2
        /               //                /
   O=C             -O—C              -O—C
        \               \                 \\
         NH2              NH2               NH2+

     I                 II                III
```

If coordination occurs through nitrogen, the contributions of structures II and III will decrease. This results in an increase of the CO stretching frequency with a decrease in the CN stretching frequency. The NH stretching frequency in this case may fall in the same range as the value for the amido complexes (Sec. III-1). If coordination occurs through oxygen, the contribution of structure I will decrease. This may result in a decrease of the CO stretching frequency but no appreciable change in the NH stretching frequency. Since the spectrum of urea itself has been analyzed completely,[863] band shifts caused by coordination can be checked immediately. The results shown in Table III-57 indicate that coordination occurs through nitrogen in the Pt(II) complex, and through oxygen in the Cr(III) complex. It was also found that Pd(II) coordinates to the nitrogen, whereas Fe(III), Zn(II), and Cu(II) coordinate to the oxygen of urea. The infrared spectra of tetramethylurea (tmu) complexes of lanthanide elements, $[Ln(tmu)_6](ClO_4)_3$, indicate the presence of O-coordination.[864]

From infrared studies on thiourea $[(NH_2)_2CS]$ complexes, Yamaguchi et al.[865] found that all the metals studied (Pt, Pd, Zn, and Ni) form M–S bonds, since the CN stretching frequency increases and the CS stretching frequency decreases upon coordination, without an appreciable change in the NH stretching frequency. On the basis of the same criterion, thiourea complexes of Fe(II),[866] Mn(II), Co(II), Cu(I), Hg(II), Cd(II), and Pb(II) were shown to be S-bonded.[867] Several investigators[868–870] studied the far-infrared spectra of thiourea complexes and assigned the MS stretching bands between 300 and 200 $cm^{-1}$. Thus far, the only metal reported to be N-bonded is Ti(IV).[871] Infrared spectra of alkylthiourea complexes have also been studied. Lane and colleagues[872] studied the infrared spectra of methylthiourea complexes and concluded that methylthiourea forms M–S bonds with Zn(II) and Cd(II) and M—N bonds with Pd(II), Pt(II), and Cu(I). For other alkylthiourea complexes, see Refs. 873 and 874. Infrared spectra of selenourea (su) complexes of Co(II), Zn(II), Cd(II), and Hg(II) exhibit $\nu$(MSe) in the 245–167 $cm^{-1}$ region.[875] The Raman spectra of

TABLE III-57. SOME VIBRATIONAL FREQUENCIES OF UREA AND ITS METAL COMPLEXES ($CM^{-1}$)[862]

| $[Pt(urea)_2Cl_2]$ | Urea | $[Cr(urea)_6]Cl_3$ | Predominant Mode |
|---|---|---|---|
| 3390, 3290 | 3500, 3350 | 3440, 3330 | $\nu(NH_2)$, free |
| 3130, 3030 | | 3190 | $\nu(NH_2)$, bonded |
| 1725 | 1683 | 1505[a] | $\nu$(C=O) |
| 1395 | 1471 | 1505[a] | $\nu_a$(CN) |

[a] $\nu$(C=O) and $\nu$(C—N) couple in the Cr complex.

$[Pd(su)_4]^{2+}$ and $[Pt(su)_4]^{2+}$ ions exhibit the $A_{1g}$ $\nu$(MSe) at 178 and 191 cm$^{-1}$, respectively.[876]

Linkage isomerism was found for the formamidopentamminecobalt (III): $[(NH_3)_5Co(—NH_2CHO)]^{3+}$ and $[(NH_3)_5Co(—OCHNH_2)]^{3+}$. Although little difference was found in the $\nu$(C═O) region, the N-isomer showed the aldehyde $\nu$(CH) at 2700 cm$^{-1}$, whereas such a band was not obvious in the O-isomer.[877]

## (2) Complexes of Sulfoxides and Related Compounds

Cotton et al.[878] studied the infrared spectra of sulfoxide complexes to see whether coordination occurs through oxygen or sulfur. The electronic structure of sulfoxides may be represented by a resonance hybrid of these structures:

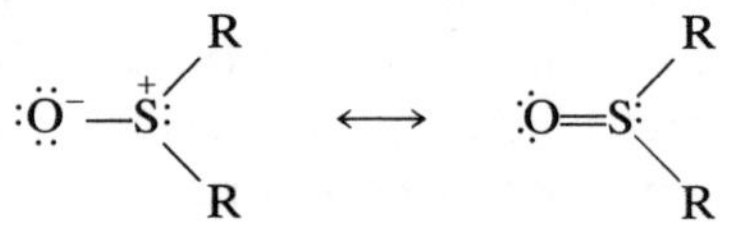

$$:\ddot{\underset{..}{O}}^{-}—\overset{+}{\underset{..}{S}}(R)_2 \quad \longleftrightarrow \quad \ddot{O}═\underset{..}{S}(R)_2$$

IV V

If coordination occurs through oxygen, the contribution of structure V will decrease and result in a decrease in $\nu$(S═O). If coordination occurs through sulfur, contribution of structure IV will decrease and may result in an increase in $\nu$(S═O). It has been concluded that coordination occurs through oxygen in the $Co(DMSO)_6^{2+}$ ion, since the $\nu$(S═O) of this ion (950 cm$^{-1}$) is lower than that of free DMSO (dimethylsulfoxide), which absorbs at 1100–1055 cm$^{-1}$. On the other hand, coordination may occur through sulfur in $PdCl_2(DMSO)_2$ and $PtCl_2(DMSO)_2$, since $\nu$(S═O) of these compounds (1157–1116 cm$^{-1}$) are higher than the value for the free ligand. Other ions such as Mn(II), Fe(II, III), Ni(II), Cu(II), Zn(II), and Cd(II) are all coordinated through oxygen, since the DMSO complexes of these metals exhibit $\nu$(S═O) between 960 and 910 cm$^{-1}$. Drago and Meek,[879] however, assigned $\nu$(S═O) of O-bonded complexes in the 1025–985 cm$^{-1}$ region, since they are metal sensitive. The bands between 960 and 930 cm$^{-1}$, which were previously assigned to $\nu$(S═O) are not metal sensitive and are assigned to $\rho_r(CH_3)$. Even so, $\nu$(S═O) of O-bonded complexes are lower than the value for free DMSO. To confirm $\nu$(S═O) assignments, it is desirable to compare the spectra of the corresponding DMSO-$d_6$ complexes since $\rho_r(CD_3)$ is outside the $\nu$(S═O) region. Table III-58 lists $\nu$(S═O) of typical compounds.

Wayland and Schramm[880] found the first example of mixed coordination of DMSO in the $[Pd(DMSO)_4]^{2+}$ ion; it exhibits two S-bonded $\nu$(S═O) at 1150 and 1140 cm$^{-1}$, and two O-bonded $\nu$(S═O) at 920 and

TABLE III-58. SO STRETCHING FREQUENCIES OF DMSO COMPLEXES ($cm^{-1}$)

| Compound | $\nu(S{=}O)$ | Bonding | Ref. |
|---|---|---|---|
| $Sn(DMSO)_2Cl_4$ | 915 | O | 880 |
| $[Cr(DMSO)_6](ClO_4)_3$ | 928 | O | 880 |
| $[Ni(DMSO)_6](ClO_4)_2$ | 955 | O | 880 |
| $[Ln(DMSO)_8](ClO_4)_3$, Ln = La, Ce, Pr, Nd | 998–992 | O | 881 |
| $[Al(DMSO)_6]X_3$, X = Cl, Br, I | 1000–1008 | O | 882 |
| $[Ru(NH_3)_5(DMSO)](PF_6)_2$ | 1045 | S | 883 |
| *trans*-$[Pd(DMSO)_2Cl_2]$ | 1116 | S | 884 |
| *cis*-$[Pt(DMSO)_2Cl_2]$ | 1134<br>1157 | S | 885 |

905 $cm^{-1}$. Thus the infrared spectrum is most consistent with a configuration in which two S-bonded and two O-bonded DMSO are in the *cis* position. The infrared and NMR spectra of $Ru(DMSO)_4Cl_2$ suggested a mixing of O- and S-coordination; $\nu(S{=}O)$ at 1120 and 1090 $cm^{-1}$ for S-coordination and at 915 $cm^{-1}$ for O-coordination.[886] X-ray analysis[887] has since shown that two Cl atoms are in the *cis*-positions of an octahedron and the remaining positions are occupied by one O-bonded and three S-bonded DMSO ligands.

Complete assignments on infrared and Raman spectra of *trans*-$Pd(DMSO)_2X_2$ (X = Cl and Br) and their deuterated analogs have been made by Tranquille and Forel.[888] Berney and Weber[889] found the order of $\nu(MO)$ in the $[M(DMSO)_6]^{n+}$ ion to be as follows:

| M = | Cr(III) | | Ni(II) | | Co(II) | | Zn(II) | | Fe(II) | | Mn(II) |
|---|---|---|---|---|---|---|---|---|---|---|---|
| $\nu(MO)$ ($cm^{-1}$) | 529 | > | 444 | > | 436 | > | 431 | > | 438<br>415 | > | 418 |

Ligands such as DPSO(diphenylsulfoxide) and TMSO (tetramethylenesulfoxide) do not exhibit the $CH_3$ rocking bands near 950 $cm^{-1}$. Thus the SO streching bands of metal complexes containing these ligands can be assigned without difficulty. Table III-59 lists the SO stretching frequencies and the magnitude of band shifts in DPSO complexes.[890] Van Leeuwen and Groeneveld[890] noted that the shift becomes larger as the electronegativity of the metal increases. In Table III-59, the metals are listed in the order of increasing electronegativity.

TABLE III-59. SHIFTS OF SO STRETCHING BANDS IN DPSO AND DMSO COMPLEXES ($cm^{-1}$)[890]

| | DPSO Complex | | DMSO Complex |
|---|---|---|---|
| Metal | ν(SO) | Shift | Shift |
| Ca(II) | 1012–1035 | 0−(−23) | — |
| Mg(II) | 1012 | −23 | — |
| Mn(II) | 983–991 | −45 | −41 |
| Zn(II) | 987–988 | −47 | — |
| Fe(II) | 987 | −48 | — |
| Ni(II) | 979–982 | −55 | −45 |
| Co(II) | 978–980 | −56 | −51 |
| Cu(II) | 1012, 948 | −23, −87 | −58 |
| Al(III) | 942 | −93 | — |
| Fe(III) | 931 | −104 | — |

In $[M(DTHO_2)_3]^{2+}$ [M = Co(II), Ni(II), Mn(II), etc], the metals are O-bonded since the ν(S=O) of free ligand (1055–1015 $cm^{-1}$) are shifted to lower frequencies by 40–22 $cm^{-1}$:

$$CH_3-S(=O)-(CH_2)_2-S(=O)-CH_3$$

2,5-Dithiahexane-2,5-dioxide ($DTHO_2$)

On the other hand, the metals are S-bonded in $M(DTHO_2)Cl_2$ [M = Pt(II) and Pd(II)] since ν(S=O) are shifted to higher frequencies by 108–77 $cm^{-1}$.[891] Dimethylselenoxide, $(CH_3)_2Se=O$, forms complexes of the $MCl_2$ $(DMSeO)_n$ type, where M is Hg(II), Cd(II), Cu(II), and so on, and $n$ is 1, $1\frac{1}{2}$, or 2. The ν(Se=O) of the free ligand (800 $cm^{-1}$) is shifted to the 770–700 $cm^{-1}$ region, indicating the O-bonding in these complexes.[892]

## References


1. K. H. Schmidt and A. Müller, *Coord. Chem. Rev.*, **19,** 41 (1976).
2. R. Plus, *J. Raman Spectrosc.*, **1,** 551 (1973).
3. N. Tanaka, M. Kamada, J. Fujita, and E. Kyuno, *Bull. Chem. Soc. Jap*, **37,** 222 (1964).
4. T. V. Long, II, and D. J. B. Penrose, *J. Am. Chem. Soc.*, **93,** 632 (1971).
5. K. H. Schmidt and A. Müller, *J. Mol. Struct*, **22,** 343 (1974).
6. K. H. Schmidt and A. Müller, *Inorg. Chem.*, **14,** 2183 (1975).
7. M. B. Fairey and R. J. Irving, *Spectrochim. Acta*, **22,** 359 (1966).

8. W. P. Griffith, *J. Chem. Soc., A*, 899 (1966).
9. T. Shimanouchi and I. Nakagawa, *Inorg. Chem.*, **3,** 1805 (1964).
10. L. Sacconi, A. Sabatini, and P. Gans, *Inorg. Chem.*, **3,** 1772 (1964).
11. K. H. Schmidt, W. Hauswirth, and A. Müller, *J. Chem. Soc.*, Dalton, 2199 (1975).
12. H. Siebert and H. H. Eysel, *J. Mol. Struct*, **4,** 29 (1969).
13. T. V. Long, II, A. W. Herlinger, E. F. Epstein, and I. Bernal, *Inorg. Chem.*, **9,** 459 (1970).
14. J. M. Terrasse, H. Poulet, and J. P. Mathieu, *Spectrochim. Acta*, **20,** 305 (1964).
15. A. Müller, K. H. Schmidt, and G. Vandrish, *Spectrochim. Acta*, **30A,** 651 (1974).
16. M. J. Nolan and D. W. James, *J. Raman Spectrosc.*, **1,** 259, 271 (1973).
17. J. Hiraishi, I. Nakagawa, and T. Shimanouchi, *Spectrochim. Acta*, **24A,** 819 (1968).
18. J. Fujita, K. Nakamoto, and M. Kobayashi, *J. Am. Chem. Soc.*, **78,** 3295 (1956).
19. K. Nakamoto, Y. Morimoto and J. Fujita, *Proc. 7th ICCC*, Stockholm, 1962, p.15.
20. T. W. Swaddle, P. J. Craig, and P. M. Boorman, *Spectrochim. Acta*, **26A,** 1559 (1970).
21. I. Nakagawa and T. Shimanouchi, *Spectrochim. Acta*, **22,** 759 (1966).
22. A. Müller, P. Christophliemk, and I. Tossidis. *J. Mol. Struct*, **15,** 289 (1973).
23. K. Nakamoto, J. Takemoto, and T. L. Chow, *Appl. Spectrosc.*, **25,** 352 (1971).
24. P. J. Hendra, *Spectrochim. Acta*, **23A,** 1275 (1967).
25. H. Poulet, P. Delorme, and J. P. Mathieu, *Spectrochim. Acta*, **20,** 1855 (1964).
26. S. J. Cyvin, B. N. Cyvin, K. H. Schmidt, W. Wiegeler, A. Müller, and J. Brunvoll, *J. Mol. Struct*, **30,** 315 (1976).
27. A. L. Geddes and G. L. Bottger, *Inorg. Chem.*, **8,** 802 (1969).
28. M. G. Miles, J. H. Patterson, C. W. Hobbs, M. J. Hopper, J. Overend, and R. S. Tobias, *Inorg. Chem.*, **7,** 1721 (1968).
29. E. P. Bertin, I. Nakagawa, S. Mizushima, T. L. Lane, and J. V. Quagliano, *J. Am. Chem. Soc.*, **80,** 525 (1958).
30. I. Nakagawa and T. Shimanouchi, *Spectrochim. Acta*, **22,** 1707 (1966).
31. T. Grzybek, J. M. Janik, A. Kulczycki, G. Pytasz, J. A. Janik, J. Ściesiński, and E. Ściesińska. *J. Raman Spectrosc.*, **1,** 185 (1973).
32. I. Nakagawa, *Bull. Chem. Soc. Jap*, **46,** 3690 (1973).
33. D. M. Adams and J. R. Hall, *J. Chem Soc.*, Dalton, 1450 (1973).
34. J. A. Janik, W. Jakob, and J. M. Janik, *Acta Phys. Pol.*, Pt. A, **38,** 467 (1970).
35. J. M. Janik, A. Magdal-Mikuli, and J. A. Janik, *Acta Phys. Pol.*, Pt. A, **40,** 741 (1971).
36. A. Adamson and T. M. Dunn, *J. Mol. Spectrosc.*, **18,** 83 (1965).
37. H. H. Eysel, *Z. Phys. Chem.* (Frankfurt), **72,** 82 (1970).
38. C. D. Flint and P. Greenough, *J. Chem. Soc.*, Faraday II, **68,** 897 (1972).
39. T. M. Loehr, J. Zinich, and T. V. Long, II, *Chem, Phys. Lett.*, **7,** 183 (1970).
40. A. F. Schreiner and J. A. McLean, *J. Inorg. Nucl. Chem.*, **27,** 253 (1965).
41. A. D. Allen and J. R. Stevens, *Can. J. Chem.*, **51,** 92 (1973).
42. M. W. Bee, S. F. A. Kettle, and D. B. Powell, *Spectrochim. Acta*, **30A,** 139 (1974).
43. C. H. Perry, D. P. Athans, E. F. Young, J. R. Durig, and B. R. Mitchell, *Spectrochim. Acta*, **23A,** 1137 (1967).
44. R. Layton, D. W. Sink, and J. R. Durig, *J. Inorg. Nucl. Chem.*, **28,** 1965 (1966).
45. J. R. Durig, R. Layton, D. W. Sink, and B. R. Mitchell, *Spectrochim. Acta*, **21,** 1367 (1965).
46. D. B. Powell, *J. Chem. Soc.*, 4495 (1956).
47. J. R. Durig and B. R. Mitchell, *Appl. Spectrosc.*, **21,** 221 (1967).
48. P. J. Hendra and N. Sadasivan, *Spectrochim. Acta*, **21,** 1271 (1965).

49. S. Mizushima, I. Nakagawa. and D. M. Sweeny, *J. Chem. Phys.*, **25,** 1006 (1956): I. Nakagawa, R. B. Penland, S. Mizushima, T. J. Lane, and J. V. Quagliano, *Spectrochim. Acta*, **9,** 199 (1957).
50. K. Niwa, H. Takahashi, and K. Higashi, *Bull. Chem. Soc. Jap*, **44,** 3010 (1971).
51. K. Brodersen and H. J. Becher, *Chem. Ber.*, **89,** 1487 (1956).
52. A. Novak, J. Portier, and P. Bouvlier, *Compt. Rend,*. **261,** 455 (1965).
53. D. C. Bradley and M. H. Gitlitz, *J. Chem. Soc.*, *A*, 980 (1969).
54. G. W. Watt, B. B. Hutchinson, and D. S. Klett, *J. Am. Chem. Soc.*, **89,** 2007 (1967).
55. Y. Y. Kharitonov, I. K. Dymina, and T. Leonova, *Izv. Akad. Nauk SSSR*, Ser. Khim., 2057 (1966).
56. M. Goldstein and E. F. Mooney, *J. Inorg. Nucl. Chem.*, **27,** 1601 (1965).
57. J. Chatt, L. A. Duncanson, and L. M. Venanzi, *J. Chem. Soc.*, 4456, 4461 (1955); 2712 (1956); *J. Inorg. Nucl. Chem.*, **8,** 67 (1958).
58. L. A. Duncanson and L. M. Venanzi, *J. Chem. Soc.*, 3841 (1960).
59. Y. Y. Kharitonov, M. A. Sarukhanov, I. B. Baranovskii, and K. U. Ikramov, *Opt, Spektrosk.*, **19,** 460 (1965).
60. L. Sacconi and A. Sabatini, *J. Inorg. Nucl. Chem.*, **25,** 1389 (1963).
61. K. Brodersen, *Z. Anorg. Allg. Chem.*, **290,** 24 (1957).
62. K. H. Linke, F. Dürholz, and P. Hädicke, *Z. Anorg. Allg. Chem.*, **356,** 113 (1968).
63. D. Nicholls and R. Swindells, *J. Inorg. Nucl. Chem.*, **30,** 2211 (1968).
64. S. Mizushima, I. Ichishima, I. Nakagawa, and J. V. Quagliano, *J. Phys. Chem.*, **59,** 293 (1955).
65. D. M. Sweeny, S. Mizushima, and J. V. Quagliano, *J. Am. Chem. Soc.*, **77,** 6521 (1955).
66. A. Nakahara, Y. Saito, and H. Kuroya, *Bull. Chem. Soc. Jap*, **25,** 331 (1952).
67. D. B. Powell and N. Sheppard, *J. Chem. Soc.*, 791 (1959); 1112 (1961); *Spectrochim. Acta.* **16,** 241 (1960); **17,** 68 (1961).
68. K. Krishnan and R. A. Plane, *Inorg. Chem.*, **5,** 852 (1966).
69. J. Gouteron-Vaissermann, *Compt. Rend.*, **275B,** 149 (1972).
70. R. W. Berg and K. Rasmussen, *Spectrochim. Acta*, **30A,** 1881 (1974).
71. R. W. Berg and K. Rasmussen, *Spectrochim. Acta*, **29A,** 319 (1973).
72. R. J. Mureinik and W. Robb, *Spectrochim. Acta*, **24A,** 837 (1968).
73. Y. Omura, I. Nakagawa, and T. Shimanouchi, *Spectrochim. Acta*, **27A,** 2227 (1971).
74. A. B. P. Lever and E. Mantovani, *Can. J. Chem.*, **51,** 1567 (1973).
75. G. W. Rayner-Canham and A. B. P. Lever, *Can. J. Chem.*, **50,** 3866 (1972).
76. A. B. P. Lever and E. Mantovani, *Inorg. Chem.*, **10,** 817 (1971).
77. M. E. Baldwin, *J. Chem. Soc.*, 4369 (1960).
78. J. A. McLean, A. F. Schreiner, and A. F. Laethem, *J. Inorg. Nucl. Chem*, **26,** 1245 (1964).
79. J. M. Rigg and E. Sherwin, *J. Inorg. Nucl. Chem.*, **27,** 653 (1965).
80. M. N. Hughes and W. R. McWhinnie, *J. Inorg. Nucl. Chem.*, **28,** 1659 (1966).
81. S. Kida, *Bull. Chem. Soc. Jap*, **39,** 2415 (1966).
82. E. B. Kipp and R. A. Haines, *Can. J. Chem.*, **47,** 1073 (1969).
83. Y. Omura, I. Nakagawa, and T. Shimanouchi, *Spectrochim. Acta*, **27A,** 1153 (1971).
84. R. W. Berg and K. Rasmussen, *Spectrochim. Acta*, **29A,** 37 (1973).
85. D. B. Powell and N. Sheppard, *J. Chem. Soc.*, 3089 (1959).
86. G. Newman and D. B. Powell, *J. Chem. Soc.*, 477 (1961); 3447 (1962).
87. K. Brodersen and T. Kahlert, *Z. Anorg. Allg. Chem.*, **348,** 273 (1966).
88. K. Brodersen, *Z. Anorg. Allg. Chem.*, **298,** 142 (1959).
89. T. Iwamoto and D. F. Shriver, *Inorg. Chem.*, **10,** 2428 (1971).

90. G. W. Watt and D. S. Klett, *Spectrochim. Acta*, **20,** 1053 (1964).
91. A. R. Gainsford and D. A. House, *Inorg. Chim. Acta*, **3,** 367 (1969).
92. H. H. Schmidtke and D. Garthoff, *Inorg. Chim. Acta*, **2,** 357 (1968).
93. K. W. Kuo and S. K. Madan, *Inorg. Chem.*, **8,** 1580 (1969).
94. J. H. Forsberg, T. M. Kubik, T. Moeller, and K. Gucwa, *Inorg. Chem.*, **10,** 2656 (1971).
95. D. A. Buckingham and D. Jones, *Inorg. Chem.*, **4,** 1387 (1965).
96. K. W. Bowker, E. R. Gardner, and J. Burgess, *Inorg. Chim. Acta*, **4,** 626 (1970).
97. R. J. H. Clark and C. S. Williams, *Inorg. Chem.*, **4,** 350 (1965).
98. Y. Saito, M. Cordes, and K. Nakamoto, *Spectrochim. Acta*, **28A,** 1459 (1972).
99. Y. Saito, C. W. Schläpfer, M. Cordes, and K. Nakamoto, *Appl. Spectrosc.*, **27,** 213 (1973).
100. M. Choca, J. R. Ferraro, and K. Nakamoto, *J. Chem. Soc.*, Dalton, 2297 (1972).
101. R. H. Nuttall, A. F. Cameron, and D. W. Taylor, *J. Chem. Soc., A*, 3103 (1971).
102. C. Postmus, J. R. Ferraro, A. Quattrochi, K. Shobatake, and K. Nakamoto, *Inorg. Chem.*, **8,** 1851 (1969).
103. J. R. Allan, D. H. Brown, R. H. Nuttall, and D. W. A. Sharp, *J. Inorg. Nucl. Chem.*, **27,** 1305 (1965).
104. N. S. Gill and H. J. Kingdon, *Aust. J. Chem.*, **19,** 2197 (1966).
105. L. Cattalini, R. J. H. Clark, A. Orio, and C. K. Poon, *Inorg. Chim. Acta*, **2,** 62 (1968).
106. J. Burgess, *Spectrochim. Acta*, **24A,** 277 (1968).
107. M. Goldstein, E. F. Mooney, A. Anderson, and H. A. Gebbie, *Spectrochim. Acta*, **21,** 105 (1965).
108. A. B. P. Lever and B. S. Ramaswamy, *Can. J. Chem.*, **51,** 1582 (1973).
109. W. R. McWhinnie, *J. Inorg. Nucl. Chem.*, **27,** 2573 (1965).
110. D. E. Billing and A. E. Underhill, *J. Inorg. Nucl. Chem.*, **30,** 2147 (1968).
111. F. Farha and R. T Iwamoto, *Inorg. Chem.*, **4,** 844 (1965).
112. D. G. Brewer and P. T. T. Wong, *Can. J. Chem.*, **44,** 1407 (1966).
113. J. Burgess, *Spectrochim. Acta*, **24A,** 1645 (1968).
114. L. El-Sayed and R. O. Ragsdale, *J. Inorg. Nucl. Chem.*, **30,** 651 (1968).
115. R. G. Garvey, J. H. Nelson, and R. O. Ragsdale, *Coord. Chem. Rev.*, **3,** 375 (1968).
116. C. P. Prabhakaran and C. C. Patel, *J. Inorg. Nucl. Chem.*, **34,** 3485 (1972).
117. I. S. Ahuja and P. Rastogi, *J. Chem. Soc., A*, 378 (1970).
118. F. A. Cotton and J. F. Gibson, *J. Chem. Soc., A*, 2105 (1970).
119. D. M. L. Goodgame, M. Goodgame, P. J. Hayward, and G. W. Rayner-Canham, *Inorg. Chem.*, **7,** 2447 (1968).
120. C. E. Taylor and A. E. Underhill, *J. Chem. Soc., A*, 368 (1969).
121. B. Cornilsen and K. Nakamoto, *J. Inorg. Nucl. Chem.*, **36,** 2467 (1974).
122. W. J. Eilbeck, F. Holmes, C. E. Taylor, and A. E. Underhill, *J. Chem. Soc., A*, 128 (1968).
123. D. M. L. Goodgame, M. Goodgame, and G. W. Rayner-Canham, *Inorg. Chim. Acta*, **3,** 399 (1969).
124. D. M. L. Goodgame, M. Goodgame, and G. W. Raymer-Canham, *Inorg. Chim. Acta*, **3,** 406 (1969).
125. W. J. Eilbeck, F. Holmes, C. E. Taylor, and A. E. Underhill, *J. Chem. Soc., A*, 1189 (1968).
126. G. A. Melson and R. H. Nuttall, *J. Mol. Struct*, **1,** 405 (1968).
127. M. M. Cordes and J. L. Walter, *Spectrochim. Acta*, **24A,** 1421 (1968).
128. B. Hutchinson, J. Takemoto, and K. Nakamoto, *J. Am. Chem. Soc.*, **92,** 3335 (1970).

129. Y. Saito, J. Takemoto, B. Hutchinson, and K. Nakamoto, *Inorg. Chem.*, **11,** 2003 (1972); J. Takemoto, B. Hutchinson, and K. Nakamoto, *Chem. Commun.*, 1007 (1971).
130. E. König and E. Lindner, *Spectrochim. Acta*, **28A,** 1393 (1972).
131. R. Wilde, T. K. K. Srinivasan, and N. Ghosh, *J. Inorg. Nucl. Chem.*, **35,** 1017 (1973).
132. J. S. Strukl and J. L. Walter, *Spectrichim. Acta*, **27A,** 223 (1971).
133. K. Krishnan and R. A. Plane, *Spectrochim. Acta*, **25A,** 831 (1969).
134. R. J. H. Clark, P. C. Turtle, D. P. Strommen, B. Streusand, J. Kincaid, and K. Nakamoto, *Inorg. Chem.*, **16,** 84 (1977).
135. J. Takemoto and B. Hutchinson, *Inorg. Nucl. Chem. Lett.*, **8,** 769 (1972).
136. E. König and K. J. Watson, *Chem. Phys. Lett.*, **6,** 457 (1970).
137. J. Takemoto and B. Hutchinson, *Inorg. Chem.*, **12,** 705 (1973).
138. J. R. Ferraro and J. Takemoto, *Appl. Spectrosc.*, **28,** 66 (1974).
139. P. F. B. Barnard, A. T. Chamberlain, G. C. Kulasingam, R. J. Dosser, and W. R. McWhinnie, *Chem. Commun.*, 520 (1970)
140. J. Takemoto, *Inorg. Chem.*, **12,** 949 (1973).
141. B. Hutchinson and A. Sunderland, *Inorg. Chem.*, **11,** 1948 (1972).
142. A. Bigotto, G. Costa, V. Galasso, and G. DeAlti, *Spectrochim. Acta*, **26A,** 1939 (1970).
143. A. Bigotto, V. Galasso, and G. DeAlti, *Spectrochim. Acta*, **27A,** 1659 (1971).
144. N. Ohkaku and K. Nakamoto, *Inorg. Chem.*, **10,** 798 (1971).
145. B. Hutchinson, A. Sunderland, M. Neal, and S. Olbricht, *Spectrochim. Acta*, **29A,** 2001 (1973).
146. Y. Murakami and K. Sakata, *Inorg. Chim. Acta*, **2,** 273 (1968).; Y. Murakami, Y. Matsuda, K. Sakata, and K. Harada, *Bull. Chem. Soc. Jap*, **47,** 458, 3021 (1974).
147. Y. Murakami, K. Sakata, Y. Tanaka, and T. Matsuo, *Bull. Chem. Soc. Jap*, **48,** 3622 (1975).
148. T. Kobayashi, F. Kurokawa, N. Uyeda, and E. Saito, *Spectrochim. Acta*, **26A,** 1305 (1970); T. Kobayashi, *ibid.*, **26A,** 1313 (1970).
149. L. J. Boucher and J. J. Katz, *J. Am. Chem. Soc.*, **89,** 1340 (1967).
150. H. Ogoshi, N. Masai, Z. Yoshida. J. Takemoto, and K. Nakamoto, *Bull. Chem. Soc. Jap*, **44,** 49 (1971).
151. H. Ogoshi, Y. Saito, and K. Nakamoto, *J. Chem. Phys.*, **57,** 4194 (1972).
152. H. Ogoshi, E. Watanabe, Z. Yoshida, J. Kincaid, and K. Nakamoto, *J. Am. Chem. Soc.*, **95,** 2845 (1973).
153. J. Kincaid and K. Nakamoto, *J. Inorg. Nucl. Chem.*, **37,** 85 (1975).
154. H. Ogoshi, E. Watanabe, Z. Yoshida, J. Kincaid, and K. Nakamoto, *Inorg. Chem.*, **14,** 1344 (1975).
155. T. G. Spiro and T. C. Strekas, *J. Am. Chem. Soc.*, **96,** 338 (1974); T. G. Spiro, *Acc. Chem. Res.*, **7,** 339 (1974).
156. T. Yamamoto, G. Palmer, D. Gill, I. T. Salmeen, and L. Rimai, *J. Biol. Chem.*, **284,** 5211 (1973).
157. T. M. Loehr and J. S. Loehr, *Biochem. Biophys. Res. Commun.*, **55,** 218 (1973).
158. A. L. Verma and H. J. Bernstein, *J. Raman Spectrosc.*, **2,** 163 (1974).
159. A. L. Verma, R. Mendelsohn, and H. J. Bernstein, *J. Chem. Phys.*, **61,** 383 (1974).
160. S. Sunder and H. J. Bernstein, *Can. J. Chem.*, **52,** 2851 (1974).
161. I. Nakagawa, T. Shimanouchi, and K. Yamasaki, *Inorg. Chem.*, **3,** 772 (1964); **7,** 1332 (1968).
162. M. Le Postolloe, J. P. Mathieu, and H. Poulet, *J. Chim. Phys.*, **60,** 1319 (1963).
163. M. J. Cleare and W. P. Griffith, *J. Chem. Soc.*, *A*, 1144 (1967).

164. M. J. Nolan and D. W. James, *Aust. J. Chem.*, **23,** 1043 (1970).
165. J. T. Huneke, B. Meister, L. Walford, and R. L. Bain, *Spectrosc. Lett.*, **7,** 91 (1974).
166. D. W. James and M. J. Nolan, *Aust. J. Chem.*, **26,** 1433 (1973).
167. P. E. Merritt and S. E. Wiberley, *J. Phys. Chem.*, **59,** 55 (1955).
168. R. B. Hagel and L. F. Druding, *Inorg. Chem.*, **9,** 1496 (1970).
169. I. Nakagawa and T. Shimanouchi, *Spectrochim. Acta*, **23A,** 2099 (1967).
170. M. J. Nolan and D. W. James, *Aust. J. Chem.*, **26,** 1413 (1973).
171. K. Nakamoto, J. Fujita, and H. Murata. *J. Am. Chem. Soc.*, **80,** 4817 (1958).
172. D. M. L. Goodgame and M. A. Hitchman, *Inorg. Chem.*, **3,** 1389 (1964).
173. W. W. Fee, C. S. Garner, and J. N. M. Harrowfield, *Inorg. Chem.*, **6,** 87 (1967).
174. D. M. L. Goodgame, M. A. Hitchman, D. F. Marsham, and C. E. Souter, *J. Chem. Soc.*, *A*, 2464 (1969).
175. D. M. L. Goodgame and M. A. Hitchman, *Inorg. Chem.*, **6,** 813 (1967).
176. L. El-Sayed and R. O. Ragsdale, *Inorg. Chem.*, **6,** 1640 (1967).
177. R. B. Penland, T. J. Lane, and J. V. Quagliano, *J. Am. Chem. Soc.*, **78,** 887 (1956).
178. I. R. Beattie and D. P. N. Satchell, *Trans. Faraday Soc.*, **52,** 1590 (1956).
179. J. L. Burmeister, *Coord. Chem. Rev.*, **3,** 225 (1968).
180. D. M. L. Goodgame and M. A. Hitchman, *Inorg. Chem.*, **4,** 721 (1965).
181. D. M. L. Goodgame and M. A. Hitchman, *J. Chem. Soc.*, *A*, 612 (1967).
182. D. M. L. Goodgame, M. A. Hitchman, and D. F. Marsham, *J. Chem. Soc.*, *A*, 1933 (1970).
183. U. Thewalt and R. E. Marsh, *Inorg. Chem.*, **9,** 1604 (1970).
184. K. Wieghardt and H. Siebert, *Z. Anorg. Allg. Chem.*, **374,** 186 (1970).
185. D. M. L. Goodgame, M. A. Hitchman, D. F. Marsham, P. Phavanantha, and D. Rogers, *Chem. Commun.*, 1383 (1969).
186. D. M. L. Goodgame, M. A. Hitchman, and D. F. Marsham, *J. Chem. Soc.*, *A*, 259 (1971).
187. F. A. Miller and C. H. Wilkins, *Anal. Chem.*, **24,** 1253 (1952).
188. M. Hass and G. B. B. M. Sutherland, *Proc. Roy. Soc.*, **A236,** 427 (1956).
189. J. van der Elsken and D. W. Robinson, *Spectrochim. Acta*, **17,** 1249 (1961).
190. K. Ichida, Y. Kuroda, D. Nakamura, and M. Kubo, *Spectrochim. Acta*, **28A,** 2433 (1972).
191. J. O. Lundgren and I. Olovsson, *Acta Crystallogr.*, **23,** 966 (1967).
192. A. C. Pavia and P. A. Giguère, *J. Chem. Phys.*, **52,** 3551 (1970).
193. J. M. Williams, *Inorg. Nucl. Chem. Lett.*, **3,** 297 (1967).
194. J. Roziere and J. Potier, *J. Inorg. Nucl. Chem.*, **35,** 1179 (1973).
195. I. Nakagawa and T. Shimanouchi, *Spectrochim, Acta*, **20,** 429 (1964).
196. D. M. Adams and P. J. Lock, *J. Chem. Soc.*, *A*, 2801 (1971).
197. H. L. Schlafer and H. P. Fritz, *Spectrochim. Acta*, **23A,** 1409 (1967).
198. T. G. Chang and D. E. Irish, *Can. J. Chem.*, **51,** 118 (1973).
199. R. E. Hester and R. A. Plane, *Inorg. Chem.*, **3,** 768 (1964).
200. H. Boutin, G. J. Safford, and H. R. Danner, *J. Chem. Phys.*, **40,** 2670 (1964).; H. J. Prask and H. Boutin, *ibid.*, **45,** 699, 3284 (1966).
201. J. J. Rush, J. R. Ferraro, and A. Walker, *Inorg. Chem.*, **6,** 346 (1967).
202. T. Dupuis, C. Duval, and J. Lecomte, *Compt. Rend.*, **257,** 3080 (1963).
203. M. Maltese and N. J. Orville-Thomas, *J. Inorg. Nucl. Chem.*, **29,** 2533 (1967).
204. J. R. Ferraro and W. R. Walker, *Inorg. Chem.*, **4,** 1382 (1965).
205. W. R. McWhinnie, *J. Inorg. Nucl. Chem.*, **27,** 1063 (1965).
206. J. R. Ferraro, R. Driver, W. R. Walker, and W. Wozniak, *Inorg. Chem.*, **6,** 1586 (1967).

207. G. Blyholder and N. Ford, *J. Phys. Chem.*, **68,** 1496 (1964).
208. V. A. Maroni and T. G. Spiro, *J. Am. Chem. Soc.*, **89,** 45 (1967).
209. R. W. Adams, R. L. Martin, and G. Winter, *Aust. J. Chem.*, **20,** 773 (1967).
210. R. C. Mehrotra and J. M. Batwara, *Inorg. Chem.*, **9,** 2505 (1970).
211. L. M. Brown and K. S. Mazdiyasni, *Inorg. Chem.*, **9,** 2783 (1970).
212. P. W. N. M. Van Leeuwen, *Rec. Trav. Chim.*, **86,** 247 (1967).
213. D. Knetsch and W. L. Groeneveld, *Inorg. Chim. Acta*, **7,** 81 (1973).
214. H. Wieser and P. J. Krueger, *Spectrochim. Acta*, **26A,** 1349 (1970).
215. G. W. A. Fowles, D. A. Rice, and R. A. Walton, *Spectrochim. Acta*, **26A,** 143 (1970).
216. W. L. Driessen and W. L. Groeneveld, *Rec. Trav. Chim.*, **88,** 977 (1969).
217. W. L. Driessen and W. L. Groeneveld, *Rec. Trav. Chim.*, **90,** 258 (1971).
218. W. L. Driessen and W. L. Groeneveld, *Rec. Trav. Chim.*, **90,** 87 (1971).
219. W. L. Driessen, W. L. Groeneveld. and F. W. Van der Wey. *Rec. Trav. Chim.*, **89,** 353 (1970).
220. K. Itoh and H. J. Bernstein, *Can. J. Chem.*, **34,** 170 (1956).
221. S. D. Robinson and M. F. Uttley, *J. Chem. Soc.*, 1912 (1973).
222. T. A. Stephenson and G. Wilkinson, *J. Inorg. Nucl. Chem.*, **29,** 2122 (1967).
223. G. Csontos, B. Heil and C. Markó. *J. Organometal. Chem.*, **37,** 183 (1972).
224. A. V. R. Warrier and P. S. Narayanan, *Spectrochim. Acta*, **23A,** 1061 (1967).
225. A. V. R. Warrier and R. S. Krishnan, *Spectrochim. Acta*, **27A,** 1243 (1971).
226. K. Nakamoto, Y. Morimoto, and A. E. Martell. *J. Am. Chem. Soc.*, **83,** 4528 (1961).
227. S. Baba and S. Kawaguchi, *Inorg. Nucl. Chem. Lett.*, **9,** 1287 (1973).
228. K. Nakamoto, P. J. McCarthy, and B. Miniatus, *Spectrochim. Acta*, **21,** 379 (1965).
229. J. Fujita, A. E. Martell, and K. Nakamoto, *J. Chem. Phys.*, **36,** 324, 331 (1962).
230. R. D. Hancock and D. A. Thornton, *J. Mol. Struct.* **6,** 441 (1970).
231. K. L. Scott, K. Wieghardt, and A. G. Sykes, *Inorg. Chem.*, **12,** 655 (1973); K. Wieghardt, *Z. Anorg. Allg. Chem.*, **391,** 142 (1972).
232. R. E. Hester and R. A. Plane, *Inorg. Chem.*, **3,** 513 (1964); E. C. Gruen and R. A. Plane, *ibid.*, **6,** 1123 (1967).
233. Y. Kuroda, M. Kato, and K. Sone, *Bull. Chem. Soc. Jap*, **34,** 877 (1961).
234. P. X. Armendarez and K. Nakamoto, *Inorg. Chem.*, **5,** 796 (1966).
235. B. B. Kedzia, P. X. Armendarez, and K. Nakamoto, *J. Inorg. Nucl. Chem.*, **30,** 849 (1968).
236. L. Cavalca, M. Nardelli, and G. Fava, *Acta Crystallogr.*, **13,** 594 (1960).
237. Y. Saito, K. Machida, and T. Uno, *Spectrochim. Acta*, **26A,** 2089 (1970).
238. K. Nakamoto, J. Fujita, S. Tanaka, and M. Kobayashi, *J. Am. Chem, Soc.*, **79,** 4904 (1957).
239. C. G. Barraclough and M. L. Tobe, *J. Chem. Soc.*, 1993 (1961).
240. R. Eskenazi, J. Rasovan, and R. Levitus, *J. Inorg. Nucl. Chem.*, **28,** 521 (1966).
241. J. E. Finholt, R. W. Anderson, J. A. Fyfe, and K. G. Caulton, *Inorg. Chem.*, **4,** 43 (1965).
242. W. R. McWhinnie, *J. Inorg. Nucl. Chem.*, **26,** 21 (1964).
243. I. S. Ahuja, *Inorg. Chim. Acta*, **3,** 110 (1969).
244. K. Wieghardt and J. Eckert, *Z. Anorg. Allg. Chem.*, **383,** 240 (1971).
245. R. Ugo, F. Conti, S. Cenini, R. Mason, and G. B. Robertson, *Chem. Commun.*, 1498 (1968).
246. R. W. Horn, E. Weissberger, and J. P. Collman, *Inorg. Chem.*, **9,** 2367 (1970).
247. J. R. Ferraro and A. Walker, *J. Chem. Phys.*, **42,** 1278 (1965).
248. N. Tanaka, H. Sugi, and J. Fujita, *Bull. Chem. Soc. Jap.* **37,** 640 (1964).

249. J. A. Goldsmith, A. Hezel, and S. D. Ross, *Spectrochim. Acta,* **24A,** 1139 (1968).
250. M. R. Rosenthal, J. Chem. Educ., **50,** 331 (1973).
251. B. J. Hathaway and A. E. Underhill, *J. Chem. Soc.,* 3091 (1961).
252. B. J. Hathaway, D. G. Holah, and M. Hudson, *J. Chem. Soc.,* 4586 (1963).
253. M. E. Farago, J. M. James, and V. C. G. Trew, *J. Chem. Soc., A,* 820 (1967).
254. A. E. Wickenden and R. A. Krause, *Inorg. Chem.,* **4,** 404 (1965).
255. L. E. Moore, R. B. Gayhart, and W. E. Bull, *J. Inorg. Nucl. Chem.,* **26,** 896 (1964).
256. S. F. Lincoln and D. R. Stranks, *Aust. J. Chem.,* **21,** 37 (1968).
257. T. A. Beech and S. F. Lincoln, *Aust. J. Chem.,* **24,** 1065 (1971).
258. R. Coomber and W. P. Griffith, *J. Chem. Soc.,* 1128 (1968).
259. S. D. Ross and N. A. Thomas, *Spectrochim. Acta,* **26A,** 971 (1970).
260. A. N. Freedman and B. P. Straughan, *Spectrochim. Acta,* **27A,** 1455 (1971).
261. P. A. Yeats, J. R. Sams, and F. Aubke, *Inorg. Chem.,* **12,** 328 (1973).
262. B. M. Gatehouse, S. E. Livingstone, and R. S. Nyholm, *J. Chem. Soc.,* 3137 (1958).
263. J. Fujita, A. E. Martell, and K. Nakamoto, *J. Chem. Phys.,* **36,** 339 (1962).
264. R. E. Hester and W. E. L. Grossman, *Inorg. Chem.,* **5,** 1308 (1966).
265. J. A. Goldsmith and S. D. Ross, *Spectrochim. Acta,* **24A,** 993 (1968).
266. H. Elliott and B. J. Hathaway, *Spectrochim. Acta,* **21,** 1047 (1965).
267. C. C. Addison, N. Logan, S. C. Wallwork, and C. D. Barner, *Quart. Rev.,* **25,** 289 (1971).
268. B. M. Gatehouse, S. E. Livingstone, and R. S. Nyholm, *J. Chem. Soc.,* 4222 (1957); *J. Inorg. Nucl. Chem.,* **8,** 75 (1958).
269. N. F. Curtis and Y. M. Curtis, *Inorg. Chem.,* **4,** 804 (1965).
270. C. C. Addison, R. Davis, and N. Logan, *J. Chem. Soc., A,* 3333 (1970).
271. C. C. Addison and W. B. Simpson, *J. Chem. Soc.,* 598 (1965).
272. J. G. Allpress and A. N. Hambly, *Aust. J. Chem.,* **12,** 569 (1959).
273. R. J. Fereday, N. Logan, and D. Sutton, *J. Chem. Soc., A,* 2699 (1969).
274. D. W. Johnson and D. Sutton, *Can. J. Chem.,* **50,** 3326 (1972).
275. N. Logan and W. B. Simpson, *Spectrochim. Acta,* **21,** 857 (1965).
276. E. J. Duff, M. N. Hughes, and K. J. Rutt, *J. Chem. Soc., A,* 2126 (1969).
277. E. M. Briggs and A. E. Hill, *J. Chem. Soc., A,* 2008 (1970).
278. A. B. P. Lever, E. Mantovani, B. S. Ramaswamy, *Can. J. Chem.,* **49,** 1957 (1971).
279. J. R. Ferraro, A. Walker, and C. Cristallini, *Inorg. Nucl. Chem. Lett.,* **1,** 25 (1965).
280. C. C. Addison, D. W. Amos, D. Sutton, and W. H. H. Hoyle, *J. Chem. Soc., A,* 808 (1967).
281. R. H. Nuttall and D. W. Taylor, *Chem. Commun.,* 1417 (1968).
282. J. I. Bullock and F. W. Parrett, *Chem. Commun.,* 157 (1969).
283. J. R. Ferraro and A. Walker, *J. Chem. Phys.,* **42,** 1273 (1965); **43,** 2689 (1965); **45,** 550 (1966).
284. D. E. Irish and G. E. Walrafen, *J. Chem. Phys.,* **46,** 378 (1967).
285. R. E. Hester and K. Krishnan, *J. Chem. Phys.,* **46,** 3405 (1967); **47,** 1747 (1967).
286. F. A. Cotton and R. Francis, *J. Am. Chem. Soc.,* **82,** 2986 (1960).
287. G. Newman and D. B. Powell, *Spectrochim. Acta,* **19,** 213 (1963).
288. M. E. Baldwin, *J. Chem. Soc.,* 3123 (1961).
289. B. Nyberg and R. Larsson, *Acta Chem. Scand.,* **27,** 63 (1973).
290. E. Lindner and G. Vitzthum, *Chem. Ber.,* **102,** 4062 (1969).
290a. G. Vitzthum and E. Lindner, *Angew. Chem. Int. Ed.,* **10,** 315 (1971).
291. K. Nakamoto and A. E. Martell, *J. Chem. Phys.,* **32,** 588 (1960).
292. M. Mikami, I. Nakagawa, and T. Shimanouchi, *Spectrochim. Acta,* **23A,** 1037 (1967).
293. S. Pinchas, B. L. Silver, and I. Laulicht, *J. Chem. Phys.,* **46,** 1506 (1967).

294. H. Musso and H. Junge, *Tetrahedron Lett.*, **33,** 4003, 4009 (1966).
295. K. Nakamoto, C. Udovich, and J. Takemoto, *J. Am. Chem. Soc.*, **92,** 3973 (1970).
296. R. C. Fay and R. N. Lowry, *Inorg. Nucl. Chem. Lett.*, **3,** 117 (1967).
297. W. D. Courrier, C. J. L. Lock, and G. Turner, *Can. J. Chem.*, **50,** 1797 (1972).
298. M. R. Caira, J. M. Haigh, and L. R. Nassimbeni, *J. Inorg. Nucl. Chem.*, **34,** 3171 (1972).
299. M. F. Richardson, W. F. Wagner, and D. E. Sands, *Inorg. Chem.*, **7,** 2495 (1968).
300. J. C. Hammel, J. A. S. Smith, and E. J. Wilkins, *J. Chem Soc.*, *A*, 1461 (1969).
301. J. C. Hammel and J. A. S. Smith, *J. Chem. Soc.*, *A*, 2883 (1969).
302. M. A. Bush, D. E. Fenton, R. S. Nyholm, M. R. Truter, *Chem. Commun.*, 1335 (1970).
303. Y. Nakamura, N. Kanehisa, and S. Kawaguchi, *Bull. Chem. Soc. Jap*, **45,** 485 (1972).
304. F. A. Cotton and R. C. Elder, *J. Am. Chem. Soc.*, **86,** 2294 (1964); *Inorg. Chem.*, **4,** 1145 (1965).
305. P. W. N. M. van Leeuwen, *Rec. Trav. Chim.*, **87,** 396 (1968).
306. Y. Nakamura and S. Kawaguchi, *Chem. Commun.*, 716 (1968).
307. S. Koda, S. Ooi, H. Kuroya, K. Isobe, Y. Nakamura, and S. Kawaguchi, *Chem. Commun.*, 1321 (1971).
308. Y. Nakamura, K. Isobe, H. Morita, S. Yamazaki, and S. Kawaguchi, *Inorg. Chem.*, **11,** 1573 (1972).
309. S. Koda, S. Ooi, H. Kuroya, Y. Nakamura, and S. Kawaguchi, *Chem. Commun.*, 280 (1971).
310. J. Lewis, R. F. Long, and C. Oldham, *J. Chem. Soc.*, 6740 (1965); D. Gibson, J. Lewis, and C. Oldham, *ibid.*, *A*, 1453 (1966).
311. G. T. Behnke and K. Nakamoto, *Inorg. Chem.*, **6,** 433 (1967).
312. G. T. Behnke and K. Nakamoto, *Inorg. Chem.*, **6,** 440 (1967).
313. G. T. Behnke and K. Nakamoto, *Inorg. Chem.*, **7,** 330 (1968).
314. F. Bonati and G. Minghetti, *Angew. Chem. Int. Ed.*, **7,** 629 (1968).
315. D. Gibson, B. F. G. Johnson, and J. Lewis, *J. Chem. Phys.*, *A*, 367 (1970).
316. S. Baba, T. Ogura, and S. Kawaguchi, *Inorg. Nucl. Chem. Lett.*, **7,** 1195 (1971).
317. G. Allen, J. Lewis, R. F. Long, and C. Oldham, *Nature*, **202,** 589 (1964).
318. G. T. Behnke and K. Nakamoto, *Inorg. Chem.*, **7,** 2030 (1968).
319. J. Lewis and C. Oldham, *J. Chem. Soc.*, *A*, 1456 (1966).
320. Y. Nakamura and K. Nakamoto, *Inorg. Chem.*, **14,** 63 (1975).
321. L. G. Hulett and D. A. Thornton, *Spectrochim. Acta*, **27A,** 2089 (1971).
322. H. Junge, *Spectrochim. Acta*, **24A,** 1957 (1968).
323. B. Hutchinson, D. Eversdyk, and S. Olbricht, *Spectrochim. Acta*, **30A,** 1605 (1974).
324. F. Sagara, H. Kobayashi, and K. Ueno, *Bull. Chem. Soc. Jap*, **45,** 794 (1972).
325. A. G. Sharp, *The Chemistry of Cyano Complexes of the Transition Metals*, Academic Press, New York, 1976.
326. W. P. Griffith, *Coord. Chem. Rev.*, **17,** 177 (1975).
327. P. Rigo and A. Turco, *Coord. Chem. Rev.*, **13,** 133 (1974).
328. M. F. A. El-Sayed and R. K. Sheline, *J. Inorg. Nucl. Chem.*, **6,** 187 (1958).
329. R. Nast and D. Rehder, *Chem. Ber.*, **104,** 1709 (1971).
330. L. H. Jones and R. A. Penneman, *J. Chem. Phys.*, **22,** 965 (1954).
331. R. A. Penneman and L. H. Jones, *J. Chem. Phys.*, **24,** 293 (1956).
332. R. A. Penneman and L. H. Jones, *J. Inorg. Nucl. Chem.*, **20,** 19 (1961).
333. H. Stammreich, B. M. Chadwick, and S. G. Frankiss, *J. Mol. Struct*, **1,** 191 (1967).
334. B. M. Chadwick and S. G. Frankiss, *J. Mol. Struct*, **2,** 281 (1968).
335. G. J. Kubas and L. H. Jones, *Inorg. Chem.*, **13,** 2816 (1974).

336. W. P. Griffith and J. R. Lane, *J. Chem. Soc.*, Dalton, 158 (1972).
337. W. P. Griffith and G. T. Turner, *J. Chem. Soc.*, *A*, 858 (1970).
338. H. Siebert and A. Siebert, *Angew. Chem. Int. Ed.*, **8,** 6009 (1969).; *Z. Anorg. Allg. Chem.*, **378,** 160 (1970).
339. G. R. Rossman. F.-D. Tsay, and H. B. Gray, *Inorg. Chem.*, **12,** 824 (1973).
340. J. L. Hoard, T. A. Hamor, and M. D. Glick, *J. Am. Chem. Soc.*, **90,** 3177 (1968).
341. H. Stammreich and O. Sala, *Z. Elektrochem.*, **64,** 741 (1960); **65,** 149 (1961).
342. K. O. Hartman and F. A. Miller, *Spectrochim. Acta*, **24A,** 669 (1968).
343. B. V. Parish, P. G. Simms, M. A. Wells, and L. A. Woodward, *J. Chem. Soc.*, 2882 (1968).
344. P. M. Kiernan and W. P. Griffith, *J. Chem. Soc.*, Dalton, 2489 (1975).
345. T. V. Long, II, and G. A. Vernon, *J. Am. Chem. Soc.*, **93,** 1919 (1971).
346. K. N. Raymond, P. W. R. Corfield, and J. A. Ibers, *Inorg. Chem.*, **7,** 1362 (1968).
347. A. Terzis, K. N. Raymond, and T. G. Spiro, *Inorg. Chem.*, **9,** 2415 (1970).
348. L. J. Basile, J. R. Ferraro, M. Choca, and K. Nakamoto, *Inorg. Chem.*, **13,** 496 (1974).
349. E. Hellner, H. Ahsbahs, G. Dehnicke, and K. Dehnicke, *Ber. Bunsen. Phys. Chem.*, **77,** 277 (1973).
350. R. L. McCullough, L. H. Jones, and R. A. Penneman, *J. Inorg. Nucl. Chem.*, **13,** 286 (1960).
351. G. W. Chantry and R. A. Plane, *J. Chem. Phys.*, **33,** 736 (1960); **34,** 1268 (1961); **35,** 1027 (1961).
352. V. Caglioti, G. Sartori, and C. Furlani, *J. Inorg. Nucl. Chem.*, **13,** 22 (1960); **8,** 87 (1958).
353. L. H. Jones, *J. Mol. Spectrosc.*, **8,** 105 (1962); *J. Chem. Phys.*, **36,** 1209 (1962).
354. L. H. Jones, *J. Chem. Phys.*, **41,** 856 (1964).
355. D. Bloor, *J. Chem. Phys.*, **41,** 2573 (1964).
356. I. Nakagawa and T. Shimanouchi, *Spectrochim. Acta*, **18,** 101 (1962).
357. L. H. Jones, *Inorg. Chem.*, **2,** 777 (1963).
358. I. Nakagawa and T. Shimanouchi, *Spectrochim. Acta*, **26A,** 131 (1970).
358a. L. H. Jones, B. I. Swanson, and G. J. Kubas, *J. Chem. Phys.*, **61,** 4650 (1974); B. I. Swanson and L. H. Jones, *Inorg. Chem.*, **13,** 313 (1974).
359. L. H. Jones *J. Chem. Phys.*, **29,** 463 (1958).
360. H. Poulet and J. P. Mathieu, *Spectrochim. Acta*, **15,** 932 (1959).
361. L. H. Jones, *Spectrochim. Acta*, **17,** 188 (1961).
362. D. M. Sweeny, I. Nakagawa, S. Mizushima, and J. V. Quagliano, *J. Am. Chem. Soc.*, **78,** 889 (1956).
363. C. W. F. T. Pistorius, *Z. Phys. Chem.*, **23,** 197 (1960).
364. R. L. McCullough, L. H. Jones, and G. A. Crosby, *Spectrochim. Acta*, **16,** 929 (1960).
365. L. H. Jones and J. M. Smith, *J. Chem. Phys.*, **41,** 2507 (1964).
366. L. H. Jones, *J. Chem. Phys.*, **27,** 665 (1957).
367. L. H. Jones, *Spectrochim. Acta*, **19,** 1675 (1963).
368. L. H. Jones, *J. Chem. Phys.*, **26,** 1578 (1957); **25,** 379 (1956).
369. L. H. Jones, *J. Chem. Phys.*, **27,** 468 (1957); **21,** 1891 (1953); **22,** 1135 (1954).
370. V. Lorenzelli and P. Delorme, *Spectrochim. Acta*, **19,** 2033 (1963).
371. L. H. Jones, *Inorg. Chem.*, **3,** 1581 (1964); **4,** 1472 (1965); J. M. Smith, L. H. Jones, I. K. Kressin, and R. A. Penneman, *ibid.*, **4,** 369 (1965).
372. J. C. Coleman, H. Peterson, and R. A. Penneman, *Inorg. Chem.*, **4,** 135 (1965).
373. J. H. Swinebart, *Coord. Chem. Rev.*, **2,** 385 (1967).

374. R. K. Khanna, C. W. Brown, and L. H. Jones, *Inorg. Chem.*, **8,** 2195 (1969); J. B. Bates and R. K. Khanna, *ibid.*, **9,** 1376 (1970).
375. D. B. Brown. *Inorg. Chim. Acta*, **5,** 314 (1971).
376. A. Poletti, A. Santucci, and G. Paliani, *Spectrochim. Acta*, **27A,** 2061 (1971).
377. E. Miki, S. Kubo, K. Mizumachi, T. Ishimori, and H. Okuno, *Bull. Chem. Soc. Jap.* **44,** 1024 (1971).
378. L. Tosi and J. Danon, *Inorg. Chem.*, **3,** 150 (1964).
379. L. H. Jones and J. M. Smith, *Inorg. Chem.*, **4,** 1677 (1965).
380. M. N. Memering, L. H. Jones, and J. C. Bailar, Jr., *Inorg. Chem.*, **12,** 2793 (1973).
381. D. F. Shriver, *J. Am. Chem. Soc.*, **84,** 4610 (1962): **85,** 1405 (1963); D. F. Shriver and J. Posner, *ibid.*, **88,** 1672 (1966).
382. D. F. Shriver, S. A. Shriver, and S. E. Anderson, *Inorg. Chem.*, **4,** 725 (1965).
383. D. B. Brown, D. F. Shriver, and L. H. Schwartz, *Inorg. Chem.*, **7,** 77 (1968).
384. H. G. Nadler, J. Pebler, and K. Dehnicke, *Z. Anorg. Allg. Chem.*, **404,** 230 (1974).
385. R. E. Wilde, S. N. Ghosh, and B. J. Marshall, *Inorg. Chem.*, **9,** 2513 (1970).
386. M. F. A. El-Sayed and R. K. Sheline, *J. Am. Chem. Soc.*, **78,** 702 (1956).
387. O. Jarchow, *Z. Anorg. Allg. Chem.*, **383,** 40 (1971).
388. W. P. Griffith and A. J. Wickham, *J. Chem. Soc.*, *A*, 834 (1969).
389. R. A. Walton, *Spectrochim. Acta*, **21,** 1795 (1965); *Can. J. Chem.*, **44,** 1480 (1966).
390. R. E. Clarke and P. C. Ford, *Inorg. Chem.*, **9,** 227 (1970).
391. J. C. Evans and G. Y.-S. Lo, *Spectrochim. Acta*, **21,** 1033 (1965).
392. J. Reedijk and W. L. Groeneveld, *Rec. Trav. Chim.*, **86,** 1127 (1967).
393. M. F. Farona and K. F. Kraus, *Inorg. Chem.*, **9,** 1700 (1970).
394. Y. Kinoshita, I. Matsubara, and Y. Saito, *Bull. Chem. Soc. Jap*, **32,** 741 (1959).
395. M. Kubota, D. L. Johnston, and I. Matsubara, *Inorg. Chem.*, **5,** 386 (1966).
396. Y. Kinoshita, I. Matsubara, and Y. Saito, *Bull. Chem. Soc. Jap*, **32,** 1216 (1959).
397. I. Matsubara, *Bull. Chem. Soc. Jap.*, **34,** 1719 (1961); *J. Chem. Phys.*, **35,** 373 (1961).
398. J. K. Brown, N. Sheppard, and D. M. Simpson, *Phil. Trans. Roy. Soc.*, **A247,** 35 (1954).
399. M. Kubota and D. L. Johnston, *J. Am. Chem. Soc.*, **88,** 2451 (1966).
400. I. Matsubara, *Bull. Chem. Soc. Jap.*, **35,** 27 (1962).
401. F. A. Cotton and F. Zingales, *J. Am. Chem. Soc.*, **83,** 351 (1961).
402. A. Sacco and F. A. Cotton, *J. Am. Chem. Soc.*, **84,** 2043 (1962).
403. J. W. Dart, M. K. Lloyd, R. Mason, J. A. McCleverty, and J. Williams, *J. Chem. Soc.*, Dalton, 1747 (1973).
404. P. M. Boorman, P. J. Craig, and T. W. Swaddle, *Can. J. Chem.*, **48,** 838 (1970).
405. J. L. Burmeister, *Coord. Chem. Rev.*, **3,** 225 (1968); **1,** 205 (1966).
406. R. A. Bailey, S. L. Kozak, T. W. Michelsen, and W. N. Mills, *Coord. Chem. Rev.*, **6,** 407 (1971).
407. A. H. Norbury, *Adv. Inorg. Chem. Radiochem.*, **17,** 231 (1975).
408. S. Ahrland, J. Chatt, and N. R. Davies, *Quart. Rev.*, **12,** 265 (1958).
409. P. C. H. Mitchell and R. J. P. Williams, *J. Chem. Soc.*, 1912 (1960).
410. A. Turco and C. Pecile, *Nature*, **191,** 66 (1961).
411. J. Lewis, R. S. Nyholm, and P. W. Smith, *J. Chem. Soc.*, 4590 (1961).
412. A. Sabatini and I. Bertini, *Inorg. Chem.*, **4,** 959 (1965).
413. C. Pecile, *Inorg. Chem.*, **5,** 210 (1966).
414. S. Fronaeus and R. Larsson, *Acta Chem. Scand.*, **16,** 1447 (1962).
415. R. A. Bailey, T. W. Michelsen, and W. N. Mills, *J. Inorg. Nucl. Chem.*, **33,** 3206 (1971).
416. R. Larsson and A. Miezis, *Acta Chem. Scand.*, **23,** 37 (1969).

417. R. J. H. Clark and C. S. Williams, *Spectrochim. Acta*, **22,** 1081 (1966).
418. D. Forster and D. M. L. Goodgame, *Inorg. Chem.*, **4,** 715 (1965).
419. M. A. Bennett, R. J. H. Clark, and A. D. J. Goodwin, *Inorg. Chem.*, **6,** 1625 (1967).
420. D. Forster and D. M. L. Goodgame, *Inorg. Chem.*, **4,** 823 (1965).
421. H. H. Schmidtke and D. Garthoff, *Helv. Chim. Acta*, **50,** 1631 (1967).
422. M. M. Chamberlain and J. C. Bailar, Jr., *J. Am. Chem. Soc.*, **81,** 6412 (1959).
423. A. B. P. Lever, B. S. Ramaswamy, S. H. Simonsen, and L. K. Thompson, *Can. J. Chem.*, **48,** 3076 (1970).
424. G. Contreras and R. Schmidt, *J. Inorg. Nucl. Chem.*, **32,** 1295, 127 (1970).
425. F. Basolo, J. L. Burmeister, and A. J. Poe, *J. Am. Chem. Soc.*, **85,** 1700 (1963).
426. J. Burmeister and F. Basolo, *Inorg. Chem.*, **3,** 1587 (1964).
427. A. Sabatini and I. Bertini, *Inorg. Chem.*, **4,** 1665 (1965).
428. D. M. L. Goodgame and B. W. Malerbi, *Spectrochim. Acta*, **24A,** 1254 (1968).
429. T. E. Sloan and A. Wojcicki, *Inorg. Chem.*, **7,** 1268 (1968).
430. I. Stotz, W. K. Wilmarth, and A. Haim, *Inorg. Chem.*, **7,** 1250 (1968).
431. R. L. Hassel and J. L. Burmeister, *Chem. Commun.*, 568 (1971).
432. L. A. Epps and L. G. Marzilli, *Inorg. Chem.*, **12,** 1514 (1973).
433. I. Bertini and A. Sabatini, *Inorg. Chem.*, **5,** 1025 (1966).
434. G. R. Clark, G. J. Palenik, and D. W. Meek, *J. Am. Chem. Soc.*, **92,** 1077 (1970).
435. D. W. Meek, P. E. Nicpon, and V. I. Meek, *J. Am. Chem. Soc.*, **92,** 5351 (1970).
436. S. M. Nelson and J. Rodgers, *Inorg. Chem.*, **6,** 1390 (1967).
437. W. Krasser and H. W. Nürnberg, *Z. Naturforsch.*, **25a,** 1394 (1970).
438. J. L. Burmeister, R. L. Hassel, and R. J. Phelan, *Inorg. Chem.*, **10,** 2032 (1971).
439. J. Chatt and L. A. Duncanson, *Nature*, **178,** 997 (1956).
440. J. Chatt, L. A. Duncanson, F. A. Hart, and P. G. Owston, *Nature*, **181,** 43 (1958).
441. P. G. Owston and J. M. Rowe, *Acta Crystallogr.* **13,** 253 (1960).
442. J. Chatt and F. A. Hart, *J. Chem. Soc.*, 1416 (1961).
443. B. R. Chamberlain and W. Moser, *J. Chem. Soc.*, A, 354 (1969).
444. G. Liptay, K. Burger, E. Papp-Molnár, and Sz. Szebeni, *J. Inorg. Nucl. Chem.*, **31,** 2359 (1969).
445. J. M. Homan, J. M. Kawamoto, and G. L. Morgan, *Inorg. Chem.* **9,** 2533 (1970).
446. R. A. Bailey and T. W. Michelsen, *J. Inorg. Nucl. Chem.*, **34,** 2671 (1972).
447. F. A. Cotton, D. M. L. Goodgame, M. Goodgame, and T. E. Haas, *Inorg. Chem.*, **1,** 565 (1962).
448. J. L. Burmeister and L. E. Williams, *Inorg. Chem.*, **5,** 1113 (1966).
449. M. E. Farago and J. M. James, *Inorg. Chem.*, **4,** 1706 (1965).
450. A. Turco, C. Pecile, and M. Nicolini, *J. Chem. Soc.*, 3008 (1962).
451. J. L. Burmeister and Y.Al-Janabi, *Inorg. Chem.*, **4,** 962 (1965).
452. D. Forster and D. M. L. Goodgame, *Inorg. Chem.*, **4,** 1712 (1965).
453. J. L. Burmeister and H. J. Gysling, *Inorg. Chim. Acta*, **1,** 100 (1967).
454. M. A. Jennings and A. Wojcicki, *Inorg. Chim. Acta*, **3,** 335 (1969).
455. J. L. Burmeister, H. J. Gysling, and J. C. Lim, *J. Am. Chem. Soc.*, **91,** 44 (1969).
456. F. A. Miller and G. L. Carlson, *Spectrochim. Acta*, **17,** 977 (1961).
456a. D. Forster and W. D. Horrocks, *Inorg. Chem.*, **6,** 339 (1967).
457. D. Forster and D. M. L. Goodgame, *J. Chem. Soc.*, 262 (1965).
458. A. R. Chugtai and R. N. Keller, *J. Inorg. Nucl. Chem.*, **31,** 633 (1969).
459. D. Forster and D. M. L. Goodgame, *J. Chem. Soc.*, 1286 (1965).
460. E. J. Peterson, A. Galliart, and J. M. Brown, *Inorg. Nucl. Chem. Lett.*, **9,** 241 (1973).
461. R. A. Bailey and S. L. Kozak, *J. Inorg. Nucl. Chem.*, **31,** 689 (1969).
462. A. H. Norbury and A. I. P. Sinha, *J. Chem. Soc.*, *A*, 1598 (1968).

463. S. J. Patel and D. G. Tuck, *J. Chem. Soc., A*, 1870 (1968).
464. S. J. Anderson and A. H. Norbury, *Chem. Commun*, 37 (1974).
465. J. Nelson and S. M. Nelson, *J. Chem. Soc., A*, 1597 (1969).
466. R. B. Saillant, *J. Organometal. Chem.*, **39,** C71 (1972).
467. W. Beck, *Chem. Ber.*, **95,** 341 (1962).
468. W. Beck, P. Swoboda, K. Feldl, and E. Schuierer, *Chem. Ber.*, **103,** 3591 (1970).
469. W. Beck and E. Schuierer, *Z. Anorg. Allg. Chem.*, **347,** 304 (1966).
470. W. Beck and E. Schuierer, *Chem. Ber.*, **98,** 298 (1965).
471. W. Beck, C. Oetker, and P. Swoboda, *Z. Naturforsch.*, **28b,** 229 (1973).
472. W. Beck, W. P. Fehlhammer, P. Pöllmann, E. Schuierer, and K. Feldl, *Chem. Ber.*, **100,** 2335 (1967); *Angew. Chem.*, **77,** 458 (1965).
473. D. Forster and W. D. Horrocks, *Inorg. Chem.*, **5,** 1510 (1966).
474. D. Seybold and K. Dehnicke, *Z. Anorg. Allg. Chem.*, **361,** 277 (1968).
475. L. F. Druding, H. C. Wang, R. E. Lohen, and F. D. Sancilio, *J. Coord. Chem.*,**3,** 105 (1973).
476. I. Agrell, *Acta Chem. Scand.*, **25,** 2965 (1971).
477. W. Beck, W. P. Fehlhammer, P. Pöllman, and R. S. Tobias, *Inorg. Chim. Acta*, **2,** 467 (1968).
478. D. R. Herrington and L. J. Boucher, *Inorg. Nucl. Chem. Lett.*, **7,** 1091 (1971).
479. G. R. Dobon, I. W. Stolz, and R. K. Sheline, *Adv. Inorg. Chem. Radiochem.*, **8,** 1 (1965).
480. E. W. Abel and F. G. A. Stone, *Quart. Rev.*, **23,** 325 (1969).
481. L. M. Haines and M. H. B. Stiddard, *Adv. Inorg. Chem. Radiochem.*, **12,** 53 (1969).
482. P. S. Braterman, *Metal Carbonyl Spectra*, Academic Press, New York, 1974.
483. J. S. Kristoff and D. F. Shriver, *Inorg. Chem.*, **13,** 499 (1974).
484. G. Bouquet and M. Bigorgne, *Spectrochim. Acta*, **27A,** 139 (1971).
485. W. F. Edgell and J. Lyford, IV, *J. Chem. Phys.*, **52,** 4329 (1970).
486. H. Stammreich, K. Kawai, Y. Tavares, P. Krumholz, J. Behmoiras, and S. Bril, *J. Chem. Phys.* **32,** 1482 (1960).
487. M. Bigorgne, *J. Organometal. Chem.*, **24,** 211 (1970).
488. L. H. Jones, R. S. McDowell, M. Goldblatt, and B. I. Swanson, *J. Chem. Phys.*, **57,** 2050 (1972).
489. L. H. Jones, R. S. McDowell, and M. Goldblatt, *Inorg. Chem.*, **8,** 2349 (1969).
490. E. W. Abel, R. A. N. McLean, S. P. Tyfield, P. S. Braterman, A. P. Walker, and P. J. Hendra, *J. Mol. Spectrosc.*, **30,** 29 (1969).
491. R. A. N. McLean, *Can. J. Chem.*, **52,** 213 (1974).
492. F. Calderazzo and F. L'Eplattenier, *Inorg. Chem.*, **6,** 1220 (1967).
493. A. Terzis and T. G. Spiro, *Inorg. Chem.*, **10,** 643 (1971).
494. P. J. Hendra and M. M. Qurashi, *J. Chem. Soc., A*, 2963 (1968).
495. S. F. A. Kettle and I. Paul, *Adv. Organometal. Chem.*, **10,** 199 (1972).
496. R. J. H. Clark and B. Crociani, *Inorg. Chim. Acta*, **1,** 12 (1967).
497. W. F. Edgell, J. Lyford, R. Wright, W. M. Risen, Jr. and A. T. Watts, *J. Am. Chem. Soc.*, **92,** 2240 (1970); W. F. Edgell and J. Lyford, *ibid.*, **93,** 6407 (1971).
498. C. D. Pribula and T. L. Brown, *J. Organometal. Chem.*, **71,** 415 (1974).
499. G. G. Summer, H. P. Klug, and L. E. Alexander, *Acta Crystallogr.*, **17,** 732 (1964).
500. F. A. Cotton and R. R. Monchamp, *J. Chem. Soc.*, 1882 (1960).
501. K. Noack, *Spectrochim. Acta*, **19,** 1925 (1963).
502. G. Bor, *Spectrochim. Acta*, **19,** 1209 (1963).
503. G. Bor and K. Noack, *J. Organometal. Chem.*, **64,** 367 (1974).
504. R. K. Sheline and K. S. Pitzer, *J. Am. Chem. Soc.*, **72,** 1107 (1950).

505. H. M. Powell and R. V. G. Ewens, *J. Chem. Soc.*, 286 (1939).
506. L. F. Dahl and R. E. Rundle, *Acta Crystallogr.*, **16,** 419 (1963).
507. D. M. Adams, M. A. Hooper, and A. Squire, *J. Chem. Soc.*, *A*, 71 (1971).
508. G. Bor, *Chem. Commun*, 641 (1969).
509. R. A. Levenson, H. B. Gray, and G. P. Ceasar, *J. Am. Chem. Soc.*, **92,** 3653 (1970).
510. I. J. Hyams, D. Jones, and E. R. Lippincott, *J. Chem. Soc. A*, 1987 (1967).
511. N. Flitcroft, D. K. Huggins, and H. D. Kaesz, *Inorg. Chem.*, **3,** 1123 (1964).
512. L. F. Dahl and J. F. Blount, *Inorg. Chem.*, **4,** 1373 (1965); C. H. Wei and L. F. Dahl, *J. Am. Chem. Soc.*, **91,** 1351 (1969).
513. N. E. Erickson and A. W. Fairhall, *Inorg. Chem.*, **4,** 1320 (1965).
514. F. A. Cotton and D. L. Hunter, *Inorg. Chim. Acta*, **11,** L9 (1974).
515. B. F. G. Johnson, *Chem. Commun.*, 703 (1976).
516. E. R. Corey and L. F. Dahl, *Inorg. Chem.*, **1,** 521 (1962).
517. D. K. Huggins, N. Flitcroft, and H. D. Kaesz, *Inorg. Chem.*, **4,** 166 (1965).
518. C. O. Quicksall and T. G. Spiro, *Inorg. Chem.*, **7,** 2365 (1968).
519. P. Corradini, *J. Chem. Phys.*, **31,** 1676 (1959).
520. H. Stammreich, K. Kawai, O. Sala, and P. Krumholz, *J. Chem. Phys.*, **35,** 2175 (1961).
521. G. Bor, *Inorg. Chim. Acta*, **3,** 196 (1969).
522. R. J. Ziegler, J. M. Burlitch, S. E. Hayes, and W. M. Risen, Jr., *Inorg. Chem.*, **11,** 702 (1972).
523. N. J. Nelson, N. E. Kime, and D. F. Shriver, *J. Am. Chem. Soc.*, **91,** 5173 (1969).
524. I. J. Hyams and E. R. Lippincott, *Spectrochim. Acta*, **25A,** 1845 (1969).
525. D. K. Ottesen, H. B. Gray, L. H. Jones, and M. Goldblatt, *Inorg. Chem.*, **12,** 1051 (1973).
526. M. J. Cleare and W. P. Griffith, *J. Chem. Soc.*, *A*, 372 (1969).
527. R. G. Denning and M. J. Ware, *Spectrochim. Acta*, **24A,** 1785 (1968).
528. K. Noack, *J. Organometal. Chem.*, **13,** 411 (1968).
529. M. F. Farona and G. R. Camp, *Inorg. Chim. Acta*, **3,** 395 (1969).
530. W. P. Anderson and T. L. Brown, *J. Organometal. Chem.*, **32,** 343 (1971).
531. M. A. El-Sayed and H. D. Kaesz, *Inorg. Chem.*, **2,** 158 (1963).
532. C. W. Garland and J. R. Wilt, *J. Chem. Phys.*, **36,** 1094 (1962).
533. L. F. Dahl, C. Martell, and D. L. Wampler, *J. Am. Chem. Soc.*, **83,** 1761 (1961).
534. B. F. G. Johnson, J. Lewis, P. W. Robinson, and J. R. Miller, *J. Chem. Soc.*, *A*, 2693 (1969).
535. F. A. Cotton and B. F. G. Johnson, *Inorg. Chem.*, **6,** 2113 (1967).
536. A. Loutellier and M. Bigorgne, *J. Chim. Phys.*, **67,** 78, 99, 107 (1970).
537. M. Bigorgne, *J. Organometal. Chem.*, **24,** 211 (1970).
538. J. Dalton, I. Paul, J. G. Smith, and F. G. A. Stone, *J. Chem. Soc.*, *A*, 1195 (1968).
539. R. J. Angelici and M. D. Malone, *Inorg. Chem.*, **6,** 1731 (1967).
540. B. Hutchinson and K. Nakamoto, *Inorg. Chim. Acta*, **3,** 591 (1969).
541. A. A. Chalmers, J. Lewis, and R. Whyman, *J. Chem. Soc.*, *A*, 1817 (1967).
542. M. F. Farona, J. G. Grasselli, and B. L. Ross, *Spectrochim. Acta*, **23A**, 1875 (1967).
543. S. Singh, P. P. Singh, and R. Rivest, *Inorg. Chem.*, **7,** 1236 (1968).
544. F. A. Cotton and C. S. Kraihanzel, *J. Am. Chem. Soc.*, **84,** 4432 (1962); *Inorg. Chem.*, **2,** 533 (1963); **3,** 702 (1964).
545. F. A. Cotton, M. Musco, and G. Yagupsky, *Inorg. Chem.*, **6,** 1357 (1967).
546. L. H. Jones, *Inorg. Chem.*, **7,** 1681 (1968); **6,** 1269 (1967).
547. F. A. Cotton, *Inorg. Chem.*, **7,** 1683 (1968).
548. M. B. Hall and R. F. Fenske, *Inorg. Chem.*, **11,** 1619 (1972).

549. A. C. Sarapu and R. F. Fenske, *Inorg. Chem.*, **14,** 247 (1975).
550. H. D. Kaesz and R. B. Saillant, *Chem. Rev.*, **72,** 231 (1972).
551. W. F. Edgell, C. Magee, and G. Gallup, *J. Am. Chem. Soc.*, **78,** 4185, 4188 (1956); W. F. Edgell and R. Summitt, *ibid.*, **83,** 1772 (1961).
552. S. J. LaPlaca, W. C. Hamilton, and J. A. Ibers, *Inorg. Chem.*, **3,** 1491 (1964); *J. Am. Chem. Soc.*, **86,** 2288 (1964).
553. D. K. Huggins and H. D. Kaesz, *J. Am. Chem. Soc.*, **86,** 2734 (1964).
554. P. S. Braterman, R. W. Harrill, and H. D. Kaesz, *J. Am. Chem. Soc.*, **89,** 2851 (1967).
555. A. Davison and J. W. Faller, *Inorg. Chem.*, **6,** 845 (1967).
556. W. F. Edgell, J. W. Fisher, G. Asato, and W. M. Risen, Jr., *Inorg. Chem.*, **8,** 1103 (1969).
557. K. Farmery and M. Kilner, *J. Chem. Soc.*, *A*, 634 (1970).
558. S. S. Bath and L. Vaska, *J. Am. Chem. Soc.*, **85,** 3500 (1963).
559. J. Chatt, N. P. Johnson, and B. L. Shaw, *J. Chem. Soc.*, 1625 (1964).
560. L. Vaska, *J. Am. Chem. Soc.*, **88,** 4100 (1966).
561. F. L'Eplattenier and F. Calderazzo, *Inorg. Chem.*, **7,** 1290 (1968).
562. D. K. Huggins, W. Fellman, J. M. Smith, and H. D. Kaesz, *J. Am. Chem. Soc.*, **86,** 4841 (1964).
563. J. M. Smith, W. Fellmann, and L. H. Jones, *Inorg. Chem.*, **4,** 1361 (1965).
564. R. G. Hayter, *J. Am. Chem. Soc.*, **88,** 4376 (1966).
565. A. P. Ginsberg and M. J. Hawkes, *J. Am. Chem. Soc.*, **90,** 5931 (1968).
566. M. J. Mays and R. N. F. Simpson, *J. Chem. Soc.*, *A*, 1444 (1968); *Chem Commun.*, 1024 (1967).
567. H. D. Kaesz, F. Fontal, R. Bau, S. W. Kirtley, and M. R. Churchill, *J. Am. Chem. Soc.*, **91,** 1021 (1969).
568. M. J. Bennett, W. A. G. Graham, J. K. Hoyano, and W. L. Hutcheon, *J. Am. Chem. Soc.*, **94,** 6232 (1972).
569. S. A. R. Knox, J. W. Koepke, M. A. Andrews, and H. D. Kaesz, *J. Am. Chem. Soc.*, **97,** 3942 (1975).
570. S. A. R. Knox and H. D. Kaesz, *J. Am. Chem. Soc.*, **93,** 4594 (1971).
571. J. W. White and C. J. Wright, *Chem. Commun.*, 971 (1970).
572. R. L. DeKock, *Inorg. Chem.*, **10,** 1205 (1971).
573. M. Moskovits and G. A. Ozin, "Characterization of the Products of Metal Atom-Molecule Condensation Reactions by Matrix Infrared and Raman Spectroscopy," in *Vibrational Spectra and Structure*, Vol. 4, J. R. Durig, ed., Elsevier, Amsterdam, 1975.
574. J. H. Darling and J. S. Ogden, *Inorg. Chem.*, **11,** 666 (1972); *J. Chem. Soc.*, Dalton, 1079 (1973).
575. H. Huber, E. P. Kündig, M. Moskovits, and G. A. Ozin, *Nature*, *Phys. Sci.*, **235**, 98 (1972); E. P. Kündig, M. Moskovits, and G. A. Ozin, *Can. J. Chem.*, **50,** 3587 (1972).
576. E. P. Kündig, D. McIntosh, M. Moskovits, and G. A. Ozin, *J. Am. Chem. Soc.*, **95,** 7234 (1973).
577. J. L. Slater, R. K. Sheline, K. C. Lin, and W. Weltner, *J. Chem. Phys.*, **55,** 5129 (1971).
578. J. L. Slater, T. C. DeVore, and V. Calder, *Inorg. Chem.*, **12,** 1918 (1973); **13,** 1808 (1974).
579. M. Poliakoff and J. J. Turner, *J. Chem. Soc.*, Dalton, 1351 (1973); 2276 (1974).
580. M. A. Graham, M. Poliakoff, and J. J. Turner, *J. Chem. Soc.*, *A*, 2939 (1971).
581. E. P. Kündig and G. A. Ozin, *J. Am. Chem. Soc.*, **96,** 3820 (1974).

582. O. Crichton, M. Poliakoff, A. J. Rest, and J. J. Turner, *J. Chem. Soc.*, Dalton, 1321 (1973).
583. M. Poliakoff and J. J. Turner, *J. Chem. Soc.*, *A*, 2403 (1971).
584. D. A. Van Leirsburg and C. W. DeKock, *J. Phys. Chem.*, **78,** 134 (1974).
585. D. Tevault and K. Nakamoto, *Inorg. Chem.*, **15,** 1282 (1976).
586. B. I. Swanson, L. H. Jones, and R. R. Ryan, *J. Mol. Spectrosc.*, **45,** 324 (1973).
587. M. Poliakoff and J. J. Turner, *J. Chem. Soc.*, *A*, 654 (1971).
588. D. Tevault and K. Nakamoto, *Inorg. Chem.*, **14,** 2371 (1975); A. Cormier, J. D. Brown, and K. Nakamoto, *ibid.*, **12,** 3011 (1973).
589. P. Gans, A. Sabatini, and L. Sacconi, *Coord. Chem. Rev.*, **1,** 187 (1966).
590. J. Masek, *Inorg. Chim. Acta Rev.*, **3,** 99 (1969).
591. B. F. G. Johnson and J. A. McCleverty, *Prog. Inorg. Chem.*, **7,** 277 (1966).
592. W. P. Griffith, *Adv. Organometal. Chem.*, **7,** 211 (1968).
593. J. A. McGinnety, *MTP, Int. Rev. Sci. Inorg. Chem.*, **5,** 229 (1972).
594. J. H. Enemark and R. D. Feltham, *Coord. Chem. Rev.*, **13,** 339 (1974).
595. C. G. Pierpont, D. G. Van Derveer, W. Durland, and R. Eisenberg, *J. Am. Chem. Soc.*, **92,** 4760 (1970).
596. C. P. Brock, J. P. Collman, G. Dolcetti, P. H. Farnham, J. A. Ibers, J. E. Lester, and C. A. Reed, *Inorg. Chem.*, **12,** 1304 (1973).
597. M. Herberhold and A. Razavi, *Angew, Chem. Int. Ed.*, **11,** 1092 (1972).
598. I. H. Sabberwal and A. B. Burg, *Chem. Commun.*, 1001 (1970).
599. L. H. Jones, R. S. McDowell, and B. I. Swanson, *J. Chem. Phys.*, **58,** 3757 (1973).
600. G. Barna and I. S. Butler, *Can. J. Spectrosc.*, **17,** 2 (1972).
601. O. Crichton and A. J. Rest, *Inorg. Nucl. Chem. Lett.*, **9,** 391 (1973).
602. B. F. G. Johnson, *J. Chem. Soc.*, *A*, 475 (1967).
603. Z. Iqbal and T. C. Waddington, *J. Chem. Soc.*, *A*, 1092 (1969).
604. E. Miki, T. Ishimori, H. Yamatera, and H. Okuno, *J. Chem. Soc. Jap*, **87,** 703 (1966).
605. J. R. Durig, W. A. McAllister, J. N. Willis, Jr., and E. E. Mercer, *Spectrochim. Acta*, **22,** 1091 (1966).
606. D. Dale and D. C. Hodgkin, *J. Chem. Soc.*, 1364 (1965).
607. D. Hall and A. A. Taggart, *J. Chem. Soc.*, 1359 (1965).
608. B. F. Hoskins, F. D. Whillans, D. H. Dale, and D. C. Hodgkin, *Chem. Commun.*, 69 (1969).
609. E. E. Mercer, W. A. McAllister, and J. R. Durig, *Inorg. Chem.*, **6,** 1816 (1967),
610. S. Pell and J. N. Armor, *Inorg. Chem.*, **12,** 873 (1973).
611. H. Mosback and K. G. Poulsen, *Chem. Commun.*, 479 (1969).
612. E. E. Mercer, W. A. McAllister, and J. R. Durig, *Inorg. Chem.*, **5,** 1881 (1966).
613. J. R. Norton, J. P. Collman, G. Dolcetti, and W. T. Robinson, *Inorg. Chem.*, **11,** 382 (1972).
614. V. J. Choy and C. H. O'Connor, *Coord. Chem. Rev.*, **9,** 45 (1972).
615. J. A. Connor and E. A. V. Ebsworth, *Adv. Inorg. Chem. Radiochem.*, **6,** 279 (1964).
616. F. Blunt, P. J. Hendra, and J. R. Mackenzie, *Chem. Commun.*, 278 (1969).
617. A. Nakamura, Y. Tatsuno, M. Yamamoto, and S. Otsuka, *J. Am. Chem. Soc.*, **93,** 6052 (1971).
618. F. Offner and J. Dehand, *Compt. Rend.*, **273,** C50 (1971).
619. C. A. Reed and W. R. Roper, *J. Chem. Soc.*, Dalton, 1370 (1973).
620. J. A. McGinnety, R. J. Doedens, and J. A. Ibers, *Inorg. Chem.*, **6,** 2243 (1967).
621. J. B. R. Dunn, D. F. Shriver, and I. M. Klotz, *Proc. Nat. Acad. Sci.*, **70,** 2582 (1973).
622. J. S. Loehr, T. B. Freedman, and T. M. Loehr, *Biochem. Biophys. Res. Commun.*, **56,** 510 (1974).
623. C. Floriani and F. Calderazzo, *J. Chem. Soc.*, 946 (1969).

624. D. L. Diemente, B. M. Hoffman, and F. Basolo, *Chem. Commun.*, 467 (1970).
625. A. L. Crumbliss and F. Basolo, *J. Am. Chem. Soc.*, **92,** 55 (1970).
625a. B. M. Hoffman, D. L. Diemente, and F. Basolo, *J. Am. Chem. Soc.*, **92,** 61 (1970).
626. S. Ikawa, K. Hasebe, and M. Kimura, *Spectrochim. Acta*, **30A,** 151 (1974).
627. W. P. Griffith and T. D. Wickins, *J. Chem. Soc.*, *A*, 397 (1968).
628. M. Mori, J. A. Weil, and M. Ishiguro, *J. Am. Chem. Soc.*, **90,** 615 (1968).
629. T. Shibahara, *Chem. Commun.*, 864 (1973).
630. T. B. Freedman, C. M. Yoshida, and T. M. Loehr, *Chem. Commun.*, 1016 (1974).
631. G. V. Jere and G. D. Gupta, *J. Inorg. Nucl. Chem.*, **32,** 537 (1970).
632. A. D. Allen and C. V. Senoff. *Chem. Commun.*, 621 (1965).
633. A. D. Allen and F. Bottomley, *Acc. Chem. Res.*, **1,** 360 (1968).
634. P. C. Ford, *Coord. Chem. Rev.*, **5,** 75 (1970).
635. R. Murray and D. C. Smith, *Coord. Chem. Rev.*, **3,** 429 (1968).
636. G. Henrici-Olive and S. Olive, *Angew. Chem. Int. Ed.*, **8,** 650 (1969).
637. A. D. Allen, F. Bottomley, R. O. Harris, V. P. Reinsalu, and C. V. Senoff, *J. Am. Chem. Soc.*, **89,** 5595 (1967).
638. S. Pell, R. H. Mann, H. Taube, and J. N. Armor, *Inorg. Chem.*, **13,** 479 (1974).
639. M. W. Bee, S. F. A. Kettle, and D. B. Powell, *Spectrochim. Acta*, **30A,** 585 (1974).
640. A. D. Allen and J. R. Stevens, *Chem. Commun.*, 1147 (1967).
641. G. Speier and L. Markó, *Inorg. Chim. Acta*, **3,** 126 (1969).
642. J. H. Enemark, B. R. Davis, J. A. McGinnety, and J. A. Ibers, *Chem. Commun.*, 96 (1968).
643. J. P. Collman, M. Kubota, F. D. Vastine, J. Y. Sun, and J. W. Kang, *J. Am. Chem. Soc.*, **90,** 5430 (1968).
644. M. Hidai, K. Tominari, Y. Uchida, and A. Misono, *Chem. Commun.*, 1392 (1969).
645. B. Bell, J. Chatt, and G. J. Leigh, *Chem. Commun.*, 842 (1970).
646. G. Speier and L. Markó, *J. Organometal. Chem.*, **21,** P46 (1970).
647. S. C. Srivastava and M. Bigorgne, *J. Organometal. Chem.*, **19,** 241 (1969).
648. D. J. Darensbourg, *Inorg. Chem.*, **11,** 1436 (1972).
649. J. N. Armor and H. Taube, *J. Am. Chem. Soc.*, **92,** 2560 (1970).
650. G. A. Ozin and A. Vander Voet, *Can. J. Chem.*, **51,** 637 (1973).
651. K. Jonas, *Angew. Chem. Int. Ed.*, **19,** 997 (1973); C. Kruger and Y.-H. Tsay, *ibid.*, **19,** 998 (1973).
652. J. Chatt, A. B. Nikolsky, R. L. Richards, and J. R. Sanders, *Chem. Commun.*, 154 (1969).
653. J. Chatt, R. C. Fay, and R. L. Richards, *J. Chem. Soc.*, *A*, 702 (1971).
654. M. Mercer, R. H. Crabtree, and R. L. Richards, *Chem. Commun.*, 808 (1973).
655. D. Sellman, A. Brandl, and R. Endell, *J. Organometal. Chem.*, **49,** C22 (1973).
656. W. P. Griffith, *Coord. Chem. Rev.*, **8,** 369 (1972).
657. M. J. Cleare and W. P. Griffith, *J. Chem. Soc.*, *A*, 1117 (1970).
658. W. Kolitsch and K. Dehnicke, *Z. Natursforsch.*, **25b,** 1080 (1970).
659. W. P. Griffith and D. Pawson, *J. Chem. Soc.*, Dalton, 1315 (1973).
660. M. Moskovits and G. A. Ozin, *Cryochemistry*, Wiley, New York, 1976.
661. L. Andrews, *J. Chem. Phys.*, **50,** 4288 (1969); *J. Phys. Chem.*, **73,** 3922 (1969).
662. H. Huber, W. Klotzbucher, G. A. Ozin, and A. Vander Voet, *Can. J. Chem.*, **51,** 2722 (1973).
663. D. McIntosh and G. A. Ozin, *Inorg. Chem.*, **15,** 2869 (1976).
664. H. Huber, E. P. Kündig, M. Moskovits, and G. A. Ozin, *J. Am. Chem. Soc.*, **95,** 332 (1973).
665. J. K. Burdett and J. J. Turner, *Chem. Commun.*, 885 (1971).

666. G. A. Ozin, M. Moskovits, P. Kündig, and H. Huber, *Can. J. Chem.*, **50,** 2385 (1972).
667. E. P. Kündig, M. Moskovits, and G. A. Ozin, *Can. J. Chem.*, **51,** 2710 (1973); D. W. Green, J. Thomas, and D. M. Gruen, *J. Chem. Phys.*, **58,** 5453 (1973).
668. G. A. Ozin and A. Vander Voet, *Can. J. Chem.*, **51,** 637 (1973).
669. G. A. Ozin and A. Vander Voet, *Can. J. Chem.*, **51,** 3332 (1973).
670. A. J. Rest, *J. Organometal. Chem.*, **40,** C76 (1972).
671. E. P. Kündig, M. Moskovits. and G. A. Ozin, *Can. J. Chem.*, **51,** 2737 (1973).
672. A. Misono, Y. Uchida, T. Saito, and K. M. Song, *Chem. Commun.*, 419 (1967).
673. A. Sacco and M. Rossi, *Chem. Commun.*, 316 (1967).
674. R. G. S. Banks and J. M. Pratt, *J. Chem. Soc.*, *A*, 854 (1968).
675. K. Krogmann and W. Binder, *Angew. Chem. Int. Ed.*, **6,** 881 (1967).
676. J. Chatt and R. G. Hayter, *J. Chem. Soc.*, 5507 (1961).
677. L. Vaska and J. W. DiLizio, *J. Am. Chem. Soc.*, **84,** 4989 (1962).
678. L. Vaska, *Chem. Commun.*, 614 (1966).
679. L. Vaska and D. L. Catone, *J. Am. Chem. Soc.*, **88,** 5324 (1966).
680. J. Chatt and R. G. Hayter, *J. Chem. Soc.*, 2605 (1961).
681. J. Chatt, L. A. Duncanson, and B. L. Shaw, *Chem.&Ind.*, 859 (1958).
682. M. J. Church and M. J. Mays, *J. Chem. Soc.*, 3074 (1968); 1938 (1970).
683. P. W. Atkins, J. C. Green, and M. L. H. Green, *J. Chem. Soc.*, *A*, 2275 (1968).
684. M. Tsuboi, K. Onishi, I. Nakagawa, T. Shimanouchi, and S. Mizushima, *Spectrochim. Acta*, **12,** 253 (1958).
685. K. Fukushima, T. Onishi, T. Shimanouchi, and S. Mizushima, *Spectrochim. Acta*, **14,** 236 (1959).
686. A. J. Stosick, *J. Am. Chem. Soc.*, **67,** 365 (1945).
687. K. Nakamoto, Y. Morimoto, and A. E. Martell, *J. Am. Chem. Soc.*, **83,** 4528 (1961).
688. R. A. Condrate and K. Nakamoto, *J. Chem. Phys.*, **42,** 2590 (1965).
689. T. J. Lane, J. A. Durkin, and R. J. Hooper, *Spectrochim. Acta*, **20,** 1013 (1964); I. Nakagawa, R. J. Hooper, J. L. Walter, and T. J. Lane, *ibid.*, **21,** 1 (1965).
690. R. J. Hooper, T. J. Lane, and J. L. Walter, *Inorg. Chem.*, **3,** 1568 (1964); J. F. Jackovitz and J. L. Walter, *Spectrochim. Acta*, **22,** 1393 (1966); J. F. Jackovitz, J. A. Durkin, and J. L. Walter, *ibid.*, **23A,** 67 (1967).
691. C. W. Rayner-Canham and A. B. P. Lever, *Spectrosc. Lett.*, **6,** 109 (1973).
692. J. Kincaid and K. Nakamoto, *Spectrochim. Acta*, **32A,** 277 (1976).
693. K. Krishnan and R. A. Plane, *Inorg. Chem.*, **6,** 55 (1967).
694. T. V. Long, II, and C. M. Yoshida, *Inorg. Chem.*, **6,** 1754 (1967).
695. A. W. Herlinger, S. L. Wenhold, and T. V. Long, II, *J. Am. Chem. Soc.*, **92,** 6474 (1970).
696. A. W. Herlinger and T. V. Long, II, *J. Am. Chem. Soc.*, **92,** 6481 (1970).
697. J. A. Kieft and K. Nakamoto, to be published.
698. J. A. Kieft and K. Nakamoto, *J. Inorg. Nucl. Chem.*, **29,** 2561 (1967).
699. J. A. Kieft and K. Nakamoto, *Inorg. Chim. Acta*, **2,** 225 (1968).
700. V. Balice and T. Theophanides, *J. Inorg. Nucl. Chem.*, **32,** 1237 (1970).
701. G. W. Watt and J. F. Knifton, *Inorg. Chem.*, **6,** 1010 (1967); **7,** 1159 (1968).
702. D. H. Busch and J. C. Bailar, Jr., *J. Am. Chem. Soc.*, **75,** 4574 (1953); **78,** 716 (1956); M. L. Morris and D. H. Busch, *ibid.*, **78,** 5178 (1956); K. Swaminathan and D. H. Busch, *J. Inorg. Nucl. Chem.*, **20,** 159 (1961); R. E. Sievers and J. C. Bailar, Jr., *Inorg. Chem.*, **1,** 174 (1962).
703. D. Chapman, *J. Chem. Soc.*, 1766 (1955).
704. Y. Tomita and K. Ueno, *Bull. Chem. Soc. Jap.* **36,** 1069 (1963).

705. K. Krishnan and R. A. Plane, *J. Am. Chem. Soc.*, **90,** 3195 (1968).
706. D. G. Brannon, R. H. Morrison, J. L. Hall, G. L. Humphrey, and D. N. Zimmerman, *J. Inorg. Nucl. Chem.*, **33,** 981 (1971).
707. S. Fronaeus and R. Larsson, *Acta Chem. Scand.*, **16,** 1433, 1447 (1962).
708. S. Fronaeus and R. Larsson, *Acta Chem. Scand.*, **14,** 1364 (1960).
709. R. Larsson, *Aota Chem. Scand.*, **19,** 783 (1965).
710. K. Nakamoto, Y. Morimoto, and A. E. Martell, *J. Am. Chem. Soc.*, **84,** 2081 (1962); **85,** 309 (1963).
711. Y. Tomita, T. Ando, and K. Ueno, *J. Phys. Chem.*, **69,** 404 (1965).
712. Y. Tomita and K. Ueno, *Bull. Chem. Soc. Jap*, **36,** 1069 (1963).
713. A. E. Martell and M. K. Kim, *J. Coord. Chem.*, **4,** 9 (1974).
714. M. K. Kim and A. E. Martell, *J. Am. Chem. Soc.*, **85,** 3080 (1963).
715. M. K. Kim and A. E. Martell, *Biochemistry*, **3,** 1169 (1964).
716. M. K. Kim and A. E. Martell, *J. Am. Chem. Soc.*, **88,** 914 (1966).
717. R. J. H. Clark, in *Halogen Chemistry*, V. Gutmann, ed., Vol. 3, Academic Press, New York, 1967. p. 85.
718. R. H. Nuttall, *Talanta*, **15,** 157 (1968).
719. A. J. Carty, *Coord. Chem. Rev.*, **4,** 29 (1969).
720. R. J. H. Clark, *Spectrochim. Acta*, **21,** 955 (1965).
721. K. Konya and K. Nakamoto, *Spectrochim. Acta*, **29A,** 1965 (1973).
722. K. Thompson and K. Carlson, *J. Chem. Phys.*, **49,** 4379 (1968).
723. K. Shobatake and K. Nakamoto, *J. Am. Chem. Soc.*, **92,** 3332 (1970).
724. C. Udovich, J. Takemoto, and K. Nakamoto, *J. Coord. Chem.*, **1,** 89 (1971).
725. P. M. Boorman and A. J. Carty, *Inorg. Nucl. Chem. Lett.*, **4,** 101 (1968).
726. I. Wharf and D. F. Shriver, *Inorg. Chem.*, **8,** 914 (1969).
727. D. F. Shriver and M. P. Johnson, *Inorg. Chem.*, **6,** 1265 (1967).
728. B. Crociani, T. Boschi, and M. Nicolini, *Inorg. Chim. Acta*, **4,** 577 (1970).
729. F. H. Herbelin, J. D. Herbelin, J. P. Mathieu, and H. Poulet, *Spectrochim. Acta*, **22,** 1515 (1966).
730. D. M. Adams, J. Chatt, J. Gerratt, and A. D. Westland, *J. Chem. Soc.*, 734 (1964).
731. T. G. Appleton, H. C. Clark, and L. E. Manzer, *Coord. Chem. Rev.*, **10,** 335 (1973).
732. R. J. Goodfellow, P. L. Goggin and D. A. Duddell, *J. Chem. Soc.*, **A,** 504 (1968).
733 M. A. Bennett, R. J. H. Clark, and D. L. Milner, *Inorg. Chem.*, **6,** 1647 (1967).
734 J. Fujita, K. Konya, and K. Nakamoto, *Inorg. Chem.*, **9,** 2794 (1970).
735 J. T. Wang, C. Udovich, K. Nakamoto, A. Quattrochi, and J. R. Ferraro, *Inorg. Chem.*, **9,** 2675 (1970).
736. B. T. Kilbourn and H. M. Powell, *J. Chem. Soc.*, *A*, 1688 (1970).
737. J. R. Ferraro, K. Nakamoto, J. T. Wang, and L. Lauer, *Chem. Commun.*, 266 (1973).
738. C. Postmus, K. Nakamoto, and J. R. Ferraro, *Inorg. Chem.*, **6,** 2194 (1967).
738a. J. R. Ferraro and G. J. Long, *Acc. Chem. Res.*, **8,** 171 (1975).
739. D. M. Adams and P. J. Chandler, *J. Chem. Soc.*, *A*, 1009 (1967).
740. N. Ohkaku and K. Nakamoto, *Inorg. Chem.*, **12,** 2440, 2446 (1973).
741. D. M. Adams and D. C. Newton, *J. Chem. Soc.*, Dalton, 681 (1972).
742. R. J. Goodfellow, P. L. Goggin, and L. M. Venanzi, *J. Chem. Soc.*, *A*, 1897 (1967).
743. D. M. Adams and P. J. Chandler, *Chem. Commun.*, 69 (1966).
744. M. Goldstein and W. D. Unsworth, *Inorg. Chim. Acta*, **4,** 342 (1970).
745. F. A. Cotton, R. M. Wing, and R. A. Zimmerman, *Inorg. Chem.*, **6,** 11 (1967).
746. R. D. Hogue and R. E. McCarley, *Inorg. Chem.*, **9,** 1354 (1970).
747. P. M. Boorman and B. P. Straughan, *J. Chem. Soc.*, 1514 (1966).

748. R. A. Mackay and R. F. Schneider, *Inorg. Chem.*, **7,** 455 (1968).
749. P. B. Fleming, J. L. Meyer, W. K. Grindstaff, and R. E. McCarley, *Inorg. Chem.*, **9,** 1769 (1970).
750. T. G. Spiro, *Prog. Inorg. Chem.*, **11,** 1 (1970).
751. K. L. Watters and W. M. Risen, Jr., *Inorg. Chim. Acta Rev.*, **3,** 129 (1969).
752. E. Maslowsky, Jr., *Chem. Rev.*, **71,** 507 (1971).
753. B. J. Bulkin and C. A. Rundell, *Coord. Chem. Rev.*, **2,** 371 (1967).
754. C. O. Quicksall and T. G. Spiro, *Inorg. Chem.*, **8,** 2363 (1969).
755. J. R. Johnson, R. J. Ziegler, and W. M. Risen, Jr., *Inorg. Chem.*, **12,** 2349 (1973).
756. K. L. Watters, J. N. Brittain, and W. M. Risen, Jr., *Inorg. Chem.*, **8,** 1347 (1969).
757. K. L. Watters, W. M. Butler, and W. M. Risen, Jr., *Inorg. Chem.*, **10,** 1970 (1971).
758. K. M. Mackay and S. R. Stobart, *J. Chem. Soc.*, Dalton, 214 (1973).
759. P. L. Goggin and R. J. Goodfellow, *J. Chem. Soc.*, Dalton, 2355 (1973).
760. C. O. Quicksall and T. G. Spiro, *Inorg. Chem.*, **8,** 2011 (1969).
761. C. O. Quicksall and T. G. Spiro, *Chem. Commun*, 839 (1967).
762. J. San Filippo, Jr., and H. J. Sniadoch, *Inorg. Chem.*, **12,** 2326 (1973).
763. B. I. Swanson, J. J. Rafalko, D. F. Shriver, J. San Filippo, Jr., and T. G. Spiro, *Inorg. Chem.*, **14,** 1737 (1975).
764. R. Mattes, *Z. Anorg. Allg. Chem.*, **357,** 30 (1968).
765. R. Mattes, *Z. Anorg. Allg. Chem.*, **364,** 279 (1969).
766. F. J. Farrell, V. A. Maroni, and T. G. Spiro, *Inorg. Chem.*, **8,** 2638 (1969).
767. R. Mattes, H. Bierbüsse, and J. Fuchs, *Z. Anorg. Allg. Chem.*, **385,** 230 (1971).
768. T. G. Spiro, D. H. Templeton, and A. Zalkin, *Inorg. Chem.*, **8,** 856 (1969).
769. T. G. Spiro, V. A. Maroni, and C. O. Quicksall, *Inorg. Chem.*, **8,** 2524 (1969).
770. A. P. Ketteringham and C. Oldham, *J. Chem. Soc.*, Dalton, 1067 (1973).
771. C. Oldham and A. P. Ketteringham, *J. Chem. Soc.*, Dalton, 2304 (1973).
772. F. A. Cotton, *Acc. Chem. Res.* **2,** 240 (1969).
773. C. L. Angell, F. A. Cotton, B. A. Frenz, and T. R. Webb, *Chem. Commun.*, 399 (1973).
774. R. J. H. Clark and M. L. Franks, *J. Am. Chem. Soc.*, **97,** 2691 (1975).
775. C. O. Quicksall and T. G. Spiro, *Inorg. Chem.*, **9,** 1045 (1970).
776. J. G. Verkade, *Coord. Chem. Rev.*, **9,** 1 (1972).
777. M. Trabelsi, A. Loutellier, and M. Bigorgne, *J. Organometal. Chem.*, **40,** C45 (1972).
778. M. Bigorgne, A. Loutellier, and M. Pańkowski, *J. Organometal. Chem.*, **23,** 201 (1970).
779. M. Trabelsi, A. Loutellier, and M. Bigorgne, *J. Organometal. Chem.*, **56,** 369 (1973).
780. Th. Kruck, *Angew. Chem. Int. Ed.*, **6,** 53 (1967).
781. Th. Kruck and K. Baur, *Z. Anorg. Allg. Chem.*, **364,** 192 (1969).
782. Th. Kruck and A. Prasch, *Z. Anorg. Allg. Chem.*, **356,** 118 (1968).
783. Th. Kruck and A. Prasch, *Z. Anorg. Allg. Chem.*, **371,** 1 (1969).
784. S. Bénazeth, A. Loutellier, and M. Bigorgne, *J. Organometal. Chem.*, **24,** 479 (1970).
785. L. A. Woodward and J. R. Hall, *Spectrochim. Acta*, **16,** 654 (1960); H. G. M. Edwards and L. A. Woodward, *ibid.*, **26A,** 897 (1970).
786. P. L. Goggin and R. J. Goodfellow, *J. Chem. Soc.*, *A*, 1462 (1966).
787. G. D. Coates and C. Parkin, *J. Chem. Soc.*, 421 (1963).
788. M. A. Bennett, R. J. H. Clark, and A. D. J. Goodwin, *Inorg. Chem.*, **6,** 1625 (1967).
789. J. H. S. Green, *Spectrochim. Acta*, **24A,** 137 (1968).
790. K. Shobatake, C. Postmus, J. R. Ferraro, and K. Nakamoto, *Appl. Spectrosc.*, **23,** 12 (1969).

791. J. Bradbury, K. P. Forest, R. H. Nuttall, and D. W. A. Sharp, *Spectrochim. Acta*, **23A,** 2701 (1967).
792. P. J. D. Park and P. J. Hendra, *Spectrochim. Acta*, **25A,** 227, 909 (1969).
793. L. M. Venanzi, *Chem. Brit.* **4,** 162 (1968).
794. J. Chatt, G. A. Gamlen, and L. E. Orgel, *J. Chem. Soc.*, 486 (1958).
795. E. O. Fischer, W. Bathelt, and J. Müller, *Chem. Ber.*, **103,** 1815 (1970).
796. R. J. Goodfellow, J. G. Evans, P. L. Goggin, and D. A. Duddell, *J. Chem. Soc.*, *A*, 1604 (1968).
797. G. B. Deacon and J. H. S. Green, *Spectrochim. Acta*, **24A,** 845 (1968).
798. D. Brown, J. Hill, and C. E. F. Richard, *J. Chem. Soc.*, *A*, 497 (1970).
799. D. M. L. Goodgame and F. A. Cotton, *J. Chem. Soc.*, 2298, 3735 (1961).
800. G. A. Rodley, D. M. L. Goodgame, and F. A. Cotton, *J. Chem. Soc.*, 1499 (1965).
801. A. P. Ginsberg and W. E. Lindsell, *Chem. Commun.*, 232 (1971).
802. A. E. Wickenden and R. A. Krause, *Inorg. Chem.*, **8,** 779 (1969).
803. J. R. Allkins and P. J. Hendra, *J. Chem. Soc.*, A, 1325 (1967); *Spectrochim. Acta*, **22,** 2075 (1966); **23A,** 1671 (1967); **24A,** 1305 (1968).
804. E. A. Allen and W. Wilkinson, *Spectrochim. Acta*, **28A,** 725 (1972).
805. R. J. H. Clark, G. Natile, U. Belluco, L. Cattalini, and C. Filippin, *J. Chem. Soc.*, *A*, 659 (1970).
806. E. A. Allen and W. Wilkinson, *Spectrochim. Acta*, **28A,** 2257 (1972).
807. B. E. Aires, J. E. Fergusson, D. T. Howarth, and J. M. Miller, *J. Chem. Soc.*, *A*, 1144 (1971).
808. E. A. Allen and W. Wilkinson, *J. Chem. Soc.*, Dalton, 613 (1972).
809. P. L. Goggin, R. J. Goodfellow, D. L. Sales, J. Stokes, and P. Woodward, *Chem. Commun.*, 31 (1968).
810. D. M. Adams and P. J. Chandler, *J. Chem. Soc.*, *A*, 588 (1969).
811. W. M. Scovell and T. G. Spiro, *Inorg. Chem.*, **13,** 304 (1974).
812. I. S. Butler and A. E. Fenster, *J. Organometal. Chem.*, **66,** 161 (1974).
813. M. C. Baird and G. Wilkinson, *Chem. Commun.*, 514 (1966); *J. Chem. Soc.*, *A*, 865 (1967).
814. M. C. Baird, G. Hartwell, and G. Wilkinson, *J. Chem. Soc.*, *A*, 2037 (1967).
815. J. D. Gilbert, M. C. Baird, and G. Wilkinson, *J. Chem. Soc.*, *A*, 2198 (1968).
816. M. P. Yagupsky and G. Wilkinson, *J. Chem. Soc.*, *A*, 2813 (1968).
817. G. R. Clark, T. J. Collins, S. M. James, W. R. Roper, and K. G. Town, *Chem. Commun.*, 475 (1976).
818. A. Efraty, R. Arneri, and M. H. A. Huang, *J. Am. Chem. Soc.*, **98,** 639 (1976).
819. B. D. Dombek and R. J. Angelici, *J. Am. Chem. Soc.*, **96,** 7568 (1974).
820. J. M. Burke and J. P. Fackler, *Inorg. Chem.*, **11,** 2744 (1972).
821. A. Cormier, K. Nakamoto, P. Christophliemk, and A. Müller, *Spectrochim. Acta*, **30A,** 1059 (1974).
822. L. Vaska and S. S. Bath, *J. Am. Chem. Soc.*, **88,** 1333 (1966).
823. L. H. Vogt, J. L. Katz, and S. E. Wiberley, *Inorg. Chem.*, **4,** 1157 (1965).
824. J. W. Moore, H. W. Baird, and H. B. Miller, *J. Am. Chem. Soc.*, **90,** 1358 (1968).
825. D. M. Byler and D. F. Shriver, *Inorg. Chem.*, **15,** 32 (1976).
826. M. R. Churchill, B. G. DeBoer, K. L. Kalra, P. Reich-Rohrwig, and A. Wojcicki, *Chem. Commun.*, 981 (1972).
826a. D. C. Moody and R. R. Ryan, *Chem. Commun.*, 503 (1976).
827. K. H. Schmidt and A. Müller, *Coord. Chem. Rev.*, **14,** 115 (1974).
828. E. Königer-Ahlborn, A Müller, A. D. Cormier, J. D. Brown, and K. Nakamoto, *Inorg. Chem.*, **14,** 2009 (1975).

829. A. Müller and H.-H. Heinsen, *Chem. Ber.*, **105,** 1730 (1972).
830. A. Müller, H.-H. Heinsen, and G. Vandrish, *Inorg. Chem.*, **13,** 1001 (1974).
831. R. G. Cavell, W. Byers, E. D. Day, and P. M. Watkins, *Inorg. Chem.*, **11,** 1598 (1972).
832. F. A. Cotton, C. Oldham, and R. A. Walton, *Inorg. Chem.*, **6,** 214 (1967).
833. M. Ikram and D. B. Powell, *Spectrochim Acta*, **28A,** 59 (1972).
834. C. W. Schläpfer, Y. Saito, and K. Nakamoto, *Inorg. Chim. Acta*, **6,** 284 (1972); C. W. Schläpfer and K. Nakamoto, *ibid.*, **6,** 177 (1972).
835. F. Bonati and R. Ugo, *J. Organometal. Chem.*, **10,** 257 (1967).
836. C. O'Connor, J. D. Gilbert, and G. Wilkinson, *J. Chem. Soc.*, *A*, 84 (1969).
837. D. C. Bradley and M. H. Gitlitz, *J. Chem. Soc.*, *A*, 1152 (1969).
838. K. A. Jensen and V. Krishnan, *Acta Chem. Scand.*, **24,** 1088 (1970).
839. K. Jensen, B. M. Dahl, P. Nielsen, and G Borch, *Acta Chem. Scand.*, **26,** 2241 (1972).
840. G. W. Watt and B. J. McCormick, *Spectrochim. Acta*, **21,** 753 (1965).
841. V. V. Savant, J. Gopalakrishnan, and C. C. Patel, *Inorg. Chem.*, **9,** 748 (1970).
842. G. A. Melson, N. P. Crawford, and B. J. Geddes, *Inorg. Chem.*, **9,** 1123 (1970).
843. M. J. M. Campbell, *Coord. Chem. Rev.*, **15,** 279 (1975).
844. M. A. Ali, S. E. Livingstone, and D. J. Phillips, *Inorg. Chim. Acta*, **5,** 119 (1971).
845. H. O. Desseyn, W. A. Jacob, and M. A. Herman, *Spectrochim. Acta*, **25A,** 1685 (1969).
846. D. Coucouvanis, N. C. Baenziger, and S. M. Johnson, *J. Am. Chem. Soc.*, **95,** 3875 (1973).
847. D. Coucouvanis and D. Piltingsrud, *J. Am. Chem. Soc.*, **95,** 5556 (1973).
848. J. A. McCleverty, *Prog. Inorg. Chem.*, **10,** 49 (1968).
849. C. W. Schläpfer and K. Nakamoto, *Inorg. Chem.*, **14,** 1338 (1975).
850. M. Cox and J. Darken, *Coord. Chem. Rev.*, **7,** 29 (1971).
851. S. E. Linvingstone, *Coord. Chem. Rev.*, **7,** 59 (1971).
852. O. Siimann and J. Fresco, *Inorg. Chem.*, **8,** 1846 (1969); *J. Chem. Phys.*, **54,** 734 (1971); *ibid.*, **54,** 740 (1971).
853. C. G. Barraclough, R. L. Martin, and I. M. Stewart, *Aust. J. Chem.*, **22,** 891 (1969); G. A. Heath and R. L. Martin, *ibid.*, **23,** 1721 (1970).
854. H. Shindo and T. L. Brown, *J. Am. Chem. Soc.*, **87,** 1904 (1965).
855. M. Chandrasekharan, M. R. Udupa, and G. Aravamudan, *Inorg. Chim. Acta*, **7,** 88 (1973).
856. C. A. McAuliffe, J. V. Quagliano, and L. M. Vallarino, *Inorg. Chem.*, **5,** 1996 (1966).
857. C. A. McAuliffe, *J. Chem. Soc.*, A, 641 (1967).
858. M. Ikram and D. B. Powell, *Spectrochim. Acta*, **27A,** 1845 (1971).
859. Y. Mido and E. Sekido, *Bull. Chem. Soc. Jap.*, **44,** 2130 (1971).
860. J. A. W. Dalziel, M. J. Hitch, and S. D. Ross, *Spectrochim. Acta*, **25A,** 1055 (1969).
861. W. W. Fee and J. D. Pulsford, *Inorg. Nucl. Chem. Lett.*, **4,** 227 (1968).
862. R. B. Penland, S. Mizushima, C. Curran, and J. V. Quagliano, *J. Am. Chem. Soc.* **79,** 1575 (1957).
863. A. Yamaguchi, T. Miyazawa, T. Shimanouchi, and S. Mizushima, *Spectrochim. Acta*, **10,** 170 (1957).
864. E. Giesbrecht and M. Kawashita, *J. Inorg. Nucl. Chem.*, **32,** 2461 (1970).
865. A. Yamaguchi, R. B. Penland, S. Mizushima, T. J. Lane, C. Curran, and J. V. Quagliano, *J. Am. Chem. Soc.*, **80,** 527 (1958).
866. R. A. Bailey and T. R. Peterson, *Can. J. Chem.*, **45,** 1135 (1967).
867. K. Swaminathan and H. M. N. H. Irving, *J. Inorg. Nucl. Chem.*, **26,** 1291 (1964).

868. C. D. Flint and M. Goodgame, *J. Chem. Soc.*, *A*, 744 (1966).
869. P. J. Hendra and Z. Jović, *J. Chem. Soc.*, *A*, 735 (1967).
870. D. M. Adams and J. B. Cornell, *J. Chem. Soc.*, *A*, 884 (1967).
871. R. Rivest, *Can. J. Chem.*, **40,** 2234 (1962).
872. T. J. Lane, A. Yamaguchi, J. V. Quagliano, J. A. Ryan, and S. Mizushima, *J. Am. Chem. Soc.*, **81,** 3824 (1959).
873. M. Schafer and C. Curran, *Inorg. Chem.*, **5,** 265 (1966).
874. R. K. Gosavi and C. N. R. Rao, *J. Inorg. Nucl. Chem.*, **29,** 1937 (1967).
875. G. B. Aitken, J. L. Duncan, and G. P. McQuillan, *J. Chem. Soc.*, Dalton, 2103 (1972).
876. P. J. Hendra and Z. Jović, *Spectrochim. Acta*, **24A,** 1713 (1968).
877. R. J. Balahura and R. B. Jordan, *J. Am. Chem. Soc.*, **92,** 1533 (1970).
878. F. A. Cotton, R. Francis, and W. D. Horrocks, *J. Phys. Chem.*, **64,** 1534 (1960).
879. R. S. Drago and D. W. Meek, *J. Phys. Chem.*, **65,** 1446 (1961): D. W. Meek, D. K. Straub, and R. S. Drago, *J. Am. Chem. Soc.*, **82,** 6013 (1960).
880. B. B. Wayland and R. F. Schramm, *Inorg. Chem.*, **8,** 971 (1969); *Chem. Commun.*, 1465 (1968).
881. V. N. Krishnamarthy and S. Soundararajan, *J. Inorg. Nucl. Chem.*, **29,** 517 (1967); S. K. Ramalingam and S. Soundararajan, *Z. Anorg. Allg. Chem.*, **353,** 216 (1967).
882. C. G. Fuentes and S. J. Patel, *J. Inorg. Nucl. Chem.*, **32,** 1575 (1970).
883. C. V. Senoff, E. Maslowsky, Jr., and R. G. Goel, *Can. J. Chem.*, **15,** 3585 (1971).
884. W. Kitching, C. J. Moore, and D. Doddrell, *Inorg. Chem.*, **9,** 541 (1970).
885. D. A. Langs, C. R. Hare, and R. G. Little, *Chem. Commun.*, 1087 (1967).
886. I. P. Evans, A. Spencer, and G. Wilkinson, *J. Chem. Soc.*, Dalton, 204 (1973).
887. A. Mercer and J. Trotter, *J. Chem. Soc.*, Dalton, 2480 (1975).
888. M. Tranquille and M. T. Forel, *Spectrochim. Acta*, **28A,** 1305 (1972).
889. C. V. Berney and J. H. Weber, *Inorg. Chem.*, **7,** 283 (1968).
890. P. W. N. M. van Leeuwen, *Rec. Trav. Chim.*, **86,** 201 (1967); P. W. N. M. van Leeuwen and W. L. Groeneveld, *ibid.*, **86,** 721 (1967).
891. S. K. Madan. C. M. Hull, and L. J. Herman, *Inorg. Chem.*, **7,** 491 (1968).
892. K. A. Jensen and K. Krishnan, *Scand. Chim. Acta*, **21,** 1988 (1967).

# Organometallic Compounds

*Part IV*

## IV-1. METAL ALKANES

The methyl group bonded to a metal (M—$CH_3$) exhibits six normal vibrations such as those shown in Fig. III-2. In addition, CMC bending and $CH_3$ torsional modes are expected for $M(CH_3)_n$ ($n \geq 2$) type compounds. Table IV-1 lists the vibrational frequencies of typical $M(CH_3)_4$ type compounds. It is seen that the $CH_3$ rocking, MC stretching, and CMC bending frequencies are most sensitive to the change in metals. Tables IV-2 and IV-3 list the MC stretching and CMC bending frequencies of various $M(CH_3)_n$ type molecules and ions. As shown in Appendix III, the number of these skeletal modes which are infrared or Raman active provides direct information about the structure of the $MC_n$ skeleton.

Some metal alkyls are polymerized in condensed phases. $Li(CH_3)$ forms a tetramer containing Li—$CH_3$—Li bridges in the solid state,[33] and its $CH_3$ frequencies are lower than those of nonbridging compounds [$\nu_a(CH_3)$ and $\nu_s(CH_3)$ are 2840 and 2780 $cm^{-1}$, respectively].[34]

Solid $Be(CH_3)_2$ and $Mg(CH_3)_2$ also form long chain polymers through $CH_3$ bridges,[35] while $Al(CH_3)_3$ is dimeric in the solid state.[36,37] The infrared spectra of $Li[Al(CH_3)_4]$ and $Li_2[Zn(CH_3)_4]$ have been interpreted on the basis of linear polymeric chains in which the Al (or Zn) atom and the Li atom are bonded alternately through two $CH_3$ groups.[38] Normal coordinate analyses have been carried out on $M(CH_3)_2$ (M = Zn, Cd, and Hg),[4,39] dimeric $Al(CH_3)_3$,[36,37] and linear $[M(CH_3)_2]^{n+}$ type cations.[23,26]

The ethyl group bonded to a metal (M—$CH_2$—$CH_3$) exhibits bands characteristic of both the $CH_3$ and $CH_2$ groups. It is difficult, however, to give complete assignments of the M—$C_2H_5$ group vibrations because of band overlapping and vibrational coupling. Table IV-4 lists the $MC_n$ skeletal frequencies of typical $M(C_2H_5)_n$ type compounds. The MC stretching frequencies of the ethyl compounds are lower than those of the corresponding methyl compounds (Table IV-2) due to the larger mass of the ethyl, relative to the methyl group.

$Li(C_2H_5)$ is hexameric in hydrocarbon solvents[54] and is polymeric in the solid state.[55] The LiC stretching bands of these polymers are at 530–300 $cm^{-1}$.[56] The vibrational spectra of such other polymeric ethyl compounds as $Be(C_2H_5)_2$ (dimer),[57] $Al(C_2H_5)_3$ (dimer),[58,59] and $Li[Al(C_2H_5)_4]$ (polymer)[60] have been reported. There are many other compounds containing higher alkyl groups. References for some typical compounds are as follows: $[Tl(n\text{-}C_3H_7)_2]Cl$ (25), $Al(n\text{-}C_3H_7)_3$ (58), $Ge(n\text{-}C_4H_9)_4$ (61), and $[Li(t\text{-}C_4H_9)]_4$ (62). Vibrational spectra have also been reported for cycloalkyl compounds such as $Zn(c\text{-}C_3H_5)_2$,[63] $M(c\text{-}C_3H_5)_4$ (M = Si, Ge, and Sn),[64] and $Sb(c\text{-}C_3H_5)_5$.[65]

TABLE IV-1. RAMAN FREQUENCIES[a] OF $M(CH_3)_4$ TYPE MOLECULES ($cm^{-1}$)[1,2]

| Compound | $\nu_a(CH_3)$ | $\nu_s(CH_3)$ | $\delta_d(CH_3)$ | $\delta_s(CH_3)$ | $\rho_r(CH_3)$ | $\nu(MC)$ | $\delta(CMC)$ |
|---|---|---|---|---|---|---|---|
| $C(CH_3)_4$ | 2959 | 2922 | (1475) | — | 926 | 733 | 418 |
| | 2963 | | 1457 | | (926) | 1260 | 332 |
| $Si(CH_3)_4$ | (2959) | 2913 | (1430) | 1271 | 870 | 593 | 239 |
| | 2964 | | 1421 | | (870) | 698 | 190 |
| | (2910) | | | | | | |
| $Ge(CH_3)_4$ | (2981) | 2920 | (1430) | 1259 | — | 561 | 196 |
| | 2982 | | 1420 | | (828) | 599 | 188 |
| $Sn(CH_3)_4$ | (2984) | 2920 | (1447) | 1211 | — | 509 | 137 |
| | 2988 | | — | | (768) | 527 | 133 |
| $Pb(CH_3)_4$ | 2996 | 2924 | 1450 | 1170 | 767 | 478 | 145 |
| | 2924 | | 1400 | 1154 | 700 | 459 | 130 |

[a] ( ): IR frequency.

## IV-2. METAL ALKENES, ALKYNES, AND PHENYLS

A vinyl group $\sigma$-bonded to a metal (M—CH═CH$_2$) exhibits CH stretching, C═C stretching, $CH_2$ scissoring, rocking, twisting and wagging, and CH in-plane and out-of-plane bending in addition to the MC stretching and CMC bending modes. Table IV-5 lists the C═C and MC stretching frequencies for typical vinyl compounds. The C═C stretching vibrations are generally strong in the Raman. Their intensity in the infrared, however, depends on the metal. Vibrational spectra of halovinyl compounds have been reported for $Hg(CH{=}CHCl)_2$[73] and $M(CF{=}CF_2)_n$ (M = Hg, As, and Sn).[74] Complete vibrational assignments are available for $\sigma$-bonded allyl compounds such as $M(CH_2{-}CH{=}CH_2)_4$ (M = Si and Sn)[75] and $Hg(CH_2{-}CH{=}CH_2)_2$.[76,77] Table IV-5 also lists the C≡C and MC stretching frequencies of acetylenic compounds. Again, the C≡C stretching vibrations are strong in the Raman, but vary from strong to weak in the infrared, depending on the metal involved.

The phenyl group when $\sigma$-bonded to a metal exhibits bands characteristic of monosubstituted benzenes.[81] The M—$C_6H_5$ type molecule exhibits 30 fundamentals, only six of which, shown in Fig. IV-1, are sensitive to the change in metals. Table IV-6 lists the observed frequencies of these six modes for typical phenyl compounds. It is seen that the *t*, *x*, and *u* modes are most sensitive to the change in metals. There are many metal–phenyl compounds containing other functional groups. The spectra of these compounds can be interpreted roughly as the overlap of

TABLE IV-2. METAL—CARBON SKELETAL FREQUENCIES OF $M(CH_3)_n$ TYPE COMPOUNDS ($cm^{-1}$)

| Compound | Structure | $\nu_a$(MC) | $\nu_s$(MC) | $\delta$(CMC) | Refs. |
|---|---|---|---|---|---|
| $Be(CH_3)_2$ | linear | 1081 | — | — | 3 |
| $Zn(CH_3)_2$ | linear | 613 | 503 | 134 | 4, 5 |
| $Cd(CH_3)_2$ | linear | 534 | 459 | 120 | 4, 5 |
| $Hg(CH_3)_2$ | linear | 540 | 515 | 161 | 4, 5 |
| $Se(CH_3)_2$[a] | bent | 604 | 589 | 233 | 6, 7 |
| $Te(CH_3)_2$[a] | bent | 528 | 528 | 198 | 6, 7 |
| $B(CH_3)_3$ | planar | 1177 | 680 | 341, 321 | 8, 9 |
| $Al(CH_3)_3$ | planar | 760 | 530 | 170 | 10 |
| $Ga(CH_3)_3$ | planar | 577 | 521.5 | 162.5 | 11, 12 |
| $In(CH_3)_3$ | planar | 500 | 467 | 132 | 12 |
| $P(CH_3)_3$ | pyramidal | 703 | 653 | 305, 263 | 13, 14 |
| $As(CH_3)_3$ | pyramidal | 583 | 568 | 238, 223 | 13, 14 |
| $Sb(CH_3)_3$ | pyramidal | 513 | 513 | 188 | 15 |
| $Bi(CH_3)_3$ | pyramidal | 460 | 460 | 171 | 15 |
| $Si(CH_3)_4$ | tetrahedral | 696 | 598 | 239, 202 | 16 |
| $Ge(CH_3)_4$ | tetrahedral | 595 | 558 | 195, 175 | 2 |
| $Sn(CH_3)_4$ | tetrahedral | 529 | 508 | 157 | 17 |
| $Pb(CH_3)_4$ | tetrahedral | 476 | 459 | 120 | 18 |
| $Ti(CH_3)_4$ | tetrahedral | 577 | 489 | 180 | 19 |
| $Sb(CH_3)_5$ | trigonal-bipyramidal | 514[b] 456[c] | 493[b] 414[c] | 213[b] 199[b] 104[c] | 20 |
| $W(CH_3)_6$ | octahedral | 482 | — | — | 21 |

[a] In addition, the torsional modes of the methyl groups have been observed for $Se(CH_3)_2$ at 201 and 175 $cm^{-1}$ and for $Te(CH_3)_2$ at 185 $cm^{-1}$ (see Ref. 7).
[b] Equatorial.
[c] Axial.

the M—$C_6H_5$ group vibrations and the vibrations of other functional groups discussed later.

## IV-3. HALOGENO AND PSEUDOHALOGENO COMPOUNDS

As discussed in Sec. III-16, terminal $\nu$(MX) bands appear in the following regions:

$$\begin{array}{ccccccc} \nu(MF) & & \nu(MCl) & & \nu(MBr) & & \nu(MI) \\ 750\text{–}500 & > & 400\text{–}200 & > & 300\text{–}200 & > & 200\text{–}100\ cm^{-1} \end{array}$$

TABLE IV-3. METAL—CARBON SKELETAL FREQUENCIES OF $[M(CH_3)_n]^{m+}$ TYPE COMPOUNDS ($cm^{-1}$)

| Compound | Structure | $\nu_a$(MC) | $\nu_s$(MC) | $\delta$(CMC) | Refs. |
|---|---|---|---|---|---|
| $[Zn(CH_3)]^+$ | — | — | 557 | — | 22 |
| $[In(CH_3)_2]^+$ | linear | 566 | 502 | — | 23 |
| $[Tl(CH_3)_2]^+$ | linear | 559 | 498 | 114 | 24, 25 |
| $[Sn(CH_3)_2]^{2+}$ | linear | 582 | 529 | 180 | 26 |
| $[Sn(CH_3)_3]^+$ | planar | 557 | 521 | 152 | 27 |
| $[Sb(CH_3)_3]^{2+}$ | planar | 582 | 536 | 166 | 28 |
| $[Se(CH_3)_3]^+$ | nonplanar | 602 | 580 | 272 | 29 |
| $[Te(CH_3)_3]^+$ | nonplanar | 534 | — | — | 30 |
| $[P(CH_3)_4]^+$ | tetrahedral | 783 | 649 | 285 | 31 |
| | | | | 170 | 32 |
| $[As(CH_3)_4]^+$ | tetrahedral | 652 | 590 | 217 | 15 |
| $[Sb(CH_3)_4]^+$ | tetrahedral | 574 | 535 | 178 | 15 |

TABLE IV-4. METAL—CARBON SKELETAL FREQUENCIES OF $M(C_2H_5)_n$ TYPE COMPOUNDS ($cm^{-1}$)

| Compound | Structure | $\nu_a$(MC) | $\nu_s$(MC) | $\delta$(MCC) | $\delta$(CMC) | Refs. |
|---|---|---|---|---|---|---|
| $Zn(C_2H_5)_2$ | linear | 561 | 484 | — | — | 40 |
| $Hg(C_2H_5)_2$ | linear | 515 | 488 | 267 | 140 | 41, 42 |
| | | | | 262 | 85 | |
| $^{10}B(C_2H_5)_3$ | planar | 1135 | — | — | 287 | 8, 43 |
| $Ga(C_2H_5)_3$ | planar | 496 | — | — | — | 44 |
| $P(C_2H_5)_3$ | pyramidal | 697 | 619 | 410—249 | — | 45 |
| | | 669 | | | | |
| $As(C_2H_5)_3$ | pyramidal | 540 | 570 | — | — | 46 |
| | | | 563 | | | |
| $Sb(C_2H_5)_3$ | pyramidal | 505 | 505 | — | — | 45, 47 |
| $Bi(C_2H_5)_3$ | pyramidal | ~450 | ~450 | 253 | 124 | 48 |
| | | | | 213 | | |
| $Si(C_2H_5)_4$ | tetrahedral | 731 | 549 | 392 | 170 | 31, 49 |
| | | | | 233 | | |
| $Ge(C_2H_5)_4$ | tetrahedral | 572 | 532 | 332 | 152 | 50, 51 |
| $Sn(C_2H_5)_4$ | tetrahedral | 508 | 490 | 272 | 132 | 52, 53 |
| | | | | | 86 | |
| $Pb(C_2H_5)_4$ | tetrahedral | 461 | 443 | 243 | 107 | 48, 52 |
| | | | | 213 | | |

TABLE IV-5. CARBON—CARBON AND METAL—CARBON STRETCHING FREQUENCIES OF VINYL AND ACETYLENIC COMPOUNDS ($cm^{-1}$)

| Compound | $\nu$(C=C) or $\nu$(C≡C) | $\nu$(MC) | Refs. |
|---|---|---|---|
| $Zn(CH{=}CH_2)_2$ | 1565 | — | 66 |
| $Hg(CH{=}CH_2)_2$ | 1603 | 541, 513 | 67 |
| $^{10}B(CH{=}CH_2)_3$ | 1604 | 1186 | 68, 69 |
| $Si(CH{=}CH_2)_4$ | 1592 | 732, 583 | 71, 70 |
| $Ge(CH{=}CH_2)_4$ | 1595 | 600, 561 | 71 |
| $Sn(CH{=}CH_2)_4$ | 1583 | 527, 513 | 71, 72 |
| $Pb(CH{=}CH_2)_4$ | 1580 | 495, 481 | 71 |
| $Si(C{\equiv}CH)_4$ | 2053 | 534, 708, or 687 | 79, 78 |
| $Ge(C{\equiv}CH)_4$ | 2057 | 507, 523 | 79, 78 |
| $Sn(C{\equiv}CH)_4$ | 2043 | 504 or 447 | 79, 78 |
| $As(C{\equiv}CH)_3$ | 2053 | 517, 526 | 79a |
| $Sb(C{\equiv}CH)_3$ | — | 474 | 79a |
| $(CH_3)_2Si(C{\equiv}CH)_2$ | 2041 | 595[a] | 79, 80 |
| $(CH_3)_2Ge(C{\equiv}CH)_2$ | 2041 | 538[a], 521[a], | 79, 80 |
| $(CH_3)_2Sn(C{\equiv}CH)_2$ | 2016 | 454,[a] 445[a] | 79, 80 |

[a] $\nu$(MC) indicates $\nu$(M—C≡C).

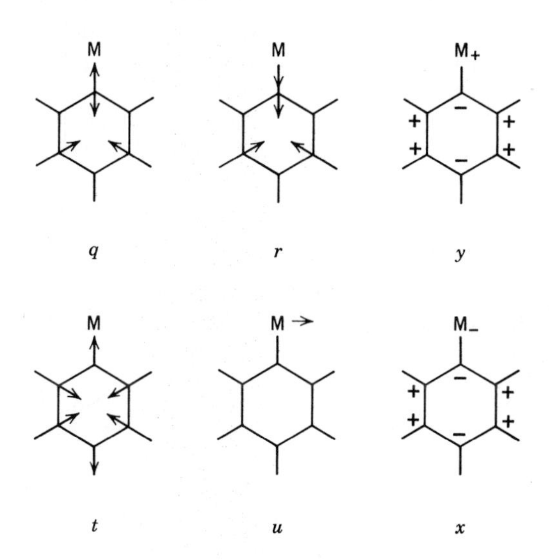


**Fig. IV-1.** Metal-sensitive modes of monosubstituted benzenes.[81]

TABLE IV-6. VIBRATIONAL FREQUENCIES OF METAL-SENSITIVE MODES OF METAL PHENYLS ($cm^{-1}$)

| Compound | *q* | *r* | *y* | *t* | *x* | *u* | Refs. |
|---|---|---|---|---|---|---|---|
| $Hg(C_6H_5)_2$ | 1067 | 661 | 456 | 258<br>252<br>248 | — | 207 | 82, 83 |
| $^{10}B(C_6H_5)_3$ | 1285<br>1248 | 893 | 600 | 650 | 245 | 408 | 84, 85 |
| $Al(C_6H_5)_3$ | 1085 | 670<br>643 | 460 | 420 | 207 | 332 | 85, 86 |
| $Ga(C_6H_5)_3$ | 1085 | 665 | 453<br>445 | 315 | 180 | 245<br>225 | 85 |
| $In(C_6H_5)_3$ | 1070 | 673 | 465 | 270 | 180 | 248<br>195 | 87, 85 |
| $Si(C_6H_5)_4$ | 1108 | 709 | 519<br>511 | 435<br>239 | 185<br>171 | 261<br>223 | 88, 89 |
| $Ge(C_6H_5)_4$ | 1091 | — | 481<br>465 | 332 | 187<br>168 | 232<br>214 | 90, 89 |
| $Sn(C_6H_5)_4$ | 1075 | 616 | 459<br>448 | 268<br>212 | 193<br>152 | 225 | 91, 89 |
| $Pb(C_6H_5)_4$ | 1061 | 645 | 450<br>440 | 223<br>201 | 147 | 181 | 89 |
| $P(C_6H_5)_3$ | 1089 | — | 501 | 428<br>398 | 248 | 209<br>190 | 92, 91 |
| $As(C_6H_5)_3$ | 1082<br>1074 | 667 | 474 | 313 | 237 | 192 | 91, 92 |
| $Sb(C_6H_5)_3$ | 1065 | 651 | 457 | 270 | 216 | 166 | 91, 92 |
| $Bi(C_6H_5)_3$ | 1055 | — | 448 | 237<br>220 | 207 | 157 | 91, 92 |

For a series with a fixed halogen, $\nu(MX)$ is higher, as M is lighter in the same family of the periodic table. Thus the $\nu(MCl)$ of the $M(CH_3)_3Cl$ series follow this order:

$$\begin{array}{lccccc} M = & Si & & Ge & & Sn \\ & 487 & > & 398 & > & 325\ cm^{-1} \end{array}$$

In general, the infrared intensity of $\nu(MX)$ decreases in the order $\nu(MF) > \nu(MCl) > \nu(MBr) > \nu(MI)$, whereas the opposite order prevails for the Raman intensity. Table IV-7 lists $\nu(MC)$ and $\nu(MX)$ of typical compounds.

In condensed phases, halogeno compounds tend to polymerize by

TABLE IV-7. METAL—CARBON AND METAL—HALOGEN STRETCHING FREQUENCIES OF TYPICAL METHYLHALOGENO COMPOUNDS ($CM^{-1}$)

| Compound | $\nu$(MC) | $\nu$(MX) | Refs. |
|---|---|---|---|
| $CH_3CdCl$ | 476 | 247 | 93 |
| $CH_3CdBr$ | 475 | 206 | 93 |
| $CH_3CdI$ | 482 | 167 | 93 |
| $(CH_3)_3SiF$ | 704, 635 | 898 | 94 |
| $(CH_3)_3SiCl$ | 704, 635 | 472 | 95 |
| $(CH_3)_3GeF$ | 623, 576 | 623 | 96 |
| $(CH_3)_3GeCl$ | 612, 569 | 378 | 97 |
| $(CH_3)_3SbF_2$ | 591, 546 | 484, 465 | 98 |
| $(CH_3)_3SbCl_2$ | 577, 538 | 282, 272 | 98 |
| $(CH_3)_3SbBr_2$ | 569, 526 | 215, 168 | 98 |
| $(CH_3)_3SbI_2$ | 559, 508 | 144, 122 | 98 |
| $[(CH_3)_2AuCl]_2$ | 571, 561 | 273 | 99 |
| $[(CH_3)_2AuBr]_2$ | 561, 550 | 181 | 99 |
| $[(CH_3)_2AuI]_2$ | 550, 545 | 141, 131 | 99 |

forming halogeno bridges between two metals. As discussed in Sec. III-16, the bridging frequencies are much lower than the terminal frequencies. Thus it is possible to distinguish monomeric and polymeric (halogen-bridged) structures by vibrational spectroscopy. It was found that $(CH_3)_2BX$ (X = F, Cl, and Br) is monomeric,[100] whereas $(CH_3)_2AlF$, $(CH_3)_2GaF$, and $(CH_3)_2InCl$ are tetrameric,[101] trimeric,[102] and dimeric,[103] respectively, in benzene solution. Alkyl silicon and germanium halides tend to be monomeric, whereas alkyl tin and lead halides tend to be polymeric, in the liquid and solid phases. For example, $(CH_3)_2SnF_2$ and $(CH_3)_3SnF$ are polymerized through the fluorine bridges:[104]

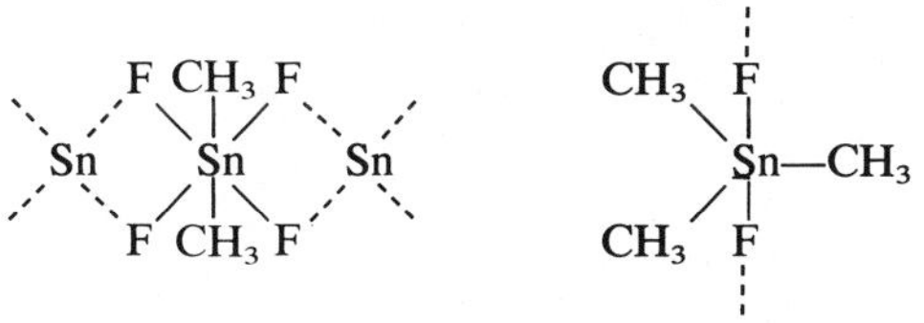

F CH3 F
Sn Sn Sn
F CH3 F

CH3 F
Sn—CH3
CH3 F

The $\nu$(SnF) for terminal bonds are at 650–625 $cm^{-1}$, whereas those for bridging bonds are at 425–335 $cm^{-1}$. In the solid state, the coordination number of tin is five or six. Dialkyl compounds prefer six-coordinate structures, while trialkyl compounds tend to form five-coordinate structures. In both cases, the favored positions of the alkyl groups are those shown in the above diagrams. Normal coordinate calculations have been made on the *trans*-$[(CH_3)_2SnX_4]^{2-}$ (X = F, Cl, and Br) series.[105]

$(CH_3)_3PbX$ (X = F, Cl, Br, and I) are monomeric in benzene but polymeric in the solid state; $\nu$(PbCl) of the monomer and polymer are at 281 and 191 cm$^{-1}$, respectively.[106] In the $[(CH_3)_2AuX]_2$ series,[99] the Au atom takes a square-planar arrangement with two methyl groups in the *cis* position:

```
CH3      X      CH3
   \   /   \   /
    Au       Au
   /   \   /   \
CH3      X      CH3
```

In this case, the bridging $\nu$(AuX) are surprisingly similar to those observed for the corresponding *cis*-$[(CH_3)_2AuX_2]^-$.[99] Fig. IV-2 shows the tetrameric structure of $[(CH_3)_3PtX]_4$ type compounds. Vibrational spectra have been reported for X = Cl, Br, I[107] and Cl, I[108] $N_3$[108a] and SCN,[108b] and for $[(CH_3)_2PtX_2]_n$ (X = Cl, Br, and I; *n* is probably 4).[109]

The vibrational spectra of coordination compounds containing pseudohalogeno groups were discussed in Sec. III-11. The vibrational spectra of azido complexes have been reported for monomeric $CH_3ZnN_3$[110] and $CH_3HgN_3$[111] and for polymeric $(CH_3)_2AlN_3$[112] and $(CH_3)_3SnN_3$.[113] The MN stretching frequencies of these compounds are in

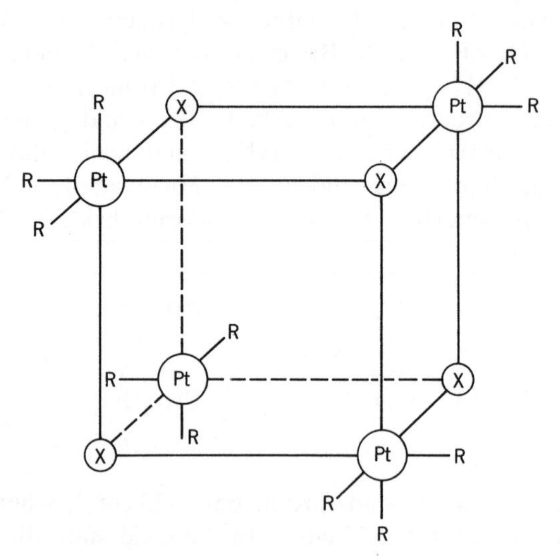


**Fig. IV-2.** Structure of $[Pt(CH_3)_3X]_4$; R denotes $CH_3$.

TABLE IV-8. VIBRATIONAL FREQUENCIES OF ISOTHIOCYANATO COMPOUNDS ($cm^{-1}$)

| Compound | Mode of coordination | $\nu$(CN) | $\nu$(CS) | $\delta$(NCS) | $\nu$(MS) or $\nu$(MN) | Ref. |
|---|---|---|---|---|---|---|
| $[CH_3Zn(SCN)]_\infty$ | (Zn)(Zn)>SCN—Zn | 2190, 2140 | 685 | 455 | — | 116 |
| $[CH_3Hg(SCN)_3]^{2-}$ | Hg—SCN | 2119 | — | — | 276 | 117 |
| $[(CH_3)_3Al(SCN)]^-$ | Al—SCN | 2097 | 845 | 485 | 335 | 118 |
| $[(CH_3)_2Al(SCN)]_3$ | (Al)(Al)>SCN | 2075 | 627 | 501, 438 | — | 119 |
| $(CH_3)_3Sn(NCS)$, solid (polymer) | Sn—NCS—Sn | 2098, 2079, 2046 | 779 | 474, 467 | — | 120 |
| $(CH_3)_3Sn(NCS)$, $CS_2$ solution (monomer) | Sn—NCS | 2050 | 781 | 485 | 478 | 120 |
| $[(CH_3)_2Au(NCS)]_2$ | >Au<(NCS)(SCN)>Au< | 2163 | 775 | 444, 430[a] | — | 99 |

[a] These bands may be assigned to $\nu$(AuN).

the 600–400 $cm^{-1}$ region. The NCO group is always bonded to the metal through the N atom (isocyanato complex). The vibrational spectra of monomeric $CH_3Hg—NCO$,[114] $(CH_3)_3Si—NCO$,[115] and $(CH_3)_3Sb(NCO)_2$[114] have been reported.

The NCS group may be bonded to a metal through the N or S atom or may form a bridge between two metals by using the N, the S, or both atoms. It is not easy to distinguish all these possible structures by vibrational spectra. Table IV-8 lists the modes of coordination and the vibrational frequencies of the M—NCS group. The NCSe group is N-bonded in $(CH_3)_3Si—NCSe$[121] and $(CH_3)_3Ge—NCSe$[121a] but is Se-bonded in $(CH_3)_3Pb—SeCN$.[122] Only a very few compounds containing the M-CNO (fulminato) group are known. The spectrum of $(CH_3)_2Tl(CNO)$ is similar to that of K[CNO], and the Tl–CNO bond may be ionic.[123] No isofulminato complexes are reported. The CN group is usually bonded to the metal through the C atom. In the case of $(CH_3)_3M(CN)$(M = Si and Ge), however, the cyano and isocyano complexes are in equilibrium in the liquid phase, although the mole fraction of the latter is rather small. The CN stretching frequencies ($cm^{-1}$) of these isomers are as follows:

| | $(CH_3)_3M—CN$ | $(CH_3)_3M—NC$ | |
|---|---|---|---|
| M = Si | 2198 | 2095 | (gas phase)[124] |
| M = Ge | 2182 | 2090 | ($CHCl_3$ solution)[125] |

## IV-4. COMPOUNDS CONTAINING OTHER FUNCTIONAL GROUPS

The $\nu$(MN) of $[CH_3HgNH_3]^+$,[126] and $[(CH_3)_3SnNH_3]^+$,[127] have been assigned at 585 and 503 $cm^{-1}$, respectively. The infrared and Raman spectra of $[(CH_3)_3Pt(NH_3)_3]^+$ have been interpreted on the basis of the *fac* structure ($\mathbf{C}_{3v}$). Its $\nu$(PtN) are at 390 $cm^{-1}$ ($A_1$, Raman) and at 410 and 377 $cm^{-1}$ ($E$, infrared).[128] The $\nu$(SnN) of alkyl tin halide adducts with bipy and phen were suggested to be close to 200 $cm^{-1}$.[129]

Compounds containing the hydroxo group exhibit $\nu$(OH), $\delta$(MOH), and $\nu$(MO) at 3760–3000, 1200–700, and 900–300 $cm^{-1}$, respectively. As expected, these frequencies depend heavily on the strength of the hydrogen bond involved References for typical hydroxo compounds are as follows: $[(CH_3)_2GaOH]_4$ (130), $(CH_3)_3SiOH$ (131), $(CH_3)_3SnOH$ (132), $(CH_3)_4SbOH$ (133), and $[(CH_3)_3PtOH]_4$ (108, 134). The vibrational spectra of aquo compounds are characterized by the bands discussed in Sec. III-5; these are some pertinent references: $[CH_3Hg(OH_2)]^+$ (135) and $[(CH_3)_3Pt(OH_2)_3]^+$ (136). As stated in Sec. III-6, the alkoxides exhibit $\nu$(CO) in the 1200–950 $cm^{-1}$ region: $(CH_3)_2Al(OCH_3)$,[137] $(CH_3)_2Si(OCH_3)_2$,[138] and $(CH_3)_2Sn(OCH_3)_2$.[139] The carboxylates, such as $(CH_3)_nSi(OCOCH_3)_{4-n}$, exhibit $\nu$(CO) at 1580 $cm^{-1}$.[140] The band assignments of the O-bonded (chelated) acac complexes were discussed in Sec. III-9. References for acac complexes of metal alkyls are as follows: $(CH_3)_2Ga(acac)$ (130), $(CH_3)_2Sn(acac)_2$ (141), $(CH_3)_2Pb(acac)_2$ (142), $(CH_3)_2SbCl_2(acac)$ (143), and $(CH_3)_2Au(acac)$ (144). The structure of $(CH_3)_2Sn(acac)_2$ has been controversial. Originally, it was suggested, on the basis of NMR and vibrational spectra, that its structure in the solid state and in solution was *trans*.[145] Later, the *cis* structure was proposed because of the large dipole moment (2.95 D) in benzene solution.[146] X-ray analysis shows that the compound is *trans* in the solid state.[147] Tobias et al.[141] suggest that the structure remains *trans* in solution and that the large dipole moment may originate in the nonplanarity of the $SnO_4$ plane with the remainder of the acac ring. The $\nu$(MS) of

$(CH_3)_3Si(SC_2H_5)$,[148] $(CH_3)_3Si(SC_6H_5)$,[149] and $(CH_3)_2Sn(SCH_3)_2$[150] are assigned at 486, 459, and 347 $cm^{-1}$, respectively.

A large number of metal alkyls containing acido groups are known. As discussed in Sec. III-8, vibrational spectra are very useful in elucidating the mode of coordination. References are given for the following acido groups: $(CH_3)_3SnNO_3$ (151), $[(CH_3)_3Sn]_2SO_4$ (152), $(C_2H_5)_2SnSO_3$ (153), $(CH_3)_2SnCO_3$ (154), and $(CH_3)_3SnClO_4$ (155).

As discussed in Sec. III-13, metal hydrido complexes exhibit sharp $\nu(MH)$ in the 2200–1700 $cm^{-1}$ region. Table IV-9 lists $\nu(MH)$ and $\nu(MC)$ of typical compounds. It is seen that $\nu(MH)$ decreases as the mass of M increases in the same family of the periodic table and as more halogens are substituted by alkyl groups. The $MH_3$ torsional modes of $CH_3MH_3$ type compounds have been assigned at 142 and 113 $cm^{-1}$ for M = Si and Ge, respectively.[163] Vibrational spectra of $B_2H_6$ type molecules were discussed in Sec. II-11. The compounds $(CH_3)_nB_2H_{6-n}$ ($n = 1$–4) exhibit the terminal and bridging $\nu(BH)$ at 2600–2500 and 2150–1525 $cm^{-1}$, respectively.[164] Dialkylaluminum hydride exists as a trimer in solution and pure liquid, and its $\nu(AlH)$ is at ca. 1800 $cm^{-1}$.[165]

As discussed in Sec. III-17, $\nu(MM')$ are generally strong in the Raman and weak in the infrared. Figure IV-3 shows the SnMn stretching bands in the far-infrared spectra of $(CH_3)_{3-n}Cl_nSn$—$Mn(CO)_5$ type compounds.[173] The vibrational spectra of $(CH_3)_3M$—$M(CH_3)_3$ (M = Si, Ge, Sn, and Pb) and $M[Si(CH_3)_3]_4$ (M = Si, Ge, and Sn) are reported in Refs. 166, 167, and Ref. 168, respectively. Table IV-10 lists the MM' stretching frequencies of metal alkylcarbonyl compounds.

TABLE IV-9. METAL—HYDROGEN AND METAL—CARBON STRETCHING FREQUENCIES OF TYPICAL HYDRIDO COMPOUNDS ($cm^{-1}$)

| Compound | $\nu(MH)$ | $\nu(MC)$ | Ref. |
|---|---|---|---|
| $CH_3SiH_3$ | 2166, 2169 | 701 | 156 |
| $(CH_3)_2SiH_2$ | 2145, 2142 | 728, 659 | 157 |
| $(CH_3)_3SiH$ | 2123 | 711, 624 | 157 |
| $CH_3GeH_3$ | 2106, 2082 | 602 | 158 |
| $(CH_3)_2GeH_2$ | 2080, 2062 | 604, 590 | 159 |
| $(CH_3)_3GeH$ | 2049 | 601, 573 | 159 |
| $CH_3SnH_3$ | 1875 | 527 | 160 |
| $(CH_3)_2SnH_2$ | 1869 | 536, 514 | 161 |
| $(CH_3)_3SnH$ | 1837 | 521, 516 | 161 |
| $(CH_3)_3PbH$ | 1709 | — | 162 |

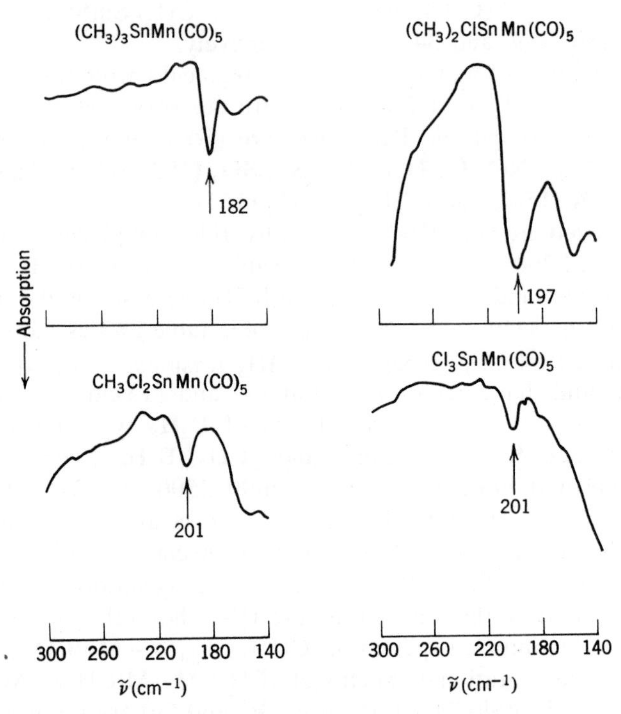


**Fig. IV-3.** Far-infrared spectra of $(CH_3)_{3-n}Cl_nSnMn(CO)_5$, where $n = 0$, 1, 2, or 3.[173] The arrow indicates the SnMn stretching band.

TABLE IV-10. METAL—METAL STRETCHING FREQUENCIES OF METAL ALKYL COMPOUNDS ($CM^{-1}$)

| Compound | $\nu(MM')$ | Ref. |
|---|---|---|
| $(CH_3)_3Si—Mn(CO)_5$ | 297 (R) | 169 |
| $(CH_3)_3Ge—Cr(CO)_3(\pi\text{-Cp})$ | 119 (R) | 170 |
| $(CH_3)_3Ge—Mn(CO)_5$ | 191 (R) | 171 |
| $(CH_3)_3Ge—Co(CO)_4$ | 192 (R) | 172 |
| $(CH_3)_3Sn—Mo(CO)_3(\pi\text{-Cp})$ | 172 (IR) | 173 |
| $(CH_3)_3Sn—Mn(CO)_5$ | 182 (IR, R) | 173 |
| $(CH_3)_3Sn—Re(CO)_5$ | 147 (R) | 171 |
| $(CH_3)_3Sn—Co(CO)_4$ | 176 (IR, R) | 172 |

The band at $284\ cm^{-1}$ in the resonance Raman spectrum of ($t$-Bu—C≡C—$t$-Bu)$_2$Fe$_2$(CO)$_4$:

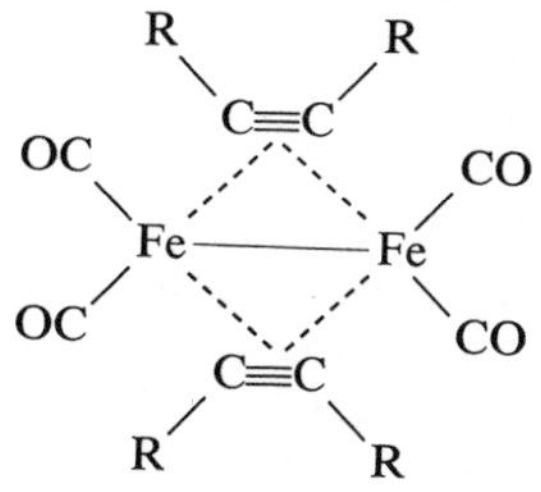

R: $t$-Bu

was assigned to $\nu$(FeFe) by Kubas and Spiro.[173a] Its FeFe stretching force constant (3.0 mdyn/Å) is about twice that of the Fe–Fe single bond (1.3 mdyn/Å), found in $Fe_2S_2(CO)_6$.[173b] Thus the Fe–Fe bond of the former compound must be close to a double bond.

## IV-5. π-COMPLEXES OF ALKENES, ALKYNES, AND RELATED LIGANDS

Vibrational spectra of $\pi$-bonded complexes of alkenes and alkynes with transition metals have been reviewed by Davidson.[174] In contrast to $\sigma$-bonded complexes (Sec. IV-2), the C═C and C≡C stretching bands of $\pi$-bonded complexes show marked shifts to lower frequencies relative to those of free ligands.

### (1) Complexes of monoolefins

Ethylene and other olefins form $\pi$-complexes with transition metals. The simplest and best-studied complex is Zeise's salt, $K[Pt(C_2H_4)Cl_3]\cdot H_2O$. Several investigators[175–177] have given complete assignments of the infrared spectra of Zeise's salt and its deuterated analog. Table IV-11 lists the observed frequencies and band assignments reported by Grogan and Nakamoto.[175] Normal coordinate analysis shows that the C═C stretching mode couples strongly with the $CH_2$ bending mode.[175–177] Thus the negative shift of the C═C stretching mode upon complexation cannot be used as a quantitative measure of the strength of the metal–olefin bond.

The assignments of the low frequency modes of Zeise's salt have been controversial. According to Chatt et al.,[178] two types of bonding are involved in the Pt–ethylene bond: (1) a $\sigma$-type bond is formed by the overlap of the filled $2p\pi$ bonding orbital of the olefin with the vacant $dsp^2$ bonding orbital of the metal, and (2) a $\pi$-type bond is formed by the

TABLE IV-11. OBSERVED FREQUENCIES OF FREE ETHYLENE, ZEISE'S SALT, AND THEIR DEUTERATED ANALOGS ($CM^{-1}$)[175]

| | $C_2H_4$ | Zeise's Salt | $C_2D_4$ | Zeise's Salt-$d_4$ | Assignment[a] |
|---|---|---|---|---|---|
| Coordinated ethylene | 3019 | 2920 | 2251 | 2115 | $\nu(CH)$ |
| | 1623 | 1526 | 1515 | 1428 | $\nu(C{=}C)+\delta_s(CH_2)$[b] |
| | 1342 | 1418 | 981 | 978 | $\delta_s(CH_2)+\nu(C{=}C)$[b] |
| | 3108 | 2975 | 2304 | 2219 | $\nu(CH)$ |
| | 1236 | 1251 | 1009 | 1021 | $\rho_r(CH_2)$ |
| | 3106 | 3098 | 2345 | 2335 | $\nu(CH)$ |
| | 810 | 844 | 586 | 536 | $\rho_r(CH_2)$ |
| | 2990 | 3010 | 2200 | 2185 | $\nu(CH)$ |
| | 1444 | 1428 | 1078 | 1067 | $\delta_a(CH_2)$ |
| | 1007 | 730 | 726 | 450 | $\rho_t(CH_2)$ |
| | 943 | 1023 | 780 | 811 | $\rho_w(CH_2)$ |
| | 949 | 1023 | 721 | 818 | $\rho_w(CH_2)$ |
| Square-planar skeleton | | 331 | | 329 | $\nu_s(PtCl)$ |
| | | 407 | | 387 | $\nu(Pt{-}C_2H_4)$ |
| | | 310 | | 305 | $\nu(PtCl_t)$[c] |
| | | 183 | | 185 | $\delta(ClPtCl)$ |
| | | 339 | | 339 | $\nu_a(PtCl)$ |
| | | 210 | | 198 | $\delta(ClPtC_2H_4)+\delta(ClPtCl)$ |
| | | 161 | | 160 | $\delta(ClPtCl_t)+\delta(ClPtC_2H_4)$ |
| | | 121 | | 117 | $\pi(C_2H_4PtCl_t)$ |
| | | 92 | | 92 | $\pi(ClPtCl)$ |

[a] Band assignments are for nondeuterated Zeise's salt.
[b] This coupling does not exist for Zeise's salt-$d_4$.
[c] $Cl_t$ denotes the Cl atom *trans* to $C_2H_4$.

overlap of the $2p\pi^*$ antibonding orbital of the olefin with a filled *dp* hybrid orbital of the metal.

H H H H C C Pt— Pt— C C H H H H

*A* *B*

If the $\sigma$-type bonding is predominant, the spectrum may be interpreted in terms of bonding scheme *A*, which predicts one Pt–olefin stretching mode. Grogan and Nakamoto[175] preferred this interpretation and assigned the 407 $cm^{-1}$ band of Zeise's salt to this mode. On the other hand, other workers[176–177] preferred bonding scheme *B*, which involves a five-coordinate Pt atom and assigned two bands at 491 and 403 $cm^{-1}$ to the symmetric and antisymmetric PtC stretching modes, respectively. The real bonding is somewhere between *A* and *B*, and the latter may become more predominant as the oxidation state of the metal becomes lower. In the case of Zeise's salt involving the Pt(II) atom, X-ray analysis[179] and M.O. calculations,[180] together with infrared studies,[175] seem to favor bonding scheme *A*. The vibrational spectra of ethylene complexes with other metals have been reported: $Fe(C_2H_4)(CO)_4$ (356 $cm^{-1}$),[181] $[Ag(C_2H_4)]^+$,[182] $[Rh(C_2H_4)_2Cl]_2$ (399 $cm^{-1}$),[183] $Ni(C_2H_4)_3$,[184] and $Ir(C_2H_4)_4Cl$ (505, 397, and 372 $cm^{-1}$).[184a] Here, the numbers in parenthesis indicate the metal–ethylene stretching frequencies.

### (2) Allyl Complexes

If the allyl group is $\pi$-bonded to a metal, the C═C stretching band characteristic of the $\sigma$-bonded allyl group (Sec. IV-2) does not appear. Instead, three bands of medium or strong intensity are observed at 1510–1375 $cm^{-1}$. In the low frequency region, the metal—olefin vibrations appear in the range from 570 to 320 $cm^{-1}$. Complete band assignments have been reported for $M(\pi\text{-}C_3H_5)_2$ (M = Ni, Pd), $M(\pi\text{-}C_3H_5)_3$ (M = Rh, Ir),[185] $[Pd(\pi\text{-}C_3H_5)X]_2$ (X = Cl and Br),[186,187] and $Mn(\pi\text{-}C_3H_5)(CO)_4$.[187a]

### (3) Complexes of Di- and Oligoolefins

Nonconjugated diolefins such as norbornadiene (NBD, $C_7H_8$) and 1,5-hexadiene ($C_6H_{10}$) form metal complexes via their C═C double bonds (Fig. IV-4*a*, *b*). Complete vibrational assignments have been made for $M(NBD)(CO)_4$ (M = Cr, Mo, and W),[188] $Cr(NBD)(CO)_4$[189] and $Pd(NBD)X_2$ (X = Cl and Br).[189] The metal–olefin vibrations are assigned in the region from 305 to 200 $cm^{-1}$. The spectrum of $K_2[(PtCl_3)_2(C_6H_{10})]$ is similar to that of the free ligand in the *trans* conformation.[190] Thus its structure may be shown as in Fig. IV-4*b*. However, the spectrum of $Pt(C_6H_{10})Cl_2$ is more complicated than that of the free ligand and suggests a chelate structure such as that shown in Fig. IV-4*a*.

Free butadiene ($C_4H_6$) is *trans*-planar. However, it takes a *cis*-planar structure in $Fe(C_4H_6)(CO)_3$[191] and $Fe(C_4H_6)_2CO$.[192] For $K_2[C_4H_6(PtCl_3)_2]$, the infrared spectrum indicates the *trans*-planar structure of the olefin.[193] In $[Rh(COT)Cl]_2$, cyclooctatetraene (COT) takes a tub conformation and coordinates to a metal via the 1,5 C═C double bonds, the 3,7 C═C double bonds being free (Fig. IV-4*c*). The C═C

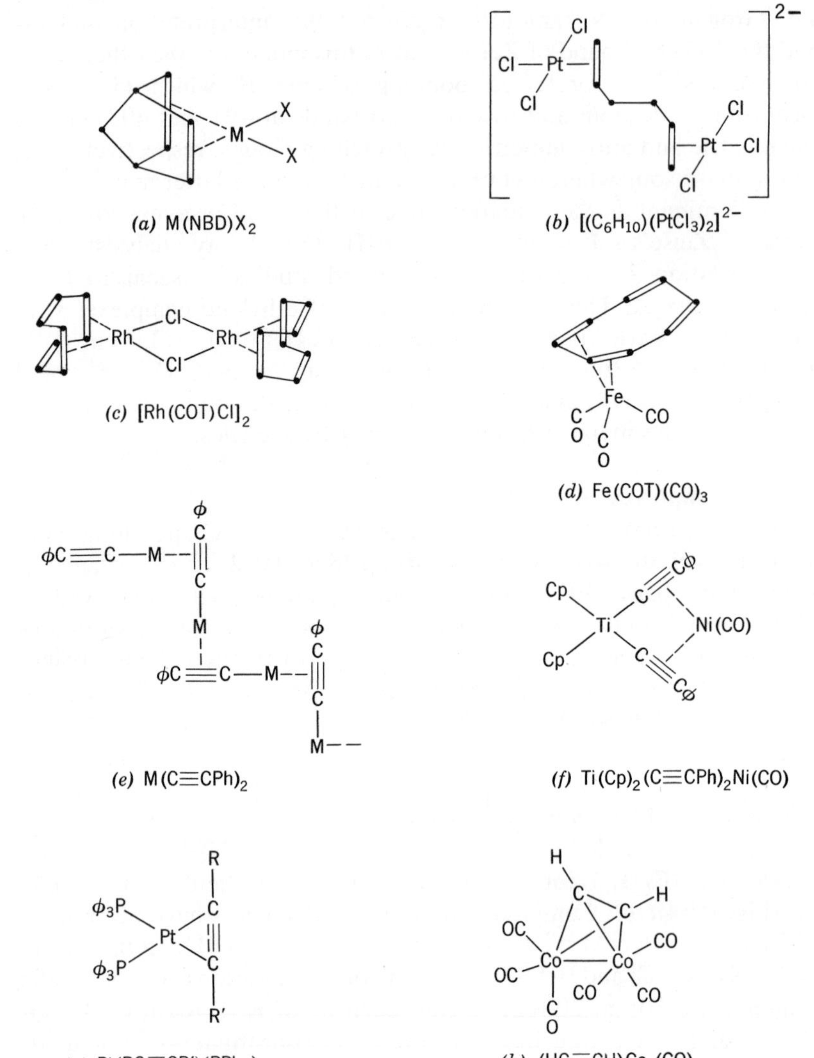

(a) $M(NBD)X_2$

(b) $[(C_6H_{10})(PtCl_3)_2]^{2-}$

(c) $[Rh(COT)Cl]_2$

(d) $Fe(COT)(CO)_3$

(e) $M(C{\equiv}CPh)_2$

(f) $Ti(Cp)_2(C{\equiv}CPh)_2Ni(CO)$

(g) $Pt(RC{\equiv}CR')(PPh_3)_2$

(h) $(HC{\equiv}CH)Co_2(CO)_6$

**Fig. IV-4.** Structures of $\pi$-complexes.

stretching bands of free COT are at 1630 and 1605 $cm^{-1}$, whereas those of the complex are at 1630 (free) and 1410 (bonded) $cm^{-1}$.[194] According to X-ray analysis,[195] only two of the four C=C double bonds of COT are bonded to the metal in $Fe(COT)(CO)_3$ (Fig. IV-4*d*). In this case, the C=C stretching band for free C=C double bonds is at 1562 $cm^{-1}$, whereas that for coordinated C=C double bonds is at 1460 $cm^{-1}$.[196]

### (4) Complexes of Alkynes

Free $HC{\equiv}C(C_6H_5)$ exhibits the $C{\equiv}C$ stretching band at $2111\ cm^{-1}$. In the case of $\sigma$-bonded complexes (Sec. IV-II), this band shifts slightly to a lower frequency ($2036$–$2017\ cm^{-1}$).[197] In $M[{-}C{\equiv}C(C_6H_5)]_2$ [$M = Cu(I)$ and $Ag(I)$], it shifts to $1926\ cm^{-1}$. This relatively large shift was attributed to the formation of both $\sigma$- and $\pi$-type bonding, shown in Fig. IV-4*e*.[198,199] $Ti[C{\equiv}C(C_6H_5)]_2(\pi\text{-}Cp)_2$ reacts with $Ni(CO)_4$ to form the complex shown in Fig. IV-4*f*. The $C{\equiv}C$ stretching band of the parent compound at $2070\ cm^{-1}$ is shifted to $1850\ cm^{-1}$ by complex formation.[200] According to Chatt and co-workers,[201] the $C{\equiv}C$ stretching bands of disubstituted alkynes ($2260$–$2190\ cm^{-1}$) are lowered to ca. $2000\ cm^{-1}$ in $Na[Pt(RC{\equiv}CR')Cl_3]$ and $[Pt(RC{\equiv}CR')Cl_2]_2$, and to ca. $1700\ cm^{-1}$ in $Pt(RC{\equiv}CR')(PPh_3)_2$. Here R and R′ denote various alkyl groups. The former represents a relatively weak $\pi$-bonding similar to that found for Zeise's salt, whereas the latter indicates strong $\pi$-bonding in which the $C{\equiv}C$ triple bond is almost reduced to the double bond (Fig. IV-4*g*). Similar results were found for $(RC{\equiv}CR')Co_2(CO)_6$, which exhibits the $C{\equiv}C$ stretching bands near $1600\ cm^{-1}$.[202] In the case of $(HC{\equiv}CH)Co_2(CO)_6$, the $C{\equiv}C$ stretching band was observed at $1402\ cm^{-1}$, which is ca. $570\ cm^{-1}$ lower than the value for free acetylene ($1974\ cm^{-1}$). The spectrum of the coordinated acetylene in this complex is similar to that of free acetylene in its first excited state, at which the molecule takes a *trans*-bent structure. Considering possible steric repulsion between the hydrogens and the $Co(CO)_3$ moiety, a structure such as that shown in Fig. IV-4*h* was proposed.[202a]

### (5) $\pi$-Complexes of Nitriles

The $C{\equiv}N$ stretching frequency of $CF_3{-}C{\equiv}N$ is $2271\ cm^{-1}$. This band is shifted to $1734\ cm^{-1}$ in $Pt(CF_3CN)(PPh_3)_2$ because of the formation of a Pt–nitrile $\pi$-bond.[202b] A similar $\pi$-bonding has been proposed for $Mn(CO)_3IL$, where L is *o*-cyanophenyldiphenylphosphine:

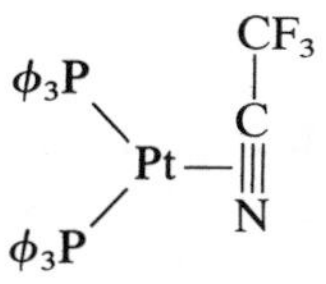

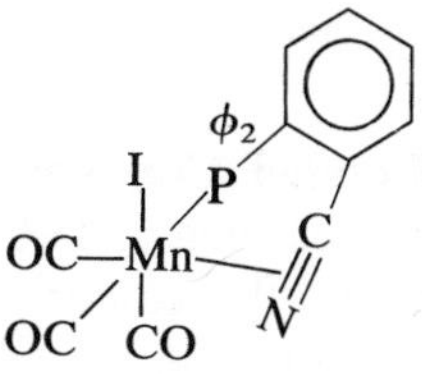

In the latter case, the $C{\equiv}N$ stretching band of the free ligand at $2225\ cm^{-1}$ is shifted to $1973\ cm^{-1}$ in the complex.[202c]

## IV-6. CYCLOPENTADIENYL COMPLEXES

The infrared spectra of cylopentadienyl ($C_5H_5$ or Cp) complexes have been reviewed extensively by Fritz.[203] According to Fritz, they are roughly classified into four groups, each of which exhibits its own characteristic spectrum.

### (1) Ionic Complexes

These are complexes such as MCp ($M = K^+$, $Rb^+$, and $Cs^+$) and $MCp_2$ ($M = Sr^{2+}$, $Ba^{2+}$, $Mn^{2+}$, and $Eu^{2+}$),[204–207] in which $M^{n+}$ and $Cp^-$ are ionically bonded. The spectra of these compounds are essentially the same as the spectrum of the $C_5H_5^-$ ion ($\mathbf{D}_{5h}$ symmetry) and consist of the following bands; $\nu$(CH), 3100–3000 $cm^{-1}$; $\nu$(CC), 1500–1400 $cm^{-1}$; $\delta$(CH), 1010–1000 $cm^{-1}$; $\pi$(CH), 750–650 $cm^{-1}$.

### (2) Centrally $\sigma$-Bonded Complexes

These are complexes such as MCp (M = Li and Na) and $MCp_2$ (M = Be, Mg, and Ca)[204,207–209] in which the metal is bonded to the center of the ring through a $\sigma$-bond. In this case, the local symmetry of the ring is regarded as $\mathbf{C}_{5v}$, and the following bands are IR active; $\nu$(CH), 3100–3000 $cm^{-1}$; $\nu$(CH), 2950–2900 $cm^{-1}$; $\nu$(CC), 1450–1400 $cm^{-1}$; $\nu$(CC), 1150–1100 $cm^{-1}$; $\delta$(CH), 1010–990 $cm^{-1}$; and two $\pi$(CH), 890–700 $cm^{-1}$. In addition, these compounds exhibit metal–ring (MR) stretching and ring–tilt vibrations below 550 $cm^{-1}$. As will be discussed in the next subsection, $MCp_2$ type compounds exhibit one MR stretching and one ring-tilt vibration in the infrared if two rings are parallel to each other. According to electron diffraction studies,[210] the two rings of $SnCp_2$ and $PbCp_2$ form angles of 45° and 55°, respectively, in the vapor state. On the assumption of angular structure in the solid state, two bands at 240 and 170 $cm^{-1}$ of $SnCp_2$ have been assigned to the antisymmetric and symmetric MR stretching modes, respectively.[211] Although the infrared spectrum of $BeCp_2$ in the vapor phase can be interpreted in terms of an ionic structure ($Cp^-Be^{2+}Cp^-$), its solid state spectrum suggests the presence of the unusual $(Cp—Be)^+Cp^-$ structure, in which one ring is covalently bonded to Be.[208]

### (3) Centrally $\pi$-Bonded Complexes*

These are complexes such as $FeCp_2$ and $RuCp_2$, in which the transition metals are bonded to the center of the ring via the $d$-$\pi$-bond. Figure IV-5 shows the infrared spectra of $FeCp_2$ and $NiCp_2$, and Table IV-12 lists the observed frequencies and band assignments for the compounds

---

* Or pentahapto ($h^5$) complexes.

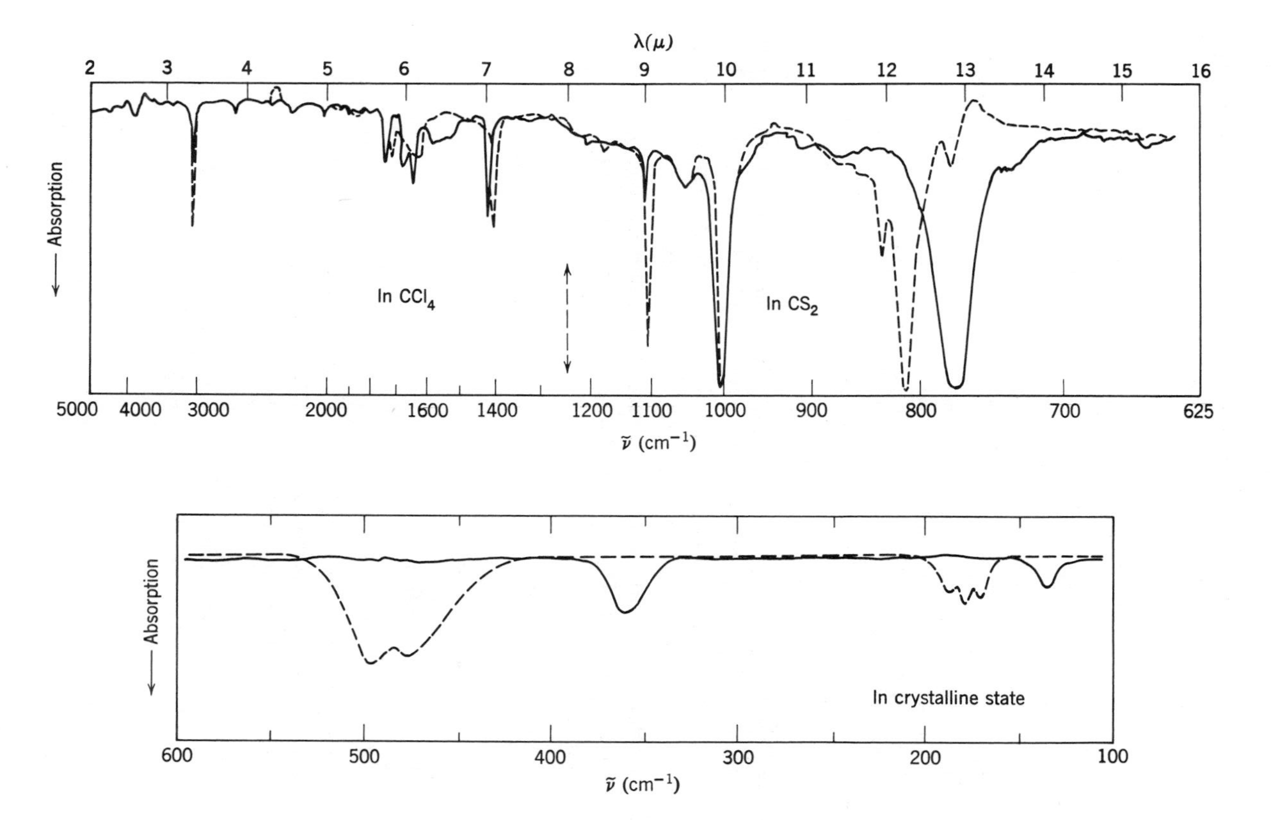


**Fig. IV-5.** Infrared spectra of $Ni(C_5H_5)_2$ (solid line) and $Fe(C_5H_5)$ (dotted line.[211a]

TABLE IV-12. OBSERVED INFRARED FREQUENCIES AND BAND ASSIGNMENTS OF CENTRALLY π-BONDED $MCp_2$ TYPE COMPOUNDS ($cm^{-1}$)

| Compound | $\nu$(CH) | | $\nu$(CC) | | $\delta$(CH) | $\pi$(CH) | | Ring Tilt | $\nu(MR)^a$ | $\delta(RMR)^a$ | Refs. |
|---|---|---|---|---|---|---|---|---|---|---|---|
| $FeCp_2$ | — | 3077 | 1110 | 1410 | 1005 | 820 | 855 | 492 | 478 | 179 | 213,211a |
| $RuCp_2$ | — | 3076 | 1095 | 1410 | 1005 | 808 | 834 | 450 | 380 | 170 | 213 |
| $OsCp_2$ | 3061 | 3061 | 1098 | 1400 | 998 | 823 | 831 | 428 | 353 | 160 | 214, 203 |
| $CoCp_2$ | 3041 | 3041 | 1101 | 1412 | 995 | 778 | 828 | 464 | 355 | — | 214, 203 |
| $NiCp_2$ | 3075 | 3075 | 1110 | 1430 | 1000 | 773 | 839 | 355 | 355 | — | 215,211a |
| $FeCp_2^+$ | 3108 | 3108 | 1116 | 1421 | 1017 | 805 | 860 | 501 | 423 | — | 216 |
| | 3100 | 3100 | 1110 | 1412 | 1001 | 779 | 841 | 490 | 405 | | |
| $CoCp_2^+$ | 3094 | 3094 | 1113 | 1419 | 1010 | 860 | 895 | 495 | 455 | 172 | 217 |
| $IrCp_2^+$ | 3077 | 3077 | 1106 | 1409 | 1009 | 818 | 862 | — | — | — | 217 |

[a] R denotes the Cp ring.

belonging to this group. In the high frequency region, these compounds exhibit seven bands, listed previously for group (2). In addition, $MCp_2$ compounds with parallel rings exhibit the six skeletal modes shown in Fig. IV-6, three of which ($\nu_3$, $\nu_5$, and $\nu_6$) are infrared active.

It is interesting to note that two rings in solid ferrocene take the staggered configuration ($\mathbf{D}_{5d}$), while those in ruthenocene take the eclipsed configuration ($\mathbf{D}_{5h}$):

Since the number of infrared or Raman active fundamentals is the same for both conformations, they cannot be distinguished on the basis of the number of fundamentals observed. According to Fritz,[203] the MR stretching frequencies and force constants follow the order:

| | Os | Fe | Ru | Cr | Co | V | Ni | Zn |
|---|---|---|---|---|---|---|---|---|
| $k$(MR) (mdyn/Å) | 2.8 | $>$2.7 | $>$2.4 | $\gg$1.6 | ~1.5 | ~1.5 | ~1.5 | ~1.5 |
| $\nu$(MR)($cm^{-1}$) | 353 | 478 | 379 | 408 | 355 | 379 | 355 | 345 |

Here, R denotes the Cp ring.

This may indicate the order of the M–R bond strength. Table IV-12 shows that the MR stretching band of $FeCp_2$ at 478 $cm^{-1}$ is shifted to a lower frequency when it is ionized to $FeCp_2^+$. Apparently, the deviation from the inert gas electronic configuration due to the ionization weakens the M–R bond.

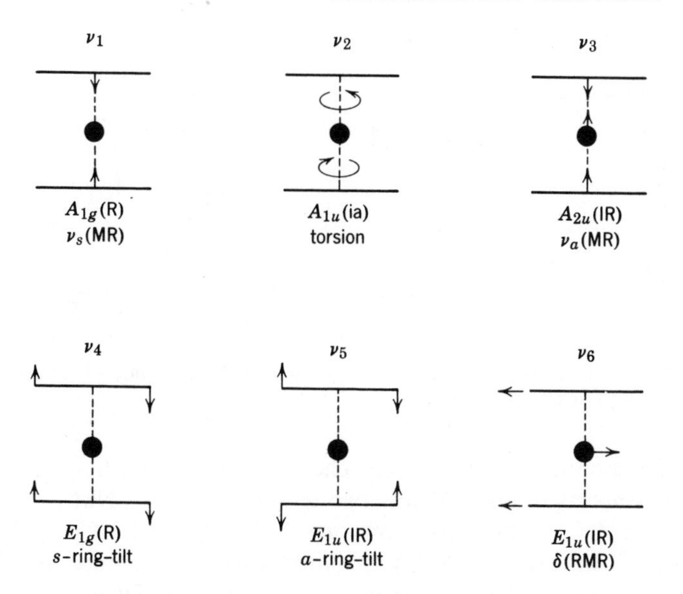


**Fig. IV-6.** Skeletal vibrations of dicyclopentadienyl metal complexes ($D_{5d}$ symmetry).

Many references are available on the vibrational spectra of $FeCp_2$ (212, 213, 218–221) and $RuCp_2$ (213, 222–223). Brunvoll et al.[224] carried out normal coordinate analysis on the whole ferrocene molecule.

### (4) Diene Type (σ-Bonded) Complexes*

These are complexes such as $HgCp_2$ and $CH_3HgCp$,[225,226] in which the metal is $\sigma$-bonded to one of the C atoms of the Cp ring:

The spectra of these compounds are similar to the spectrum of $C_5H_6$ (cyclopentadiene), and markedly different from those of the other groups discussed previously. Figure IV-7 shows the infrared spectrum of $HgCp_2$.[226a] Band assignments of these compounds can be made based on those obtained for $C_5H_6$.[227] Infrared and NMR evidence suggests the presence of diene type bonding for $(Cp)M(CH_3)_3$ (M = Si, Ge, and Sn).[228]

There are many other complexes in which the $\pi$-bonded (type 3) and the $\sigma$-bonded (type 4) cyclopentadienyl groups are mixed. As expected,

* Or monohapto ($h^1$) complexes.

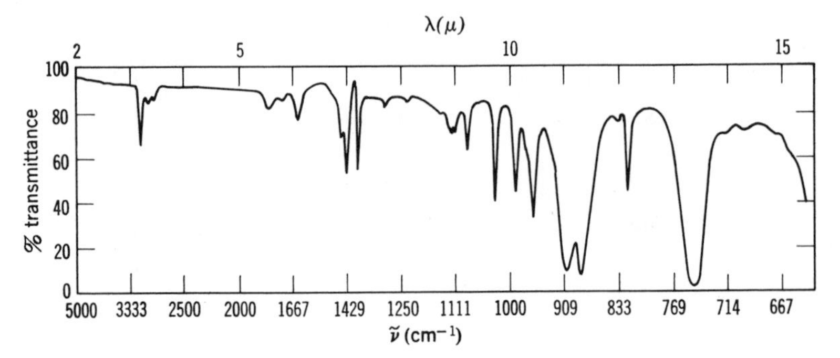


**Fig. IV-7.** Infrared spectrum of $Hg(C_5H_5)_2$, in $CS_2$ (2–6μ and 7.1–15.5μ), in $CHCl_3$ (6–6.6μ), and in $CCl_4$ (6.6–7.1μ).[226a]

these compounds exhibit bands characteristic of both groups. Typical examples are as follows: $VCp_3$ (two π and one σ),[229] $NbCp_4$ (two π and two σ),[230] $ZrCp_4$ (three π(?) and one σ),[231,232] and $MoCp_4$ (three π and one σ).[233] Infrared,[230] X-ray,[234] and NMR[235] evidence indicates the presence of two π-and two σ-bonded Cp rings in $TiCp_4$.

## (5) Complexes of Other Types

In addition to the complexes discussed above, recent X-ray studies have revealed the existence of other types. For example, an allylic (or a trihapto, $h^3$) bonding was found in $[Ni(h^3-Cp)(C_3H_4)]_2$, whose structure is shown in Fig. IV-8*a*.[236] In $TiCp_3$, two rings are π-bonded while the third is bonded to the metal through only two adjacent C atoms.[237] It is rather difficult, however, to distinguish these structures from other types by vibrational spectroscopy. In the case of $NbCp_2$, the very unusual

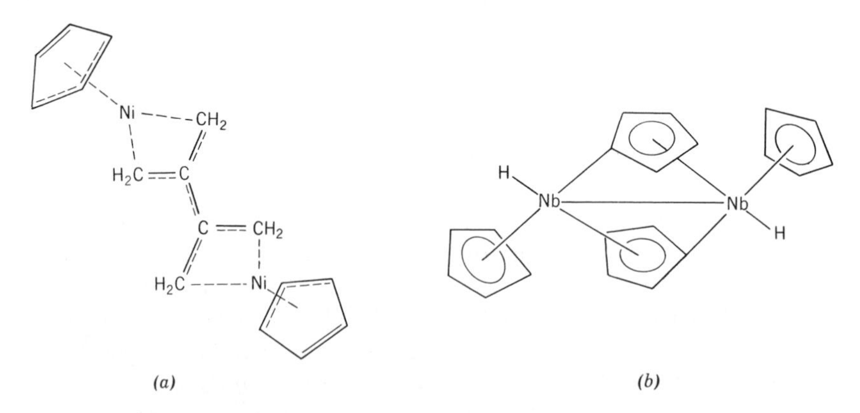


**Fig. IV-8.** Structures of some cyclopentadienyl compounds.

structure shown in Fig. IV-8*b* was suggested from NMR and other evidence.[238] Here the Cp ring serves as a bridge between two metal atoms by forming one $\pi$- and one $\sigma$-bond.

## IV-7. CYCLOPENTADIENYL COMPOUNDS CONTAINING OTHER GROUPS

### (1) Carbonyl Compounds

The vibrational spectra of carbonyl compounds were discussed in Sec. III-12. Here we discuss only those containing cyclopentadienyl rings.* It has been well established that the number of CO stretching bands observed in the infrared depends on the local symmetry of the $M(CO)_n$ group in $M(Cp)_m(CO)_n$ type compounds.[203,239] For example, only two CO stretching bands have been observed for the following compounds, in accordance with the prediction from local symmetry:

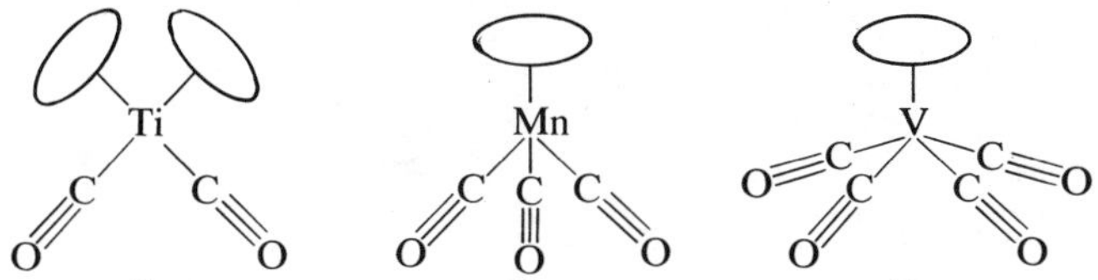

| $C_{2v}$ | $C_{3v}$ | $C_{4v}$ |
|---|---|---|
| 1890 ($B_2$) | 1938 ($E$) | 1916 ($E$) |
| 1969 ($A_1$) | 2025 ($A_1$) | 2016 ($A_1$) |

For more recent studies of these compounds, see the references cited: $Mn(Cp)(CO)_3$ (240–242), $Re(Cp)(CO)_3$ (243), and $V(Cp)(CO)_4$ (244). In $M(Cp)(CO)_3$ type compounds,[245] the CO stretching frequencies increase in the order: $V^{-1} < Cr^0 < Mn^{+1} < Fe^{2+}$. This indicates that the higher the oxidation state of the metal, the less the M–C $\pi$-back bonding and the higher the CO stretching frequency.

Originally, $Fe(Cp)_2(CO)_2$ was thought to contain two $\pi$-bonded Cp rings.[246] However, an infrared and NMR study[247] showed that one ring is $\pi$-bonded and the other $\sigma$-bonded to the metal. Later, X-ray analysis confirmed this structure.[248] The structure of $Fe_2(Cp)_2(CO)_4$ has been studied extensively. In the solid state, it takes a *trans*-bridged structure (Fig. IV-9*a*),[249] or a *cis*-bridged structure (Fig. IV-9*b*) if crystallized in polar solvents at low temperatures.[250] The *cis*-isomer exhibits two terminal (1975 and 1933 $cm^{-1}$) and two bridging (1801 and 1766 $cm^{-1}$) bands. Although the *trans*-isomer also exhibits two terminal (1956 and 1935 $cm^{-1}$) and two bridging (1769 and 1755 $cm^{-1}$) bands, these splittings are probably due to the crystal field effect.

* Hereafter, Cp denotes a $\pi$-bonded $C_5H_5$ group.

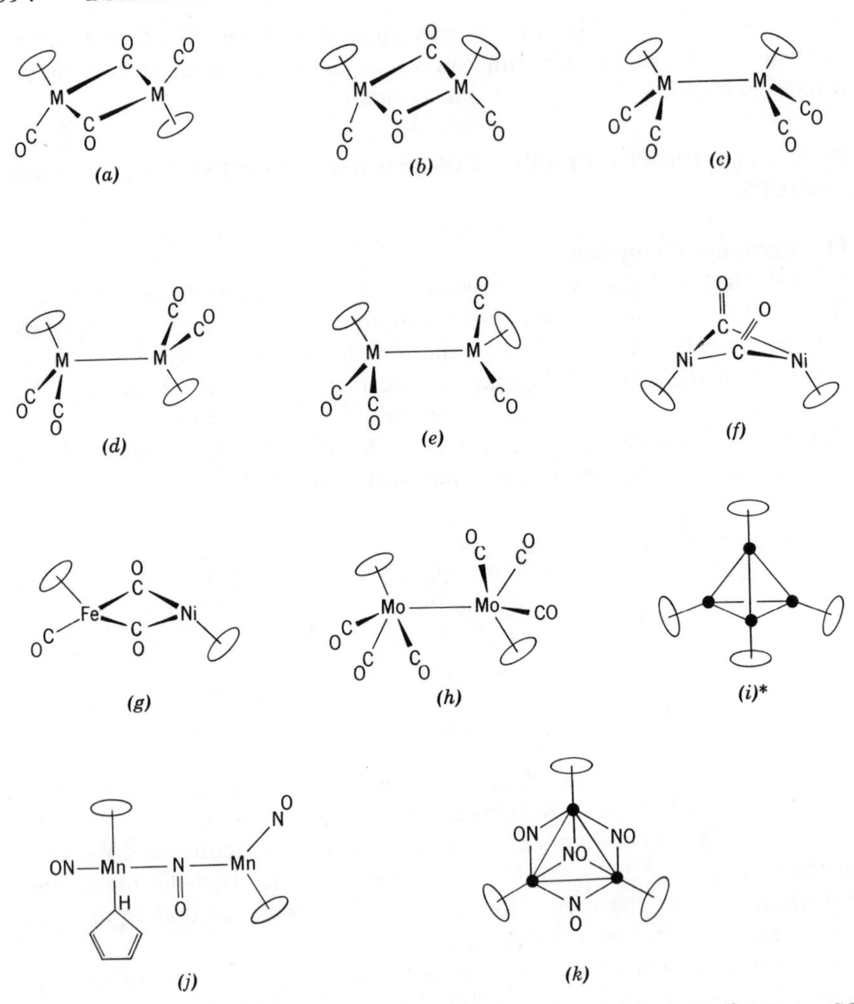


**Fig. IV-9.** Structures of cyclopentadienyl carbonyl and nitrosyl compounds. (*:bridging CO groups are not shown.)

The structure of $Fe_2(Cp)_2(CO)_4$ in solution has been controversial. Early infrared studies[251,252] suggested the presence of the *cis*-bridged structure (Fig. IV-9*b*) mixed with a trace of noncentrosymmetric, nonbridging isomer (Fig. IV–9*c*). Manning[253] proposed, however, an equilibrium involving the three isomers, *a*, *b*, and *c* of Fig. IV-9. This was confirmed by Bullitt et al.,[254] who gave the following assignments for the spectrum in a $CS_2$—$C_6D_5CD_3$ solution: *trans*-isomer (*a*)—1954 and 1781 $cm^{-1}$; *cis*-isomer (*b*)—1998, 1954, 1810, and 1777 $cm^{-1}$. The frequencies of nonbridged species could not be determined because of their very low concentration. In the case of $Ru_2(Cp)_2(CO)_4$, Bullitt et al.[254] proposed an equilibrium containing four isomers; *a*, *b*, *d*, and *e* of Fig. IV-9.

It is interesting to note that the bridging CO groups of $Fe_2(Cp)_2(CO)_4$ form an adduct with trialkylaluminum[255] (see Sec. III-12):

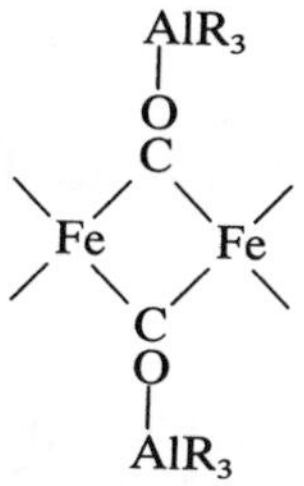

```
        AlR3
         |
         O
         C
   \   /   \   /
    Fe       Fe
   /   \   /   \
         C
         O
         |
        AlR3
```

This indicates that the basicity of the bridging CO group is greater than that of the terminal CO group. The CO stretching bands of the parent compound (R: isobutyl) are at 2005 and 1962 (terminal) and 1794 (bridging) $cm^{-1}$ in heptane solution. These bands are shifted to 2041 and 2003 (terminal) and 1680 (bridging) $cm^{-1}$ by adduct formation. X-ray analysis has been carried out on $[Fe_2(Cp)_2(CO)_4][Al(C_2H_5)_3]_2$.[256] Formation of adducts such as $[Fe_2(Cp)_2(CO)_4]BX_3$ (X = Cl and Br) and $[Fe(Cp)(CO)]_4 \cdot BX_3$ (X = F, Cl, and Br) has also been confirmed.[257] These compounds exhibit bands at 1470–1290 $cm^{-1}$ for bridging CO groups, which are bonded to a Lewis acid via the O atom.

$Ni_2(Cp)_2(CO)_2$ exhibits two bridging CO stretching bands at 1854 and 1896 $cm^{-1}$ in heptane solution. The structure shown in Fig. IV-9*f* with a puckered $Ni(CO)_2Ni$ bridge was proposed for this compound.[258] In heptane solution $FeNi(Cp)_2(CO)_3$ shows a strong terminal CO stretching at 2004 $cm^{-1}$ and two bridging CO stretching bands at 1855 and 1825 $cm^{-1}$. Since the 1855 $cm^{-1}$ band (symmetric type) is very weak, the $Ni(CO)_2Fe$ bridge in this compound was thought to be virtually planar, as shown in Fig. IV-9*g*.[258]

According to X-ray analysis,[259] the structure of $Mo_2(Cp)_2(CO)_6$ is *trans*-centrosymmetric, as shown in Fig. IV-9*h*. The infrared spectrum in the CO stretching region is consistent with this structure, both in the solid state and in solution.[260] In solvents of high dielectric constants, however, the *trans*-rotamer is rearranged into the *gauche*-rotamer.[261] For the infrared spectra of analogous tungsten compounds, see Refs. 241 and 262. According to X-ray analysis,[263] $Fe_4(Cp)_4(CO)_4$ takes a regular tetrahedral structure such as that shown in Fig. IV-9*i*. It exhibits a bridging CO stretching band at 1649 $cm^{-1}$,[255] in the infrared and a FeFe stretching band at 214 $cm^{-1}$,[264] in the Raman.

**(2) Halogeno Compounds**

Cyclopentadienyl complexes containing metal–halogen bonds exhibit metal–halogen vibrations (Sec. III-16), together with those of the cyclopentadienyl rings. The low frequency spectra of these compounds are complicated[265,266] because metal–ring skeletal modes couple with metal-halogen modes. The infrared spectra of $M(Cp)_2X_2$ type compounds (M = Ti, Zr, and Hf; X = Cl, Br, and I) have been studied by several investigators.[265–268] Infrared spectra have been reported for Mo(Cp)-$(CO)_3X$[240] and $Mo(Cp)(CO)_2X_3$ (X = Cl, Br, and I).[269]

**(3) Nitrosyl Compounds**

Vibrational spectra of nitrosyl compounds were discussed in Sec. III-12. The vibrational spectra of Ni(Cp)(NO) and its deuterated and $^{15}N$ species have been assigned completely[270]: the NO stretching, NiN stretching, NiCp stretching, and NiCp tilt vibrations are at 1809, 649, 322, and 290 $cm^{-1}$, respectively. If this compound is irradiated by UV light in an Ar matrix, the bands near 1830 $cm^{-1}$ disappear and a new band emerges at 1390 $cm^{-1}$.[271] This has been interpreted as indicating the following photoionization: $Ni(Cp)NO \xrightarrow{h\nu} Ni(Cp)^+(NO)^-$. $Mn_2(Cp)_3(NO)_3$ exhibits two NO stretching bands at 1732 and 1510 $cm^{-1}$.[272] With the former attributed to the terminal and the latter to the bridging NO, the structure shown in Fig. IV-9*j* was proposed. The infrared spectrum of Mo(Cp)-$(CO)_2(NO)$ has been reported.[273,240] Figure IV-9*k* shows the structure of $Mn_3(Cp)_3(NO)_4$, containing doubly and triply bridging NO groups. The bands at 1530 and 1480 $cm^{-1}$ were assigned to the doubly bridged NO groups, whereas the 1320 $cm^{-1}$ band was attributed to the triply bridged NO group.[274]

**(4) Hydrido Complexes**

Vibrational spectra of hydrido complexes were reviewed in Sec. III-13. The metal-hydrogen stretching bands for $Mo(Cp)_2H_2$,[275] $Re(Cp)_2H_2$, and $W(Cp)_2H_2$[276,277] have been observed in the 2100–1800 $cm^{-1}$ region. X-ray analysis on $Mo(Cp)_2H_2$[278] suggests that the coordination around the Mo atom is approximately tetrahedral. In polymeric $[Zr(Cp)_2H_2]_n$,[279] the bridging ZnH stretching vibration is observed as a strong, broad band at 1540 $cm^{-1}$.[280] A similar bridging TiH vibration is found at 1450 $cm^{-1}$ for $[Ti(Cp)_2H]_2$.[281] In $[\{Rh(Cp')\}_2HCl_3]$, where Cp′ denotes the pentamethyl-Cp group, the bridging RhH vibration was assigned at 1151 $cm^{-1}$.[282]

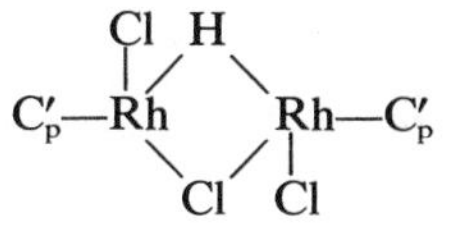

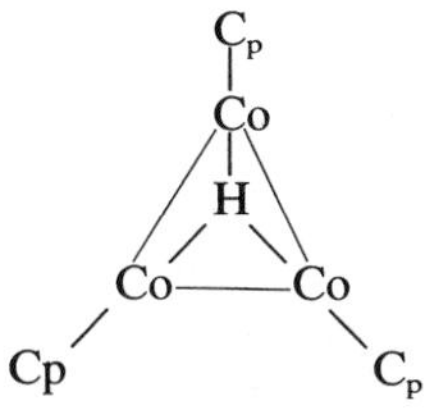

```
          Cl  H                              Cp
          |  / \                             |
   C'p—Rh      Rh—C'p                        Co
         \    / |                           /|\
          Cl   Cl                          / H \
                                          / / \ \
                                        Co———————Co
                                        /          \
                                      Cp            Cp
```

An extremely low CoH stretching frequency (950 $cm^{-1}$), together with an unusually high field proton chemical shift observed for $[Co(Cp)H]_4$, was attributed to the triply bridged structure shown above (only one face of the tetrahedron is shown).[283]

### (5) Complexes Containing Other Groups

As discussed in Sec. III-11, the mode of coordination of the pseudohalide ion can be determined by vibrational spectroscopy. Burmeister et al.[284] found that all NCS and NCSe groups are N-bonded in $M(Cp)_2X_2$ type compounds, where M is Ti, Zr, Hf, or V, and X is NCS or NCSe. In the case of analogous NCO complexes, Ti, Zr, and Hf form O-bonded complexes, whereas V forms an N-bonded complex. Later, Jensen et al.[285] suggested the N-bonded structure for the titanium complex.

A strong N≡N stretching band is observed at 1910 $cm^{-1}$ in the Raman spectrum of $L_2(Cp)Mo—N{\equiv}N—Mo(Cp)L_2$ (L: $PPh_3$).[286] Thiocarbonyl complexes of the $(Cp)Mn(CO)_{3-n}(CS)_n$ ($n = 1, 2, 3$) type exhibit the C=S stretching bands at 1340—1235 $cm^{-1}$.[287] In $(Cp)Nb(S_2)X$ type compounds (X = Cl, Br, I, and SCN), the $S_2$ is probably coordinated to the metal in a side-on fashion, and its SS stretching band may be assigned at 540 $cm^{-1}$.[288]

## IV-8. COMPLEXES OF OTHER CYCLIC UNSATURATED LIGANDS

The vibrational spectra of a cyclobutadiene complex, $Fe(C_4H_4)(CO)_3$, have been measured,[289] and the FeR stretching and tilt vibrations have been assigned at ca. 397 and 471 $cm^{-1}$, respectively. The infrared spectra of $Ni(C_4(CH_3)_4)Cl_2$ and $M(C_4(C_6H_5)_4)X_2$ (M = Ni and Pd; X = Cl, Br, and I) have been reported.[290]

Dibenzene chromium, $Cr(C_6H_6)_2$, takes a ferrocene-like sandwich structure, and its vibrational spectra have been assigned by Fritz and Fischer.[291] Table IV-13 lists the vibrational frequencies of $Cr(C_6H_6)_2$ and

TABLE IV-13. INFRARED FREQUENCIES OF DIBENZENE–METAL COMPLEXES ($cm^{-1}$)[203]

| Complex | $\nu$(CH) | | $\nu$(CC) | $\delta$(CH) | $\nu$(CC) | $\pi$(CH) | | Ring Tilt | $\nu(MR)^a$ | $\delta(RMR)^a$ |
|---|---|---|---|---|---|---|---|---|---|---|
| $Cr(C_6H_6)_2$ | 3037 | — | 1426 | 999 | 971 | 833 | 794 | 490 | 459 | (140) |
| $Cr(C_6H_6)_2^+$ | 3040 | — | 1430 | 1000 | 972 | 857 | 795 | 466 | 415 | (144) |
| $Mo(C_6H_6)_2$ | 3030 | 2916 | 1425 | 995 | 966 | 811 | 773 | 424 | 362 | — |
| $W(C_6H_6)_2$ | 3012 | 2898 | 1412 | 985 | 963 | 882 | 798 | 386 | 331 | — |
| $V(C_6H_6)_2$ | 3058 | — | 1416 | 985 | 959 | 818 | 739 | 470 | 424 | — |

$^a$ R denotes the $C_6H_6$ ring.

its analogs. Figure IV-10 shows the infrared spectrum of $Cr(C_6H_6)_2$.[291a] On the basis of the three-body approximation, the Cr—R (ring) stretching force constant of $Cr(C_6H_6)_2$ was estimated to be 2.39 mdyn/Å, which is smaller than that of ferrocene (2.7 mdyn/Å). For complete normal coordinate analysis of $Cr(C_6H_6)_2$, see Ref. 292. As expected, the infrared spectra of $M(C_6H_6)(C_5H_5)$ type complexes show bands characteristic of both $C_6H_6$ and $C_5H_5$ rings.[293] Complete assignments are available for $Cr(Ar)(CO)_3$ type compounds, where Ar represents $C_6H_6$, $C_6H_5CH_3$, and so on.[294,295] The local $\mathbf{C}_{3v}$ symmetry of the $Cr(CO)_3$ group does not hold if Ar is a monosubstituted benzene. Thus the *E* mode CO stretching band splits into two, even in solution.

Originally, two CH stretching bands of $M(C_5H_6)(C_5H_5)$ (M = Co and Rh) near 2750 and 2945 $cm^{-1}$ were assigned to $\nu(CH_{endo})$ and $\nu(CH_{exo})$, respectively:[296]

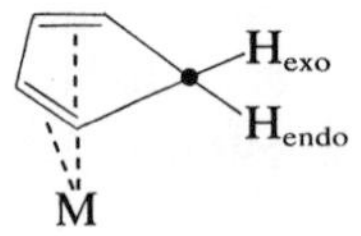

A later study[297] shows, however, that the lower frequency band near 2750 $cm^{-1}$ must be assigned to $\nu(CH_{exo})$, since replacement of the exo hydrogen by the phenyl or perfluorophenyl group results in the disappearance of this band. In the case of $Mn(C_6H_7)(CO)_3$, the bands at 2970 and 2830 $cm^{-1}$ were assigned to $\nu(CH_{endo})$ and $\nu(CH_{exo})$, respectively.[298]

Vibrational frequencies have been reported for $MCl_2(C_7H_7)_2$,[299] $MCl_2(C_8H_7)_2$ (M = Ti and Zr),[300] and $M(C_8H_8)_2$ (M = Th and U).[301] The indenyl group may coordinate to the metal through a $\sigma$- or a $\pi$-bond:

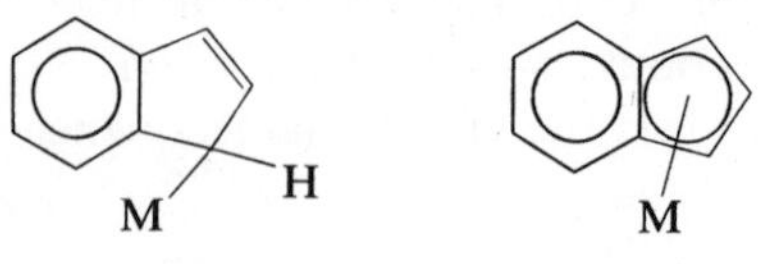

$\sigma$-Complex

$\pi$-Complex

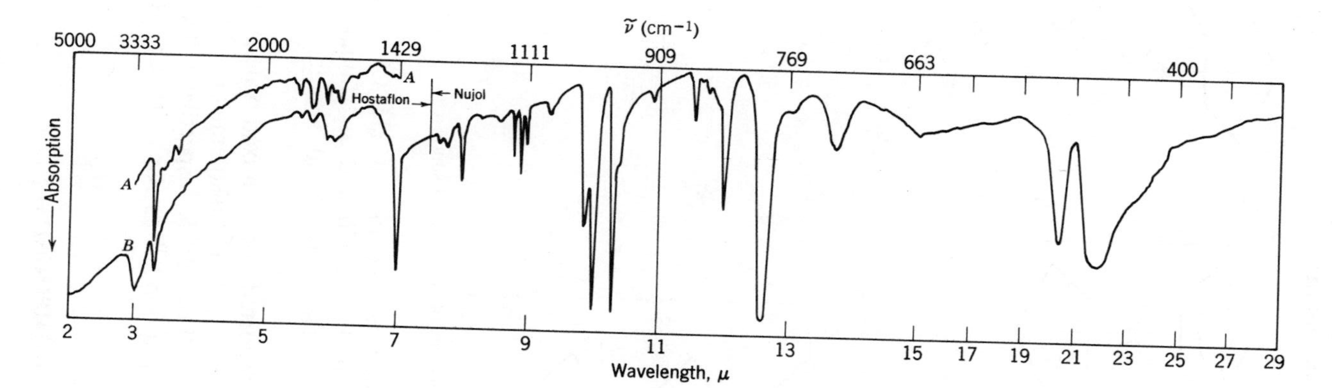


**Fig. IV-10.** Infrared spectrum of crystalline $Cr(C_6H_6)_2$: (*a*) KBr pellet; (*b*) Hostaflon-oil suspension (2–7.5 μ) and Nujol mull suspension (7.5–29 μ).[291a]

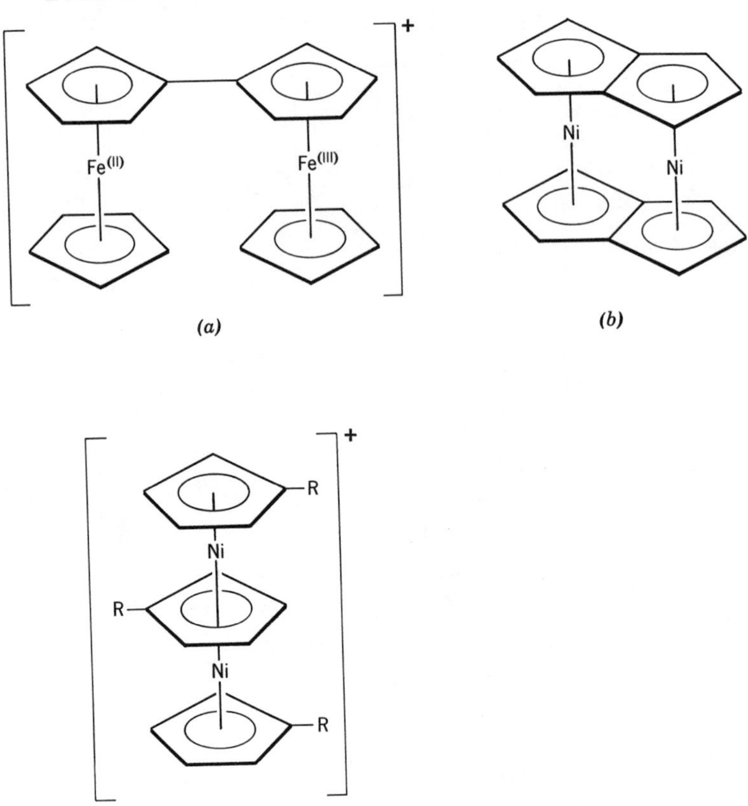


**Fig. IV-11.** Structures of some metal sandwich compounds.

An example of the former is seen in $Hg(C_9H_7)Cl$, which exhibits an aromatic CH stretching at 3060–3050 and an aliphatic CH stretching band at 2920–2850 $cm^{-1}$. The latter band should be absent in the $\pi$-bonded complex.[302]

Infrared spectra are reported for a mixed valence state complex, biferrocene ($Fe^{2+}$, $Fe^{3+}$) picrate[303] and bis(pentalenyl)Ni,[304] whose structures are shown in Fig. IV-11*a* and *b*, respectively. The spectrum of a triple decker compound, $[Ni_2(Cp')_3]BF_4$ (Cp′: $CH_3$-Cp) (Fig. IV-11*c*), is similar to that of $Ni(Cp')_2$.[305]

## IV-9. MISCELLANEOUS COMPOUNDS

There are many other organometallic compounds which have not been covered in the preceding sections. For these, the reader should consult general references such as the following:

1. N. N. Greenwood, ed., *Spectroscopic Properties of Inorganic and Organometallic Compounds*, Vols. 1–9, The Chemical Society, London, 1968–1975.
2. M. Dub, ed., *Organometallic Compounds: Methods of Synthesis, Physical Constants and Chemical Reactions*, Vols. 1–3, Springer-Verlag, New York, 1966.
3. E. Maslowsky, Jr., *Vibrational Spectra of Organometallic Compounds*, Wiley, New York, 1976.
4. K. Nakamoto, "Characterization of Organometallic Compounds by Infrared Spectroscopy", in *Characterization of Organometallic Compounds*, Part I, M. Tsutsui ed., Wiley–Interscience, New York, 1969.

Other review articles on specific groups of compounds are as follows:

*Alkyldiboranes*: W. J. Lehmann and I. Shapiro, *Spectrochim. Acta*, **17,** 396 (1961).
*Organoaluminum Compounds*: E. G. Hoffman, *Z. Elektrochem.*, **64,** 616 (1960).
*Organosilicon Compounds*: A. L. Smith, *Spectrochim. Acta*, **16,** 87 (1960).
*Organogermanes*: R. J. Cross and R. Glockling, *J. Organometal. Chem.*, **3,** 146 (1965).
*Organotin Compounds*: R. Okawara and W. Wada, *Adv. Organometal. Chem.*, **5,** 137 (1967).
*Organophosphorus Compounds*: D. E. C. Corbridge, *The Structural Chemistry of Phosphorus*, Elsevier, Amsterdam, 1974; L. C. Thomas, *Interpretation of the Infrared Spectra of Organophosphorus Compounds*, Heyden, London, 1974.
*Organometallic Compounds of P, As, Sb, and Bi*: E. Maslowsky, Jr., *J. Organometal. Chem.*, **70,** 153 (1974).

## References

1. A. M. Pyndyk, M. R. Aliev, and V. T. Aleksanyan, *Opt Spectrosc.* (English), **36,** 393 (1974).
2. E. R. Lippincott and M. C. Tobin, *J. Am. Chem. Soc.*, **75,** 4141 (1953).
3. R. A. Kovar and G. L. Morgan, *Inorg. Chem.*, **8,** 1099 (1969).
4. A. M. W. Bakke, *J. Mol. Spectrosc.*, **41,** 1 (1972).
5. J. R. Durig and S. C. Brown, *J. Mol. Spectrosc.*, **45,** 338 (1973).
6. J. R. Allkins and P. J. Hendra, *Spectrochim. Acta*, **22,** 2075 (1966).
7. J. R. Durig, C. M. Plater, Jr., J. Bragin, and Y. S. Li, *J. Chem. Phys.*, **55,** 2895 (1971).
8. W. J. Lehmann, C. O. Wilson, and I. Shapiro, *J. Chem. Phys.*, **31,** 1071 (1959).
9. L. A. Woodward, J. R. Hall, R. N. Nixon, and N. Sheppard, *Spectrochim. Acta*, **15,** 249 (1959).
10. R. J. O'Brien and G. A. Ozin, *J. Chem. Soc.*, **A,** 1136 (1971).
11. G. E. Coates and A. J. Downs, *J. Chem. Soc.*, 3353 (1964).
12. J. R. Hall, L. A. Woodward, and E. A. V. Ebsworth, *Spectrochim. Acta*, **20,** 1249 (1964).
13. G. Bouquet and M. Bigorgne, *Spectrochim. Acta*, **23A,** 1231 (1967).
14. P. J. D. Park and P. J. Hendra, *Spectrochim. Acta*, **24A,** 2081 (1968).
15. H. Siebert, *Z. Anorg. Allg. Chem.*, **273,** 161 (1953).
16. S. Sportouch, C. Lacoste and R. Gaufrès, *J. Mol. Struct.*, **9,** 119 (1971).
17. W. F. Edgell and C. H. Ward, *J. Am. Chem. Soc.*, **77,** 6486 (1955).

18. G. A. Crowder, G. Gorin, F. H. Kruse, and D. W. Scott, *J. Mol. Spectrosc.*, **16,** 115 (1965).
19. H. H. Eysel, H. Siebert, G. Groh, and H. J. Berthold, *Spectrochim. Acta*, **26A,** 1595 (1970).
20. A. J. Downs, R. Schmutzler, and I. A. Steer, *Chem. Commun.*, 221 (1966).
21. A. J. Shortland and G. Wilkinson, *J. Chem. Soc.*, Dalton, 872 (1973).
22. J. W. Nibler and T. H. Cook, *J. Chem. Phys.*, **58,** 1596 (1973).
23. C. W. Hobbs and R. S. Tobias, *Inorg. Chem.*, **9,** 1998 (1970).
24. P. L. Goggin and L. A. Woodward, *Trans. Faraday Soc.*, **56,** 1591 (1960).
25. G. B. Deacon and J. H. S. Green, *Spectrochim. Acta*, **24A,** 885 (1968).
26. M. G. Miles, J. H. Patterson, C. W. Hobbs, M. J. Hopper, J. Overend, and R. S. Tobias, *Inorg. Chem.*, **7,** 1721 (1968).
27. H. Kriegsmann and S. Pischtschan, *Z. Anorg. Allg. Chem.*, **308,** 212 (1961).
28. A. J. Downs and I. A. Steer, *J. Organometal. Chem.*, **8,** P21 (1967).
29. K. J. Wynne and J. W. George, *J. Am. Chem. Soc.*, **91,** 1649 (1969).
30. M. T. Chen and J. W. George, *J. Am. Chem. Soc.*, **90,** 4580 (1968).
31. J. A. Creighton, G. B. Deacon, and J. H. S. Green, *Aust. J. Chem.*, **20,** 583 (1967).
32. R. Baumgärtner, W. Sawodny, and J. Goubeau, *Z. Anorg. Allg. Chem.*, **333,** 171 (1964).
33. E. Weiss and E. A. Lucken, *J. Organometal. Chem.*, **2,** 197 (1964).
34. R. West and W. Glaze, *J. Am. Chem. Soc.*, **83,** 3580 (1961).
35. P. Krohmer and J. Goubeau, *Z. Anorg. Allg. Chem.*, **369,** 238 (1969).
36. T. Ogawa, K. Hirota, and T. Miyazawa, *Bull. Chem. Soc. Jap.*, **38,** 1105 (1965).
37. T. Ogawa, *Spectrochim. Acta*, **23A,** 15 (1968).
38. J. Yamamoto and C. A. Wilkie, *Inorg. Chem.*, **10,** 1129 (1971).
39. J. Mink and B. Gellai, *J. Organometal. Chem.*, **66,** 1 (1974).
40. S. Inoue and T. Yamada, *J. Organometal. Chem.*, **25,** 1 (1970).
41. J. Mink and Y. A. Pentin, *J. Organometal. Chem.*, **23,** 293 (1970).
42. J. L. Bribes and R. Gaufrès, *Spectrochim. Acta*, **27A,** 2133 (1971).
43. W. J. Lehmann, C. O. Wilson, and I. Shapiro, *J. Chem. Phys.*, **28,** 781 (1958).
44. J. Chouteau, G. Davidovics, F. D'Amato, and L. Savidan, *Compt. Rend.*, **260,** 2759 (1965).
45. J. H. S. Green, *Spectrochim. Acta*, **24A,** 137 (1968).
46. A. E. Borisov, N. V. Novikova, N. A. Chumaevskii, and E. B. Shkirtil, *Dokl. Akad. Nauk SSSR*, **173,** 855 (1967).
47. R. L. McKenney and H. H. Sisler, *Inorg. Chem.*, **6,** 1178 (1967).
48. J. A. Jackson and R. J. Nielson, *J. Mol. Spectrosc.*, **14,** 320 (1964).
49. M. I. Batuev, A. D. Petrov, V. A. Ponomarenko, and A. D. Mateeva, *Izv. Akad. Nauk SSSR, Otd. Kim. Nauk*, 1070 (1956).
50. L. A. Leites, Y. P. Egorov, J. Y. Zueva, and V. A. Ponomarenko, *Izv. Akad. Nauk SSSR, Otd. Kim. Nauk*, 2132 (1961).
51. W. R. Cullen, G. B. Deacon, and J. H. S. Green, *Can. J. Chem.*, **43,** 3193 (1965).
52. P. Taimasalu and J. L. Wood, *Trans. Faraday Soc.*, **59,** 1754 (1963).
53. C. R. Dillard and J. R. Lawson, *J. Opt. Soc. Am.*, **50,** 1271 (1960).
54. T. L. Brown, R. L. Gerteis, D. A. Bafus, and J. A. Ladd, *J. Am. Chem. Soc.*, **86,** 2134 (1964); T. L. Brown, J. A. Ladd, and C. N. Newmann, *J. Organometal. Chem.*, **3,** 1 (1965).
55. H. Dietrich, *Acta Crystallogr.*, **16,** 681 (1963).
56. T. L. Brown in *Advances in Organometallic Chemistry*, F. G. A. Stone and R. West, eds., Vol. III, Academic Press, New York, 1965, p. 374.

57. C. N. Atam, H. Müller, and K. Dehnicke, *J. Organometal. Chem.*, **37,** 15 (1972).
58. E. G. Hoffmann, *Z. Elektrochem.*, **64,** 616 (1960).
59. O. Yamamoto, *Bull. Chem. Soc. Jap.*, **35,** 619 (1962).
60. C. A. Wilkie, *J. Organometal. Chem.*, **32,** 161 (1971).
61. R. J. Cross and F. Glocking, *J. Organometal. Chem.*, **3,** 146 (1965).
62. W. M. Scovell, B. Y. Kimura, and T. G. Spiro, *J. Coord. Chem.*, **1,** 107 (1971).
63. K. H. Thiele, S. Wilcke, and M. Ehrhardt, *J. Organometal. Chem.*, **14,** 13 (1968).
64. B. Busch and K. Dehnicke, *J. Organometal. Chem.*, **67,** 237 (1974).
65. A. H. Cowley, J. L. Mills, T. M. Leohr, and T. V. Long, *J. Am. Chem. Soc.*, **93,** 2150 (1971).
66. H. D. Kaesz and F. G. A. Stone, *Spectrochim. Acta*, **15,** 360 (1959).
67. J. Mink and Y. A. Pentin, *Acta Chim. Acad. Sci. Hung.*, **66,** 277 (1970).
68. A. K. Holliday, W. Reade, K. R. Seddon, and I. A. Steer, *J. Organometal. Chem.*, **67,** 1 (1974).
69. J. D. Odom, L. W. Hall, S. Riethmiller, and J. R. Durig, *Inorg. Chem.*, **13,** 170 (1974).
70. G. Davidson, *Spectrochim. Acta*, **27A,** 1161 (1971).
71. G. Masetti and G. Zerbi, *Spectrochim. Acta*, **26A,** 1891 (1970).
72. U. Kunze, E. Lindner, and J. Koola, *J. Organometal. Chem.*, **57,** 319 (1973).
73. A. N. Nesmeyanov, A. E. Borisov, N. V. Novikova, and E. I. Fedin, *J. Organometal. Chem.*, **15,** 279 (1968).
74. S. L. Stafford and F. G. A. Stone, *Spectrochim. Acta*, **17,** 412 (1961).
75. G. Davidson, P. G. Harrison, and E. M. Riley, *Spectrochim. Acta*, **29A,** 1265 (1973).
76. J. Mink and Y. A. Pentin, *J. Organometal. Chem.*, **23,** 293 (1970).
77. C. Souresseau and B. Pasquier, *J. Organometal. Chem.*, **39,** 51 (1972).
78. R. E. Sacher, D. H. Lemmon, and F. A. Miller, *Spectrochim. Acta*, **32A,** 1169 (1967).
79. D. I. Maclean and R. E. Sacher, *J. Organometal. Chem.*, **74,** 197 (1974).
79a. W. M. A. Smit and G. Dijkstra, *J. Mol. Struct.*, **8,** 263 (1971).
80. R. E. Sacher, W. Davidsohn, and F. A. Miller, *Spectrochim. Acta*, **26A,** 1011 (1970).
81. D. H. Whiffen, *J. Chem. Soc.*. 1350 (1956).
82. J. H. S. Green, *Spectrochim. Acta*, **24A,** 863 (1968).
83. J. Mink, G. Végh, and Y. A. Pentin, *J. Organometal. Chem.*, **35,** 225 (1972).
84. G. Costa, A. Camus, N. Marsich, and L. Gatti, *J. Organometal. Chem.*, **8,** 339 (1967).
85. A. N. Rodionov, N. I. Rucheva, I. M. Viktorova, D. N. Shigorin, N. I. Sheverdina, and K. A. Kocheshkov, *Izv. Akad. Nauk SSSR, Ser. Khim.*, 1047 (1969).
86. H. F. Shurvell, *Spectrochim. Acta*, **23A,** 2925 (1967).
87. N. Kumar, B. L. Kalsotra, and R. K. Multani, *J. Inorg. Nucl. Chem.*, **35,** 3019 (1973).
88. A. L. Smith, *Spectrochim. Acta*, **23A,** 1075 (1967).
89. A. L. Smith, *Spectrochim. Acta*, **24A,** 695 (1968).
90. J. R. Durig, C. W. Sink, and J. B. Turner, *Spectrochim. Acta*, **26A,** 557 (1970).
91. D. H. Brown, A. Mohammed, and D. W. A. Sharp, *Spectrochim. Acta*, **21,** 663 (1965).
92. K. Shobatake, C. Postmus, J. R. Ferraro, and K. Nakamoto, *Appl. Spectrosc.*, **23,** 12 (1969).
93. K. Cavanaugh and D. F. Evans, *J. Chem. Soc.*, A, 2890 (1969).
94. H. Kriegsmann, *Z. Anorg. Allg. Chem.*, **294,** 113 (1958).
95. H. Bürger, *Spectrochim. Acta*, **24A,** 2015 (1968).
96. K. Licht and P. Koehler, *Z. Anorg. Allg. Chem.*, **383,** 174 (1971).

97. J. R. Durig, K. K. Lau, J. B. Turner, and J. Bragin, *J. Mol. Spectrosc.*, **31,** 419 (1971).
98. B. A. Nevett and A. Perry, *J. Organometal. Chem.*, **71,** 399 (1974).
99. W. M. Scovell, G. C. Stocco, and R. S. Tobias, *Inorg. Chem.*, **9,** 2682 (1970); W. M. Scovell and R. S. Tobias, *ibid.*, **9,** 945 (1970).
100. H. J. Becher, *Z. Anorg. Allg. Chem.*, **294,** 183 (1958).
101. J. Weidlein and V. Krieg, *J. Organometal. Chem.*, **11,** 9 (1968).
102. H. Schmidbaur, J. Weidlein, H. F. Klein, and K. Eiglmeier, *Chem. Ber.*, **101,** 2268 (1968).
103. H. C. Clark and A. L. Pichard, *J. Organometal. Chem.*, **8,** 427 (1967).
104. L. E. Levchuk, J. R. Sams, and F. Aubke, *Inorg. Chem.*, **11,** 43 (1972).
105. C. W. Hobbs and R. S. Tobias, *Inorg. Chem.*, **9,** 1037 (1970).
106. E. Amberger and R. Honigschmid-Grossich, *Chem. Ber.*, **98,** 3795 (1965).
107. D. E. Clegg and J. R. Hall, *J. Organometal. Chem.*, **22,** 491 (1970).
108. P. A. Bulliner, V. A. Maroni, and T. G. Spiro, *Inorg. Chem.*, **9,** 1887 (1970).
108a. K.-H. von Dahlen and J. Lorberth, *J. Organometal. Chem.*, **65,** 267 (1974).
108b. G. C. Stocco and R. S. Tobias, *J. Coord. Chem.*, **1,** 133 (1971).
109. J. R. Hall and G. A. Swile, *J. Organometal. Chem.*, **56,** 419 (1973).
110. J. Müller and K. Dehnicke, *J. Organometal. Chem.*, **10,** P1 (1967).
111. K. Dehnicke and D. Seybold, *J. Organometal. Chem.*, **11,** 227 (1968).
112. J. Müller and K. Dehnicke, *J. Organometal. Chem.*, **12,** 37 (1968).
113. J. S. Thayer and D. P. Strommen, *J. Organometal. Chem.*, **5,** 383 (1966).
114. H. Leimeister and K. Dehnicke, *J. Organometal. Chem.*, **31,** C3 (1971).
115. R. G. Goel and D. R. Ridley, *Inorg. Chem.*, **13,** 1252 (1974).
116. J. E. Förster, M. Vargas, and H. Müller, *J. Organometal. Chem.*, **59,** 97 (1973).
117. J. Relf, R. P. Cooney, and H. F. Henneike, *J. Organometal. Chem.*, **39,** 75 (1972).
118. F. Weller, I. L. Wilson, and K. Dehnicke, *J. Organometal. Chem.*, **30,** C1 (1971).
119. K. Dehnicke, *Angew. Chem.*, **79,** 942 (1967).
120. M. Wada and R. Okawara, *J. Organometal. Chem.*, **8,** 261 (1967).
121. H. Bürger and U. Goetze, *J. Organometal. Chem.*, **10,** 380 (1967).
121a. J. S. Thayer, *Inorg. Chem.*, **7,** 2599 (1968).
122. E. E. Aynsley, N. N. Greenwood, G. Hunter, and M. J. Sprague, *J. Chem. Soc.*, *A*, 1344 (1966).
123. W. Beck and E. Schuierer, *J. Organometal. Chem.*, **3,** 55 (1965).
124. M. R. Booth and S. G. Frankiss, *Spectrochim. Acta*, **26A,** 859 (1970).
125. J. R. Durig, Y. S. Li, and J. B. Turner, *Inorg. Chem.*, **13,** 1495 (1974).
126. N. Q. Dao and D. Breitinger. *Spectrochim. Acta*, **27A,** 905 (1971).
127. H. C. Clark, R. J. O'Brien, and A. L. Pickard, *J. Organometal. Chem.*, **4,** 43 (1965).
128. D. E. Clegg and J. R. Hall, *Spectrochim. Acta*, **23A,** 263 (1967).
129. R. J. H. Clark, A. G. Davies, and R. J. Puddenphatt, *J. Chem. Soc.*, **A,** 1828 (1968).
130. R. S. Tobias, M. J. Sprague, and G. E. Glass, *Inorg. Chem.*, **7,** 1714 (1968).
131. J. Rouviere, V. Tabacik, and G. Fleury, *Spectrochim. Acta*, **29A,** 229 (1973).
132. J. M. Brown, A. C. Chapman, R. Harper, D. J. Mowthorpe, A. G. Davies, and P. J. Smith, *J. Chem. Soc.*, Dalton 338 (1972).
133. H. Schmidbauer, J. Weidlein, and K. H. Mitschke, *Chem. Ber.*, **102,** 4136 (1969).
134. P. A. Bulliner and T. G. Spiro, *Inorg. Chem.*, **8,** 1023 (1969).
135. P. L. Goggin and L. A. Woodward, *Trans. Faraday Soc.*, **58,** 1495 (1962).
136. D. E. Clegg and R. J. Hall, *J. Organometal. Chem.*, **17,** 175 (1969).; *Spectrochim. Acta*, **21,** 357 (1965).
137. G. Mann, A. Haaland, and J. Weidlein, *Z. Anorg. Allg. Chem.*, **398,** 231 (1973).
138. T. Tanaka. *Bull. Chem. Soc. Jap.*, **33,** 446 (1960).
139. J. Lorberth and M. R. Kula, *Chem. Ber.*, **97,** 3444 (1964).
140. R. Okawara, *J. Am. Chem. Soc.*, **82,** 3287 (1960).
141. V. B. Ramos and R. S. Tobias, *Spectrochim. Acta.* **29A,** 953 (1973).

142. Y. Kawasaki, T. Tanaka, and R. Okawara, *Spectrochim. Acta*, **22,** 1571 (1966).
143. H. A. Meinema, A. Mackor, and J. G. Noltes, *J. Organometal. Chem.*, **37,** 285 (1972).
144. M. G. Miles, G. E. Glass, and R. S. Tobias, *J. Am. Chem. Soc.*, **88,** 5738 (1966).
145. M. M. McGrady and R. S. Tobias, *Inorg. Chem.*, **3,** 1161 (1964); *J. Am. Chem. Soc.*, **87,** 1909 (1965).
146. C. Z. Moore and W. H. Nelson, *Inorg. Chem.*, **8,** 138 (1969).
147. G. A. Miller and E. O. Schlemper, *Inorg. Chem.*, **12,** 677 (1973).
148. E. W. Abel, *J. Chem. Soc.*, 4406 (1960).
149. K. A. Hooton and A. L. Allred, *Inorg. Chem.*, **4,** 671 (1965).
150. P. G. Harrison and S. R. Stobart, *J. Organometal. Chem.*, **47,** 89 (1973).
151. D. Potts, H. D. Sharma, A. J. Carty, and A. Walker, *Inorg. Chem.*, **13,** 1205 (1974).
152. H. C. Clark and R. G. Goel, *Inorg. Chem.*, **4,** 1428 (1965).
153. U. Kunze, E. Lindner, and J. Koola, *J. Organometal. Chem.*, **38,** 1 (1972).
154. H. C. Clark and R. G. Goel, *J. Organometal. Chem.*, **7,** 263 (1967).
155. B. J. Hathaway and A. E. Underhill, *J. Chem. Soc.*, 3091 (1961).
156. D. F. Ball, T. Carter, D. C. McKean, and L. A. Woodward, *Spectrochim. Acta*, **20,** 1721 (1964).
157. D. F. Ball, P. L. Goggin, D. C. McKean, and L. A. Woodward, *Spectrochim. Acta*, **16,** 1358 (1960).
158. J. E. Griffiths, *J. Chem. Phys.*, **38,** 2879 (1963).
159. D. F. Van de Vondel and G. P. Van der Kelen, *Bull. Soc. Chim. Belg.*, **74,** 467 (1965).
160. H. Kimmel and C. R. Dillard, *Spectrochim. Acta*, **24A,** 909 (1968).
161. C. R. Dillard and L. May, *J. Mol. Spectrosc.*, **14,** 250 (1964).
162. E. Amberger, *Angew. Chem.*, **72,** 494 (1960).
163. J. R. Durig and C. W. Hawley, *J. Phys. Chem.*, **75,** 3993 (1971).
164. W. J. Lehmann, C. O. Wilson, and I. Shapiro, *J. Chem. Phys.*, **32,** 1088, 1786 (1960); **33,** 590 (1960); **34,** 476, 783 (1961).
165. E. G. Hoffmann and G. Schomburg, *Z. Elektrochem.*, **61,** 1101 (1957).
166. B. Fontal and T. G. Spiro, *Inorg. Chem.*, **10,** 9 (1971).
167. R. J. H. Clark, A. G. Davies, R. J. Puddenphatt, and W. McFarlane, *J. Am. Chem. Soc.*, **91,** 1334 (1969).
168. H. Bürger and U. Goetze, *Spectrochim. Acta*, **26A,** 685 (1970).
169. R. A. Burnham and S. R. Stobart, *J. Chem. Soc.*, Dalton, 1269 (1973).
170. D. J. Cardin, S. A. Keppie, and M. F. Lappert, *Inorg. Nucl. Chem. Lett.*, **4,** 365 (1968).
171. A. Terzis, T. C. Strekas, and T. G. Spiro, *Inorg. Chem.*, **13,** 1346 (1974).
172. G. F. Bradley and S. R. Stobart, *J. Chem. Soc.*, Dalton, 264 (1974).
173. N. A. D. Carey and H. C. Clark, *Chem. Commun.*, 292 (1967); *Inorg. Chem.*, **7,** 94 (1968).
173a. G. J. Kubas and T. G. Spiro, *Inorg. Chem.*, **12,** 1797 (1973).
173b. W. M. Scovell and T. G. Spiro, *Inorg. Chem.*, **13,** 304 (1974).
174. G. Davidson, *Organometal. Chem. Rev.*, A, 303 (1972).
175. M. J. Grogan and K. Nakamoto, *J. Am. Chem. Soc.*, **88,** 5454 (1966); **90,** 918 (1968).
176. J. P. Sorzano and J. P. Fackler, *J. Mol. Spectrosc.*, **22,** 80 (1967).
177. J. Hiraishi, *Spectrochim. Acta*, **25A,** 749 (1969).
178. J. Chatt, L. A. Duncanson, and R. G. Guy, *Nature*, **184,** 526 (1959).
179. J. A. J. Jarvis, B. T. Kilbourn, and P. G. Owston, *Acta Crystallogr.*, **B27,** 366 (1971).
180. N. Rösch, R. P. Messmer, and K. H. Johnson, *J. Am. Chem. Soc.*, **96,** 3855 (1974).
181. D. C. Andrews and G. Davidson, *J. Organometal. Chem.*, **35,** 161 (1972).
182. D. P. Powell, J. G. V. Scott, and N. Sheppard, *Spectrochim. Acta*, **28A,** 327 (1972).
183. M. A. Bennett, R. J. H. Clark, and D. L. Miller, *Inorg. Chem.*, **6,** 1647 (1967).
184. K. Fischer, K. Jonas, and G. Wilke, *Angew. Chem. Int. Ed.*, **12,** 565 (1973).

184a. A. L. Onderdelinden and A. van der Ent, *Inorg. Chim. Acta*, **6,** 420 (1972).
185. D. C. Andrews and G. Davidson, *J. Organometal. Chem.*, **55,** 383 (1973).
186. K. Shobatake and K. Nakamoto, *J. Am. Chem. Soc.*, **92,** 3339 (1970).
187. D. M. Adams and A. Squire, *J. Chem. Soc.*, A, 1808 (1970).
187a. G. Davidson and D. C. Andrews, *J. Chem. Soc.*, Dalton, 126 (1972).
188. I. S. Butler and G. G. Barna, *J. Raman Spectrosc.*, **1,** 141 (1973).
189. D. M. Adams and W. S. Fernando, *Inorg. Chim. Acta*, **7,** 277 (1973).
190. P. J. Hendra and D. B. Powell, *Spectrochim. Acta*, **17,** 909 (1961).
191. G. Davidson, *Inorg. Chim. Acta*, **3,** 596 (1969).
192. G. Davidson and D. A. Duce, *J. Organometal. Chem.*, **44,** 365 (1972).
193. M. J. Grogan and K. Nakamoto, *Inorg. Chim. Acta*, **1,** 228 (1967).
194. M. A. Bennett and J. D. Saxby, *Inorg. Chem.*, **7,** 321 (1968).
195. B. Dickens and W. N. Lipscomb, *J. Am. Chem. Soc.*, **83,** 4062 (1961); *J. Chem. Phys.*, **37,** 2084 (1962).
196. R. T. Bailey, E. R. Lippincott, and D. Steele, *J. Am. Chem. Soc.*, **87,** 5346 (1965).
197. M. A. Coles and F. A. Hart, *J. Organometal. Chem.*, **32,** 279 (1971).
198. R. Nast and H. Schindel, *Z. Anorg. Allg. Chem.*, **326,** 201 (1963).
199. I. A. Garbusova, V. T. Alexanjan, L. A. Leites, I. R. Golding, and A. M. Sladkov, *J. Organometal. Chem.*, **54,** 341 (1973).
200. K. Yasufuku and H. Yamazaki, *Bull. Chem. Soc. Jap.*, **45,** 2664 (1972).
201. J. Chatt, G. A. Rowe, and A. A. Williams, *Proc. Chem. Soc.*, 208 (1957); J. Chatt, R. Guy, and L. A. Duncanson, *J. Chem. Soc.*, 827 (1961).
202. Y. Iwashita, A. Ishikawa, and M. Kainosho, *Spectrochim. Acta*, **27A,** 271 (1971).
202a. Y. Iwashita, F. Tamura, and A. Nakamura, *Inorg. Chem.*, **8,** 1179 (1969).
202b. W. J. Bland, R. D. Kemmitt, and R. D. Moore, *J. Chem. Soc.*, Dalton, 1292 (1973).
202c. D. H. Payne, Z. A. Payne, R. Rohmer, and H. Frye, *Inorg. Chem.*, **12,** 2540 (1973).
203. H. P. Fritz, *Adv. Organometal. Chem.*, **1,** 239 (1964).
204. H. P. Fritz and L. Schäfer, *Chem. Ber.*, **97,** 1827 (1964).
205. E. O. Fischer and H. Fischer, *J. Organometal. Chem.*, **3,** 181 (1965).
206. E. O. Fischer and G. Stölzle, *Chem. Ber.*, **94,** 2187 (1961).
207. E. R. Lippincott, J. Xavier, and D. Steele, *J. Am. Chem. Soc.*, **83,** 2262 (1961).
208. G. B. McVicker and G. L. Morgan, *Spectrochim. Acta*, **26A,** 23 (1970).
209. K. A. Allan, B. G. Gowenlock, and W. E. Lindsell, *J. Organometal. Chem.*, **55,** 229 (1973).
210. A. Almenningen, A. Haaland, and T. Motzfeldt, *J. Organometal. Chem.*, **7,** 97 (1967).
211. P. G. Harrison and M. A. Healy, *J. Organometal. Chem.*, **51,** 153 (1973).
211a. G. Wilkinson, P. L. Pauson, and F. A. Cotton, *J. Am. Chem. Soc.*, **76,** 1970 (1954).
212. R. T. Bailey, *Spectrochim. Acta*, **27A,** 199 (1971).
213. J. S. Bodenheimer and W. Low, *Spectrochim. Acta*, **29A,** 1733 (1973).
214. B. V. Lokshin, V. T. Aleksanian, and E. B. Rusach, *J. Organometal. Chem.*, **86,** 253 (1975).
215. E. R. Lippincott and R. D. Nelson, *Spectrochim. Acta*, **10,** 307 (1958).
216. I. Pavlík and J. Klilorka, *Collect. Czech. Chem. Commun.*, **30,** 664 (1965).
217. D. Hartley and M. J. Ware, *J. Chem. Soc.*, A, 138 (1969).
218. F. Rocquet, L. Berreby, and J. P. Marsault, *Spectrochim. Acta*, **29A,** 1101 (1973).
219. I. J. Hyams, *Spectrochim. Acta*, **29A,** 839 (1973).
220. K. Nakamoto, C. Udovich, J. R. Ferraro, and A. Quattrochi, *Appl. Spectrosc.*, **24,** 606 (1970).
221. L. Schäfer, J. Brunvoll, and S. J. Cyvin, *J. Mol. Struct.*, **11,** 459 (1972).
222. D. M. Adams and W. S. Fernado, *J. Chem. Soc.*, Dalton, 2507 (1972).
223. J. Brunvoll, S. J. Cyvin, and L. Schäfer, *Chem. Phys. Lett.*, **13,** 286 (1972).
224. J. Brunvoll, S. J. Cyvin, and L. Schäfer, *J. Organometal. Chem.*, **27,** 107 (1971).
225. E. Maslowsky, Jr., and K. Nakamoto, *Inorg. Chem.*, **8,** 1108 (1969).

226. F. A. Cotton and T. J. Marks, *J. Am. Chem. Soc.*, **91,** 7281 (1969).
226a. G. Wilkinson and T. S. Piper, *J. Inorg. Nucl. Chem.*, **2,** 32 (1956).
227. E. Gallinella, B. Fortunato, and P. Mirone, *J. Mol. Spectrosc.*, **24,** 345 (1967).
228. A. Davison and P. E. Rakita, *Inorg. Chem.*, **9,** 289 (1970).
229. F. W. Siegert and H. J. de Liefde Meijer, *J. Organometal. Chem.*, **15,** 131 (1968).
230. F. W. Siegert and H. J. de Liefde Meijer, *J. Organometal. Chem.*, **20,** 141 (1969).
231. V. I. Kulishov, E. M. Brainina, N. G. Bokiy, and Yu. T. Struchkov, *Chem. Commun.* 475 (1970).
232. J. L. Calderon, F. A. Cotton, B. G. DeBoer, and J. Takats, *J. Am. Chem. Soc.*, **93,** 3592 (1971).
233. E. O. Fischer and Y. Hristidu, *Chem. Ber.*, **95,** 253 (1962).
234. J. L. Calderon, F. A. Cotton, B. G. DeBoer, and J. Takats, *J. Am. Chem. Soc.*, **93,** 3592 (1971).
235. J. L. Calderon, F. A. Cotton, and J. Takats, *J. Am. Chem. Soc.*, **93,** 3587 (1971).
236. A. E. Smith, *Inorg. Chem.*, **11,** 165 (1972).
237. R. A. Forder and K. Prout, *Acta Crystallogr.*, **B30,** 491 (1974).
238. F. N. Tebbe and G. W. Parshall, *J. Am. Chem. Soc.*, **93,** 3793 (1971).
239. H. P. Fritz and E. F. Paulus, *Z. Naturforsch.*, **18b,** 435 (1963).
240. D. J. Parker, *J. Chem. Soc., A*, 1382 (1970).
241. I. J. Hyams, R. T. Bailey, and E. R. Lippincott, *Spectrochim. Acta*, **23A,** 273 (1967).
242. D. M. Adams and A. Squire, *J. Organometal. Chem.*, **63,** 381 (1973).
243. B. V. Lokshin, Z. S. Klemmenkova, and Yu. V. Makarov, *Spectrochim. Acta*, **28A,** 2209 (1972)
244. J. R. Durig, R. B. King, L. W. Houk, and A. L. Marston, *J. Organometal, Chem.*, **16,** 425 (1969).
245. A. Davison, M. L. H. Green, and G. Wilkinson, *J. Chem. Soc.*, 3172 (1961).
246. B. F. Hallam and P. L. Pauson, *Chem. & Ind.*, **23,** 653 (1955).
247. T. S. Piper and G. Wilkinson, *Chem. & Ind.*, **23,** 1296 (1955); *J. Inorg. Nucl. Chem.*, **3,** 104 (1956).
248. M. J. Bennett, F. A. Cotton, A. Davison, J. W. Faller, S. J. Lippard, and S. M. Morehouse, *J. Am. Chem. Soc.*, **88,** 4371 (1966).
249. O. S. Mills, *Acta Crystallogr.*, **11,** 620 (1958); R. F. Bryan and P. T. Greene, *J. Chem. Soc., A*, 3064 (1970).
250. R. F. Bryan, P. T. Greene, M. J. Newlands, and D. S. Field, *J. Chem. Soc.*, **A,** 3068 (1970).
251. F. A. Cotton and G. Yagupsky, *Inorg. Chem.*, **6,** 15 (1967).
252. R. D. Fischer, A. Vogler, and K. Noack, *J. Organometal. Chem.*, **7,** 135 (1967).
253. A. R. Manning, *J. Chem. Soc., A*, 1319 (1968).
254. J. G. Bullitt, F. A. Cotton, and T. J. Marks, *Inorg. Chem.*, **11,** 671 (1972).
255. A. Alich, N. J. Nelson, D. Strope, and D. F. Shriver, *Inorg. Chem.*, **11,** 2976 (1972); N. J. Nelson, N. E. Kime, and D. F. Shriver, *J. Am. Chem. Soc.*, **91,** 5173 (1969).
256. N. E. Kim, N. J. Nelson, and D. F. Shriver, *Inorg. Chim. Acta*, **7,** 393 (1973).
257. J. S. Kristoff and D. F. Shriver, *Inorg. Chem.*, **13,** 499 (1974).
258. P. McArdle and A. R. Manning, *J. Chem. Soc., A*, 717 (1971).
259. F. C. Wilson and D. P. Shoemaker, *J. Chem. Phys.*, **27,** 809 (1957).
260. G. Davidson and E. M. Riley, *J. Organometal. Chem.*, **51,** 297 (1973).
261. R. D. Adams and F. A. Cotton, *Inorg. Chim. Acta*, **7,** 153 (1973).
262. A. Davison, W. McFarlane, E. Pratt, and G. Wilkinson, *J. Chem. Soc.*, 3653 (1962).
263. M. A. Neuman, Trinh-Toan, and L. F. Dahl, *J. Am. Chem. Soc.*, **94,** 3382 (1972).
264. A. Terzis and T. G. Spiro, *Chem. Commun.*, 1160 (1970).

265. E. Maslowsky. Jr., and K. Nakamoto, *Appl. Spectrosc.*, **25,** 187 (1971).
266. E. Samuel, R. Ferner, and M. Bigorgne, *Inorg. Chem.*, **12,** 881 (1973).
267. P. M. Druce, B. M. Kingston, M. F. Lappert, and R. C. Srivastava, *J. Chem. Soc.*, *A*, 2106 (1969).
268. P. M. Druce, B. M. Kingston, M. F. Lappert, R. C. Srivastava, M. J. Frazer, and W. E. Newton, *J. Chem. Soc.*, *A*, 2814 (1969).
269. R. J. Haines, R. S. Nyholm, and M. H. B. Stiddard, *J. Chem. Soc.*, *A*, 1606 (1966).
270. G. Paliani, R. Cataliotti, A. Poletti, and A. Foffani, *J. Chem. Soc.*, Dalton, 1741 (1972).
271. O. Crichton and A. J. Rest, *Chem. Commun.*, 407 (1973).
272. T. S. Piper and G. Wilkinson, *J. Inorg. Nucl. Chem.*, **2,** 38 (1956).
273. H. Brunner, *J. Organometal. Chem.*, **16,** 119 (1969).
274. R. C. Elder, F. A. Cotton, and R. A. Schunn, *J. Am. Chem. Soc.*, **89,** 3645 (1967).
275. M. J. D'Aniello, Jr., and E. K. Barefield, *J. Organometal. Chem.*, **76,** C50 (1974).
276. R. L. Cooper, M. L. H. Green, and J. T. Moelwyn-Hughes, *J. Organometal. Chem.*, **3,** 261 (1965).
277. M. P. Johnson and D. F. Shriver, *J. Am. Chem. Soc.*, **88,** 301 (1966).
278. M. Gerloch and R. Mason, *J. Chem. Soc.*, 296 (1965).
279. B. D. James, R. K. Nanda, and M. G. H. Wallbridge, *Inorg. Chem.*, **6,** 1979 (1967).
280. L. Banford and G. E. Coates, *J. Chem. Soc.*, 5591 (1964).
281. J. E. Bercaw and H. H. Brintzinger, *J. Am. Chem. Soc.*, **91,** 7301 (1969).
282. C. White, D. S. Gill, J. W. Kang, H. B. Lee, and P. M. Maitlis, *Chem. Commun.*, 734 (1971).
283. J. Müller and H. Dorner, *Angew. Chem. Int. Ed.*, **12,** 843 (1973).
284. J. L. Burmeister, E. A. Deardorff, A. Jensen, and V. H. Christiansen, *Inorg. Chem.*, **9,** 58 (1970).
285. A. Jensen, V. H. Christiansen, J. F. Hansen, T. Likowski, and J. L. Burmeister, *Acta Chem, Scand.*, **26,** 2898 (1972).
286. M. L. H. Green and W. E. Silverthorn, *Chem. Commun.*, 557 (1971).
287. A. E. Fenster and I. S. Butler, *Can. J. Chem.*, **50,** 598 (1972).
288. P. M. Treichel and G. P. Werber, *J. Am. Chem. Soc.*, **90,** 1753 (1968).
289. D. C. Andrews and G. Davidson, *J. Organometal. Chem.*, **36,** 349 (1972); **76,** 373 (1974).
290. H. P. Fritz, *Z. Naturforsch.*, **16b,** 415 (1961).
291. H. P. Fritz, and E. O. Fischer, *J. Organometal. Chem.*, **7,** 121 (1967).
291a. H. P. Fritz, W. Lüttke, H. Stammreich, and R. Forneris, *Spectrochim. Acta*, **17,** 1068 (1961).
292. S. J. Cyvin, J. Brunvoll, and L. S. Schäfer, *J. Chem. Phys.*, **54,** 1517 (1971).
293. H. P. Fritz and J. Manchot, *J. Organometal. Chem.*, **2,** 8 (1964).
294. G. Davidson and E. M. Riley, *Spectrochim. Acta*, **27A,** 1649 (1971).
295. R. Cataliotti, A. Poletti, and A. Santucci, *J. Mol. Struct.*, **5,** 215 (1970).
296. M. L. H. Green, L. Pratt, and G. Wilkinson, *J. Chem. Soc.*, 3753 (1959).
297. P. M. Treichel and R. L. Shubkin, *Inorg. Chem.*, **6,** 1328 (1967).
298. G. Winkhaus, L. Pratt, and G. Wilkinson, *J. Chem. Soc.*, 3807 (1961).
299. K. M. Sharma, S. K. Anand, R. K. Multani, and B. D. Jain, *J. Organometal. Chem.*, **23,** 173 (1970).
300. K. M. Sharma, S. K. Anand, R. K. Multani, and B. D. Jain, *J. Organometal. Chem.*, **25,** 447 (1970).

301. J. Goffart, J. Fuger, B. Gilbert, B. Kanellakopulos, and G. Duyckaerts, *Inorg. Nucl. Chem. Lett.*, **8,** 403 (1972).
302. E. Samuel and M. Bigorgne, *J. Organometal. Chem.*, **19,** 9 (1969); **30,** 235 (1971).
303. F. Kaufman and D. O. Cowan, *J. Am. Chem. Soc.*, **92,** 6198 (1970).
304. T. J. Katz and N. Acton, *J. Am. Chem. Soc.*, **94,** 3281 (1972).
305. A. Salzer and H. Werner, *Angew. Chem. Int. Ed.*, **11,** 930 (1972).

# Appendices

# APPENDIX I

## POINT GROUPS AND THEIR CHARACTER TABLES

The following are the character tables of the point groups that appear frequently in this book. The species (or the irreducible representations) of the point group are labeled according to the following rules: $A$ and $B$ denote nondegenerate species (one-dimensional representation). $A$ represents the symmetric species (character $=+1$) with respect to rotation about the principal axis (chosen as $z$ axis), whereas $B$ represents the antisymmetric species (character $=-1$) with respect to rotation about the principal axis; $E$ and $F$ denote doubly degenerate (two-dimensional representation) and triply degenerate species (three-dimensional representation), respectively. If two species in the same point group differ in the character of $C$ (other than the principal axis), they are distinguished by subscripts 1, 2, 3, . . . . If two species differ in the character of $\sigma$ (other than $\sigma_v$), they are distinguished by $'$ and $''$. If two species differ in the character of $i$, they are distinguished by subscripts $g$ and $u$. If these rules allow several different labels, $g$ and $u$ take precedence over 1, 2, 3, . . . , which in turn take precedence over $'$ and $''$. The labels of species of point groups $\mathbf{C}_{\infty v}$ and $\mathbf{D}_{\infty h}$ (linear molecules) are exceptional and are taken from the notation for the component of the electronic orbital angular momentum along the molecular axis.

| $\mathbf{C}_s$ | $I$ | $\sigma(xy)$ | | |
|---|---|---|---|---|
| $A'$ | $+1$ | $+1$ | $T_x, T_y, R_z$ | $\alpha_{xx}, \alpha_{yy}, \alpha_{zz}, \alpha_{xy}$ |
| $A''$ | $+1$ | $-1$ | $T_z, R_x, R_y$ | $\alpha_{yz}, \alpha_{xz}$ |

| $\mathbf{C}_2$ | $I$ | $C_2(z)$ | | |
|---|---|---|---|---|
| $A$ | $+1$ | $+1$ | $T_z, R_z$ | $\alpha_{xx}, \alpha_{yy}, \alpha_{zz}, \alpha_{xy}$ |
| $B$ | $+1$ | $-1$ | $T_x, T_y, R_x, R_y$ | $\alpha_{yz}, \alpha_{xz}$ |

| $\mathbf{C}_i$ | $I$ | $i$ | | |
|---|---|---|---|---|
| $A_g$ | $+1$ | $+1$ | $R_x, R_y, R_z$ | all components of $\alpha$ |
| $A_u$ | $+1$ | $-1$ | $T_x, T_y, T_z$ | |

| $\mathbf{C}_{2v}$ | $I$ | $C_2(z)$ | $\sigma_v(xz)$ | $\sigma_v(yz)$ | | |
|---|---|---|---|---|---|---|
| $A_1$ | +1 | +1 | +1 | +1 | $T_z$ | $\alpha_{xx}, \alpha_{yy}, \alpha_{zz}$ |
| $A_2$ | +1 | +1 | −1 | −1 | $R_z$ | $\alpha_{xy}$ |
| $B_1$ | +1 | −1 | +1 | −1 | $T_x, R_y$ | $\alpha_{xz}$ |
| $B_2$ | +1 | −1 | −1 | +1 | $T_y, R_x$ | $\alpha_{yz}$ |

| $\mathbf{C}_{3v}$ | $I$ | $2C_3(z)$ | $3\sigma_v$ | | |
|---|---|---|---|---|---|
| $A_1$ | +1 | +1 | +1 | $T_z$ | $\alpha_{xx}+\alpha_{yy}, \alpha_{zz}$ |
| $A_2$ | +1 | +1 | −1 | $R_z$ | |
| $E$ | +2 | −1 | 0 | $(T_x, T_y), (R_x, R_y)$ | $(\alpha_{xx}-\alpha_{yy}, \alpha_{xy}), (\alpha_{yz}, \alpha_{xz})$ |

| $\mathbf{C}_{4v}$ | $I$ | $2C_4(z)$ | $C_4^2 \equiv C_2''$ | $2\sigma_v$ | $2\sigma_d$ | | |
|---|---|---|---|---|---|---|---|
| $A_1$ | +1 | +1 | +1 | +1 | +1 | $T_z$ | $\alpha_{xx}+\alpha_{yy}, \alpha_{zz}$ |
| $A_2$ | +1 | +1 | +1 | −1 | −1 | $R_z$ | |
| $B_1$ | +1 | −1 | +1 | +1 | −1 | | $\alpha_{xx}-\alpha_{yy}$ |
| $B_2$ | +1 | −1 | +1 | −1 | +1 | | $\alpha_{xy}$ |
| $E$ | +2 | 0 | −2 | 0 | 0 | $(T_x, T_y), (R_x, R_y)$ | $(\alpha_{yz}, \alpha_{xz})$ |

$C_p^n$ (or $S_p^n$) denotes that the $C_p$ (or $S_p$) operation is carried out successively $n$ times.

| $\mathbf{C}_{\infty v}$ | $I$ | $2C_\infty^\phi$ | $2C_\infty^{2\phi}$ | $2C_\infty^{3\phi}$ | ... | $\infty\sigma_v$ | | |
|---|---|---|---|---|---|---|---|---|
| $\Sigma^+$ | +1 | +1 | +1 | +1 | ... | +1 | $T_z$ | $\alpha_{xx}+\alpha_{yy}, \alpha_{zz}$ |
| $\Sigma^-$ | +1 | +1 | +1 | +1 | ... | −1 | $R_z$ | |
| $\Pi$ | +2 | $2\cos\phi$ | $2\cos 2\phi$ | $2\cos 3\phi$ | ... | 0 | $(T_x, T_y), (R_x, R_y)$ | $(\alpha_{yz}, \alpha_{xz})$ |
| $\Delta$ | +2 | $2\cos 2\phi$ | $2\cos 2\cdot 2\phi$ | $2\cos 3\cdot 2\phi$ | ... | 0 | | $(\alpha_{xx}-\alpha_{yy}, \alpha_{xy})$ |
| $\Phi$ | +2 | $2\cos 3\phi$ | $2\cos 2\cdot 3\phi$ | $2\cos 3\cdot 3\phi$ | ... | 0 | | |
| ... | ... | ... | ... | ... | ... | ... | | |

| $\mathbf{C}_{2h}$ | $I$ | $C_2(z)$ | $\sigma_h(xy)$ | $i$ | | |
|---|---|---|---|---|---|---|
| $A_g$ | +1 | +1 | +1 | +1 | $R_z$ | $\alpha_{xx}, \alpha_{yy}, \alpha_{zz}, \alpha_{xy}$ |
| $A_u$ | +1 | +1 | −1 | −1 | $T_z$ | |
| $B_g$ | +1 | −1 | −1 | +1 | $R_x, R_y$ | $\alpha_{yz}, \alpha_{xz}$ |
| $B_u$ | +1 | −1 | +1 | −1 | $T_x, T_y$ | |

| $\mathbf{D}_3$ | $I$ | $2C_3(z)$ | $3C_2$ | | |
|---|---|---|---|---|---|
| $A_1$ | +1 | +1 | +1 | | $\alpha_{xx}+\alpha_{yy}, \alpha_{zz}$ |
| $A_2$ | +1 | +1 | −1 | $T_z, R_z$ | |
| $E$ | +2 | −1 | 0 | $(T_x, T_y), (R_x, R_y)$ | $(\alpha_{xx}-\alpha_{yy}, \alpha_{xy}), (\alpha_{yz}, \alpha_{xz})$ |

| $\mathbf{D}_{2d} \equiv \mathbf{V}_d$ | $I$ | $2S_4(z)$ | $S_4^2 \equiv C_2''$ | $2C_2$ | $2\sigma_d$ | | |
|---|---|---|---|---|---|---|---|
| $A_1$ | +1 | +1 | +1 | +1 | +1 | | $\alpha_{xx}+\alpha_{yy}, \alpha_{zz}$ |
| $A_2$ | +1 | +1 | +1 | −1 | −1 | $R_z$ | |
| $B_1$ | +1 | −1 | +1 | +1 | −1 | | $\alpha_{xx}-\alpha_{yy}$ |
| $B_2$ | +1 | −1 | +1 | −1 | +1 | $T_z$ | $\alpha_{xy}$ |
| $E$ | +2 | 0 | −2 | 0 | 0 | $(T_x, T_y), (R_x, R_y)$ | $(\alpha_{yz}, \alpha_{xz})$ |

| $\mathbf{D}_{3d}$ | $I$ | $2S_6(z)$ | $2S_6^2 \equiv 2C_3$ | $S_6^3 \equiv S_2 \equiv i$ | $3C_2$ | $3\sigma_d$ | | |
|---|---|---|---|---|---|---|---|---|
| $A_{1g}$ | +1 | +1 | +1 | +1 | +1 | +1 | | $\alpha_{xx}+\alpha_{yy}, \alpha_{zz}$ |
| $A_{1u}$ | +1 | −1 | +1 | −1 | +1 | −1 | | |
| $A_{2g}$ | +1 | +1 | +1 | +1 | −1 | −1 | $R_z$ | |
| $A_{2u}$ | +1 | −1 | +1 | −1 | −1 | +1 | $T_z$ | |
| $E_g$ | +2 | −1 | −1 | +2 | 0 | 0 | $(R_x, R_y)$ | $(\alpha_{xx}-\alpha_{yy}, \alpha_{xy}), (\alpha_{yz}, \alpha_{xz})$ |
| $E_u$ | +2 | +1 | −1 | −2 | 0 | 0 | $(T_x, T_y)$ | |

| $\mathbf{D}_{4d}$ | $I$ | $2S_8(z)$ | $2S_8^2 \equiv 2C_4$ | $2S_8^3$ | $S_8^4 \equiv C_2''$ | $4C_2$ | $4\sigma_d$ | | |
|---|---|---|---|---|---|---|---|---|---|
| $A_1$ | +1 | +1 | +1 | +1 | +1 | +1 | +1 | | $\alpha_{xx}+\alpha_{yy}, \alpha_{zz}$ |
| $A_2$ | +1 | +1 | +1 | +1 | +1 | −1 | −1 | $R_z$ | |
| $B_1$ | +1 | −1 | +1 | −1 | +1 | +1 | −1 | | |
| $B_2$ | +1 | −1 | +1 | −1 | +1 | −1 | +1 | $T_z$ | |
| $E_1$ | +2 | $+\sqrt{2}$ | 0 | $-\sqrt{2}$ | −2 | 0 | 0 | $(T_x, T_y)$ | |
| $E_2$ | +2 | 0 | −2 | 0 | +2 | 0 | 0 | | $(\alpha_{xx}-\alpha_{yy}, \alpha_{xy})$ |
| $E_3$ | +2 | $-\sqrt{2}$ | 0 | $+\sqrt{2}$ | −2 | 0 | 0 | $(R_x, R_y)$ | $(\alpha_{yz}, \alpha_{xz})$ |

| $\mathbf{D}_{2h} \equiv \mathbf{V}_h$ | $I$ | $\sigma(xy)$ | $\sigma(xz)$ | $\sigma(yz)$ | $i$ | $C_2(z)$ | $C_2(y)$ | $C_2(x)$ | | |
|---|---|---|---|---|---|---|---|---|---|---|
| $A_g$ | +1 | +1 | +1 | +1 | +1 | +1 | +1 | +1 | | $\alpha_{xx}, \alpha_{yy}, \alpha_{zz}$ |
| $A_u$ | +1 | −1 | −1 | −1 | −1 | +1 | +1 | +1 | | |
| $B_{1g}$ | +1 | +1 | −1 | −1 | +1 | +1 | −1 | −1 | $R_z$ | $\alpha_{xy}$ |
| $B_{1u}$ | +1 | −1 | +1 | +1 | −1 | +1 | −1 | −1 | $T_z$ | |
| $B_{2g}$ | +1 | −1 | +1 | −1 | +1 | −1 | +1 | −1 | $R_y$ | $\alpha_{xz}$ |
| $B_{2u}$ | +1 | +1 | −1 | +1 | −1 | −1 | +1 | −1 | $T_y$ | |
| $B_{3g}$ | +1 | −1 | −1 | +1 | +1 | −1 | −1 | +1 | $R_x$ | $\alpha_{yz}$ |
| $B_{3u}$ | +1 | +1 | +1 | −1 | −1 | −1 | −1 | +1 | $T_x$ | |

| $\mathbf{D}_{3h}$ | $I$ | $2C_3(z)$ | $3C_2$ | $\sigma_h$ | $2S_3$ | $3\sigma_v$ | | |
|---|---|---|---|---|---|---|---|---|
| $A_1'$ | +1 | +1 | +1 | +1 | +1 | +1 | | $\alpha_{xx}+\alpha_{yy}, \alpha_{zz}$ |
| $A_1''$ | +1 | +1 | +1 | −1 | −1 | −1 | | |
| $A_2'$ | +1 | +1 | −1 | +1 | +1 | −1 | $R_z$ | |
| $A_2''$ | +1 | +1 | −1 | −1 | −1 | +1 | $T_z$ | |
| $E'$ | +2 | −1 | 0 | +2 | −1 | 0 | $(T_x, T_y)$ | $(\alpha_{xx}-\alpha_{yy}, \alpha_{xy})$ |
| $E''$ | +2 | −1 | 0 | −2 | +1 | 0 | $(R_x, R_y)$ | $(\alpha_{yz}, \alpha_{xz})$ |

| $\mathbf{D}_{4h}$ | $I$ | $2C_4(z)$ | $C_4^2 \equiv C_2''$ | $2C_2$ | $2C_2'$ | $\sigma_h$ | $2\sigma_v$ | $2\sigma_d$ | $2S_4$ | $S_2 \equiv i$ | | |
|---|---|---|---|---|---|---|---|---|---|---|---|---|
| $A_{1g}$ | +1 | +1 | +1 | +1 | +1 | +1 | +1 | +1 | +1 | +1 | | $\alpha_{xx}+\alpha_{yy}, \alpha_{z}$ |
| $A_{1u}$ | +1 | +1 | +1 | +1 | +1 | −1 | −1 | −1 | −1 | −1 | | |
| $A_{2g}$ | +1 | +1 | +1 | −1 | −1 | +1 | −1 | −1 | +1 | +1 | $R_z$ | |
| $A_{2u}$ | +1 | +1 | +1 | −1 | −1 | −1 | +1 | +1 | −1 | −1 | $T_z$ | |
| $B_{1g}$ | +1 | −1 | +1 | +1 | −1 | +1 | +1 | −1 | −1 | +1 | | $\alpha_{xx}-\alpha_{yy}$ |
| $B_{1u}$ | +1 | −1 | +1 | +1 | −1 | −1 | −1 | +1 | +1 | −1 | | |
| $B_{2g}$ | +1 | −1 | +1 | −1 | +1 | +1 | −1 | +1 | −1 | +1 | | $\alpha_{xy}$ |
| $B_{2u}$ | +1 | −1 | +1 | −1 | +1 | −1 | +1 | −1 | +1 | −1 | | |
| $E_g$ | +2 | 0 | −2 | 0 | 0 | −2 | 0 | 0 | 0 | +2 | $(R_x, R_y)$ | $(\alpha_{yz}, \alpha_{xz})$ |
| $E_u$ | +2 | 0 | −2 | 0 | 0 | +2 | 0 | 0 | 0 | −2 | $(T_x, T_y)$ | |

| $\mathbf{D}_{5h}$ | $I$ | $2C_5(z)$ | $2C_5^2$ | $\sigma_h$ | $5C_2$ | $5\sigma_v$ | $2S_5$ | $2S_5^3$ | | |
|---|---|---|---|---|---|---|---|---|---|---|
| $A_1'$ | +1 | +1 | +1 | +1 | +1 | +1 | +1 | +1 | | $\alpha_{xx}+\alpha_{yy}, \alpha_{zz}$ |
| $A_1''$ | +1 | +1 | +1 | −1 | +1 | −1 | −1 | −1 | | |
| $A_2'$ | +1 | +1 | +1 | +1 | −1 | −1 | +1 | +1 | $R_z$ | |
| $A_2''$ | +1 | +1 | +1 | −1 | −1 | +1 | −1 | −1 | $T_z$ | |
| $E_1'$ | +2 | 2 cos 72° | 2 cos 144° | +2 | 0 | 0 | +2 cos 72° | +2 cos 144° | $(T_x, T_y)$ | |
| $E_1''$ | +2 | 2 cos 72° | 2 cos 144° | −2 | 0 | 0 | −2 cos 72° | −2 cos 144° | $(R_x, R_y)$ | $(\alpha_{yz}, \alpha_{xz})$ |
| $E_2'$ | +2 | 2 cos 144° | 2 cos 72° | +2 | 0 | 0 | +2 cos 144° | +2 cos 72° | | $(\alpha_{xx}-\alpha_{yy}, \alpha_{xy})$ |
| $E_2''$ | +2 | 2 cos 144° | 2 cos 72° | −2 | 0 | 0 | −2 cos 144° | −2 cos 72° | | |

| $\mathbf{D}_{6h}$ | $I$ | $2C_6(z)$ | $2C_6^2 \equiv 2C_3$ | $C_6^3 \equiv C_2''$ | $3C_2$ | $3C_2'$ | $\sigma_h$ | $3\sigma_v$ | $3\sigma_d$ | $2S_6$ | $2S_3$ | $S_6^3 \equiv S_2 \equiv i$ | | |
|---|---|---|---|---|---|---|---|---|---|---|---|---|---|---|
| $A_{1g}$ | +1 | +1 | +1 | +1 | +1 | +1 | +1 | +1 | +1 | +1 | +1 | +1 | | $\alpha_{xx}+\alpha_{yy}, \alpha_{zz}$ |
| $A_{1u}$ | +1 | +1 | +1 | +1 | +1 | +1 | −1 | −1 | −1 | −1 | −1 | −1 | | |
| $A_{2g}$ | +1 | +1 | +1 | +1 | −1 | −1 | +1 | −1 | −1 | +1 | +1 | +1 | $R_z$ | |
| $A_{2u}$ | +1 | +1 | +1 | +1 | −1 | −1 | −1 | +1 | +1 | −1 | −1 | −1 | $T_z$ | |
| $B_{1g}$ | +1 | −1 | +1 | −1 | +1 | −1 | −1 | −1 | +1 | +1 | −1 | +1 | | |
| $B_{1u}$ | +1 | −1 | +1 | −1 | +1 | −1 | +1 | +1 | −1 | −1 | +1 | −1 | | |
| $B_{2g}$ | +1 | −1 | +1 | −1 | −1 | +1 | −1 | +1 | −1 | +1 | −1 | +1 | | |
| $B_{2u}$ | +1 | −1 | +1 | −1 | −1 | +1 | +1 | −1 | +1 | −1 | +1 | −1 | | |
| $E_{1g}$ | +2 | +1 | −1 | −2 | 0 | 0 | −2 | 0 | 0 | −1 | +1 | +2 | $(R_x, R_y)$ | $(\alpha_{yz}, \alpha_{xz})$ |
| $E_{1u}$ | +2 | +1 | −1 | −2 | 0 | 0 | +2 | 0 | 0 | +1 | −1 | −2 | $(T_x, T_y)$ | |
| $E_{2g}$ | +2 | −1 | −1 | +2 | 0 | 0 | +2 | 0 | 0 | −1 | −1 | +2 | | $(\alpha_{xx}-\alpha_{yy}, \alpha_{xy})$ |
| $E_{2u}$ | +2 | −1 | −1 | +2 | 0 | 0 | −2 | 0 | 0 | +1 | +1 | −2 | | |

| $\mathbf{D}_{\infty h}$ | $I$ | $2C_\infty^{\phi}$ | $2C_\infty^{2\phi}$ | $2C_\infty^{3\phi}$ | ... | $\sigma_h$ | $\infty C_2$ | $\infty\sigma_v$ | $2S_\infty^{\phi}$ | $2S_\infty^{2\phi}$ | ... | $S_2 \equiv i$ | | |
|---|---|---|---|---|---|---|---|---|---|---|---|---|---|---|
| $\Sigma_g^+$ | +1 | +1 | +1 | +1 | ... | +1 | +1 | +1 | +1 | +1 | ... | +1 | | $\alpha_{xx}+\alpha_{yy}, \alpha_{zz}$ |
| $\Sigma_u^+$ | +1 | +1 | +1 | +1 | ... | −1 | −1 | +1 | −1 | −1 | ... | −1 | $T_z$ | |
| $\Sigma_g^-$ | +1 | +1 | +1 | +1 | ... | +1 | −1 | −1 | +1 | +1 | ... | +1 | $R_z$ | |
| $\Sigma_u^-$ | +1 | +1 | +1 | +1 | ... | −1 | +1 | −1 | −1 | −1 | ... | −1 | | |
| $\Pi_g$ | +2 | $2\cos\phi$ | $2\cos 2\phi$ | $2\cos 3\phi$ | ... | −2 | 0 | 0 | $-2\cos\phi$ | $-2\cos 2\phi$ | ... | +2 | $(R_x, R_y)$ | $(\alpha_{yz}, \alpha_{xz})$ |
| $\Pi_u$ | +2 | $2\cos\phi$ | $2\cos 2\phi$ | $2\cos 3\phi$ | ... | +2 | 0 | 0 | $+2\cos\phi$ | $+2\cos 2\phi$ | ... | −2 | $(T_x, T_y)$ | |
| $\Delta_g$ | +2 | $2\cos 2\phi$ | $2\cos 4\phi$ | $2\cos 6\phi$ | ... | +2 | 0 | 0 | $+2\cos 2\phi$ | $+2\cos 4\phi$ | ... | +2 | | $(\alpha_{xx}-\alpha_{yy}, \alpha_{xy})$ |
| $\Delta_u$ | +2 | $2\cos 2\phi$ | $2\cos 4\phi$ | $2\cos 6\phi$ | ... | −2 | 0 | 0 | $-2\cos 2\phi$ | $-2\cos 4\phi$ | ... | −2 | | |
| $\Phi_g$ | +2 | $2\cos 3\phi$ | $2\cos 6\phi$ | $2\cos 9\phi$ | ... | −2 | 0 | 0 | $-2\cos 3\phi$ | $-2\cos 4\phi$ | ... | +2 | | |
| $\Phi_u$ | +2 | $2\cos 3\phi$ | $2\cos 6\phi$ | $2\cos 9\phi$ | ... | +2 | 0 | 0 | $+2\cos 3\phi$ | $+2\cos 4\phi$ | ... | −2 | | |
| ... | ... | ... | ... | ... | ... | ... | ... | ... | ... | ... | ... | ... | | |

| $\mathbf{T}_d$ | $I$ | $8C_3$ | $6\sigma_d$ | $6S_4$ | $3S_4^2 \equiv 3C_2$ | | |
|---|---|---|---|---|---|---|---|
| $A_1$ | +1 | +1 | +1 | +1 | +1 | | $\alpha_{xx}+\alpha_{yy}+\alpha_{zz}$ |
| $A_2$ | +1 | +1 | −1 | −1 | +1 | | |
| $E$ | +2 | −1 | 0 | 0 | +2 | | $(\alpha_{xx}+\alpha_{yy}-2\alpha_{zz}, \alpha_{xx}-\alpha_{yy})$ |
| $F_1$ | +3 | 0 | −1 | +1 | −1 | $(R_x, R_y, R_z)$ | |
| $F_2$ | +3 | 0 | +1 | −1 | −1 | $(T_x, T_y, T_z)$ | $(\alpha_{xy}, \alpha_{yz}, \alpha_{xz})$ |

| $\mathbf{O}_h$ | $I$ | $8C_3$ | $6C_2$ | $6C_4$ | $3C_4^2 \equiv 3C_2''$ | $S_2 \equiv i$ | $6S_4$ | $8S_6$ | $3\sigma_h$ | $6\sigma_d$ | | |
|---|---|---|---|---|---|---|---|---|---|---|---|---|
| $A_{1g}$ | +1 | +1 | +1 | +1 | +1 | +1 | +1 | +1 | +1 | +1 | | $\alpha_{xx}+\alpha_{yy}+\alpha_{zz}$ |
| $A_{1u}$ | +1 | +1 | +1 | +1 | +1 | −1 | −1 | −1 | −1 | −1 | | |
| $A_{2g}$ | +1 | +1 | −1 | −1 | +1 | +1 | −1 | +1 | +1 | −1 | | |
| $A_{2u}$ | +1 | +1 | −1 | −1 | +1 | −1 | +1 | −1 | −1 | +1 | | |
| $E_g$ | +2 | −1 | 0 | 0 | +2 | +2 | 0 | −1 | +2 | 0 | | $(\alpha_{xx}+\alpha_{yy}-2\alpha_{zz}, \alpha_{xx}-\alpha_{yy})$ |
| $E_u$ | +2 | −1 | 0 | 0 | +2 | −2 | 0 | +1 | −2 | 0 | | |
| $F_{1g}$ | +3 | 0 | −1 | +1 | −1 | +3 | +1 | 0 | −1 | −1 | $(R_x, R_y, R_z)$ | |
| $F_{1u}$ | +3 | 0 | −1 | +1 | −1 | −3 | −1 | 0 | +1 | +1 | $(T_x, T_y, T_z)$ | |
| $F_{2g}$ | +3 | 0 | +1 | −1 | −1 | +3 | −1 | 0 | −1 | +1 | | $(\alpha_{xy}, \alpha_{yz}, \alpha_{xz})$ |
| $F_{2u}$ | +3 | 0 | +1 | −1 | −1 | −3 | +1 | 0 | +1 | −1 | | |

# APPENDIX II

## GENERAL FORMULAS FOR CALCULATING THE NUMBER OF NORMAL VIBRATIONS IN EACH SPECIES

These tables are quoted from G. Herzberg, *Molecular Spectra and Molecular Structure*, Vol. II (Ref. I-1).

### A. Point Groups Including Only Nondegenerate Vibrations

| Point Group | Total Number of Atoms | Species | Number of Vibrations[a] |
|---|---|---|---|
| $\mathbf{C}_2$ | $2m + m_0$ | $A$<br>$B$ | $3m + m_0 - 2$<br>$3m + 2m_0 - 4$ |
| $\mathbf{C}_s$ | $2m + m_0$ | $A'$<br>$A''$ | $3m + 2m_0 - 3$<br>$3m + m_0 - 3$ |
| $\mathbf{C}_i \equiv \mathbf{S}_2$ | $2m + m_0$ | $A_g$<br>$A_u$ | $3m - 3$<br>$3m + 3m_0 - 3$ |

| | | | |
|---|---|---|---|
| $\mathbf{C}_{2v}$ | $4m+2m_{xz}$ $+2m_{yz}+m_0$ | $A_1$ | $3m+2m_{xz}+2m_{yz}+m_0-1$ |
| | | $A_2$ | $3m+m_{xz}+m_{yz}-1$ |
| | | $B_1$ | $3m+2m_{xz}+m_{yz}+m_0-2$ |
| | | $B_2$ | $3m+m_{xz}+2m_{yz}+m_0-2$ |
| $\mathbf{C}_{2h}$ | $4m+2m_h$ $+2m_2+m_0$ | $A_g$ | $3m+2m_h+m_2-1$ |
| | | $A_u$ | $3m+m_h+m_2+m_0-1$ |
| | | $B_g$ | $3m+m_h+2m_2-2$ |
| | | $B_u$ | $3m+2m_h+2m_2+2m_0-2$ |
| $\mathbf{D}_{2h}\equiv\mathbf{V}_h$ | $8m+4m_{xy}+4m_{xz}$ $+4m_{yz}+2m_{2x}$ $+2m_{2y}+2m_{2z}+m_0$ | $A_g$ | $3m+2m_{xy}+2m_{xz}+2m_{yz}+m_{2x}+m_{2y}+m_{2z}$ |
| | | $A_u$ | $3m+m_{xy}+m_{xz}+m_{yz}$ |
| | | $B_{1g}$ | $3m+2m_{xy}+m_{xz}+m_{yz}+m_{2x}+m_{2y}-1$ |
| | | $B_{1u}$ | $3m+m_{xy}+2m_{xz}+2m_{yz}+m_{2x}+m_{2y}+m_{2z}+m_0-1$ |
| | | $B_{2g}$ | $3m+m_{xy}+2m_{xz}+m_{yz}+m_{2x}+m_{2z}-1$ |
| | | $B_{2u}$ | $3m+2m_{xy}+m_{xz}+2m_{yz}+m_{2x}+m_{2y}+m_{2z}+m_0-1$ |
| | | $B_{3g}$ | $3m+m_{xy}+m_{xz}+2m_{yz}+m_{2y}+m_{2z}-1$ |
| | | $B_{3u}$ | $3m+2m_{xy}+2m_{xz}+m_{yz}+m_{2x}+m_{2y}+m_{2z}+m_0-1$ |

[a] Note that $m$ is always the number of sets of equivalent nuclei not on any element of symmetry; $m_0$ is the number of nuclei lying on all symmetry elements present; $m_{xy}$, $m_{xz}$, $m_{yz}$ are the numbers of sets of nuclei lying on the $xy$, $xz$, $yz$ plane, respectively, but not on any axes going through these planes; $m_2$ is the number of sets of nuclei on a twofold axis but not at the point of intersection with another element of symmetry; $m_{2x}$, $m_{2y}$, $m_{2z}$ are the numbers of sets of nuclei lying on the $x$, $y$, $z$ axis if they are twofold axes, but not on all of them; $m_h$ is the number of sets of nuclei on a plane $\sigma_h$ but not on the axis perpendicular to this plane.

## B. Point Groups Including Degenerate Vibrations

| Point Group | Total Number of Atoms | Species | Number of Vibrations[a] |
|---|---|---|---|
| $\mathbf{D}_3$ | $6m+3m_2+2m_3+m_0$ | $A_1$ | $3m+m_2+m_3$ |
| | | $A_2$ | $3m+2m_2+m_3+m_0-2$ |
| | | $E$ | $6m+3m_2+2m_3+m_0-2$ |
| $\mathbf{C}_{3v}$ | $6m+3m_v+m_0$ | $A_1$ | $3m+2m_v+m_0-1$ |
| | | $A_2$ | $3m+m_v-1$ |
| | | $E$ | $6m+3m_v+m_0-2$ |
| $\mathbf{C}_{4v}$ | $8m+4m_v+4m_d+m_0$ | $A_1$ | $3m+2m_v+2m_d+m_0-1$ |
| | | $A_2$ | $3m+m_v+m_d-1$ |
| | | $B_1$ | $3m+2m_v+m_d$ |
| | | $B_2$ | $3m+m_v+2m_d$ |
| | | $E$ | $6m+3m_v+3m_d+m_0-2$ |
| $\mathbf{C}_{\infty v}$ | $m_0$ | $\Sigma^+$ | $m_0-1$ |
| | | $\Sigma^-$ | 0 |
| | | $\Pi$ | $m_0-2$ |
| | | $\Delta, \Phi, \ldots$ | 0 |
| $\mathbf{D}_{2d} \equiv \mathbf{V}_d$ | $8m+4m_d+4m_2+2m_4+m_0$ | $A_1$ | $3m+2m_d+m_2+m_4$ |
| | | $A_2$ | $3m+m_d+2m_2-1$ |
| | | $B_1$ | $3m+m_d+m_2$ |
| | | $B_2$ | $3m+2m_d+2m_2+m_4+m_0-1$ |
| | | $E$ | $6m+3m_d+3m_2+2m_4+m_0-2$ |

| | | | |
|---|---|---|---|
| $\mathbf{D}_{3d}$ | $12m+6m_d+6m_2+2m_6+m_0$ | $A_{1g}$ | $3m+2m_d+m_2+m_6$ |
| | | $A_{1u}$ | $3m+m_4+m_2$ |
| | | $A_{2g}$ | $3m+m_d+2m_2-1$ |
| | | $A_{2u}$ | $3m+2m_d+2m_2+m_6+m_0-1$ |
| | | $E_g$ | $6m+3m_d+3m_2+m_6-1$ |
| | | $E_u$ | $6m+3m_d+3m_2+m_6+m_0-1$ |
| $\mathbf{D}_{4d}$ | $16m+8m_d+8m_2+2m_8+m_0$ | $A_1$ | $3m+2m_d+m_2+m_8$ |
| | | $A_2$ | $3m+m_d+2m_2-1$ |
| | | $B_1$ | $3m+m_d+m_2$ |
| | | $B_2$ | $3m+2m_d+2m_2+m_8+m_0-1$ |
| | | $E_1$ | $6m+3m_d+3m_2+m_8+m_0-1$ |
| | | $E_2$ | $6m+3m_d+3m_2$ |
| | | $E_3$ | $6m+3m_d+3m_2+m_8-1$ |
| $\mathbf{D}_{3h}$ | $12m+6m_v+6m_h+3m_2+2m_3+m_0$ | $A_1'$ | $3m+2m_v+2m_h+m_2+m_3$ |
| | | $A_1''$ | $3m+m_v+m_h$ |
| | | $A_2'$ | $3m+m_v+2m_h+m_2-1$ |
| | | $A_2''$ | $3m+2m_v+m_h+m_2+m_3+m_0-1$ |
| | | $E'$ | $6m+3m_v+4m_h+2m_2+m_3+m_0-1$ |
| | | $E''$ | $6m+3m_v+2m_h+m_2+m_3-1$ |
| $\mathbf{D}_{4h}$ | $16m+8m_v+8m_d+8m_h+4m_2+4m_2'+2m_4+m_0$ | $A_{1g}$ | $3m+2m_v+2m_d+2m_h+m_2+m_2'+m_4$ |
| | | $A_{1u}$ | $3m+m_v+m_d+m_h$ |
| | | $A_{2g}$ | $3m+m_v+m_d+2m_h+m_2+m_2'-1$ |
| | | $A_{2u}$ | $3m+2m_v+2m_d+m_h+m_2+m_2'+m_4+m_0-1$ |
| | | $B_{1g}$ | $3m+2m_v+m_d+2m_h+m_2+m_2'$ |
| | | $B_{1u}$ | $3m+m_v+2m_d+m_h+m_2'$ |
| | | $B_{2g}$ | $3m+m_v+2m_d+2m_h+m_2+m_2'$ |
| | | $B_{2u}$ | $3m+2m_v+m_d+m_h+m_2$ |
| | | $E_g$ | $6m+3m_v+3m_d+2m_h+m_2+m_2'+m_4-1$ |
| | | $E_u$ | $6m+3m_v+3m_d+4m_h+2m_2+2m_2'+m_4+m_0-1$ |

| | | | |
|---|---|---|---|
| $\mathbf{D}_{5h}$ | $20m+10m_v+10m_h+5m_2+2m_5+m_0$ | $A_1'$ | $3m+2m_v+2m_h+m_2+m_5$ |
| | | $A_1''$ | $3m+m_v+m_h$ |
| | | $A_2'$ | $3m+m_v+2m_h+m_2-1$ |
| | | $A_2''$ | $3m+2m_v+m_h+m_2+m_5+m_0-1$ |
| | | $E_1'$ | $6m+3m_v+4m_h+2m_2+m_5+m_0-1$ |
| | | $E_1''$ | $6m+3m_v+2m_h+m_2+m_5-1$ |
| | | $E_2'$ | $6m+3m_v+4m_h+2m_2$ |
| | | $E_2''$ | $6m+3m_v+2m_h+m_2$ |
| $\mathbf{D}_{6h}$ | $24m+12m_v+12m_d+12m_h+6m_2+6m_2'+2m_6+m_0$ | $A_{1g}$ | $3m+2m_v+2m_d+2m_h+m_2+m_2'+m_6$ |
| | | $A_{1u}$ | $3m+m_v+m_d+m_h$ |
| | | $A_{2g}$ | $3m+m_v+m_d+2m_h+m_2+m_2'-1$ |
| | | $A_{2u}$ | $3m+2m_v+2m_d+m_h+m_2+m_2'+m_6+m_0-1$ |
| | | $B_{1g}$ | $3m+m_v+2m_d+m_h+m_2'$ |
| | | $B_{1u}$ | $3m+2m_v+m_d+2m_h+m_2+m_2'$ |
| | | $B_{2g}$ | $3m+2m_v+m_d+m_h+m_2$ |
| | | $B_{2u}$ | $3m+m_v+2m_d+2m_h+m_2+m_2'$ |
| | | $E_{1g}$ | $6m+3m_v+3m_d+2m_h+m_2+m_2'+m_6-1$ |
| | | $E_{1u}$ | $6m+3m_v+3m_d+4m_h+2m_2+2m_2'+m_6+m_0-1$ |
| | | $E_{2g}$ | $6m+3m_v+3m_d+4m_h+2m_2+2m_2'$ |
| | | $E_{2u}$ | $6m+3m_v+3m_d+2m_h+m_2+m_2'$ |
| $\mathbf{D}_{\infty h}$ | $2m_\infty+m_0$ | $\Sigma_g^+$ | $m_\infty$ |
| | | $\Sigma_u^+$ | $m_\infty+m_0-1$ |
| | | $\Sigma_g^-, \Sigma_u^-$ | 0 |
| | | $\Pi_g$ | $m_\infty-1$ |
| | | $\Pi_u$ | $m_\infty+m_0-1$ |
| | | $\Delta_g, \Delta_u$ | 0 |
| | | $\Phi_g, \Phi_u, \ldots$ | 0 |

| | | | |
|---|---|---|---|
| $\mathbf{T}_d$ | $24m+12m_d$ $+6m_2+4m_3+m_0$ | $A_1$ | $3m+2m_d+m_2+m_3$ |
| | | $A_2$ | $3m+m_d$ |
| | | $E$ | $6m+3m_d+m_2+m_3$ |
| | | $F_1$ | $9m+4m_d+2m_2+m_3-1$ |
| | | $F_2$ | $9m+5m_d+3m_2+2m_3+m_0-1$ |
| $\mathbf{O}_h$ | $48m+24m_h+24m_d$ $+12m_2+8m_3$ $+6m_4+m_0$ | $A_{1g}$ | $3m+2m_h+2m_d+m_2+m_3+m_4$ |
| | | $A_{1u}$ | $3m+m_h+m_d$ |
| | | $A_{2g}$ | $3m+2m_h+m_d+m_2$ |
| | | $A_{2u}$ | $3m+m_h+2m_d+m_2+m_3$ |
| | | $E_g$ | $6m+4m_h+3m_d+2m_2+m_3+m_4$ |
| | | $E_u$ | $6m+2m_h+3m_d+m_2+m_3$ |
| | | $F_{1g}$ | $9m+4m_h+4m_d+2m_2+m_3+m_4-1$ |
| | | $F_{1u}$ | $9m+5m_h+5m_d+3m_2+2m_3+2m_4+m_0-1$ |
| | | $F_{2g}$ | $9m+4m_h+5m_d+2m_2+2m_3+m_4$ |
| | | $F_{2u}$ | $9m+5m_h+4m_d+2m_2+m_3+m_4$ |

[a] Note that $m$ is the number of sets of nuclei not on any element of symmetry; $m_0$ is the number of nuclei on all elements of symmetry; $m_2$, $m_3$, $m_4$, . . . are the numbers of sets of nuclei on a twofold, threefold, fourfold, . . . axis but not on any other element of symmetry that does not wholly coincide with that axis; $m_2'$ is the number of sets of nuclei on the twofold axis called $C_2'$ in the preceding character tables; $m_v$, $m_d$, $m_h$ are the numbers of sets of nuclei on planes $\sigma_v$, $\sigma_d$, $\sigma_h$, respectively, but not on any other element of symmetry.

# APPENDIX III

## NUMBER OF INFRARED AND RAMAN ACTIVE STRETCHING VIBRATIONS FOR $MX_nY_m$ TYPE MOLECULES

| Compound | Structure | Point group | Infrared or Raman | MX Stretching | MY Stretching |
|---|---|---|---|---|---|
| $MX_6$ | octahedral | $\mathbf{O}_h$ | IR | $F_{1u}$ | |
| | | | R | $A_{1g}, E_g$ | |
| $MX_5Y$ | octahedral | $\mathbf{C}_{4v}$ | IR | $2A_1, E$ | $A_1$ |
| | | | R | $2A_1, B_1, E$ | $A_1$ |
| *trans*-$MX_4Y_2$ | octahedral | $\mathbf{D}_{4h}$ | IR | $E_u$ | $A_{2u}$ |
| | | | R | $A_{1g}, B_{1g}$ | $A_{1g}$ |
| *cis*-$MX_4Y_2$ | octahedral | $\mathbf{C}_{2v}$ | IR | $2A_1, B_1, B_2$ | $A_1, B_1$ |
| | | | R | $2A_1, B_1, B_2$ | $A_1, B_1$ |
| *mer*-$MX_3Y_3$ | octahedral | $\mathbf{C}_{2v}$ | IR | $2A_1, B_2$ | $2A_1, B_1$ |
| | | | R | $2A_1, B_2$ | $2A_1, B_1$ |
| *fac*-$MX_3Y_3$ | octahedral | $\mathbf{C}_{3v}$ | IR | $A_1, E$ | $A_1, E$ |
| | | | R | $A_1, E$ | $A_1, E$ |
| $MX_5$ | trigonal-bipyramidal | $\mathbf{D}_{3h}$ | IR | $A''_2, E'$ | |
| | | | R | $2A'_1, E'$ | |
| $MX_5$ | tetragonal-pyramidal | $\mathbf{C}_{4v}$ | IR | $2A_1, E$ | |
| | | | R | $2A_1, B_1, E$ | |
| $MX_4$ | tetrahedral | $\mathbf{T}_d$ | IR | $F_2$ | |
| | | | R | $A_1, F_2$ | |
| $MX_3Y$ | tetrahedral | $\mathbf{C}_{3v}$ | IR | $A_1, E$ | $A_1$ |
| | | | R | $A_1, E$ | $A_1$ |
| $MX_2Y_2$ | tetrahedral | $\mathbf{C}_{2v}$ | IR | $A_1, B_1$ | $A_1, B_2$ |
| | | | R | $A_1, B_1$ | $A_1, B_2$ |
| Polymeric $MX_2Y_2$[a] | octahedral | $\mathbf{C}_i$ | IR | $2A_u$ | $A_u$ |
| | | | R | $2A_g$ | $A_g$ |
| $MX_4$ | square-planar | $\mathbf{D}_{4h}$ | IR | $E_u$ | |
| | | | R | $A_{1g}, B_{1g}$ | |
| $MX_3Y$ | planar | $\mathbf{C}_{2v}$ | IR | $2A_1, B_1$ | $A_1$ |
| | | | R | $2A_1, B_1$ | $A_1$ |
| *trans*-$MX_2Y_2$ | planar | $\mathbf{C}_{2h}$ | IR | $B_u$ | $B_u$ |
| | | | R | $A_g$ | $A_g$ |
| *cis*-$MX_2Y_2$ | planar | $\mathbf{C}_{2v}$ | IR | $A_1, B_2$ | $A_1, B_2$ |
| | | | R | $A_1, B_2$ | $A_1, B_2$ |
| $MX_3$ | pyramidal | $\mathbf{C}_{3v}$ | IR | $A_1, E$ | |
| | | | R | $A_1, E$ | |
| $MX_3$ | planar | $\mathbf{D}_{3h}$ | IR | $E'$ | |
| | | | R | $A'_1, E'$ | |

[a] Bridging through X atoms.

# APPENDIX IV

## DERIVATION OF EQUATION 11.3 (PART I)

Using the rectangular coordinates, we write the kinetic energy as

$$2T = \tilde{\dot{\mathbf{X}}}\mathbf{M}\dot{\mathbf{X}} \tag{1}$$

where

$$\mathbf{X} = \begin{bmatrix} x_1 \\ y_1 \\ z_1 \\ x_2 \\ \cdot \\ \cdot \\ \cdot \\ z_N \end{bmatrix} \quad \text{and} \quad \mathbf{M} = \begin{bmatrix} m_1 & & & & & & & \\ & m_1 & & & & & & \\ & & m_1 & & & & & \\ & & & m_2 & & & & \\ & & & & \cdot & & & \\ & & & & & \cdot & & \\ & & & & & & \cdot & \\ & & & & & & & m_N \end{bmatrix}$$

By definition, the momentum $p_{x1}$ conjugated with $x_1$ is given by

$$p_{x_1} = \frac{\partial T}{\partial \dot{x}_1} = m_1 \dot{x}_1$$

and $p_{y_1} \ldots p_{z_N}$ take similar forms. Using the conjugate momenta, we write $T$ as

$$\begin{aligned} 2T &= \frac{1}{m_1} p_{x_1}^2 + \frac{1}{m_1} p_{y_1}^2 + \cdots + \frac{1}{m_N} p_{z_N}^2 \\ &= \tilde{\mathbf{P}}_x \mathbf{M}^{-1} \mathbf{P}_x \end{aligned} \tag{2}$$

where

$$\mathbf{P}_x = \begin{bmatrix} p_{x_1} \\ p_{y_1} \\ \cdot \\ \cdot \\ \cdot \\ p_{z_N} \end{bmatrix} \quad \text{and} \quad \mathbf{M}^{-1} = \begin{bmatrix} \mu_1 & & & & & \\ & \mu_1 & & & & \\ & & \cdot & & & \\ & & & \cdot & & \\ & & & & \cdot & \\ & & & & & \mu_N \end{bmatrix}$$

The column matrix $\mathbf{P}_x$ can be expressed as

$$\mathbf{P}_x = \mathbf{M}\dot{\mathbf{X}} \tag{3}$$

Define a set of conjugate momenta $\mathbf{P}$ associated with internal coordinates $\mathbf{R}$. As is shown at the end of this appendix, we have

$$\mathbf{P}_x = \tilde{\mathbf{B}}\mathbf{P} \tag{4}$$

Equations 3 and 4 give

$$\mathbf{M}\dot{\mathbf{X}} = \tilde{\mathbf{B}}\mathbf{P} \tag{5}$$

Equation 11.8 in the text gives

$$\mathbf{R} = \mathbf{B}\mathbf{X} \qquad \text{and} \qquad \dot{\mathbf{R}} = \mathbf{B}\dot{\mathbf{X}} \tag{6}$$

By inserting Eq. 5 into Eq. 6, we obtain

$$\dot{\mathbf{R}} = \mathbf{B}\mathbf{M}^{-1}\tilde{\mathbf{B}}\mathbf{P} \tag{7}$$

Using Eq. 4, we write Eq. 2 as

$$2T = \tilde{\mathbf{P}}\mathbf{B}\mathbf{M}^{-1}\tilde{\mathbf{B}}\mathbf{P} \tag{8}$$

If we define

$$\mathbf{G} = \mathbf{B}\mathbf{M}^{-1}\tilde{\mathbf{B}} \qquad \text{(11.7), text}$$

Eq. 8 is written as

$$2T = \tilde{\mathbf{P}}\mathbf{G}\mathbf{P} \tag{9}$$

If Eq. 11.7 is combined with Eq. 7, we obtain

$$\dot{\mathbf{R}} = \mathbf{G}\mathbf{P}$$

or

$$\mathbf{G}^{-1}\dot{\mathbf{R}} = \mathbf{G}^{-1}\mathbf{G}\mathbf{P} = \mathbf{P} \tag{10}$$

Using Eq. 10, we can write Eq. 9 as

$$\begin{aligned} 2T &= \tilde{\dot{\mathbf{R}}}\tilde{\mathbf{G}}^{-1}\mathbf{G}\mathbf{G}^{-1}\dot{\mathbf{R}} \\ &= \tilde{\dot{\mathbf{R}}}\mathbf{G}^{-1}\dot{\mathbf{R}} \end{aligned} \qquad \text{(11.3), text}$$

**Derivation of Eq. 4**

The momentum, $p_{R_k}$ conjugated with the internal coordinate, $R_k$, is given by

$$p_{R_k} = \frac{\partial T}{\partial \dot{R}_k} \qquad k = 1, 2, \ldots, s$$

If we denote the coordinates corresponding to the translational and rotational motions of the molecule by $R_j^0$ and its conjugate momentum by

$p^0_{R_j}$, we have

$$p^0_{R_j} = \frac{\partial T}{\partial \dot{R}^0_j} \qquad j = 1, 2, \ldots, 6$$

Then the momentum, $p_{x_1}$, in terms of rectangular coordinates is written as

$$p_{x_1} = \frac{\partial T}{\partial \dot{x}_1} = \sum_k^s \frac{\partial T}{\partial \dot{R}_k}\frac{\partial R_k}{\partial x_1} + \sum_j^6 \frac{\partial T}{\partial \dot{R}^0_j}\frac{\partial R^0_j}{\partial x_1}$$
$$= \sum_k^s p_{R_k} B_{k,x_1} + \sum_j^6 p^0_{R_j}\frac{\partial R^0_j}{\partial x_1}$$

The seond term becomes zero since the momenta corresponding to the translational and rotational motions are zero. Thus we have

$$p_{x_1} = \sum_k^s p_{R_k} B_{k,x_1}$$
$$p_{y_1} = \sum p_{R_k} B_{k,y_1}$$
$$\vdots \qquad \vdots$$
$$p_{z_N} = \sum p_{R_k} B_{k,z_N}$$

In matrix form, this is written as

$$\mathbf{P}_x = \tilde{\mathbf{B}}\mathbf{P} \tag{4}$$

## APPENDIX V

## THE G AND F MATRIX ELEMENTS OF TYPICAL MOLECULES

In the following equations, $F$ represents **F** matrix elements in the GVF field, whereas $F^*$ denotes those in the UBF field. In the latter, $F' = -\frac{1}{10}F$ was assumed for all cases, and the *molecular tension* (see Ref. I-41) was ignored.

### (1) Bent $XY_2$ Molecules ($C_{2v}$)

$A_1$ species—infrared and Raman active

$$G_{11} = \mu_y + \mu_x(1 + \cos\alpha)$$

$$G_{12} = -\frac{\sqrt{2}}{r}\mu_x \sin\alpha$$

$$G_{22} = \frac{2}{r^2}[\mu_y + \mu_x(1 - \cos\alpha)]$$

$$F_{11} = f_r + f_{rr}$$

$$F_{12} = (\sqrt{2})rf_{r\alpha}$$

$$F_{22} = r^2 f_\alpha$$

$$F^*_{11} = K + 2F\sin^2\frac{\alpha}{2}$$

$$F^*_{12} = (0.9)(\sqrt{2})rF\sin\frac{\alpha}{2}\cos\frac{\alpha}{2}$$

$$F^*_{22} = r^2\left[H + F\left\{\cos^2\frac{\alpha}{2} + (0.1)\sin^2\frac{\alpha}{2}\right\}\right]$$

$B_2$ species—infrared and Raman active

$$G = \mu_y + \mu_x(1 - \cos\alpha)$$

$$F = f_r - f_{rr}$$

$$F^* = K - (0.2)F\cos^2\frac{\alpha}{2}$$

**(2) Pyramidal $XY_3$ Molecules ($C_{3v}$)**

$A_1$ species—infrared and Raman active

$$G_{11} = \mu_y + \mu_x(1 + 2\cos\alpha)$$

$$G_{12} = -\frac{2}{r}\frac{(1 + 2\cos\alpha)(1 - \cos\alpha)}{\sin\alpha}\mu_x$$

$$G_{22} = \frac{2}{r^2}\left(\frac{1 + 2\cos\alpha}{1 + \cos\alpha}\right)[\mu_y + 2\mu_x(1 - \cos\alpha)]$$

$$F_{11} = f_r + 2f_{rr}$$

$$F_{12} = r(2f_{r\alpha} + f'_{r\alpha})$$

$$F_{22} = r^2(f_\alpha + 2f_{\alpha\alpha})$$

$$F^*_{11} = K + 4F\sin^2\frac{\alpha}{2}$$

$$F^*_{12} = (1.8)rF\sin\frac{\alpha}{2}\cos\frac{\alpha}{2}$$

$$F^*_{22} = r^2\left[H + F\left\{\cos^2\frac{\alpha}{2} + (0.1)\sin^2\frac{\alpha}{2}\right\}\right]$$

$E$ species—infrared and Raman active

$$G_{11} = \mu_y + \mu_x(1 - \cos\alpha)$$

$$G_{12} = \frac{1}{r}\frac{(1-\cos\alpha)^2}{\sin\alpha}\mu_x$$

$$G_{22} = \frac{1}{r^2(1+\cos\alpha)}[(2+\cos\alpha)\mu_y + (1-\cos\alpha)^2\mu_x]$$

$$F_{11} = f_r - f_{rr}$$

$$F_{12} = r(-f_{r\alpha} + f'_{r\alpha})$$

$$F_{22} = r^2(f_\alpha - f_{\alpha\alpha})$$

$$F^*_{11} = K + \left[\sin^2\frac{\alpha}{2} - (0.3)\cos^2\frac{\alpha}{2}\right]F$$

$$F^*_{12} = -(0.9)rF\sin\frac{\alpha}{2}\cos\frac{\alpha}{2}$$

$$F^*_{22} = r^2\left[H + F\left\{\cos^2\frac{\alpha}{2} + (0.1)\sin^2\frac{\alpha}{2}\right\}\right]$$

Here $f_{r\alpha}$ denotes interaction between $\Delta r$ and $\Delta\alpha$ having a common bond (e.g., $\Delta r_1$ and $\Delta\alpha_{12}$ or $\Delta\alpha_{31}$); $f'_{r\alpha}$ denotes interaction between $\Delta r$ and $\Delta\alpha$ having no common bonds (e.g., $\Delta r_1$ and $\Delta\alpha_{23}$); see Fig. I-11*c*.

**(3) Planar $XY_3$ Molecules ($D_{3h}$)**

$A'_1$ species—Raman active

$$G = \mu_y$$

$$F = f_r + 2f_{rr}$$

$$F^* = K + 3F$$

$A''_2$ species—infrared active

$$G = \frac{9}{4r^2}(\mu_y + 3\mu_x)$$

$$F = F^* = r^2 f_\theta$$

$E'$ species—infrared and Raman active

$$G_{11} = \mu_y + \tfrac{3}{2}\mu_x$$

$$G_{12} = \frac{3\sqrt{3}}{2r}\mu_x$$

$$G_{22} = \frac{3}{2r^2}(2\mu_y + 3\mu_x)$$

$$F_{11} = f_r - f_{rr}$$
$$F_{12} = r(f'_{r\alpha} - f_{r\alpha})$$
$$F_{22} = r^2(f_\alpha - f_{\alpha\alpha})$$
$$F^*_{11} = K + 0.675F$$
$$F^*_{12} = -(0.9)\frac{\sqrt{3}}{4}\,rF$$
$$F^*_{22} = r^2(H + 0.325F)$$

The symbols $f_{r\alpha}$ and $f'_{r\alpha}$ are defined in (2); $f_\theta$ denotes the force constant for the out-of-plane mode (see Fig. I-11$f$).

### (4) Tetrahedral $XY_4$ Molecules ($T_d$)

$A_1$ species—Raman active

$$G = \mu_y$$
$$F = f_r + 3f_{rr}$$
$$F^* = K + 4F$$

$E$ species—Raman active

$$G = \frac{3\mu_y}{r^2}$$
$$F = r^2(f_\alpha - 2f_{\alpha\alpha} + f'_{\alpha\alpha})$$
$$F^* = r^2(H + 0.37F)$$

$F_2$ species—infrared and Raman active

$$G_{11} = \mu_y + \tfrac{4}{3}\mu_x$$
$$G_{12} = -\frac{8}{3r}\,\mu_x$$
$$G_{22} = \frac{1}{r^2}\left(\tfrac{16}{3}\mu_x + 2\mu_y\right)$$
$$F_{11} = f_r - f_{rr}$$
$$F_{12} = (\sqrt{2})r(f_{r\alpha} - f'_{r\alpha})$$
$$F_{22} = r^2(f_\alpha - f'_{\alpha\alpha})$$
$$F^*_{11} = K + \tfrac{6}{5}F$$
$$F^*_{12} = \tfrac{3}{5}rF$$
$$F^*_{22} = r^2(H + \tfrac{1}{2}F)$$

where $f_{\alpha\alpha}$ denotes interaction between two $\Delta\alpha$ having a common bond, and $f'_{\alpha\alpha}$ denotes interaction between two $\Delta\alpha$ having no common bond.

**(5) Square-Planar $XY_4$ Molecules ($D_{4h}$)**

$A_{1g}$ species—Raman active

$$G = \mu_y$$
$$F = f_r + 2f_{rr} + f'_{rr}$$
$$F^* = K + 2F$$

$B_{1g}$ species—Raman active

$$G = \mu_y$$
$$F = f_r - 2f_{rr} + f'_{rr}$$
$$F^* = K - 0.2F$$

$B_{2g}$ species—Raman active

$$G = \frac{4\mu_y}{r_2}$$
$$F = r^2(f_\alpha - 2f_{\alpha\alpha} + f'_{\alpha\alpha})$$
$$F^* = r^2(H + 0.55F)$$

$E_u$ species—infrared active

$$G_{11} = 2\mu_x + \mu_y$$
$$G_{12} = -\frac{2\sqrt{2}}{r}\mu_x$$
$$G_{22} = \frac{2}{r^2}(\mu_y + 2\mu_x)$$
$$F_{11} = f_r - f'_{rr}$$
$$F_{12} = (\sqrt{2})r(f_{r\alpha} - f'_{r\alpha})$$
$$F_{22} = r^2(f_\alpha - f'_{\alpha\alpha})$$
$$F^*_{11} = K + 0.9F$$
$$F^*_{12} = -(\sqrt{2})r(0.45)F$$
$$F^*_{22} = r^2(H + 0.55F)$$

The symbol $f_{rr}$ denotes interaction between two $\Delta r$ perpendicular to each other; $f'_{rr}$ denotes interaction between two $\Delta r$ on the same straight line. In addition, a square-planar $XY_4$ molecule has two out-of-plane vibrations in the $A_{2u}$ and $B_{2u}$ species.

### (6) Octahedral $XY_6$ Molecules ($O_h$)

$A_{1g}$ species—Raman active

$$G = \mu_y$$
$$F = f_r + 4f_{rr} + f'_{rr}$$
$$F^* = K + 4F$$

$E_g$ species—Raman active

$$G = \mu_y$$
$$F = f_r - 2f_{rr} + f'_{rr}$$
$$F^* = K + 0.7F$$

$F_{1u}$ species—infrared active

$$G_{11} = \mu_y + 2\mu_x$$
$$G_{12} = -\frac{4}{r}\mu_x$$
$$G_{22} = \frac{2}{r^2}(\mu_y + 4\mu_x)$$
$$F_{11} = f_r - f'_{rr}$$
$$F_{12} = 2rf_{r\alpha}$$
$$F_{22} = r^2(f_\alpha + 2f_{\alpha\alpha})$$
$$F^*_{11} = K + 1.8F$$
$$F^*_{12} = 0.9rF$$
$$F^*_{22} = r^2(H + 0.55F)$$

$F_{2g}$ species—Raman active

$$G = \frac{4\mu_y}{r^2}$$
$$F = r^2(f_\alpha - 2f'_{\alpha\alpha})$$
$$F^* = r^2(H + 0.55F)$$

$F_{2u}$ species—inactive

$$G = \frac{2\mu_y}{r^2}$$

$$F = r^2(f_\alpha - 2f_{\alpha\alpha})$$

$$F^* = r^2(H + 0.55F)$$

The symbol $f_{rr}$ denotes interaction between two $\Delta r$ perpendicular to each other, whereas $f'_{rr}$ denotes that between two $\Delta r$ on the same straight line; $f_{\alpha\alpha}$ denotes interaction between two $\Delta\alpha$ perpendicular to each other, whereas $f'_{\alpha\alpha}$ denotes that between two $\Delta\alpha$ on the same plane. Only the interaction between two $\Delta\alpha$ having a common bond was considered.

# APPENDIX VI

## GROUP FREQUENCY CHARTS

The data cited in this book were used in the preparation of the following group frequency charts. Each section of Part III gives a number of group frequencies that are not included here. For the physical meaning of "group frequency," see Sec. I-16.

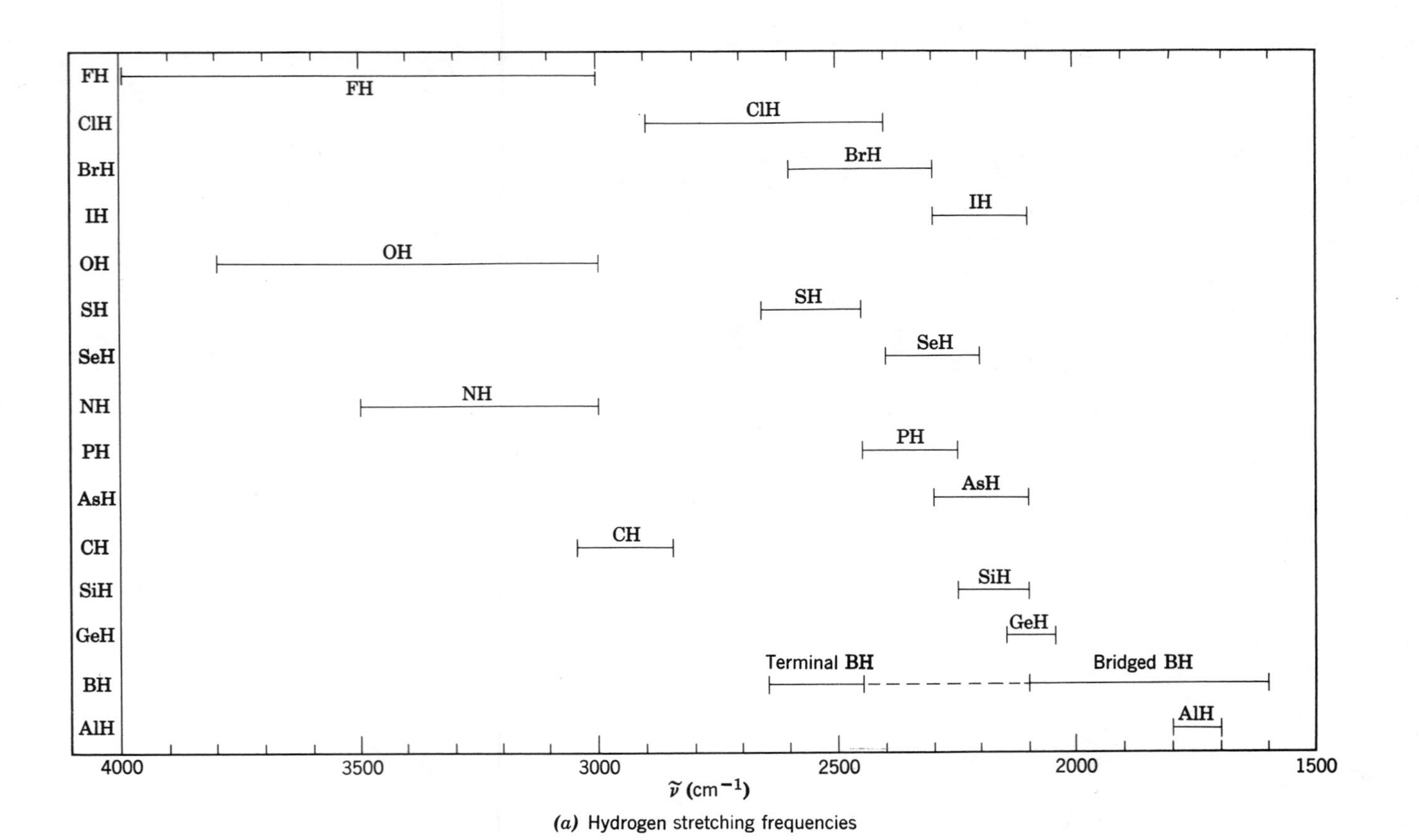


(*a*) Hydrogen stretching frequencies

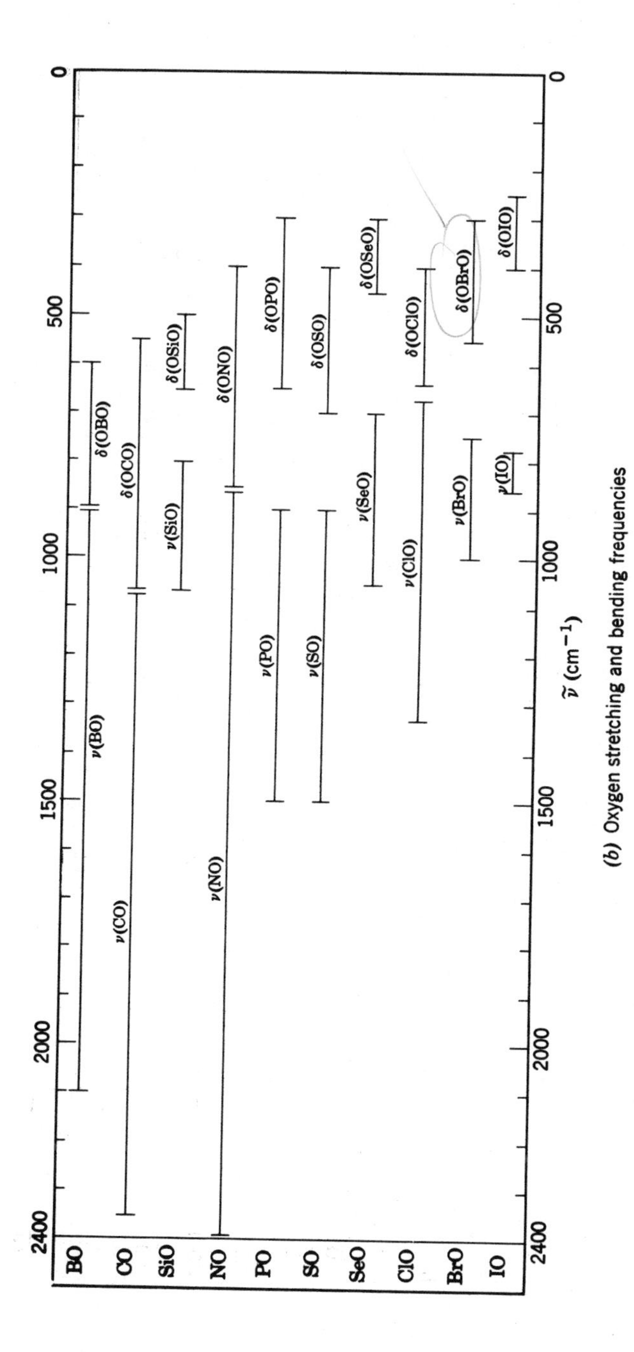


(*b*) Oxygen stretching and bending frequencies

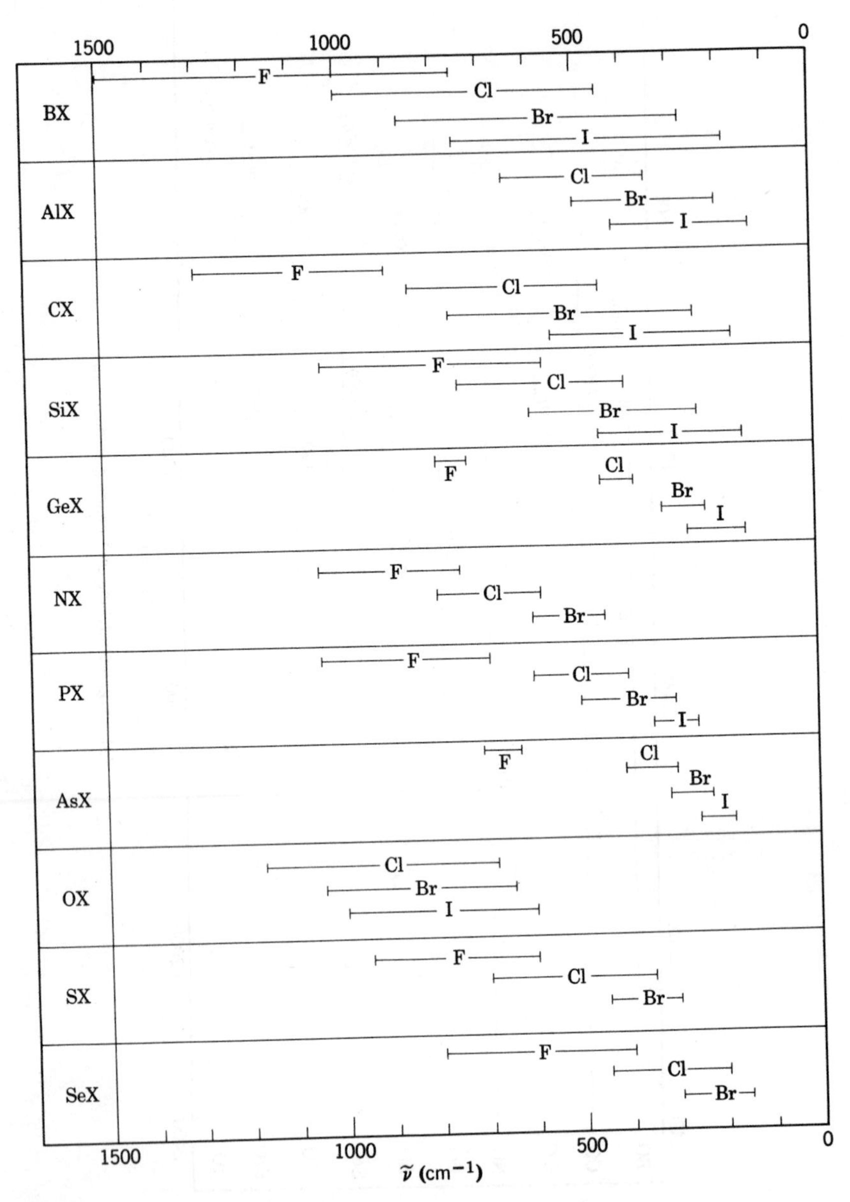


*(c)* Halogen (X) stretching frequencies

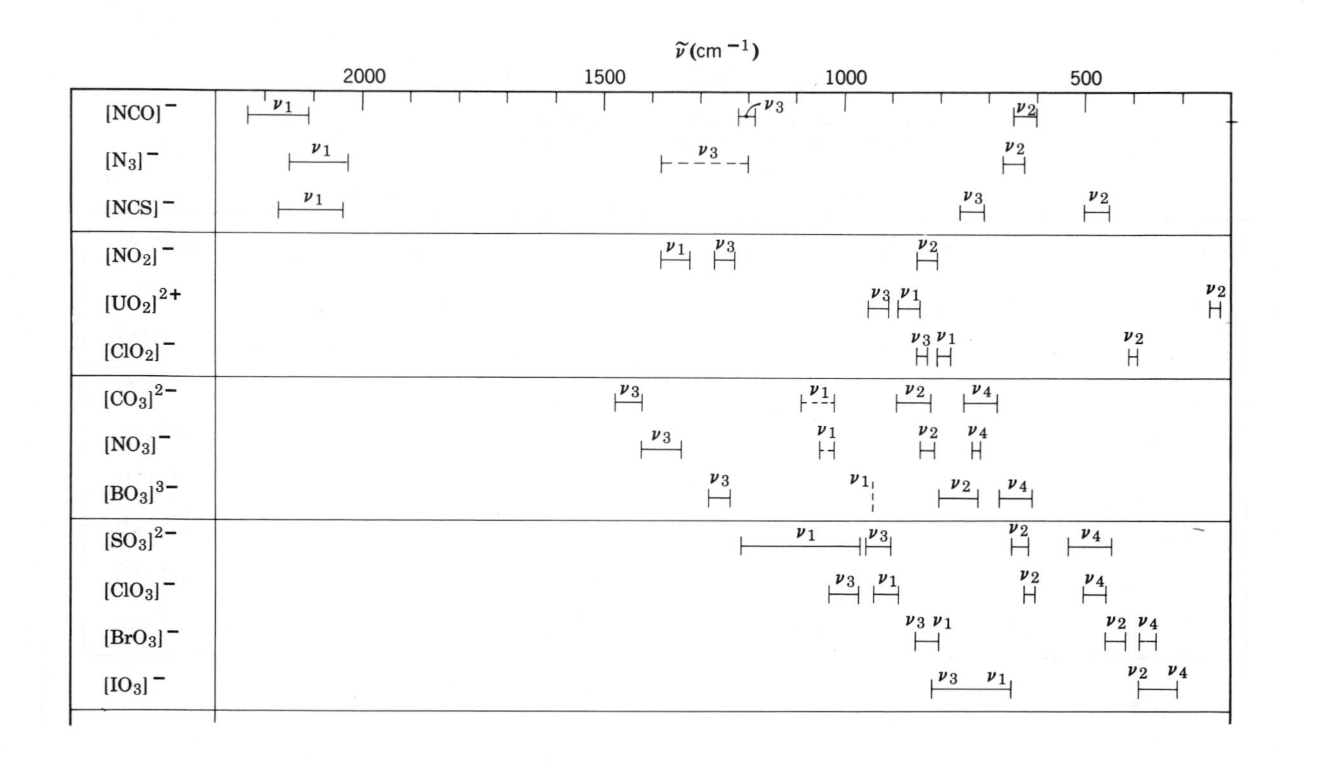

$\tilde{\nu}$ (cm$^{-1}$)
2000
1500
1000
500
$[NCO]^-$
$[N_3]^-$
$[NCS]^-$
$[NO_2]^-$
$[UO_2]^{2+}$
$[ClO_2]^-$
$[CO_3]^{2-}$
$[NO_3]^-$
$[BO_3]^{3-}$
$[SO_3]^{2-}$
$[ClO_3]^-$
$[BrO_3]^-$
$[IO_3]^-$
$\nu_1$
$\nu_2$
$\nu_3$
$\nu_4$

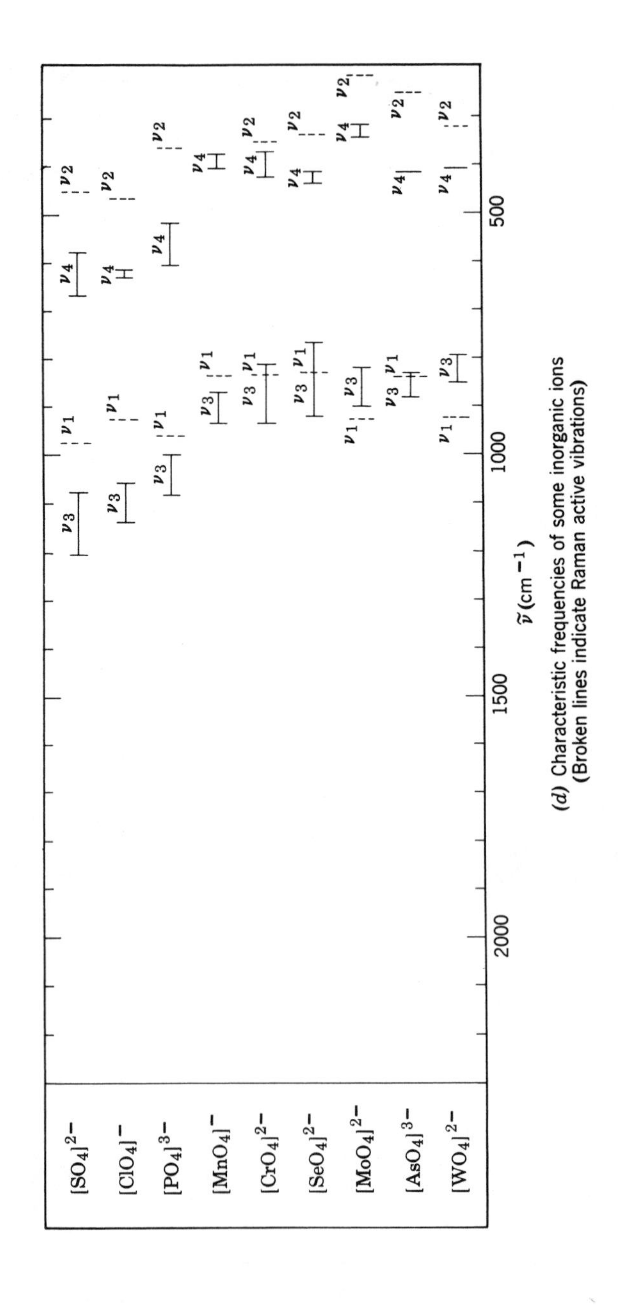


*(d)* Characteristic frequencies of some inorganic ions
(Broken lines indicate Raman active vibrations)

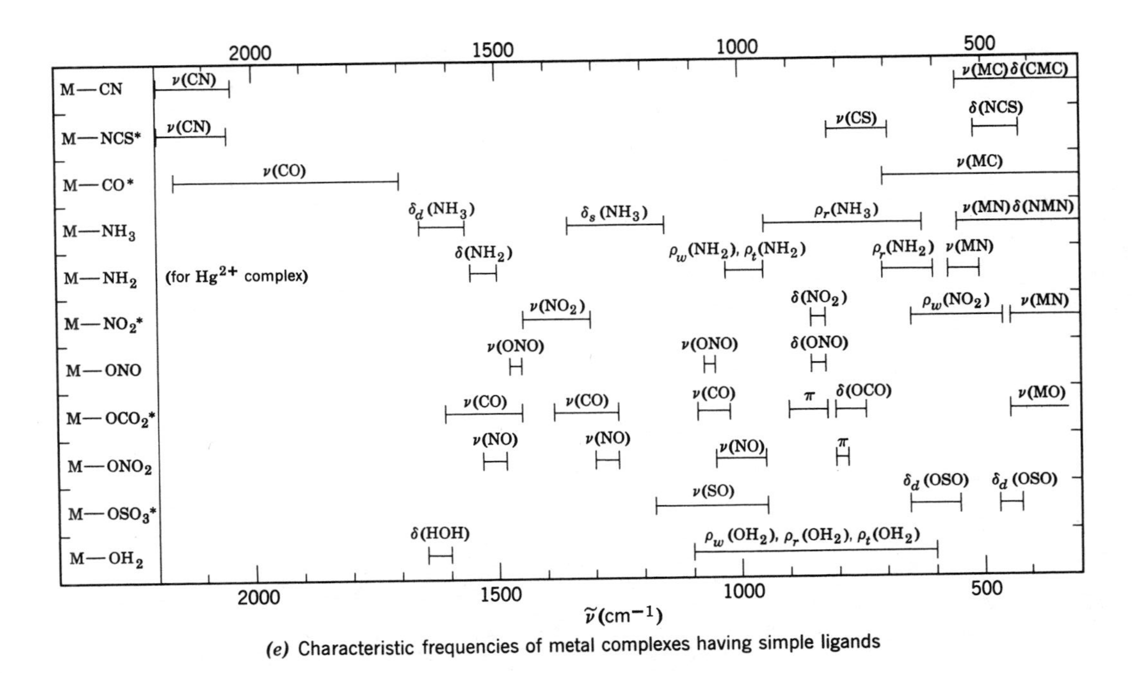


*(e)* Characteristic frequencies of metal complexes having simple ligands

(Frequency ranges include bidentate and bridged complexes for the ligands marked by an asterisk)

# Index

Since the number of compounds included in this book is numerous, entries are given for only a few compounds appearing in the text. The majority of other compounds may be reached through general entries such as diatomic molecules, $XY_4$ molecules and ammine complexes. **Bold face** page numbers refer to figures.